MATHEMATICS OF MICROSTRUCTURE EVOLUTION

MATHEMATICS OF MICROSTRUCTURE EVOLUTION

A volume in the Electronic, Magnetic and Photonic Materials Division (EMPMD) Monograph Series sponsored by the TMS MSD-ASM Thermodynamics and Phase Equilibria Committee and the MSD-ASM Atomic Transport Committee.

This symposium was held during
Materials Week '95, October 29 - November 2, 1995
in Cleveland, Ohio.

Edited by

Long-Qing Chen • Brent Fultz • John W. Cahn
John R. Manning • John E. Morral • John Simmons

A Publication of

TMS   siam
Minerals • Metals • Materials

A Publication of

The Minerals, Metals & Materials Society
420 Commonwealth Drive
Warrendale, Pennsylvania 15086
(412) 776-9000

Society for Industrial and Applied Mathematics
3600 University City Science Center
Philadelphia, PA 19104-2688
(215) 382-9800

The Minerals, Metals & Materials Society is not responsible for statements or opinions and is absolved of liability due to misuse of information contained in this publication.

Printed in the United States of America
Library of Congress Catalog Number 96-078157
TMS ISBN Number 0-87339-351-1
SIAM ISBN Number 0-89871-386-2

Authorization to photocopy items for internal or personal use, or the internal or personal use of specific clients, is granted by The Minerals, Metals & Materials Society for users registered with the Copyright Clearance Center (CCC) Transactional Reporting Service, provided that the base fee of $3.00 per copy is paid directly to Copyright Clearance Center, 27 Congress Street, Salem, Massachusetts 01970. For those organizations that have been granted a photocopy license by Copyright Clearance Center, a separate system of payment has been arranged.

© 1996

If you are interested in purchasing a copy of this book, or if you would like to receive the latest TMS publications catalog, please telephone 1-800-759-4867.

CONTENTS

PREFACE ix

SURFACE EVOLVER AS A TOOL FOR MATERIALS SCIENCE RESEARCH 1
    W. C. Carter

THE SIMULATION OF TWO-DIMENSIONAL GRAIN
GROWTH BY THE SURFACE EVOLVER 15
    K. Marthinsen, O. Hunderi, and N. Ryum

EDGE-ENERGY MINIMIZING SURFACES 23
    M. J. Kelley

ZENER PINNING OF NORMAL GRAIN GROWTH 31
    J. Gao, R. G. Thompson, and B. R. Patterson

THE THERMAL STABILITY OF ALLOYS HAVING PERIODIC
MINIMAL SURFACE MICROSTRUCTURES 39
    B. Fultz and L. Anthony

MODELING MAGNETOSTRICTIVE MICROSTRUCTURE
UNDER LOADING 49
    R. James, D. Kinderlehrer, and L. Ma

MICROLOCAL FORCES AND PRECURSOR OSCILLATIONS 59
    D. Brandon and R. C. Rogers

REAL-SPACE LATTICE STATICS DESCRIPTION
OF A RANDOM LATTICE 71
    J. R. Willis and T. J. Gosling

MODULATED PATTERNS AND PINNING EFFECT
IN PHASE-SEPARATING ALLOYS 87
    A. Onuki

DISCRETE ATOM METHOD FOR MORPHOLOGICAL
EVOLUTION OF COHERENT PARTICLES 101
    J. H. Choy, S. A. Hackney, and J. K. Lee

COMPUTER SIMULATIONS OF PHASE DECOMPOSITION
IN REAL ALLOY SYSTEMS BASED ON A DISCRETE
TYPE DIFFUSION EQUATION 111
    T. Miyazaki and T. Koyama

EQUILIBRIUM PARTICLE MORPHOLOGIES
IN ELASTICALLY STRESSED SOLIDS 125
    M. E. Thompson and P. W. Voorhees

ANISOTROPIC INTERFACE MOTION 135
    J. E. Taylor

THERMODYNAMIC DRIVING FORCES AND EQUILIBRIUM IN
MULTICOMPONENT SYSTEMS WITH ANISOTROPIC SURFACES 149
    J. W. Cahn and J. E. Taylor

ANALYSIS OF ANISOTROPIC CHARACTERISTICS
OF DISLOCATION LOOP MORPHOLOGY 161
    D. Schwendeman and K. Rajan

ON THE PHYSICS OF SELF-ORGANIZING
MICROSTRUCTURAL EVOLUTION 173
    J. S. Kirkaldy

THE GEOMETRICAL PHASE TRANSFORMATIONS
DURING EVOLUTION IN FINITE MEDIA 187
    V. Ya. Shur, E. L. Rumyantsev, and S. D. Makarov

PHASE-FIELD MODELING OF CRYSTAL GROWTH: NEW
ASYMPTOTICS AND COMPUTATIONAL LIMITS 195
    A. Karma and W.-J. Rappel

PHASE FIELD INSTABILITIES AND ADAPTIVE MESH REFINEMENT 205
    R. F. Almgren and A. S. Almgren

MICROSTRUCTURAL EVOLUTION AND GRAIN GROWTH KINETICS
IN A TWO-PHASE SOLID WITH QUADRIJUNCTIONS 215
    D. Fan and L. Q. Chen

ANISOTROPIC INTERFACES AND ORDERING IN FCC ALLOYS:
A MULTIPLE-ORDER-PARAMETER CONTINUUM THEORY 225
    R. J. Braun, J. W. Cahn, J. Hagedorn, G. B. McFadden, and A. A. Wheeler

DIFFUSIONAL PHASE TRANSFORMATIONS ON THE
ATOMIC SCALE: EXPERIMENT AND MODELING 245
    J. M. Hyde, A. Cerezo, M. K. Miller, R. P. Setna, and G. D. W. Smith

NON-CLASSICAL NUCLEATION AND GROWTH OF
ORDERED DOMAINS IN A DISORDERED MATRIX 269
    R. Poduri and L. Q. Chen

THE MATHEMATICS OF PROCESSING FOR
MATERIAL MICROSTRUCTURE 281
    J. A. Simmons

APPEAL TO MATHEMATICIANS ON THE
HIERARCHY IN CVM AND PPM 311
    R. Kikuchi

MEAN-FIELD EQUATIONS FOR CONFIGURATION KINETICS OF
ALLOYS AT ARBITRARY DEGREE OF NONEQUILIBRIUM 321
    V. G. Vaks, S. V. Beiden, and V. Yu. Dobretsov

MODELING AND CONTROL OF ADVANCED CHEMICAL
VAPOR DEPOSITION PROCESSES 327
    H. T. Banks, K. Ito, J. S. Scroggs, H. T. Tran, N. Dietz, and K. J. Bachmann

PATTERN FORMATION IN NONUNIFORM MEDIA: CO OXIDATION
ON MICROSTRUCTURED AND COMPOSITE PT SURFACES 345
    M. Bär, A. K. Bangia, I. G. Kevrekidis, G. Haas, and H.-H. Rotermund

STRUCTURAL STABILITY OF SPINODAL DECOMPOSITION 355
    Y. Oono

CASCADES OF SPINODAL DECOMPOSITIONS IN
THE TERNARY CAHN-HILLIARD EQUATIONS 367
    D. J. Eyre

CHEMICAL POTENTIAL APPROACH TO DIFFUSION-LIMITED
MICROSTRUCTURE EVOLUTION 379
    M. A. Fradkin, J. Goldak, and R. C. Reed

SUBJECT INDEX 387

AUTHOR INDEX 391

PREFACE

This proceedings contains 31 of the 70 papers presented at the symposium "Modern Modeling of Microstructural Evolution" which was held from October 29 - November 1 at the TMS/ASM fall meeting in Cleveland, Ohio. The Symposium was jointly sponsored by ASM International (formerly the American Society for Metals), SIAM (Society for Industrial & Applied Mathematics) and TMS (The Minerals, Metals & Materials Society). The purpose of the symposium was to bring together mathematicians, computer scientists, physicists and materials scientists who model and characterize the evolution of microstructure. Of special interest were models and computer simulations of the kinetics of phase transitions, the evolution of shapes towards equilibrium, the comparisons of models with sharp or diffuse interfaces, and effects of stress or electromagnetic fields, comparisons between lattice (large-scale ODE) and continuum (PDE) theories, and the various theories and methods for dealing with strong interracial or elastic anisotropies.

The symposium was large, lively, and brought new interactions between the mathematics and materials science communities. The editors hope that this book catches some of the spirit of this symposium. The symposium organizers are indebted to the referees of the papers, the speakers, the session chairpersons and the attendees for the success of this symposium and these proceedings.

The organizers acknowledge the generous financial support from ARPA (the Defense Department's Advanced Research Projects Agency), NIST (the National Institute of Standards and Technology) and the Advanced Micro Devices Corporation.

Organizers:

L. Q. Chen
B. Fultz
J. W. Cahn
J. R. Manning
J. E. Morral
J. A. Simmons

SURFACE EVOLVER AS A TOOL FOR MATERIALS SCIENCE RESEARCH

W. Craig Carter

Ceramics Division, Materials Science and Engineering Laboratory,
National Institute of Standards and Technology, Gaithersburg, MD 20899

Abstract

Several examples of the *Surface Evolver* code for computational problems in materials sciences are demonstrated. Particular attention is given to the mechanics of setting up and running *Evolver* code and details are presented from which new applications could be extrapolated. The three examples(the microstructure of second phases wetting a three-dimensional grain boundary network; the stability of a solder bridge connecting two leads; and the forces and potentials that obtain in a powder compact with a small amount of fluid phase) not only are pertinent calculations by themselves, they demonstrate the wide range of problems accessible to *Evolver* solution

Introduction

In materials science, many phenomena depend on the shapes of interfaces and surfaces. Many properties depend on interface shape, stability, and associated potentials. Calculating the shape of an interface is not usually a very easy exercise. When it is necessary to find the shape of an interface, a materials scientist can sometimes obtain serviceable information by judiciously reducing the dimensionality of the problem, or make some other simplifying approximation. However, sometimes such approximations are not warranted.

In some cases, the pertinent interface shapes are those A which minimize a total energy which depends, in part, on the interfaces. Finding the functional form of the minimizing interface is typically posed as a problem in the calculus of variations. The forces due to the interface, the interface potential and stability are can also be cast in the calculus of variations. Solutions have been worked out for simple geometries[1]; however, general problems are typically intractable except to numerical methods.

In some cases, numerical algorithms are written to solve a particular class of such interface problems and this can be an effective yet time-consuming endeavor.

A general program to solve minimal interface problems (and many others where a function is desired which minimizes other quantities) exists: Ken Brakke's Surface Evolver[2] [1] is public domain software[4] which is flexible and useful in materials where such problems often arise.

Below, examples will be shown where: 1) the surface shape is the object of interest, 2) the surface stability is the quantity of interest, and 3) the potentials are the quantity of interest.

Each of these *Evolver* calculations are pertinent to the field of materials science and each is a valid subject for study. However, the purpose is to demonstrate the utility of *Evolver* –hopefully with enough detail that they can be extended by similarity to other materials science problems. A library or working evolver codes is also available in the public domain. [5].

Since *Evolver* gives a representation of a surface as its result, those surface representation may be utilized as input for other stand-alone software. For instance, the surface may be extended into a volume mesh on which finite element techniques for determination of stress concentrations could be applied to solder joint geometries. In such cases, solder volumes and wetting conditions are either variable or not completely known, systematic variation of such parameters could give a materials scientist methods for determining the robustness of a process; or, via a combination of bootstrap[6] methods and finite element methods, means to determine reliability and lifetime predictions.

Brief Description of *SURFACE EVOLVER*

Evolver comes with a manual which describes the use and technical details of the code in great depth[4]. A short (and incomplete) description is presented here to put the following examples in context. An example follows which may clarify some of this description.

---

[1]The *Surface Evolver* is most often referred to as simply *Evolver* . The name sometimes causes confusion in the materials science community as evolution in the sense of *kinetic* evolution is not usually implied. Rather the function is evolved by iteratively finding a sequence of discretized shapes which continuously minimize the energy while satisfying imposed constraints. The particular path is determined by the energy minimization algorithm, of which there are several in *Evolver* . In some cases however, such as grain growth[3], the minimizing algorithm can be chosed so that it simulates the kinetics of motion by mean curvature properly; but this is not the usual case.

Evolver is a vertex-edge-facet element model of surfaces. Surfaces can be associated with bodies, but there are no volume elements. Energy is calculated by associating quantities, often vector quantities, to subsets of the facets, edges and vertices. The simplest contribution to the energy is from isotropic surface tension; in this case energy depends only on the area associated with a facet.

Volume elements do not exist, but volumes and volume energy are included by an integration technique which utilizes the divergence theorem as discussed below.

Facets always have three edges; any facets which are input to *Evolver* which have four or more edges get divided into coplanar triangles. Edges are directed quantities: they have a head and a tail vertex. Edges are assembled onto facets so that their direction obeys the right-hand-rule for the facet outward normal.

There are different types of constraints. The first are spatial constraints given as mathematical surfaces or boundaries to which vertices, edges or facets can be restricted. A second type of constraint acts in the same manner as a Lagrangian multiplyer. For instance, the volume of a body may be fixed. A body is the interior of the union of all its facets and regions bounded by edges on surface constraints (the example below will clarify this).

For facets which are supposed to change shape and orientation, fixing them to a spatial constraint is a bad idea (but still allowed as a last resort) since the vertices such facets will have gradients which typically lie normal to the constraint surface. Since they must be constantly projected back to the surface, those vertices tend not to move and become a bottleneck to further minimization. In order to avoid this, it is possible to associate line integrals with edges on constraints so as to account for the energy and volume associated with those surfaces so the need for placing facets there is obviated.

Once a model has been assembled, *Evolver* iterates towards a minimum in energy (plus possible Lagrange multiplyer terms) by methods which can be selected by the user. The default method simultaneously moves each vertex in the direction of gradient energy associated with the vertex coordinates (i.e., in the direction of the force on the vertex). Alternatively, one can use variants of conjugate gradient schemes or higher order methods or seeking minima. The mesh can be refined at any time, the most effortless scheme is to simply divide each triangle into four by placing a vertex in the middle of each edge by typing the single-letter command r.

The use of *Evolver* may be a daunting at first, but it really is not very complicated. One of the best things about *Evolver* (and this is a direct result of the thoughtfulness of its author) is that it is so flexible. It is flexible because *Evolver* is really a programming language and like all languages its utility depends on the user. There are several standard methods of solution which frequently are found in *Evolver* codes. With the benefit that one can solve many different kinds of problems comes the compromise is that much of the burden for the construction of a working model resides with the user. It is often productive to start from a working model. The following example is both useful and instructive.

An Example: The Shapes of Grain Boundary Second Phases

The code for this particular application (which can be found in Appendix 1) was written in less than one hour; a fairly accurate solution is obtained in about 30 seconds on a medium speed workstation.

The object is to calculate the shape of a fixed volume second phase which lies along the triple junctions of grains composed of a primary phase. The second phase forms a dihedral angle at the grain boundaries. If the second phase is gas phase (say, a vapor in equilibrium with the primary phase), this would be a good geometric model of a sintering body.

However, the arrangement of grains is somewhat unrealistic. The unit cell (primary phase in center plus grain boundary phase along the edges) for each grain was chosen as a rectangular prism to simplify the *Evolver* code. It is reasonably straightforward to set up the case where the grains are tetrakaidecahedra.

The method by which the code in Appendix 1 was developed can be briefly summarized. The model variables were determined. Account was made of various surfaces which were known to be important; in this case, the grain boundary and primary-secondary interface. Symmetries which were available were used to simplify the coding. Some vector potentials (described below in detail) were determined by "intelligent guessing" them and verifying the result. For the initial mesh, a picture was sketched by hand and the vertices, edges and facets where assigned. Several *Evolver* runs are made and adjustments are made to the surface mesh by visual inspection.

Please refer to Figure 1 for a graphic description of the code in Appendix 1. Note that *Evolver* comments have the C++ syntax (e.g, they are preceded by \\ or begin with * and end with *\

The *Evolver* code initializes parameters which may considered to make up a three-parameter family of interesting boundary phase shapes: the volume fraction of the primary phase, the dihedral angle, and the cell height. All these parameters can be changed any time during an *Evolver* session.

Next, various functions or useful quantities are defined. The functions can depend on the parameters; adjusting a parameter changes the value of a function during an *Evolver* session. The width and depth of the cell are defined so that total cell volume is unity; this could be easily generalized by defining more parameters. The effective tension is the difference in energy per unit area created by uncovering area on the constraint. The surface tension of the primary/secondary interface is unity. The offset is a parameter which is useful in setting up the intial mesh, described below.

The symmetry of the rectangular prism unit cell is used to advantage. [2] Only the upper octant is modeled. Generators of the symmetry group are introduced only to view the entire cell graphically.

Six constraints are specified in this *Evolver* code: three cell walls, where the grain boundaries are located, and three mirror planes.

Consider `constraint 1` (Figure 1). The constraint is given as the formula, `z = height`. Since the grain boundary is flat and since we would like to avoid putting facets on the grain boundary (see reasons given above), a method for accounting for the energy is required. the following stratagem is used (this trick is ubiquitous in *Evolver*): The energy of the grain boundary on `constraint 1` is given by:

$$E_{gb} = \gamma_{gb} \int_{A_{gb}} dA = \gamma_{gb} \int_{A_{gb}} -\mathbf{e_3} \cdot \mathbf{dA} \qquad (1)$$

where $\mathbf{e_3}$ is the unit vector parallel to the z direction and $\mathbf{dA}$ is an element of area on the constraint.

---

[2] It is also important to note that the imposed symmetry can also impose stability on a system which is unstable to asymmetric deformations of the surface. An example of this follows. When questions of stability are important, it is essential to allow the system to break symmetry. The current example is easily extended for such cases.

If the closed curve which surrounds A_{gb} (call it ∂A_{gb}, the boundary curve) is known, then it follows that A_{gb} should be determined from that curve. This is just a statement of a fundamental theorem from vector calculus:

$$\int_A \nabla \times \mathbf{v} \cdot \mathbf{dA} = \int_{\partial A} \mathbf{v} \cdot \mathbf{ds} \tag{2}$$

where $\mathbf{ds}$ is a directed line element on the boundary curve ∂A.

The task at hand is to find any $\mathbf{v}$ such that $\nabla \times \mathbf{v} = -\gamma_{gb}\mathbf{e_3}$. For planar problems like this, one choice is $\mathbf{v} = \gamma_{gb}\, y\, \mathbf{e_1}$. That vector potential becomes input for the energy: definition of that constraint and we leave it up to *Evolver* to use those edges on constraint 1 as the $\mathbf{ds}$ parts of a discrete approximation to an integral of $\mathbf{v} \cdot \mathbf{ds}$ over all the edges on the constraint.

There is also volume content associated with constraint 1. One way that *Evolver* calculates can calculate volume is simply account for the volume between a facet and its projection on the $z = 0$ plane. However there are no facets on constraint 1, so we must account for the volume by using the edges lying on the constraint.

The same trick is applied yet again to find the volume under the grain boundary:

$$V_{gb} = \int_V dV = \int_V \nabla \cdot \mathbf{F} dV = \int_{\partial V} \mathbf{F} \cdot \mathbf{dA} \tag{3}$$

where the vector field $\mathbf{F}$ is chosen so that $\nabla \cdot \mathbf{F} = 1$. There are many such choices for $\mathbf{F}$, a simple choice is: $\mathbf{F} = z\mathbf{e_3}$, in fact this is the one that *Evolver* uses as the default for the volume below a facet. [3] The last integral of Equation 3 looks like the first integral of Equation 2, so we should try to find vector field $\boldsymbol{\omega}$ such that $\nabla \times \boldsymbol{\omega} = \mathbf{F}$. However, it is not so straightforward since for any vector field $\boldsymbol{\omega}$ $\nabla \cdot (\nabla \times \boldsymbol{\omega}) = 0$, and we have specifically picked $\mathbf{F}$ so that $\nabla \cdot \mathbf{F} = 1$! The trick is to convert $\mathbf{F} = z\,\mathbf{e_3}$ into an equivalent divergenceless form on the constraint: $\mathbf{F} = (\texttt{height})\, \mathbf{e_3}$. Therefore, an $\boldsymbol{\omega}$ can be found, $\boldsymbol{\omega} = -(\texttt{height})\, x\, \mathbf{e_2}$ and that is what is written into the *Evolver* file as (c_1, c_2, c_3).

The other grain boundaries, constraint 2 and constraint 3, have similar terms for the energy vector integrand, but do not contribute any extra volume content since each one has a normal which is perpendicular to $\mathbf{e_3}$.

The other mirror planes contribute neither energy nor content. Note that contributing no energy is equivalent to setting the dihedral angle to π: where it intersects, the surface is perpendicular to the mirror plane.

Next, the geometry and connectivity of an initial surface is delineated. Referring to Figure 1, the vertices are "named" with a integer index. The index can be any positive integer (but smaller that the largest integer that your computer will read, of course) and it is useful to give elements indices that helps the user associate them. The coordinates of the vertices are given and the constraints on which they lie.

The edges are also given by an index and the vertices that define their beginning and end. The constraints not only specify a surface that the edge will lie upon (as well as any new edges which inherit that constraint as the edge is subdivided) but also specify that vector integrands be performed as required by the constraint. Here the edge indices are chosen here in an obvious manner which is not only useful, but aid in debugging their ordering on the face list.

---

[3]Note that for a facet, the there is no contribution to the volume from the three planar parts of the bounding surface ∂V which are the projections of the edges of the facets to $z = 0$ since $\mathbf{F}$ is normal to those surfaces, and there is no contribution from the "floor" $z = 0$.

Faces are given by an index plus an ordered list of oriented edges. It is useful to color the faces so that they can be graphically distinguished. Faces that have more than three edges are automatically subdivided by *Evolver* .

The bodies are given by an index plus a list of all of its faces and a volume may also be specified. The additional surfaces which bound the body are implied by the edges which lie on the constraint which are also members of faces in the face list.

Finally the program can be run. If the *Evolver* code given above is in a file called boundary_phase.fe it is typically executed by typing evolver boundary_phase.fe at a computer prompt. It is essential to use graphics of some sort. Refer to the manual that comes with the *Evolver* source for details for particular computer systems.

To produce the result given in Figure 1, the following *Evolver* commands were given r; g30 ; u ; V ; g30 ; r ; g30 ; quadratic ; U ; g50; which means: 1) r refine the surface mesh by subdividing the edges, 2) g30 go 30 iterations by moving each vertex in a direction parallel to its local gradient, 3) u adjust the mesh by making edge switches to make facets more equilateral, 4) V adjust the mesh by vertex averaging, this tends to tighten the distribution of facet areas, 5) g30 ; r ; g30 go another 30, refine again, and go another 30, 6) quadratic switch to a higher-order approximation, 7) U turn on conjugate gradient as the minimizing scheme, 8) g30 go 30 conjugate gradient iterations.

The parameters can be changed at any time by typing A.

The entire unit cell can be viewed by utilizing all the symmetry operations which were part of the boundary_phase.fe by typing the *Evolver* command transform_expr "abc" which generates the (mmm) group.

Questions of convergence naturally arise and there are some built-in methods in *Evolver* to judge how close a current mesh is to its minimal configuration. As an initial guide, one uses the relative decrease of each iteration's relative decrease in energy judge the magnitude of the global gradient in energy. It is useful to switch between minimization schemes as the minimizing solution is approached. However, care must be taken to insure that the mesh does not develop irregularities such as very small edges or facets at the same time–such irregularities *globally* decrease the distance that *Evolver* allows each iteration to move the vertices. Keeping the mesh regular is an art which is developed by experience and careful inspection of the graphical output as well as a useful continuously displayed quantity called the "motion scale".

It is also useful to perturb vertices see if they are restored to their positions through further iteration. This method is described in the next session.

Using *SURFACE EVOLVER* to Determine Stability

Another example of a useful *Evolver* calculation is the determination of the critical volume of solder which can cause a "short" between two flat leads in a chip array[7].

In Figure 2, two such leads are modeled as constraints for an evolver calculation. The leads are separated by a gap, over which the solder droplet sits. The *Evolver* file for this calculation is not very different from the one given above, except that the entire droplet is modeled instead of using the two possible mirror planes.

The wetting angle for the solder on the lead was arbitrarily chosen as $\pi/4$. The gap distance was 1/2 (gravity was ignored, so the length units can be ignored, but then calculation then only applies for small capillary lengths). The initial mesh was symmetric across the gap. The volume as manually adjusted until a critical amount was obtained that gave a "stable" *Evolver* solution. The amount of volume depicted in Figure 2 was

4.875 (in units of gap distance cubed) and it is slightly above the critical volume at which the droplet breaks symmetrically into two isolated droplets.

Evolver can calculate the Hessian of this current solution and one finds that it is not positive definite; so this solution is not a good determination of the critical bridging volume.

It is instructive to examine the stability using the following scheme. It may be conjectured that the solder drop will go unstable by breaking the symmetry across the gap; this can be tested by giving the vertices a "kick" in that direction. This is executed by typing the *Evolver* command set vertices x x+.05, which moves all vertices to one side, except those constrained from such motions. Subsequent iteration clearly shows (Figure 2c) that the system was in fact unstable to an asymmetric mode. Considerably more investigation is required to find the actual critical volume; note that merely calculating the droplet sizes resulting from asymetric break-up would be insufficient since hysteresis should be expected.

Imposing a mirror plane on the original calculation would lead to an unphysical estimation for the critical volume. If the lead is finite, instead of semi-infinite as in Figure 2, the stability will also be affected as the droplet contacts the other lead boundaries. Such contact problems are handled in a straightforward manner[8].

Using *SURFACE EVOLVER* to Determine Forces and Potentials

As a final example of a useful *Evolver* calculation, consider the case of an assembly of small spheres which have a liquid phase wetting the contact points and interstices.

This is a reasonable idealization of a drying powder compact or liquid-phase sintering. In each case, the forces which the liquid menisci apply to the assembly of spheres can cause local rearrangement, shrinkage, or result in crack-like features where the menisci become unstable and break. In addition, the meniscus curvature will depend on the wetting angle, liquid volume, and the local arrangement of the solid particle. Thus, the local vapor pressures within the compact may vary considerably and influence drying behavior.

Evolver can be used to calculate such forces and curvatures[9]. Consider three spheres with a fixed amount of volume of fluid located in the interstice as in the top of Figure 3. If the total energy E of such a system can be calculated as the upper two spheres in Figure 3 are rotated away from each other by some angle θ, then the torque T that the meniscus applies to the spheres can be calculated by $T = -(dE/d\theta)_{fixed\ volume}$. Similarly, if the energy is calculated as the volume of fluid is increased, then the mean curvature can be calculated as $K = (dE/dV)_{fixed\ \theta}$ and the vapor pressure can be inferred through the Gibbs-Thomsen equation.

The *Evolver* code in Appendix 2 illustrates how to do such a calculation. Parameters are defined for the wetting angle, the sphere opening angle, and the meniscus volume. To model the entire system, three constraints must be defined and vector integrands must be obtained for the energy and volume content on each; the results are given in the code which follows.

Macro functions are defined at the end of the file in Appendix 2 which allows the program to run on its own. These may provide some useful examples for other similar problems. The macro run_this given at the end of the file does the following: increments θ; iterates toward a solution while doing some mesh cleaning operations; stops iterating when the energy decrease falls below a chosen tolerance; writes data to a file; saves the surface every ten increments; repeats.

Some results are plotted in Figure 3 for two $V_f \equiv = V_{meniscus}/V_{sphere}$. The end of the curves signify opening angles where no *Evolver* solution could be found and are thus an estimate for the critical opening angle for each V_f

The torques are such that the particles tend to be drawn back together; this is probably not the case for $\phi > \pi/2$. The solid vertical curve in the plot can be used to calculate the curvatures in the collapsed configuration.

Such calculations could be done for many different values of V_f and θ and placed in a look-up table for a simulation of a many particle compact. The solutions for bridges between two spheres are well-known and have been worked out.[10]. An *Evolver* calculation for a meniscus in a tetrahedral interstice would be a nice contribution to the literature.

The same method of taking derivatives can be applied to the first example of the grain boundary phase. Using that same program and varying the height of the cell, the force exerted by the top grain boundary constraint could not be numerically distinguished from 0. However, the second derivative of energy with respect the height is negative–indicating that this particular symmetric geometry is unstable. The total surface potential, $\partial E/\partial V$, was also numerically determined to be $0.37 \gamma_{lv}/(cell\ volume)^{(1/3)}$ and the potential is a decreasing function of density; so the system will be unstable to coarsening.

Summary

The three calculations using *Evolver* which have been demonstrated are fairly disparate examples of useful computations in materials science. There are many more which will be described in a future publication. The purpose in this paper was to introduce *Evolver* to members of the materials science community as new research tool–hopefully with enough working (and hopefully motivating) examples that new challenging materials science problems can be tackled with *Evolver* .

Ken Brakke, the mathematician who wrote *Evolver*, deserves an enormous amount of credit. Since materials science tends to be such an eclectic discipline, it is very useful to absorb progress from different disciplines. There is a richness of mathematical ideas in *Evolver*, after all it was written with problems in mathematics in mind. Nevertheless, materials scientists will use new tools where they can find them.

One reason that *Evolver* is so useful is that it is a *programming language*. This gives it useful features; *Evolver* is:

Flexible It can be used to solve an enormous range of problems. The concept of an *Evolver* mesh need not even be geometric, but any data structure that can mapped onto an *Evolver* mesh could be minimized. Because parameters need not be fixed, but can be changed on the fly, the code is re-usable.

Progammable One can query variables, calculate expressions and then have *Evolver* act in a controlled fashion according to the result. New macros and functions can be written *in-situ*. There are very many ways to add energies to elements; the *Evolver* manual describes about 40 and more are always being added.

Quantitative It is possible to get real numbers out of *Evolver* such as those which were used to make the plots in Figure 3. One can also obtain quite general information, for instance in one application it was important to find how a contact angle was changing on a constraint. The following *Evolver* command was able to extract the information: `foreach edge ee where on_constraint 1 do { foreach ee.vertex do`

```
{printf "%g %g %g\n",x,y,z}; foreach ee.facet fff do
{ foreach fff.vertex do {printf "%g %g %g\n",x,y,z}}}
```
. This printed the positions of the ends of each edge on a constraint, and then printed the positions of the corners of the facet belonging to that edge. In turn, this later served as input to some auxiliary code which calculated the contact angle as a function of position.

Visual Seeing the result is very important and certainly one of the best ways to learn more about the problem being solved.

Extensible Instructions for adding source code to *Evolver* is included with the distribution. Invariably, one wants to try something new and since the source is free it's possible to try anything. A good example is an *Evolver* calculation on line energies referred to in these proceedings by Jennifer Kelley.

The bottom line is that one doesn't have to write a minimization algorithm. It becomes necessary only to concentrate on the problem at hand. However, there is still some work involved. An initial mesh needs to be constructed. There is some art in maintaining a good mesh by judiciously deleting and refining elements. And there is some cleverness required in the application of the Stoke's and the divergence theorems.

One of the most useful ways to get started is to adapt some working *Evolver* code; such working codes, including the ones described in this paper are available on the net[5].

Acknowledgements: Ken Brakke not only wrote *Evolver* but also patiently gives astoundingly great answers to all of my questions. Jim Warren provided useful dialogue and suggestions as well as produced code for the bridging example. Carol Handwerker has inspired much of my *Evolver* related work. NSF and the Geometry Center hosted a workshop called, "Computational Crystal Growth" which was organized by Jean Taylor where I was introduced to *Evolver* and many other important applications of mathematics in materials science. Long-Qing Chen made useful and encouraging suggestions and has my gratitude for his patience.

References

[1] W.C. Carter. The forces and behavior of fluids constrained by solids. *Acta. Metall.*, 36(8):2283–2292, 1988.

[2] K. A. Brakke. The surface evolver. *Experimental Mathematics*, 1(2):141–165, 1992.

[3] K.A. Brakke. Grain growth with the surface evolver. In J.A. Taylor, editor, *Video Proceedings of the Workshop on Computational Crystal Growth*, Providence, RI, 1991. American Mathematical Society.

[4] Available at (http://www.geom.umn.edu/software/download/evolver.html), or via anonymous ftp (geom.umn.edu: pub/software/evolver).

[5] Available at (http://www.ctcms.nist.gov/programs/solder/archive_solder.html).

[6] B. Efron and R. Tibshirani. Bootstrap Methods for Standard Errors, Confidence Intervals, and Other Measures of Statistical Accuracy. *Stat. Sci.*, 1:54–77, 1986.

[7] J.A. Warren, W.C. Carter, D. Josell, C.A. Handwerker, K-W. Moon. Moldeling solder interconnects: A nist program. In *I.P.C. Proceedings*, 1996. In Press.

[8] J.A. Warren, W.C. Carter, D. Josell. Stability of the gull-wing solder geometry. In Preparation, 1996.

[9] W.C. Carter and T. Shaw. Rearrangement forces in liquid phase sintering. Unpublished Research.

[10] W. C. Carter. *Capillary Induced Microstructural Evolution in Porous Materials*. PhD thesis, University of California, Berkeley, 1989.

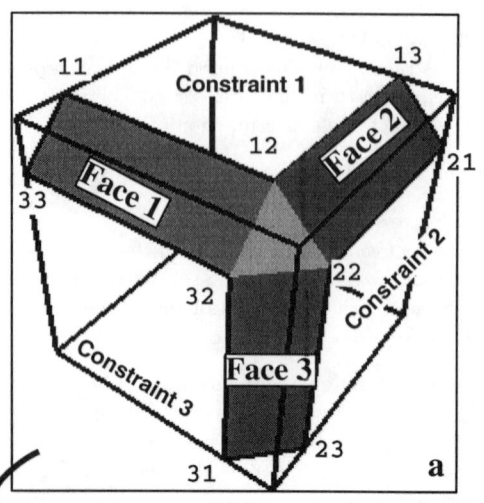
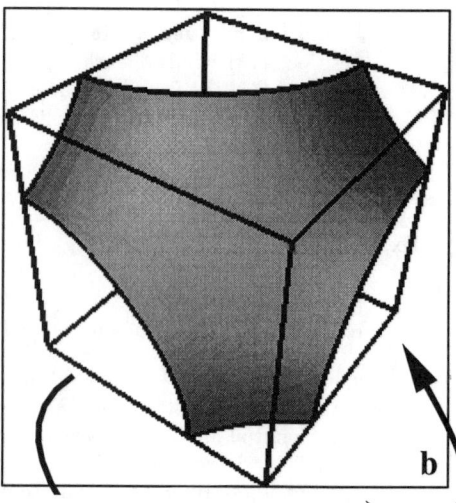

refine and iterate with Evolver (`r;g30;u;V;g30;r;g30;quadratic;U;g50`)

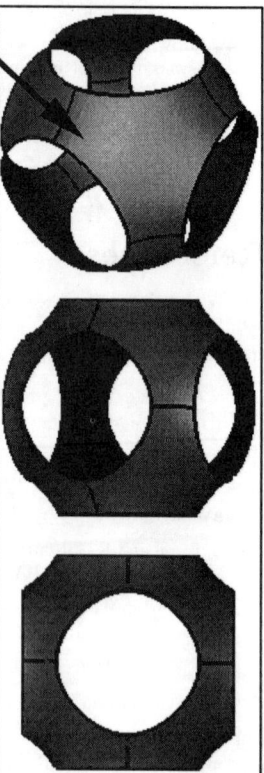

Dihedral Angle 120°
Volume Fraction of Boundary Phase, 25%
Figure 1: Example of *Evolver* calculation of the shape of a second phase wetting a grain boundary. The first panel shows how the initial mesh was constructed.

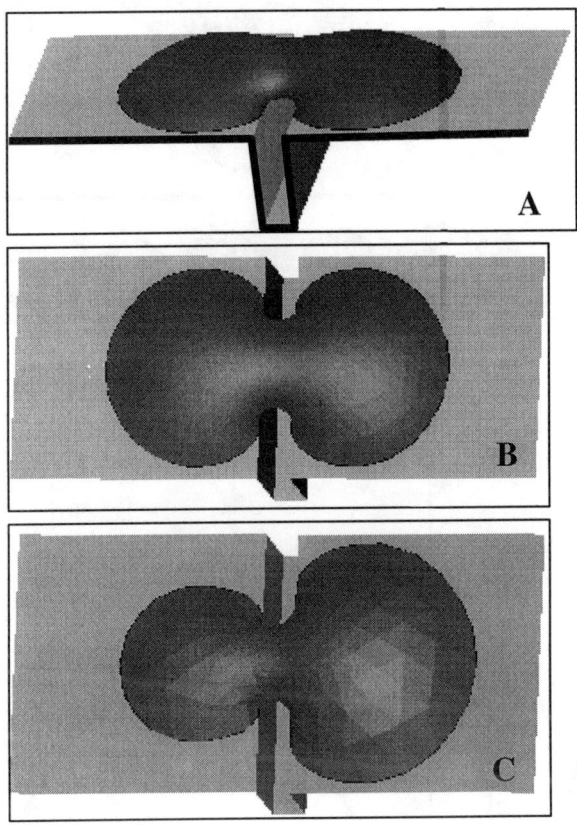

Figure 2: Evolver calculation for a the minimun critical volume for a symmetric solder bridge spanning two leads. (C) shows the onset of asymmetric instability when the vertices are perturbed to the right

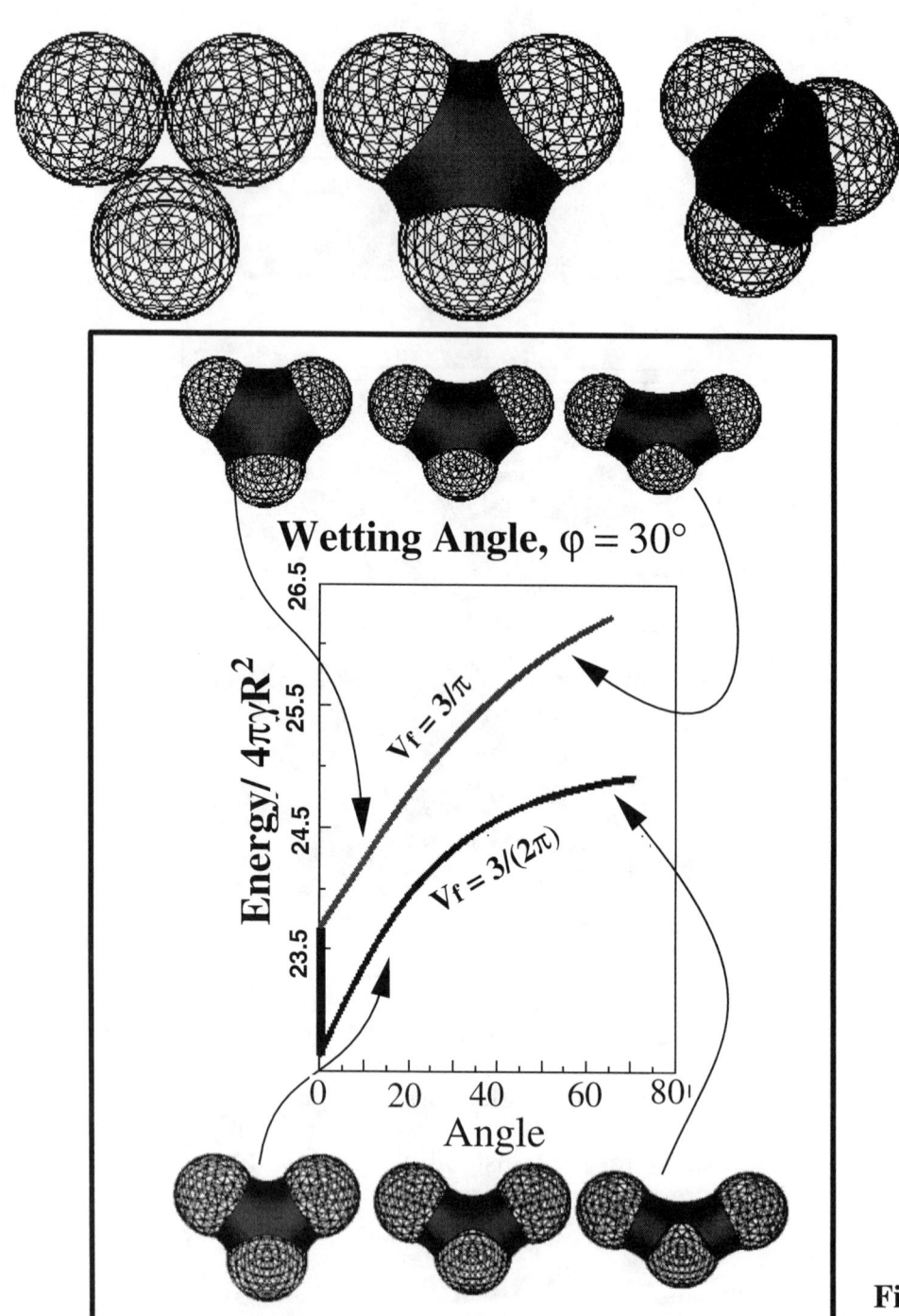

Fig. 3

Appendix 1: boundary_phase.fe

```
/* boundary_phase.fe
   ─────────────
Rectangular prism of volume 1 and height="height"
Grain Boundaries at the prism faces.
Second Phase at prism edges,
with dihedral angle = "dihedral_angle"
Volume of Primary Phase = "theoretical_density"
All PARAMETERS adjustable with
the "A" single-letter evolver command.
Only upper octant of prism modeled,
other parts can be viewed
using the group generators included here (mmm).

Have Fun, Right to use unlimited.
W. Craig Carter, NIST, wcraig@pruffle.nist.gov
6 Feb. 1996

*/

PARAMETER theoretical_density = 0.75
PARAMETER dihedral_angle = 120
PARAMETER height = 1

#define width sqrt(1/height)
#define depth sqrt(1/height)
#define effective_tension (cos(dihedral_angle*pi/360))
#define offset 0.8

/** symmetry stuff, ignore this if you are new to Evolver *

view_transform_generators 3 //three mirror planes
//to generate full rectangular symmetry
//only the 3x3 upper submatrix matters
//for our purposes

-1 0 0 0  //a: mirror across X
 0 1 0 0
 0 0 1 0
 0 0 0 1

 1 0 0 0  //b: mirror across Y
 0 -1 0 0
 0 0 1 0
 0 0 0 1

 1 0 0 0  //c: mirror across Z
 0 1 0 0
 0 0 -1 0
 0 0 0 1
/***** end of symmetry stuff *************************

constraint 1 //top of prism
formula: z = height  //normal -k
energy: // vector integrand for effective energy
e1: effective_tension*y // components are: (e1,e2,e3)
e2: 0
e3: 0
content: //vector integrand for volume
c1: 0 // from constraint surface
c2: -height*x
c3: 0

constraint 2 //right side
formula: y = width//normal -j
energy:
e1: 0
e2: 0
e3: effective_tension*x

constraint 3 //front side
formula: x = depth //normal -i
energy:
e1: 0
e2: effective_tension*z
e3: 0

constraint 11 //mirror plane
formula: z = 0

constraint 12 //mirror plane
formula: y = 0

constraint 13 //mirror plane
formula: x = 0

vertices //given as: vertex_id x y x and which constraint
11 offset 0 1 constraint 1 12
12 offset offset 1 constraint 1
13 0 offset 1 constraint 1 13

21 0 1 height*offset constraint 2 13
22 offset 1 height*offset constraint 2
23 offset 1 0 constraint 2 11

31 1 offset 0 constraint 3 11
32 1 offset height*offset constraint 3
33 1 0 height*offset constraint 3 12

edges //given as: id head_vertex_id tail_vertex_id and constraint
// Evolver will do line integrals for edges on constraints
// to account for energy and volume
1133 11 33 constraint 12
3332 33 32 constraint 3
3212 32 12
1211 12 11 constraint 1

1222 12 22
2221 22 21 constraint 2
2113 21 13 constraint 13
1312 13 12 constraint 1

3123 31 23 constraint 11
2322 23 22 constraint 2
2232 22 32
3231 32 31 constraint 3

faces // given as list of oriented edges
1 1133 3332 3212 1211 color red //it is useful to use colors
2 1222 2221 2113 1312 color green //to identify which is which
3 3123 2322 2232 3231 color blue
4 -1222 -3212 -2232

bodies
1 1 2 3 4 volume theoretical_density
/* We have one body made up of four faces
the faces have edges which lie on constraints
which encloses the body and add volume through
line integral of content vector field.

read //i.e., read the rest as Evolver commands
//P 8 /*
Commented out, use only for fancy graphics, see manual for
graphics commands for the system you use.
*/
//r; g30 ; u ; V ; g30 ; r ; g30 ; quadratic ; U ; g50
/*
Suggested method to produce
decent result and one illustrated in figure.
*/

//transform_expr "abc"
/*If graphics are on, this generates,
by symmetry, the whole prism.*/
```

Appendix 2: three_spheres.fe

```
/* Evolver file for three spheres,
modeled fluid volume sitting in
interstice.
Bottom one with center at origin,
is fixed, the other rotate around
the center amounts +- theta/2
W. Craig Carter, 1996
*/

SYMMETRIC_CONTENT

PARAMETER phi = 30 // liquid wetting angle
PARAMETER theta = 0 // sphere opening angle
PARAMETER vol = 1.5 // liquid volume
// !!! NEED volconst = -2*pi

#define Rs 1 // sets units

#define gamsl (-cos(phi*pi/180)) // effective energy

/* a bunch of functions below, used for vector integrands */

#define ct2 cos( 60*pi/180 - theta*pi/360 )
#define st2 sin( 60*pi/180 - theta*pi/360 )
#define ct3 cos( 120*pi/180 + theta*pi/360 )
#define st3 sin( 120*pi/180 + theta*pi/360 )

/*radius^2 of bottom sphere*/
#define r1sq (x)^2 + (y)^2 + z^2

/*upper right sphere */
#define r2sq (x - 2*Rs*ct2)^2 + (y - 2*Rs*st2)^2 + z^2

/*upper left sphere */
#define r3sq (x - 2*Rs*ct3)^2 + (y - 2*Rs*st3)^2 + z^2

#define w1num y
#define w1den (x^2 + z^2)*sqrt(r1sq)
#define w1 w1num/d1den

#define w2num x*ct2 + y*st2 - 2*Rs
#define w2den ((st2*x - ct2*y)^2 + z^2)*sqrt(r2sq)
#define w2 w2num/d2den

#define w3num x*ct3 + y*st3 - 2*Rs
#define w3den ((st3*x - ct3*y)^2 + z^2)*sqrt(r3sq)
#define w3 w3num/d3den

constraint 1 //bottom sphere
formula: r1sq = Rs^2
energy:
e1: gamsl*(Rs^2)*w1*z
e2: 0
e3: -gamsl*(Rs^2)*w1*x
content:
c1: ((Rs^3)/3)*w1*z
c2: 0
c3: -((Rs^3)/3)*w1*x

constraint 2 //northeast sphere theta = Pi/6
formula: r2sq = Rs^2
energy:
e1: (gamsl*(Rs^2)*w2*(z*st2))
e2: -(gamsl*(Rs^2)*w2*(z*ct2))
e3: -(gamsl*(Rs^2)*w2*(x*st2 - y*ct2))
content:
c1: -(-(((Rs^3)/3)*w2*(z*st2)) + z*(2*Rs)*st2/3)
c2: -((((Rs^3)/3)*w2*(z*ct2)) - z*(2*Rs)*ct2/3)
c3: -(((Rs^3)/3)*w2*(x*st2 - y*ct2))

constraint 3
formula: r3sq = Rs^2
energy:
e1: (gamsl*(Rs^2)*w3*(z*st3))
e2: -(gamsl*(Rs^2)*w3*(z*ct3))
e3: -(gamsl*(Rs^2)*w3*(x*st3 - y*ct3))
content:
c1: -(-(((Rs^3)/3)*w3*(z*st3)) + z*(2*Rs)*st3/3)
c2: -((((Rs^3)/3)*w3*(z*ct3)) - z*(2*Rs)*ct3/3)
c3: -(((Rs^3)/3)*w3*(x*st3 - y*ct3))

vertices
//:
//:
edges
//:
//:
faces
//:
//:
bodies
//:

//: volconst -2*pi

read
checksmall := {delete edges where length < 0.0001; \
delete facets where area < 0.005; u; \
delete edges where length < 0.0001; u;g3}

edgemax := 0.4
echeck := {while max(edges,length) > edgemax do \
{ refine edges where length > edgemax } ; \
checksmall }

grind := { ig:=1 ; change := 1.; olde := total_energy ;\
while (ig < 10000 && abs(change) > .00001) \
do {ig := ig+1; \
if ig%10 == 1 then checksmall ; \
g -= "cat > null" ; \
if ig%3 == 1 then \
{ change := total_energy - olde ;\
olde := total_energy;\
printf "sum of 3 dE: %12.10f\n", change}}}

goprint := \
{ printf "%f %f %20.12f %f " \
,phi, theta, total_energy, change -- "cat >> results";\
foreach body do { printf "%f ", volume -- "cat >> results" } ;\
printf "%f ", total_area -- "cat >> results" ;\
printf "%g ", facet_count -- "cat >> results" ; \
printf "%f \n", max(edges,length) -- "cat >> results" }

dump_to_file := \
{filename := sprintf "dump.v_%f_p_%f_t_%f",vol,phi,theta;\
dump filename}

inc := {echeck; D; D ; O ; o ;u; theta := theta + 1 ; \
printf "%f", theta ;u;g5; biggrind ; \
if theta%10 == 0 then dump_to_file; goprint }

run_this := {dump_to_file ; while (theta < 120) do {inc}}
```

THE SIMULATION OF TWO-DIMENSIONAL GRAIN GROWTH

BY THE SURFACE EVOLVER

Knut Marthinsen[1], Ola Hunderi[2], and Nils Ryum[3]

[1]SINTEF Materials Technology
and
[2]Department of Physics and [3]Department of Metallurgy,
Norwegian University of Science and Technology,
N-7034 Trondheim, NORWAY

Abstract

The present work describes an interesting an important application of the program Surface Evolver, namely the dynamic process of normal grain growth in two dimensions. The main focus of the present work is on the evolution of the grain size distribution and the existence and characterization of grain size correlations. Two-dimensional grain structures with a total of more than 10^4 Poisson Voronoi cells have been coarsened until approximately 2000 grains remained. It is demonstrated that a size correlation between neighboring grains exists in such two-dimensional grain structures, and a spatial grain size correlation function that relates the size of a grain to the average size of its neighbors is defined. It is demonstrated that both the distribution function and the correlation function approach a quasi stationary state when the mean grain area has roughly doubled compared to its initial value. The results are related to similar investigations in one dimension and also to results obtained by the Monte Carlo technique in two dimensions.

Introduction

Although grain growth has been extensively studied for several decades, not all fundamental aspects of the process are fully understood (see e.g., [1] and references therein). In particular, there is still no fully adequate analytical theory for normal grain growth in two or three dimensions, partly because of the complexity of the topological interactions involved. The interest in grain growth is related to the importance of grain size in controlling the physical and mechanical properties of polycrystalline materials.

The driving force for the process of normal grain growth is the reduction of the total grain boundary free energy. The process is thus well suited for simulation by the program Surface Evolver [2]. The Surface Evolver is an interactive computer program that minimizes the energy of a surface subjected to constraints. The energy can include surface energy, gravitational energy and other energies. The minimization is carried out by evolving the system towards a minimal total surface energy by a gradient descent method. In the present work the Surface Evolver has been used to simulate normal grain growth in two dimensions with focus on the evolution of the grain size distribution and in particular the evolution of spatial grain size correlations. The objective has been to demonstrate the application of the Surface Evolver to two-dimensional grain growth and in particular to demonstrate and characterize the existence of spatial size correlations during two-dimensional grain growth.

In a recent work by Hunderi and Ryum [3], it is demonstrated by computer simulations that a size correlation exists between neighboring grains in one-dimensional grain growth, and it was also found that by including the correlation effect in a modified mean field theory a very accurate analytical description could be given for the coarsening process in one dimension. A correlation effect in the coarsening of two-dimensional network structures has also been discussed in relation to two-dimensional soap-froth experiments and in corresponding computer simulations (e.g., [4-8]). Usually correlations in two-dimensional grain growth have been discussed in terms of the Aboav-Weaire's law [9,10]. Aboav-Weaire's law relates the number of sides n of a grain to the average sides of its neighbors, m(n). In the present work we focus on **size** correlations, and a spatial grain size correlation function is introduced which relates the size of a grain, R, and the average size if its neighbors. In a separate publication this size correlation effect is further discussed and used to develop a modified mean field theory, and its effect on the growth kinetics is discussed and compared to simulation results [11].

Simulation of two-dimensional grain growth by the Surface Evolver

The basic operation of the Surface Evolver is the evolution of an initial surface toward minimal energy by a gradient descent method [2]. In general, the Surface Evolver operates on two-dimensional surfaces in three-dimensional space, but strings in two dimensions can also be handled. This is referred to as the string model. It is the string model that is the relevant one for two-dimensional grain growth.

The starting structure for the evolution is an arbitrary grain structure in a two-dimensional unit square with periodic boundary conditions. A two-dimensional grain structure is defined by a set of vertices, edges and faces. The vertices are points defined by their coordinates, the edges are straight lines joining pair of vertices and each face (the two-dimensional "grains") is defined by a chain of directed edges that make up its boundary. To approximate curved grain boundaries each boundary is made up of many straight line segments (edges). The total energy of the grain structure is the total length of grain boundaries times the surface tension γ (i.e., in two dimensions, the grain boundary free energy per unit length). The driving force for the grain structure evolution is the minimization of total grain boundary energy where each grain boundary is assumed to move towards its center of curvature with a velocity V proportional to mean curvature, 1/R, i.e.,

$$V = M\gamma \frac{1}{R} \qquad (1)$$

where M is the grain boundary mobility, and both M and γ are constant. Computationally the grain structure is changed by moving the vertices. This is done by calculating the velocity at each vertex and move the vertex in the given direction. The velocity is calculated from the force on each vertex which is the negative gradient of the total energy of the system as a function of the position of that vertex. Motion by mean curvature is approximated by operating the Surface Evolver in the "area normalization" mode where the force on each vertex is divided by half the length of the adjacent edges. By the command "effective_area" in effect it is further guaranteed that the rate of change of area of a grain is proportional to (n-6), where n is the number of grains, i.e., in accordance with the von-Neumann-Mullin's law [12,13]:

$$\frac{dA}{dt} = \frac{\pi}{3} M(n-6) \qquad (2)$$

One of the challenges of modeling grain growth is the continual change of topology as edges and grains disappear and new edges appear. Surface Evolver automatically detects an edge that shrinks to less than a critical value and eliminates it by merging its endpoints. If this forms a vertex with more than three edges, that vertex is split into several triple vertices with short edges between them.

Simulation results

Two-dimensional grain growth has been investigated from initial grain structures of 10^3 Poisson Voronoi cells that are subjected to extensive grain growth. The results presented are averages over more than ten runs with different initial Voronoi cell structures. The program to generate the two-dimensional Voronoi cell structures of a format suitable for input to the Surface Evolver has been provided by Brakke [14].

Fig. 1 shows the growth kinetics as expressed by the variation of <A>/<A>(t=0) versus time. It is seen that the growth rate (d<A>/dt) initially is low, according to the correlation effect demonstrated below (Fig. 6), and then increases until it reaches a constant value, i.e., in the long term regime there is a linear relationship between <A> and time in accordance with theoretical predictions (e.g., [15,16]) and recent similar simulation results (e.g., [8,17]).

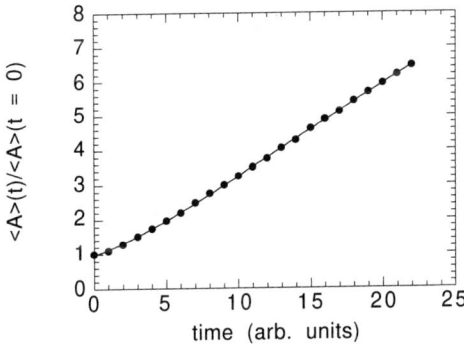

Fig. 1   The variation of the mean grain size <A> with time t.

The initial normalized grain size distribution function F(R/<R>) as well as its evolution with time are shown in Fig. 2. After an initial period with a general broadening, the size distribution approaches a quasi stationary shape, i.e., a scaling regime, after a relative short period of time.

The time to reach the scaling regime corresponds roughly to a doubling of the mean grain size <A> and is equal to the time needed to reach the linear growth region in Fig. 1. This is in agreement with what was observed in the one-dimensional case as discussed by Hunderi and Ryum [3]. However, in contrast to the latter case it is interesting to note that except for a general broadening, the initial symmetry of the distribution function around the mean grain size is largely retained during the grain growth process. The final quasi stationary size distribution function, however, is very similar to the one obtained for the one-dimensional case. The results in Fig. 2 are based on an average over 11 runs, and the final distribution function corresponds to an increase of the mean grain size, <A>, by a factor of 4 - 5.

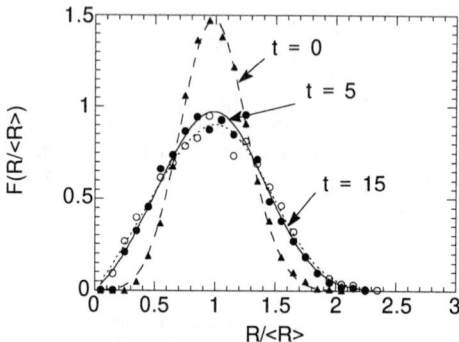

Fig. 2   The normalized grain size distribution function F(R/<R>) at three different coarsening time steps. The corresponding coarsening factors can be found from Fig. 1.

Several expressions have been proposed to describe this quasi stationary grain size distribution (e.g., [18-20]). In Fig. 3 the present simulation result is compared to Hillert's mean field result, which is given by [19]

$$F(\rho) = (2e)^2 \frac{2\rho}{(2-\rho)^4} \exp(-\frac{4}{2-\rho}), \qquad (3)$$

and the generalized Louat's function [20], given by

$$F(\rho) = 2\alpha\rho \exp(-\alpha\rho^2) \qquad (4)$$

Here $\rho = R/<R>$, and α is a parameter. It is clear that none of these theoretical distribution functions gives an appropriate fit to the simulation results, although the Louat function gives a somewhat better fit than Hillert's mean field result. However, this is in agreement with previous findings, and the present simulation results seem to agree fairly well with previous Monte Carlo Q-Potts model simulations [8,17,21].

Correlation effects in two-dimensional grain growth are usually discussed in terms of Aboav-Weaire's law, which in a general form, as suggested by Weaire [10], is given by

$$m(n) = 6 - a + \frac{6a + \mu_2}{n} \qquad (5)$$

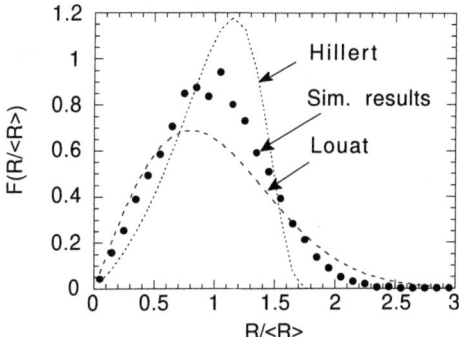

Fig. 3 A comparison of the quasi stationary distribution function as obtained from the present simulation (back dots) and those proposed by Hillert [19] (dotted) and Louat [20] (dashed).

Here m(n) is the average number of sides of the neighboring grains to grains with n sides, μ_2 is the second moment (variance) of the side distribution, and a is a constant with a value close to 1. The grain side distribution function, i.e., the frequency of occurrence of grains with a certain number of sides as a function of the number of sides, corresponding to t = 15 in Fig. 1, is shown in Fig. 4. It is seen to have a maximum at n = 6, and the mean value, <n>, and the second moment, μ_2, can be calculated to be 6.00 and 1.62, respectively. These results agree reasonably well with recent Monte Carlo Q-Potts model simulations [8]. The reason why we get <n> = 6 is a direct consequence of the fact that von-Neumann-Mullin's law is an intrinsic part of the present simulation model [2]. However, it is not a neccessary consequence of von-Neumann-Mullin's law that the side distribution should have a maximum at n = 6. In the Q-Potts model simulations [8] and in recent simulations using a continuum field model [22,23] it was observed that the maximum was shifted from n = 6 to n = 5 in the scaling regime. The reason why we do not observe the same result in our simulations may be due to the fact that the side distribution function converges more slowly to a quasi stationary state than the size distribution function. The corresponding m(n) values as obtained from our simulation data are shown in Fig. 5. These data are very well fitted by eqn. (5) with parameters a = 1.3 and μ_2 = 2.4 (solid line)

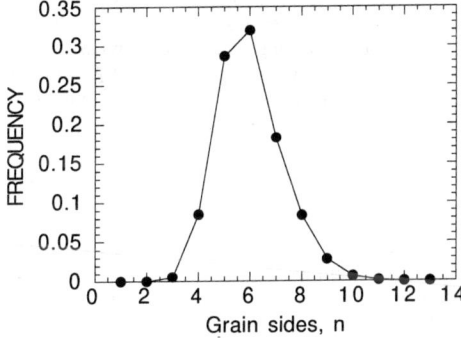

Fig. 4 The grain side distribution at t = 15., i.e., at a time where the grain size distribution function has reached its scaling regime.

in agreement with the previous simulation results [8,22,23]. The fact that the side distribution (Fig. 4) seems to predict a lower value for µ2 is not clear, but a possible explanation may be related to an inadequate description of the upper tail of the grain side distribution. The value of µ2 is very sensitive to this tail, and since the number of grains which constitute the upper tail in Fig. 4 is very small, it is not very well defined and it may be erroneous, and this may be the reason for an underestimation of the value of µ2 as compared to its actual value.

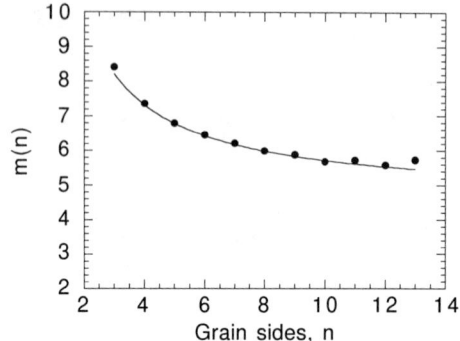

Fig. 5   The neighbor side correlation function m(n) at t = 15, i.e., at the same time as in Fig. 4.

In order to further characterize the grain size correlation effect and its time evolution, a grain size correlation function is defined. In a two-dimensional grain structure each grain has several nearest neighbors. For one particular grain of size R, we call the mean size of the neighbors R_n. The different grains of size R will of course have different values of R_n. The grain size correlation function $G(R/<R>)$ is now defined in the following way

$$G(\frac{R}{<R>}) = \frac{<R_n>}{R_{nm}} \qquad (6)$$

i.e., $G(R/<R>)$ is the average size of all the neighbors to the grains in class R, $<R_n>$, scaled by the arithmetic mean of all neighboring grains, R_{nm}. It should be noted that R_{nm} is not simply equal to $<R>$, the arithmetic mean of all the grains in the ensemble. This is due to the fact that a large grain in most cases is neighbor to more grains than a smaller grain. This means that on an average a large grain is averaged more frequent than a smaller grain in the calculation of R_{nm}. Consequently R_{nm} is somewhat larger than $<R>$. From the present two-dimensional simulations it was found that $R_{nm}/<R> = 1.22$ in the scaling regime.

The grain size correlation function for the initial random Voronoi structure and its time evolution are shown in Fig. 6. As in Fig. 2 these results are based on an average over more than 10 runs. As can be seen the initial correlation effect is quite strong and, on an average, small grains tend to be surrounded by small grains and vice versa. However, it is seen that the correlation function changes quite dramatically during grain growth. After a relative short period of time, there is a tendency for the small grains to be surrounded by large grains and vice versa. This is just the opposite of the correlation that existed in the initial grain structure. It is further seen that also the grain size correlation function seems to approach a quasi stationary state after a transient period comparable to the one observed for the grain size distribution function. The results in Fig. 6 are in agreement with the corresponding results obtained for one-dimensional grain growth [3].

However, both the initial and the final quasi stationary size correlation seem to be somewhat weaker in the two-dimensional than in the one-dimensional case. This is not unexpected, since in the one-dimensional case, the average size of the neighbors to a particular grain is always an average of only two grains, while the corresponding average in the two-dimensional case is an average of typically 3-9 grains.

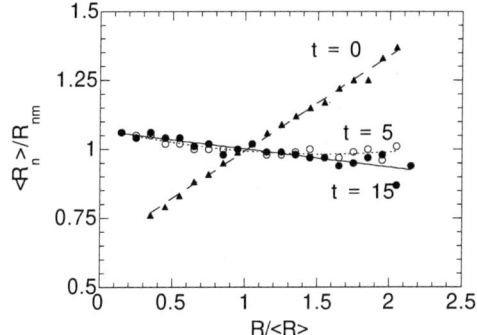

Fig. 6   The correlation function ($G(R/<R>)$) at three different time steps corresponding to Fig. 2. The corresponding coarsening factors can be found from Fig. 1.

Discussion and conclusions

It is clear from the present work that the program Surface Evolver is capable of simulating idealized two-dimensional grain growth in a fully adequate way. Both as far as it concerns the growth kinetics, the size distribution and the grain side correlation in the scaling regime they are in agreement with previous theoretical predictions and similar simulation results (e.g., [8,15,17]). After an initial transient the average grain size scales linearly with time, and after about the same period of time, which roughly corresponds to a doubling of the average grain size, the grain size distribution function approaches a quasi stationary state. The neighbor grain side correlation in the scaling regime is further found to be in agreement with the Aboav-Weaire's law [10].

In addition to these results it is further demonstrated that a spatial grain **size** correlation exists in two-dimensional grain structures, and a spatial grain size correlation function which relates the size of a grain to the average size of its neighbors is defined. It is shown that this grain size correlation function, as the size distribution function, approaches a quasi stationary state when the grain size has roughly doubled. In this quasi stationary state the small grains tend to be surrounded by large grains and vice versa. The existence of such a spatial size correlation implies that the assumption on which Hillert's mean field theory [19] is based, i.e. that all grains have the same environment, is not correct. In a separate publication we have developed a modified mean field theory that incorporates the size correlation effect that is demonstrated and quantified here [11]. It is found that as far as growth rate and the quasi stationary grain size distribution are concerned, the modified mean field theory partly accounts for the differences between Hillert's theoretical predictions and the present simulation results. The residual differences are believed to be due to local topological constraints that are very difficult to include in an analytical theory.

Acknowledgments. Professor Kenneth A. Brakke is gratefully acknowledged for many helpful hints about how to implement two-dimensional grain growth with the Surface Evolver. One of the authors (K.M.) also wants to express his gratitude to NTH/SINTEF through the "Strong Point Center for Light Metals" for financial support.

References

[1] G. Abbrusseze and P. Brozzo, eds., Material Science Forum vols. 94-96: Grain Growth in Polycrystalline Materials (Zurich, Switzerland: Trans. Tech. Publications, (1992).

[2] K.A. Brakke, "The Surface Evolver," Experimental mathematics 1 (2) (1992), 141-165.

[3] O. Hunderi and N. Ryum, "The influence of spatial grain size correlation on normal grain growth in one dimension," Acta Met. (1996), in press.

[4] D.A. Aboav, "The arrangement of Cells in a Net," Metallography 13 (1980), 43-58.

[5] D. Weaire and J.P. Kermode, "Computer simulation of two-dimensional soap froth," Phil. Mag. B 50 (1984), 379-395.

[6] J.A. Glazier, S.P. Gross, and J. Stavans, "Dynamics of two-dimensional soap froths," Phys. Rev. A 36 (1987), 306-312.

[7] C.W.J. Beenakker, "Numerical simulation of a coarsening two-dimensional network," Phys. Rev. A 37 (5) (1988), 1697-1702.

[8] J.A. Glazier, M.P. Anderson, and G.S. Grest, "Coarsening in the two dimensional soap froth and the large Q Potts model: A detailed comparison," Phil. Mag. B 62 (1990), 615-645.

[9] D.A. Aboav, "The arrangement of Grains in a Polycrystal", Metallography 3 (1970), 383-390.

[10] D. Weaire, "Some Remarks on the Arrangement of Grains in a Polycrystal," Metallography 7 (1974), 157-160.

[11] K. Marthinsen, O. Hunderi, and N. Ryum, "The influence of spatial grain size correlation and topology on normal grain growth in two dimensions," Acta Met. (1996), in press.

[12] J. von Neumann, "Discussion" in Metal Interfaces, (Cleveland, OH: American Society for Metals, 1952), 108-110.

[13] W.W. Mullins, "Two-dimensional Motion of Idealized Grain Boundaries," J. Appl. Phys. 37 (8) (1956), 900-904.

[14] K.A. Brakke, private communications (1993).

[15] M. Marder, "Soap-bubble growth," Phys. Rev. A 36 (1987), 438-440.

[16] J.A. Glazier and J. Stavans, "Non-Ideal Effects in Two-Dimensional Soap Froth," Phys. Rev. A 40 (1989), 7398-7401.

[17] B. Radhakrishnan and T. Zacharia, "Simulation of Curvature-Driven Grain Growth by Using a Modified Monte Carlo Algorithm," Metall. and Mater. Trans. A 26 (1995), 167-180.

[18] F. Feltham, "Grain growth in Metals," Acta Met. 5 (1957), 97-105.

[19] M. Hillert, "On the theory of normal and abnormal grain growth," Acta Met. 13, (1965), 227-238.

[20] N.P. Louat, "On the Theory of Normal Grain Growth," Acta Met. 22 (1974), 721-724.

[21] D.J. Srolovitz, M.P. Anderson, P.S. Sahni, and G.S. Grest, "Computer simulation of grain growth -- II. Grain Size Distr, Topology, and Local Dynamics," Acta Met. 32 (1984), 793-802.

[22] D.F. Fan and L.Q. Chen, "Computer simulation of grain growth using a continuum model," submitted to Acta Met. (1995).

[23] D.F. Fan, C. Geng, and L.Q. Chen, "Computer simulation of grain growth using a diffuse-interface model: Local kinetics and topology," submitted to Acta Met. (1995).

EDGE ENERGY-MINIMIZING SURFACES

AND CRYSTAL SHAPE

M. Jeannette Kelley

Mathematics Department
Rutgers University
New Brunswick, NJ 08903
mkelley@math.rutgers.edu

Abstract

For polyhedral Wulff shapes a complete catalog of surface energy-minimizing tangent cones is known. However, assigning an energy to edges formed by intersecting planar facets can create edge and surface energy-minimizing structures that are not included in the original classification. A simple model for edge energy gives each edge the same energy per unit length, and another model assigns a weighted energy based on the angle between the normals to the edge's two adjacent plane segments. Attempts to identify local structures that minimize edge energy are discussed along with the effects of each model on the ideal shape for small crystals.

This work was partially funded by the Advanced Research Projects Agency through the National Institute of Standards and Technology and by the National Science Foundation.

Introduction

The Wulff shape W is the shape that minimizes the surface free energy of a crystal of fixed volume [1,2,3]. If the surface energy is sufficiently anisotropic, W is composed of planar facets intersecting in sharp straight edges. A model for the free energy of such a material might add to the surface energy an energy associated with the edges. There is initial experimental evidence that such an energy exists and has an impact on the shape of small-scale crystals. Physicists have studied the influence of an ultrathin Pd film on the faceting of W(111), and using an ultrahigh vacuum scanning tunneling microscope they have observed that upon annealing the surface becomes completely faceted to {211} pyramids. The facet sizes increase with annealing temperature, and because this increase does not theoretically reduce the surface energy, this effect appears to be a consequence of an energy associated with the edges of the pyramids [4]. If an edge energy is included in the free energy of a material, its edge plus surface energy-minimizing shape I_V may depend on the volume V of the crystal as well as the method of assigning edge energy. A comparison of the minimizing shapes under different edge energy models to the shapes of small natural crystals may lead to an understanding of which model best describes their energy.

Two Models and Effects

The first edge energy model we will consider gives each edge the same energy $\rho_1 > 0$ per unit length. A more general model weights the energy of each edge according to some function ω of the dihedral angle θ between the normals to its adjacent planes; i.e., the energy per unit length is $\rho_2 \omega(\theta)$, with θ in radians in the range 0 to π. The second model we consider allows the energy to vary linearly with θ, so ω is the identity function and an edge has energy $\rho_2 \theta$ times its length. We make the initial assumption that planes that do not appear in W will not appear in I_V; such planes may be infinitesimally corrugated and thus high in edge energy. We also assume that I_V is convex and symmetric.

In order to demonstrate the possible effects of these two models on crystal shape, we calculate the evolution of I_V with increasing volume for a simple Wulff shape. Let W be a truncated cube [Fig. 1], so that for either edge energy model and for any V, I_V will be a cube or a truncated cube. For the purpose of calculation we define f to be the fraction of one of the cube's edges removed when one corner is removed. For any V we compute the total energy of a crystal as a function of f, and by locating the function's minimum, we find the truncation of I_V. For our first model, where edge energy is proportional to length, the total energy $E_L(f)$ of the evolving crystal is its surface energy plus ρ_1 times its edge length. First, for small volumes, the function E_L is strictly increasing [Fig. 2(a)] and

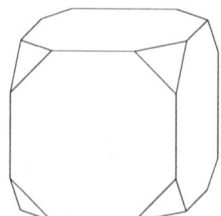

Figure 1: A truncated cube.

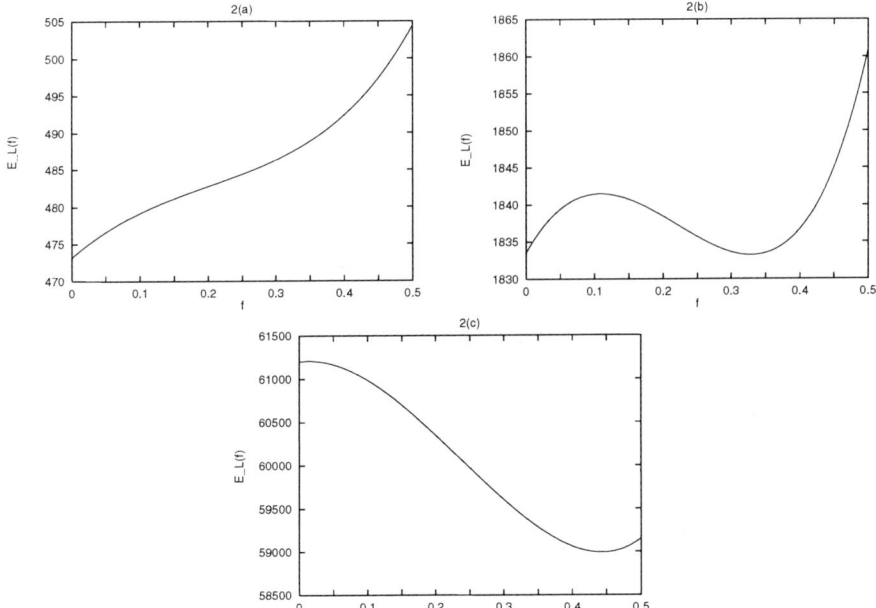

Figure 2: Evolution of E_L with increasing volume. (a) For small volumes, E_L takes its minimum at $f = 0$. (b) At some critical volume, a significant corner truncation becomes minimizing. (c) At large volumes, the minimizing f approaches approximately 0.46077, which corresponds to the truncation of W.

thus takes its minimum at $f = 0$. As V increases, a local minimum develops away from zero, but the global minimum is still fixed at zero. As the volume continues to grow, the value of E_L at the local minimum decreases toward $E_L(0)$ until at some critical volume V_0 we have $E_L(0) = E_L(f_0)$ for some $f_0 > 0$ [Fig. 2(b)]. Thus until V_0 is reached, the minimizing shape I_V is a cube with no corner truncation at all, whereas for V larger than V_0, some truncation is minimizing. In fact as V grows without bound, the surface energy term dominates the total energy, so that the truncation of I_V approaches that of W [Fig. 2(c)].

For the second model, in which edge energy is proportional to both length and angle, the evolution of I_V is strikingly different. In this case for a given V the total energy $E_A(f)$ of a crystal is its surface energy plus the sum of $\rho_2 \arccos(1/\sqrt{3})$ times the length of the edges that border triangular facets and $(\rho_2)(\pi/2)$ times the remaining edge length. For this model the behavior is simple: at small volumes the minimizing f is large [Fig. 3(a)], and as V grows, the minimizing truncation again approaches that of W [Fig. 3(b)]. Thus at small volumes I_V is a cube with a large corner truncation, and as volume increases the minimizing shape becomes similar to the shape of W.

Such different evolutions of I_V under the two models suggest that observation of the shapes of natural crystals at varying scales may help determine which model better describes edge energy. For a more detailed discussion of the calculation described see [5].

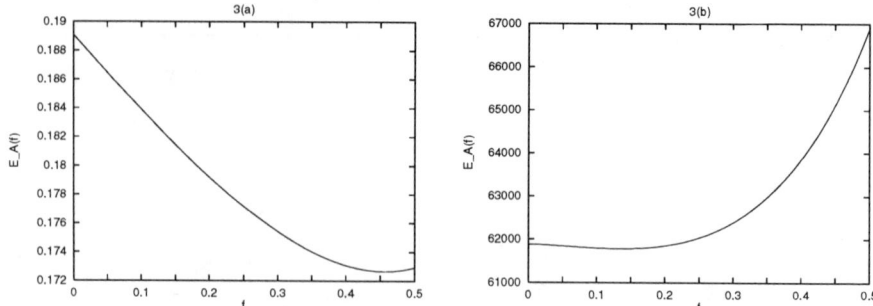

Figure 3: Evolution of E_A with increasing volume. (a) For small volumes, the value of f that minimizes E_A is large. (b) At large volumes, the minimizing f approaches approximately 0.114359, which corresponds to the truncation of W.

Truncation Results

Given that the two models yield distinct results with regard to corner truncation in the example above, we now address the general question of when truncating a corner causes a decrease in edge energy. We have seen already that in the first model truncation may actually increase edge energy; truncating a cube's corner replaces three edges of length t with three edges of length $\sqrt{2}t$. For edge energy proportional to length and angle, however, computational results indicate that truncation of any convex corner by any plane causes a reduction in edge energy. This assertion is based on three properties of this edge energy. Demonstrating each of the properties involves comparison of complicated functions of five or six variables. Although no analytic proofs exist currently, the comparisons have been made by computer on a fine grid of points located throughout the domains of the functions, and the data indicate that the properties hold for edge energy that is proportional to angle as well as length.

The first property is that the edge energy of an arbitrary three-plane corner is greater than the energy of the edges created when the corner is truncated by an arbitrary plane [Fig. 4]. The second property is that the edge energy of a four-plane convex corner is greater than the energy of the edges created upon its truncation by a plane, provided that the edges created form a parallelogram [Fig. 5]. The third property treats convex n-plane corners having three adjacent plane segments such that the two outer segments may be extended to intersect in front of the middle plane segment. This extension creates an $(n-1)$-plane convex corner. The property requires that the difference in edge energy between the n-plane corner and its truncation be greater than the difference in edge energy between the $(n-1)$-plane corner and its truncation by the same plane [Fig. 6].

Theorem 1 *If all three properties hold true for a certain edge energy, then any n-plane convex corner has more edge energy than its truncation by any plane.*

Proof: For $n = 3$ the result holds true by property one. We proceed by induction on n. Assume the result true for $n - 1$, if $n > 3$. We have two cases. Either there exist three adjacent plane segments such that the two outer segments may be extended to intersect in front of the middle plane segment, or $n = 4$ and the edges created by truncation form a

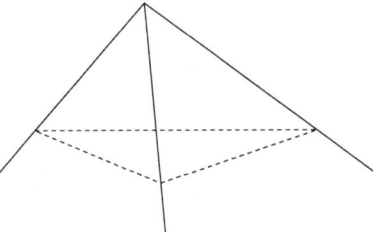

Figure 4: Property one states that the energy of the solid edges that are truncated is greater than the energy of the dotted edges, which are created by the truncation.

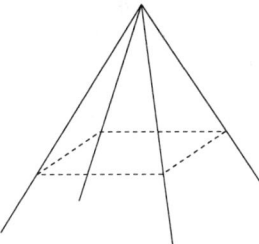

Figure 5: Property two states that the energy of the solid edges that are truncated is greater than the energy of the dotted edges created by the truncation. The dotted edges form a parallelogram in the truncating plane.

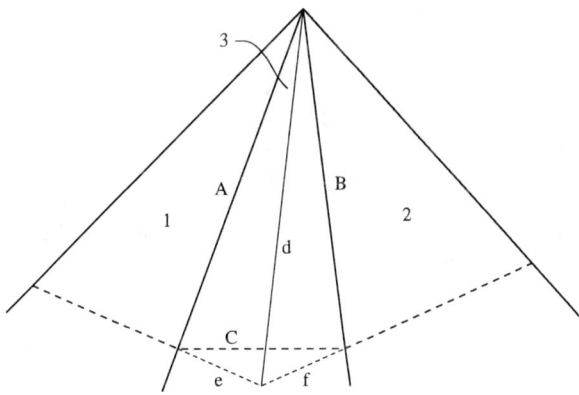

Figure 6: Shown in dark solid and dotted lines is the front view of a convex corner. Planes 1 and 2 may be extended to meet in front of plane 3. Property three states that the difference in energy between the dark solid edges A and B and the dark dotted edge C is greater than the difference in energy between the light solid edge d and the light dotted edges e and f.

parallelogram. In the first case the result holds true by property three and our inductive hypothesis, and in the second case property two guarantees the result.

This result can be extended to address the truncation of an arbitrary convex surface by a plane.

Theorem 2 *If a surface is composed of plane segments intersecting only in convex edges, and if the energy of the edges is a linear function of length, is a concave function of dihedral angle, and satisfies all three properties, then truncating the surface by any plane causes a reduction in edge energy.*

The proof uses property three and the previous theorem.

The assumption of convexity in these results is necessary; for edge energy proportional to length and angle it is possible to construct a four-plane non-convex corner that has less edge energy than its truncation by certain planes.

Conjectures and Evidence

It is interesting to contemplate the shape of crystals at the smallest scale, where edge energy so dominates surface energy that we may ignore the surface's contribution (W plays no role in determining the shape). In this case if edge energy is proportional only to length, the problem becomes one of finding the polyhedron of least edge length enclosing some fixed volume. A conjecture posed by Melzak [6] and promoted by Frank Morgan [7] suggests that the "shortest polyhedron" is a triangular prism with an equilateral base. Thus in this model small crystals would resemble triangular prisms. Joint work with W. Craig Carter at the National Institute of Standards and Technology gives computational evidence in support of this conjecture. We wrote a new energy function for the *Surface Evolver* software [8] in order to search for length-minimizing polyhedra. The new function assigns to each edge of a polyhedron the energy given by its length times $1 - e^{-\alpha^2 \theta^2}$, where θ is dihedral angle and α is a user-defined parameter. As α increases, the functions converge pointwise to the function that gives true edges the energy associated with their lengths but assigns zero energy to edges whose dihedral angle is zero [Fig. 7]. With this energy we were able to evolve bodies to decrease their edge length while maintaining their volume, and by carefully adding noise and increasing α at certain places in the evolution, we succeeded in demonstrating the evolution of a (triangulated) sphere into the conjectured prism. In numerous trials the initial shapes never evolved into polyhedra with less edge length than the prism; therefore our evidence supports Melzak's conjecture.

Another conjecture suggests that if edge energy is proportional to length and angle, the ideal shape may actually be a sphere [5]. This conjecture rests on the idea that the edge energy of any convex corner is greater than the energy of the edges formed when the corner is truncated by any plane, as discussed above. This assertion implies that any candidate for the minimizing polyhedron can be reduced in energy by making a small truncation of any of its corners and scaling the body up again to unit volume. Thus the minimizer does not exist in the class of polyhedra; the conjecture states that minimizing sequences approach the sphere. Therefore in this model, as in nature, small crystals resemble spheres, and for this reason edge energy proportional to length and angle may yield more reasonable results than edge energy proportional to length alone. Again, the theory has been tested by adding

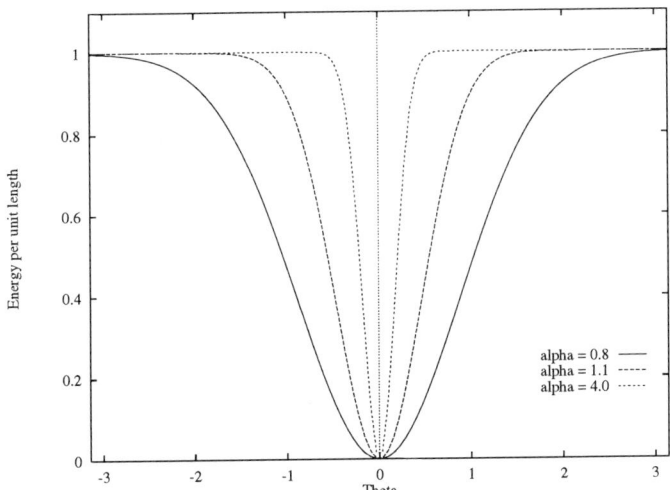

Figure 7: Several functions $1 - e^{-\alpha^2 \theta^2}$. As α increases, the functions converge pointwise to the function that gives each edge an energy equal to its length.

to the *Surface Evolver* code an energy function that approximates edge energy proportional to dihedral angle. In trials with this energy function regulating the evolution, spheres did not change shape significantly, and rectangular prisms evolved toward the sphere. Thus the initial computer evidence supports the conjecture.

References

[1] G. Wulff. Zur Frage der Geschwindigkeit des Wachsthums und der Auflösung der Krystallflachen. *Zeitschrift für Krystallographie und Mineralogie*, 34:449-530, 1901.

[2] D. Dinghas. *Zeitschrift für Krystallographie und Mineralogie*, 105:304, 1944.

[3] H. Busemann. *Am. J. Math.*, 71:743, 1949.

[4] C.-Z. Dong, et al. Microscopic aspects of faceting induced by ultrathin metal films: a comparison of Pd/W(111) and Pd/Mo(111). In X. Xie et al., editors, *The Structure of Surfaces IV*, 1994.

[5] M. Jeannette Kelley. Two models for edge energy and their effects on edge and surface energy-minimizing shapes. *Scripta Metallurgica et Materialia*, 33(9):1493-1496, 1995.

[6] Z. A. Melzak. Problems connected with convexity. *Canad. Math. J.*, 8:565-573, 1965.

[7] Frank Morgan. Surfaces minimizing area plus length of singular curves. *Proceedings of the American Mathematical Society*, 112(4):1153-1161, 1994.

[8] Kenneth A. Brakke. *Surface Evolver Manual*. The Geometry Center, Minneapolis, MN, March 1995.

ZENER PINNING OF NORMAL GRAIN GROWTH

Jinhua Gao, Raymond G. Thompson and Burton R. Patterson

Department of Materials Science and Engineering
University of Alabama at Birmingham
Birmingham, AL 35294-4461

Abstract

By introducing a new parameter, the degree of contact between grain boundaries and second phase particles, a modified two-dimensional Zener pinning model is proposed. The modified Zener pinning model can be expressed as: $\frac{D}{r} = \frac{K}{Rf}$, where D is the pinned grain size, r is the mean size of second phase particles, K is a constant, f is the volume fraction (or the area fraction in 2-D) of second phase, and R is the degree of contact between grain boundaries and second phase particles. In 2-D Monte Carlo simulations of grain growth with second phase particles, the ratio of pinned grain size to second phase particle size is proportional to $f^{-0.5}$. It was found that if the degree of contact is taken into account, this ratio is proportional to $(Rf)^{-1}$.

Introduction

Second phase particles in materials retard normal grain growth by pinning the boundaries. Zener[1] developed the first analytical model of such pinning for three dimensional (3-D) grain growth by assuming that second phase particles were monosized and spherical, second phase particles were distributed randomly in the material, and grain boundary curvature radius was equal to the grain size:

$$\frac{D}{r} = \frac{4}{3}f^{-1} \qquad (1)$$

where D was the grain size, r was the mean radius of second phase particles and f was the volume fraction of second phase in the material. This model has been discussed in many textbooks for several decades.[2-4] However, in real materials the exponent of the f term is usually different from -1. The reason for such a difference may arise from the assumption of the model that precipitates are distributed randomly in the material during grain growth. If the second phase particles prohibit grain growth by pinning boundaries, then it is expected that they would accumulate on grain boundaries. Such an accumulation would make the placement of precipitates not random but preferred at boundaries. Thus the volume fraction of the particles on grain boundaries should be greater than that in the material. Therefore, it seems more reasonable to replace the volume fraction of second phase in the material (f) in Equation (1) by the volume fraction of second phase on grain boundaries (f_{gb}). Equation (1) can be rewritten to reflect this change as:

$$\frac{D}{r} = \frac{4}{3}f_{gb}^{-1} \qquad (2)$$

It is shown in the following section that the expression for the 2-D Zener pinning model has the same form as that for the 3-D model. Monte Carlo simulations of 2-D grain growth containing precipitates were used to demonstrate the influence of second phase particles on grain growth.

2-D Zener Pinning Model and Its Modification

2-D Model

A 2-D pinning model can be obtained in a fashion similar to that used by Zener to derive Equation (1). The complete derivation will be presented elsewhere. The driving force of normal grain growth per unit length of grain boundary in 2-D is:

$$F_d = \frac{T}{D} \qquad (3)$$

where T is grain boundary line tension and D is the radius of grain boundary curvature which is assumed to be equal to the mean intercept of the grains, i.e. grain size.

The retarding force per unit length of grain boundary is:

$$F_r = \frac{4Tf}{\pi r} \qquad (4)$$

The grain boundaries may break through the drag of some second phase particles if the driving force is greater than the retarding force. That is, if $F_d > F_r$ the grains can continue to grow. When $F_D = F_r$, the grain will stop growing and the grain size becomes stable and pinned. The balance of the driving and retarding force yields the relationship of pinned grain size, particle size and area fraction of second phase:

$$\frac{D}{r} = \frac{\pi}{4} f^{-1} \qquad (5)$$

From Equations (1) and (5), it can be seen that both 3-D and 2-D Zener pinning models have the same form. The difference between the two solutions is the coefficient of the f^{-1} term.

Modification

A modification to the Zener pinning model is required as suggested by Liu and Patterson[5,6] which accounts for the nonrandomness of the precipitate distribution. The preference for precipitate accumulation on grain boundaries is obtained for the modified 2-D Zener pinning model by replacing the area fraction of the second phase particles in the material, f, in Equation (5) with the area fraction on grain boundaries f_{gb} :

$$\frac{D}{r} = \frac{\pi}{4} f_{gb}^{-1} \qquad (6)$$

By defining R, the degree of contact between grain boundaries and second phase particles, as the ratio of second phase area fraction on the grain boundaries, f_{gb}, to that in the material, f,

$$R = \frac{f_{gb}}{f} \qquad (7)$$

Equation (6) can be rewritten as:

$$\frac{D}{r} = \frac{\pi}{4} (Rf)^{-1} \qquad (8)$$

Calculation of Degree of Contact Between Grain Boundaries and Second Phase Particles, R

In 2-D, the area fraction of second phase particles on grain boundaries is

$$f_{gb} = \frac{(N^{ppt})_{gb} \times (\pi r^2)}{2r \times L_A^{\alpha\alpha} \times A}$$
$$= \frac{(N^{ppt})_{gb} \times (\pi r^2)}{2r \times \frac{\pi}{2}(\frac{1}{D}) \times A} \qquad (9)$$
$$= \frac{(N^{ppt})_{gb} \times r \times D}{A}$$

where $(N^{ppt})_{gb}$ is the total number of second phase particles on grain boundaries, r is the mean radius of second phase particles, (L_A^{aa}) is the grain boundary length in unit area of material, A is the total area of the material and D is the mean intercept of grains.

By substituting Equation (9) into Equation (7), the degree of contact between grain boundaries and second phase particles can be calculated as

$$R = \frac{(N^{ppt})_{gb} \times r \times D}{f \times A} \qquad (10)$$

Simulation Procedures

The Monte Carlo simulations of grain growth were performed in a 500×500 2-D triangular matrix, as shown in Figure 1. Each lattice site in the simulation matrix represented a hexagonal area of the material. This hexagonal area was referred to as a cell. The orientation numbers of each lattice site ranged from 1 to 100. A special negative orientation number, -20, was assigned to the lattice site where the second phase particles were embedded. In order to accelerate the simulations, a modified Monte Carlo simulation algorithm was applied.[7] The switching probability from one orientation number to another was given by:

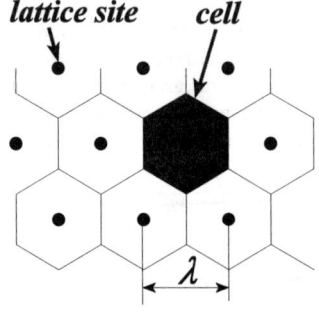

Figure 1. Simulation Matrix Configuration

$$P = \begin{cases} 0 & (\Delta E > 0) \\ 1 & (\Delta E \leq 0) \end{cases} \qquad (11)$$

where ΔE was the energy change if switching of the orientation number was accepted. The switching of a grain orientation number, either from the orientation number assigned to second phase particles (-20) or to this number was prohibited. This means the second phase particles were stable and immobile during the simulations. The mean grain size and degree of contact between grain boundaries and second phase particles were measured during the simulations.

The simulations were performed in following steps:
- Run the grain growth simulation without second phase particles until mean grain size reached 5λ, where λ was the lattice site spacing,
- Put second phase particles into the simulation matrix. The area fraction of second phase particles was 0.005, 0.01, 0.05 and 0.1. The particle size (mean particle area) was 1, 2 and 3 cells. The second phase particles were originally located either only on grain boundaries or throughout the material.
- Continue grain growth simulation with second phase particles until mean grain size remains constant.

Simulation Results and Discussion

Figure 2 and Figure 3 show the simulated grain structures. The degree of contact (R) for particles randomly placed in the microstructure, Figure 2(a), was R = 2. This can be compared to R = 10 for the microstructure of Figure 2(b). These R values show that the second phase particles were, as intended, distributed uniformly throughout the material prior to grain growth. However, during grain growth the particles began to accumulate in the grain boundaries as the boundaries became pinned. If, on the other hand, all second phase particles were originally located on grain boundaries as in Figure 3(a), the grain boundaries would break through the pinning of some particles, as shown in Figure 3(b). It should be noted that both Figure 2 and Figure 3 show only a small part (4%) of the total simulation matrix size. This was small area presented so that the detail of the precipitate - grain boundary contact could be enlarged for visual examination. However, all calculations of the microstructure were preformed on the entire simulation area.

The grain growth kinetics in the simulations are shown in Figure 4, where MCS refers to Monte Carlo simulation iterations. The grain growth kinetics with no second phase particles is also shown in the figure for reference. It is interesting to note that no matter how the second phase particles were originally distributed in the matrix, the grain growth kinetics were similar if the particle size and area fraction were the same. It can be also seen from the figure that both the simulation time required for a pinned structure and the pinned grain size increase with particle size for a given area fraction of second phase. As expected, larger particles have less pinning effect. For same particle size, the larger the area fraction, the greater is the pinning effect. Figure 5 shows the plot of log(D/r) against log(f). By linear regression analysis, the relationship between log(D/r) and log(f) can be expressed as

$$\log\left(\frac{D}{r}\right) = 0.152 - 0.53 \log(f) \tag{12}$$

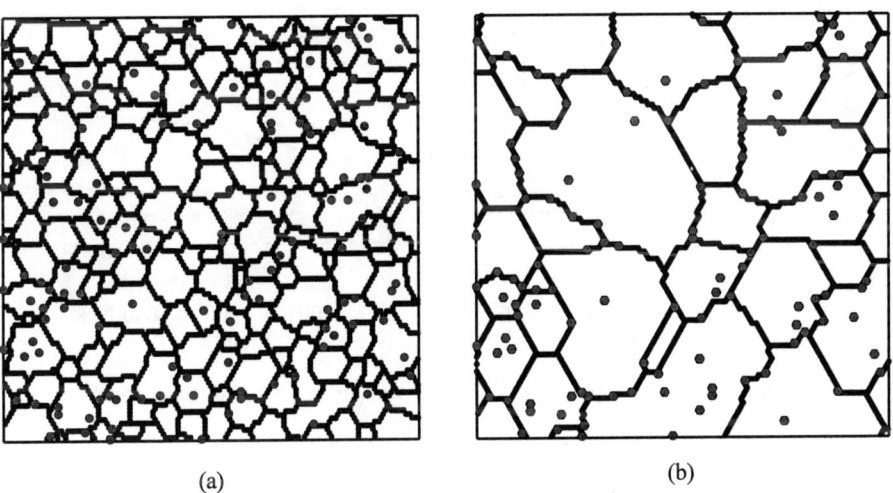

(a) (b)

Figure 2 Simulated (a) original and (b) pinned grain structures, f=0.01, with the second phase particles originally located throughout the material

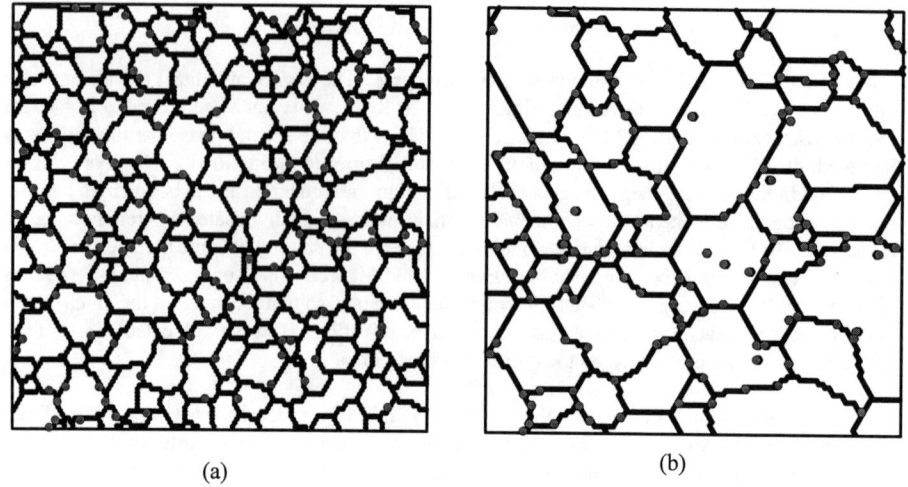

Figure 3 Simulated (a) original and (b) pinned grain structures, f=0.01, with the second phase particles originally located on grain boundaries

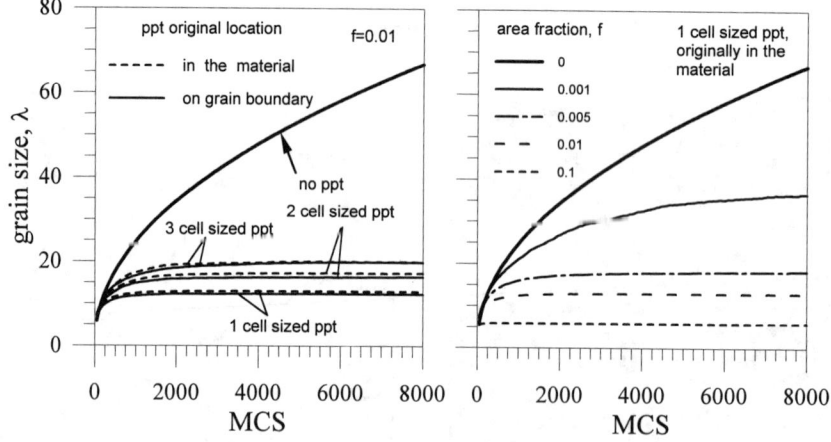

Figure 4 The effect of particle size and area fraction on grain growth kinetics

i.e.

$$\frac{D}{r} = 1.15 f^{-0.53} \qquad (13)$$

The exponent of f is similar to Srolovitz's[8], but significantly different from Equation (5). However,

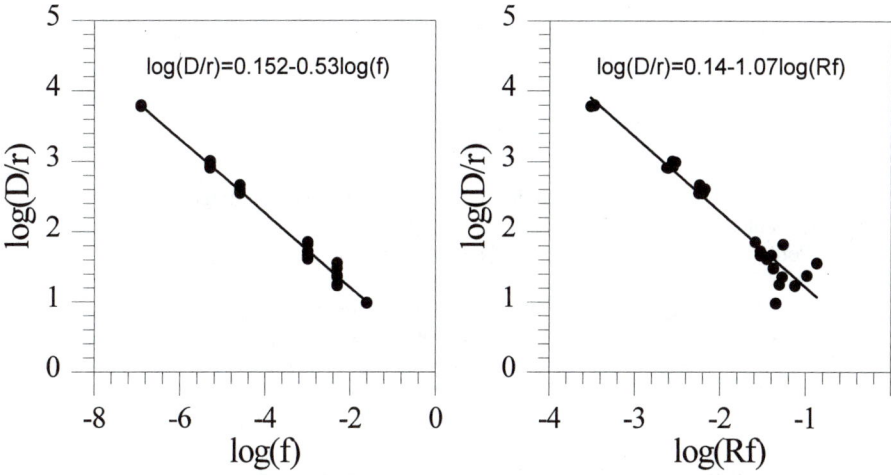

Figure 5 Relationship between log(D/r) and log(f)

Figure 6 Relationship between log(D/r) and log(Rf)

when log(D/r) was plotted against log(Rf), as shown in Figure 6, a relationship similar to Equation (8) is obtained.

$$\frac{D}{r} = 1.15(Rf)^{-1.07} \tag{14}$$

The coefficient in Equation (14), i.e. 1.15, is different from that in Equation (8), $\pi/4$. This may arise from the assumptions used in the derivation of Equation (11). There are two possible factors that contribute to the value of the coefficient. One is the shape factor of the second phase particles and the other is the relation of grain boundary curvature radius to mean grain size. In the derivation of Equation (8), we assumed that the second phase particles were circular and the grain boundary curvature radius was equal to the mean grain intercept. However, the exponent value of the (Rf) term, -1.07, in Equation (14) suggests that the simulation results confirm the modified Zener pinning model.

Conclusions

1. By introducing the degree of contact between grain boundaries and second phase particles, the effect of Zener pinning on normal grain growth can be modified as $\frac{D}{r} = K(Rf)^{-1}$.

2. 2-D Monte Carlo simulation of normal grain growth in the presence of second phase particles confirmed the behavior predicted by the modified pinning model.

3. The original location of second phase particles does not significantly change their pinning effect.

Acknowledgements

The authors gratefully acknowledge the financial support from the National Science Foundation under grant No. DMR-9417326.

References

1. C. Zener: quoted by C.S.Smith, "Grains, Phases, and Interfaces: An Interpretation of Microstructure", Trans. TMS-AIME, 175(1949), 15-51.
2. D. A. Porter and K. E. Easterling, Phase Transformations in Metals and Alloys, Second Edition, (London: Chapman & Hill, 1992), 141.
3. P. Hassen, Physical Metallurgy, (New York: Van Norstrand, 1978), 362.
4. P. G. Shewmon, Transformations in Metals, (New York: McGraw-Hill, 1969), 122.
5. Y. Liu and B. R. Patterson, "Grain Growth Inhibition by Porosity", Acta Metall. Mater., 41(1993), 2651-2656.
6. Y. Liu and B. R. Patterson, "A Stereological Model of the Degree of Grain Boundary-Pore Contact during Sintering", Metall. Trans., 24A(1993), 1497-1505.
7. B. Radhakrishnan and T. Zacharia, "Simulation of Curvature-Driven Grain Growth by Using a Modified Monte Carlo Algorithm", Metall. Trans. A, 26A(1994),167-180.
8. D. J. Srolovitz, M. P. Anderson, G. S. Grest and P. S. Sahni, "Computer Simulation of Grain Growth – III. Influence of a Particle Dispersion", 32(9)(1984), 1429-1438.

The Thermal Stability of Alloys having Periodic Minimal Surface Microstructures

Brent Fultz* and Lawrence Anthony†

* California Institute of Technology, Mail 138-78, Pasadena, California 91125, USA
† University of Toledo, Dept. of Physics and Astronomy, Toledo, OH 43606-3390, USA.

Abstract

An alloy microstructure comprising two domains separated by a boundary configured as a periodic minimal surface (a three-dimensional structure with zero mean curvature) was created, and its stability was studied at several temperatures with atomistic Monte Carlo simulations. After a quick relaxation of short range order, the periodic minimal surface microstructure (PMSM) was stable for modest times. The instability of the PMSM began with the coalescence of two adjacent cells in the structure, after which rapid domain growth occurred. This instability occurred by thermal fluctuations of the domain shapes, but with a surprisingly low increase in surface area. Surfaces with low mean curvature may be favored in nature partly for reasons of kinetics.

Acknowledgment

This work was supported by the U. S. Department of Energy under contract DE-FG03-86ER45270.

Mathematics of Microstructure Evolution
Edited by L. Q. Chen, B. Fultz, J. W. Cahn,
J. R. Manning, J. E. Morral, and J. A. Simmons
The Minerals, Metals & Materials Society, 1996

1. Introduction

Most analytical or Monte Carlo simulational studies on the kinetics of ordering or clustering in a crystalline alloy begin with quenched-in disorder [1-10]. It is known, however, that the kinetics of domain growth in a material is sensitive to the initial morphology of the domain microstructure. Further insight into kinetic processes can be obtained by studying the kinetic evolution from different initial microstructural states, including those that may seem artificial (e.g., [3,4]).

The present work examines the thermal stabilities of ordered alloys having antiphase domain boundaries arranged as a periodic minimal surface microstructure (PMSM). The salient feature of this PMSM is that the domain boundary has zero mean curvature everywhere. This condition is impossible in two dimensions, except when antiphase domain boundaries are straight lines. In three dimensions the mean curvature, κ, defined as:

$$\kappa = \frac{1}{2}\left(\frac{1}{r_1} + \frac{1}{r_2}\right) , \qquad (1)$$

can be zero if the two principal curvatures at a point on the surface, $1/r_1$ and $1/r_2$, are of equal magnitude but of opposite sign. It is possible to construct a surface that has zero mean curvature everywhere by translation of a specially-designed unit cell. Such a surface is termed a "periodic minimal surface". Nearly twenty types of periodic minimal surfaces have been found [11]. One of the simplest was reported by Schwarz in 1890 [12], termed the Schwarz P-type periodic minimal surface, and depicted in Figure 1.

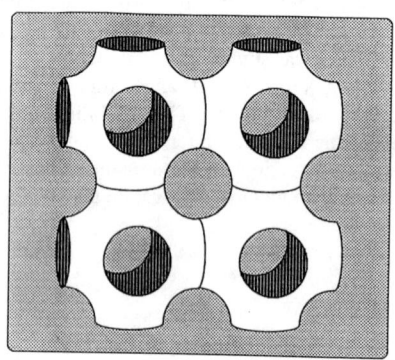

Fig. 1) Four cells of a Schwarz P-type periodic minimal surface.

Intuitively, an alloy with a periodic minimal surface microstructure (PMSM) is expected to be long-lived if domain growth is driven by curvature [13-17], since neither domain is favored over the other on any part of the domain boundary. A long lifetime is not a certainty, however. It is known that the PMSM is unstable [18], although it can be stabilized by boundary effects that are of importance in block copolymers [19,20]. Unlike the case for block copolymers, however, for metallic alloys we do not expect a conservation of domain boundary area. In particular, we expect the PMSM to eventually succumb to thermal fluctuations of its boundary shapes [3,4].

We have reported results from Monte Carlo simulations showing the Schwarz P-type PMSM to be relatively stable compared to other initial microstructural states [21]. The present paper expands on these results in two ways. First, we show that similar results are obtained for a pair interchange mechanism and for a vacancy mechanism of atom movements. Second, we calculate the relaxation of the free energy function in the Monte Carlo alloys, obtained by a hybrid cluster - Monte Carlo method. The free energy shows a kinetic arrest when the domain structure of the alloy is arranged as a PMSM. This kinetic arrest is similar for either the vacancy or pair interchange mechanisms of atom movements, and we argue that this kinetic arrest originates with a saddle point of the alloy free energy function.

2. Monte Carlo Method

The present Monte Carlo simulations of the kinetic stabilities of B2 ordered domains were performed for equiatomic bcc alloys. The alloys had 524,288 sites and periodic boundaries. (The crystal was a cube with 64 unit cells on an edge, and 2 atoms per unit cell.) Two algorithms were used for atom movement.

For the vacancy algorithm, atom movements occurred when a solitary vacancy on the lattice changed sites with a first-nearest-neighbor atom [22,23]. The characteristic rate for an interchange of the vacancy and the jth neighboring atom is ω_j, but the jth atom is competing with the other neighbors of the vacancy for an atom-vacancy interchange. One of these atoms was selected to move by picking a random number, using the $\{\omega_i\}$ as weights for the different movements. An activated state rate theory was used to calculate the characteristic transition rates, $\{\omega_i\}$. To interchange sites with the vacancy, the candidate atom must surmount a barrier of height E^*. The energy required is the difference between E^* and E_j, the energy of the jth atom in its initial state. We determine E_j with only the first-nearest-neighbor (1nn) environment of the candidate j^{th} atom. If the candidate atom is of type "A", for example:

$$E_j = n_{AA}V_{AA} + n_{AB}V_{AB} = 2V(n_{AA} - n_{AB}) \quad , \quad (2)$$

leading to a characteristic jump frequency of:

$$\omega_j \propto \exp\left(-\frac{(E^* - E_j)}{k_BT}\right) \quad . \quad (3)$$

Our candidate A-atom has n_{AA} A-atoms in its 1nn shell, and n_{AB} B-atoms. We have defined a parameter V in Eq. 2, which could be defined in an analogous way for the chemical preference of a candidate B-atom. In the work reported here, we have set the parameter V in a symmetrical way so that it is the same for both species of atoms. The time scale for the Monte Carlo simulation was obtained as a running sum of the times for the individual jumps, taken as the inverse of the ω_j for each jump.

In our second kinetic algorithm, the pair interchange algorithm, atom movements occurred via site exchanges of two nearest-neighbor atoms, selected at random from all possible nearest-neighbor pairs in the alloy. The energy change was computed for the local environments of the two atoms before (E_1) and after (E_2) the possible swap, using the local energies given by Eq. 2. The probability of the swap occurring was then computed using the formula:

$$P = \frac{1}{1 + \exp\left(\frac{E_2 - E_1}{kT}\right)} \quad . \quad (4)$$

The time scale for the pair interchange mechanism was set by the frequency of all attempts at a pair interchange. We took this frequency, without loss of generality, to be unity per atom per unit time.

Warren-Cowley short-range order (SRO) parameters were obtained from Monte Carlo simulations by counting all the A-B pairs, N_{AB}, in the equiatomic crystal containing N atoms. The 1nn Warren Cowley parameter, denoted SRO(1), was obtained as:

$$\text{SRO}(1) = 1 - \frac{4 N_{AB}}{z N}, \quad (5)$$

where the coordination number of the bcc lattice, z, is 8. Long range order (LRO) was measured by first obtaining the diffracted wave, $\psi(\mathbf{k})$, as the three-dimensional Fourier transform of the alloy, and then calculating $\psi^*\psi$. The degree of B2 order was obtained by integrating $\psi^*\psi$ in a cubical volume of edge length Δk centered around the (100) reciprocal lattice point[*] :

$$I(100) = \int_{k_x=2\pi/a-\Delta k/2}^{2\pi/a+\Delta k/2} \int_{k_y=-\Delta k/2}^{+\Delta k/2} \int_{k_z=-\Delta k/2}^{+\Delta k/2} \left| \sum_{\text{all sites}} f(\mathbf{r}) \exp(-i \mathbf{k}\cdot\mathbf{r}) \right|^2 dk_z\, dk_y\, dk_x \quad . \quad (6)$$

The atomic form factor, $f(\mathbf{r})$, was taken as unity for A-atoms and zero for B-atoms. Measuring LRO requires a decision on the spatial extent of the LRO, which is not infinite in either Monte Carlo simulations or in real alloys. We found it most meaningful to choose a Δk approximately equal to twice $2\pi/a_M$, where a_M is the primitive translation vector of the P-type periodic minimal surface. (Most results presented here are for $\Delta k = 7/4 \; 2\pi/a_M$.)

We periodically sampled the evolving crystal and calculated its free energy with a hybrid cluster - Monte Carlo method, developed by one of the authors [24]. The internal energy of the alloy, E, was obtained easily and exactly by summing the energy of all pairs of atoms in the alloy. The challenge was to calculate an entropy for use in the free energy expression. This was done by a method inspired by the cluster variation method (CVM) [25]. In this method a base cluster is chosen, and the results presented here used an octahedron-cube cluster described by Sanchez and de Fontaine [26]. The entropy calculation algorithm identified all octahedron-cube configurations on the bcc lattice of the alloy, and counted the frequencies of all chemically distinct clusters. These frequencies were then used in the well-known CVM entropy expressions [26] to obtain the entropy of the alloy. The free energy, F, was then obtained as usual: $F = E - TS$. We note that the free energy is not quite the same as that in the CVM in the octahedron-cube approximation because the internal energy for our alloy is exact.

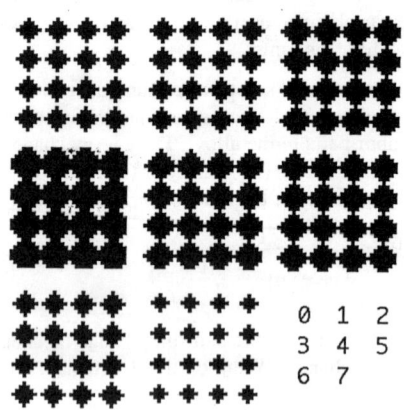

Fig. 2) Eight planes of a discretized version of the Schwarz P-type minimal surface; white squares denote A-atoms, black denote B-atoms. The planes are stacked vertically out of the plane of the paper in the sequence given in the lower right corner. Each small square in the figure represents one bcc unit cell; the cell size of this PMSM is eight bcc unit cells.

---

[*] An LRO parameter could be obtained by normalizing by an equivalent integral around the fundamental (110) reciprocal lattice point, and taking the square root. For convenience, however, we present the diffracted intensity directly, denoted as "I(100)".

For comparison, Monte Carlo simulations were performed with alloys in various initial states. A conventional reference state was a disordered solid solution, prepared with a random number generator. We compared the kinetic stability of this disordered alloy to an alloy with an initial microstructure that was a discretized approximation to the Schwarz P-type periodic minimal surface (translation vectors of PMSM cell were 4, 8 (Fig. 2), 16, 32, and 64 bcc unit cells). This microstructure is depicted in Fig. 2, which shows eight layers of unit cells. This discretized structure should be compared to Fig. 1, which it was intended to emulate.

Since the initial configuration of Fig. 2 has sharp features not present in Fig. 1, a perturbation on the Schwarz P-type minimal surface was studied unintentionally. We found that the internal energy of the alloy underwent a quick early decrease as the alloy found ways to reduce its domain boundary area. This is seen in the results for SRO kinetics.

3. Results

Figure 3 presents typical results, averaged over 6 to 10 runs, for LRO kinetics in alloys that were prepared initially with two different domain morphologies. For the disordered solid solution, ordering begins at early times and increases monotonically until the equilibrium state of order is achieved. Figure 3 also presents the data on the ordering of alloys that were prepared initially with the PMSM of Fig. 2. Although the two domains separated by the periodic minimal surfaces had perfect B2 order, the LRO parameter was nearly zero initially, but the LRO increased once the domain structure coarsened to a size greater than the PMSM cell size.

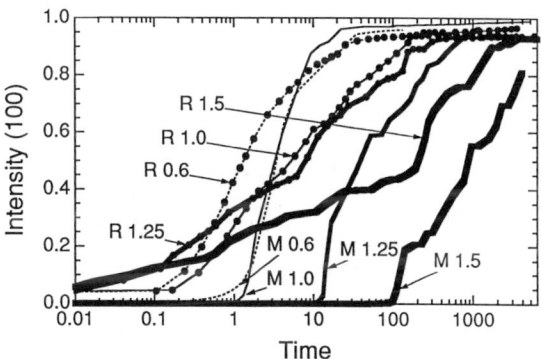

Fig. 3) Vacancy mechanism kinetics of LRO for two microstructures, one initially arranged as a PMSM (M), and the other as a random disordered solid solution (R). The curves are labeled with the potentials, 2V, where $V_{AA} = V_{BB} = 2V\ k_BT$, and $V_{AB} = 0$. Results from four temperatures are presented; the potential for the highest temperature is +0.6, the lowest is +1.5.

The LRO for the PMSM remains near zero for short and intermediate times. In this time interval, the cells of the PMSM were observed to undulate in their sizes and shapes, but their periodic arrangement was retained. Figure 4 presents one plane from an alloy that began as a PMSM as in Fig. 2, at a time shortly after the LRO parameter was observed to increase. In every case studied, it was found that the onset of coarsening of the PMSM began with the coalescence of two adjacent cells in the structure. After coalescence, the growth of the LRO occurred quickly. In the Discussion below, we therefore assume the lifetime of the PMSM is the same as the time for coalescence of two cells in the PMSM.

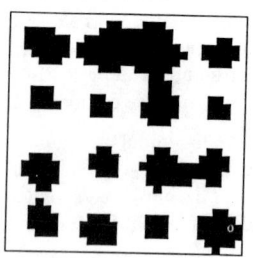

Fig. 4) Typical coalescence of cells in the PMSM, shortly after the onset of long range ordering; white squares denote A-atoms, black denote B-atoms. Top plane of crystal is shown (plane 0 of Fig. 2).

Kinetics of SRO and LRO are shown in Fig. 5. The initial state of Fig. 2 is not quite a periodic minimal surface, and there is an early relaxation of the microstructure into a less-angular morphology. This is seen as a quick change in the SRO that occurs in tens of jumps per atom. After this quick relaxation of SRO, the state of the alloy changes more slowly while the LRO remains near zero. Both the SRO and the LRO evolve towards their equilibrium values after a time of 200 (V = 1.0) and 400 (V=1.5) Monte Carlo steps. The evolution of the free energy function, shown in Fig. 6, shows a strong arrest from times of 20 to about 300 Monte Carlo steps. While the free energy tends to track the SRO, and hence the internal energy of the alloy, there is a large increase in entropy during the early relaxation that relaxes the free energy more quickly.

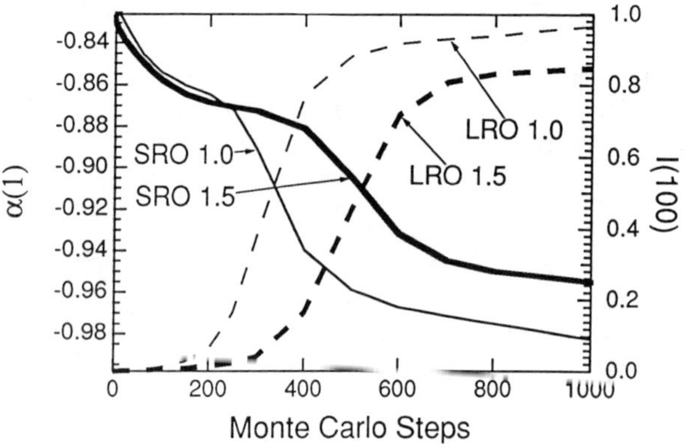

Fig. 5) Pair interchange mechanism kinetics of SRO and LRO evolution in cell of PM3M with cell size of 16 unit cells in a crystal of 64 cells on a side.

Figure 7 compares the time scales for instability of the PMSM with both the vacancy mechanism and the pair interchange mechanism for atom movements. Not surprisingly, these time scales are different. More surprisingly, however, are the results for the kinetic paths of the evolution of these two alloys through the space spanned by the SRO and LRO parameters. The kinetic paths of Fig. 8 are identical, at least within the statistics of our simulation.

Fig. 6) Pair interchange evolution of the free energy, F, and the LRO for the PMSM of Fig. 5. Results with three potentials are shown: V = 1.0, V = 1.25, and V = 1.50 (in units of kT).

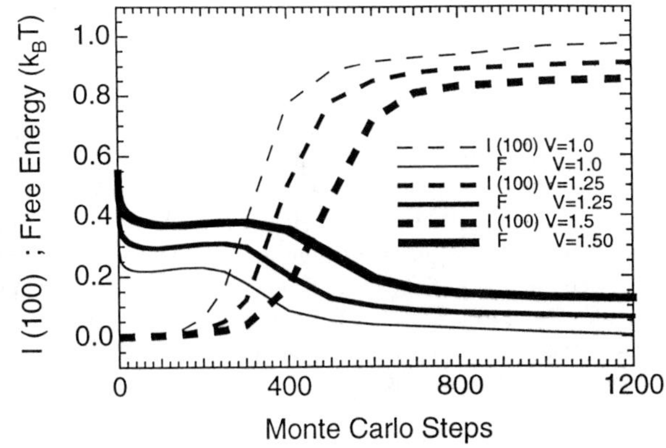

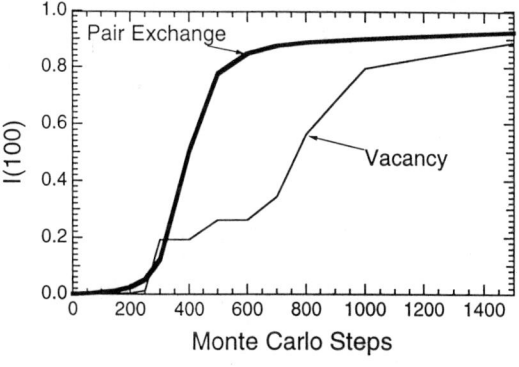

Fig. 7) LRO evolution of the PMSM of Fig. 5 for vacancy mechanism and pair interchange mechanism.

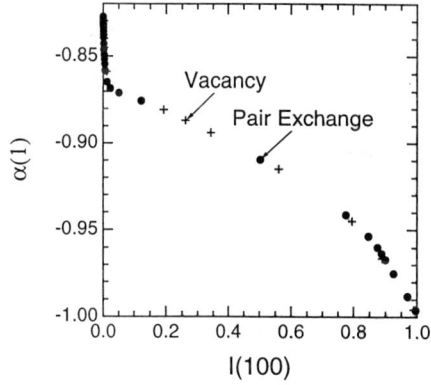

Fig. 8) Kinetic paths through SRO and LRO parameters for the PMSM of Fig. 5 for vacancy mechanism, and for pair interchange mechanism.

4. Discussion

We seek to understand the lifetime of the periodic minimal surface microstructure (PMSM). Some discussion on how the lifetime depended on the size of the cells in the PMSM was presented previously [21]. Here we present new results on the temperature dependence of the lifetime. This analysis is facilitated by the use of the pair interchange mechanism. The previous work with the vacancy mechanism for atom movements suffered from a peculiarity of vacancy mobility at low temperatures. With very weak thermal activations, the vacancy often becomes "trapped" between lattice sites that are mutually accessible through a low energy activation [22,27]. If a greater activation energy is required for the vacancy to leave these mutually-accessible sites than to jump between them, at low temperatures the vacancy will remain trapped between these sites for long times. The kinetic evolution of the alloy therefore stalls at low temperatures. Fortunately, the pair interchange mechanism does not suffer from this problem, since the algorithm picks pairs randomly throughout the alloy, and not from the limited set of vacancy neighbors. The kinetic evolution with the pair interchange mechanism does not stall at low temperatures, and in terms of Monte Carlo steps, the kinetic relaxation actually becomes slightly more efficient at low temperatures.

Figure 6 shows that with lower temperature (higher potential, V), the lifetime of the PMSM becomes larger. To understand the lifetime, we need to consider the mechanism of collapse of the PMSM, which is the coalescence of two adjacent cells in the structure. The coalescence of two cells requires the enlargement of at least one of them. Since a minimal surface has minimum surface area, coalescence must require an increase in the antiphase domain boundary area as one of the cells expands at the expense of its neighbors. Since such a local fluctuation carries an energy cost, we expect that the lifetime of the PMSM could be suppressed at low temperature by a reduction in the thermal probability of a critical fluctuation.

Assuming that the change in area during the critical fluctuation is ΔA (in units of the face of a unit cell, a_0^2), and the energy cost per unit area of the fluctuation is approximately V, the temperature dependence of a critical fluctuation, $P(T)$, is:

$$P(T) = \exp\left(-\frac{E - TS}{kT}\right) = \exp\left(\frac{-V \Delta A}{kT}\right) \exp(S/k) \quad . \tag{7}$$

We expect the entropy factor, $\exp(S/k)$, to be temperature-independent. The lifetime of the PMSM will depend on the inverse of the probability of a critical fluctuation. We can therefore use the ratio of the lifetimes for the cases of V = 1.0 kT and 1.5 kT, $\tau_{1.0}/\tau_{1.5}$, to determine ΔA for the critical fluctuation:

$$\frac{\tau_{1.0}}{\tau_{1.5}} = \frac{\exp(-1.5 \, \Delta A) \exp(S/k)}{\exp(-1.0 \, \Delta A) \exp(S/k)} \tag{8}$$

Figure 6 shows a lifetime ratio of at most 2, which indicates that $\Delta A \cong 1.4$. The critical fluctuation corresponds to only about 3 atom areas at the interface. This is a surprisingly small increase in area, considering that a cell of our PMSM had a interface area of several hundred atomic areas. Only very small fractional changes (of order 0.01) in the antiphase domain boundary area occur during a critical fluctuation, at least much smaller than our intuition would suggest.[*] The small change in area, ΔA, during the critical fluctuation is probably a self-assembling feature of the

---

[*] This small area change is consistent with the evolution of short range order parameter, shown in Fig. 5; the SRO decreases continuously. However, these types of SRO measurements could detect fluctuations of only a few tens of atoms per PMSM cell.

structure. The kinetic evolution evidently chooses a path that does not require a large increase in surface area.

We can use some observations from the Monte Carlo simulations to deduce features of a nonequilibrium free energy, F, that is a function of two variables, S (SRO) and L (LRO), and is defined over a time consistent with the initial coalescence of the PMSM. As just discussed, there is only a small energy cost when cell shapes are perturbed during thermal fluctuations. These LRO perturbations are unbiased towards one domain or the other owing to the symmetry of the Schwarz P-type periodic minimal surface. The instability of F with respect to L, the weak dependence of F on small amplitude LRO perturbations, ΔL [21], and the symmetry of ΔF with respect to perturbations of either $+\Delta L$ or $-\Delta L$ indicates that for a PMSM:

$$\frac{dF}{dL} = 0 \quad \text{and} \quad \frac{d^2F}{dL^2} < 0 \quad . \tag{9}$$

On the other hand, over short times and short-range length scales such as 1nn distances, the PMSM was found to be stable against perturbations. This was seen in the early adjustment of SRO when the alloy began with the initial state of Fig. 2, for example. Also, when SRO perturbations were imposed on the PMSM, the SRO perturbations were relaxed in a short time, and the kink in the kinetic path remained unchanged [21]. Given a particular value of L, within a short time the SRO relaxes into a value of S that is stable against perturbations. Evidently, for a PMSM:

$$\frac{dF}{dS} = 0 \quad \text{and} \quad \frac{d^2F}{dS^2} > 0 \quad . \tag{10}$$

From Eqs. 9 and 10 it appears that the PMSM is at a saddle point of F(S,L). Unstable states at saddle points of free energy functions were previously termed "pseudostable", since they are unstable, but kinetically long-lived [28,29].

The kinetic paths in Fig. 8 show a distinct kink near L = 0 and S = –0.87. The shape of the kinetic path is determined largely by the position of its kink. The branch of the path to the right occurs after coalescence of two cells in the PMSM. The time scales for kinetic evolution with the vacancy mechanism and the pair interchange mechanism are quite different, as indicated in Fig. 7, for example. An impressive feature about the kink in the kinetic path is that it is at the same position for both the vacancy and pair interchange mechanisms. Because the kink does not depend on the kinetic mechanism, we suggest that the origin of the kink is better understood in terms of a saddle point in a free energy function than in terms of kinetic rate equations.

It is interesting to speculate if thermal fluctuations like those in the shape of the cells of the PMSM could be responsible for setting the time scale and morphology in the kinetic evolution of other microstructures. At low temperatures the local energy variations during the undulations must be small, otherwise the microstructure will quickly select morphological features having lower energies. Domain structures with low mean curvatures allow for small local energy variations during diffusion-driven undulation. When an evolving microstructure takes on a low-curvature domain structure, its kinetics could be retarded. Finally, it is interesting to note the similarities between the microstructures of the PMSM and the microstructures observed by position sensitive atom probe microscopy in spinodally-decomposed Fe-Cr alloys [30,31]. The translational periodicity is not present in these experimental images, but a low mean curvature is seen. Further studies on the distributions of local mean curvatures in evolving microstructures would be interesting.

5. Conclusions

At low temperatures, microstructures with domain boundaries arranged as periodic minimal surface microstructures (PMSM) were found to be relatively stable. Only small variations in surface area occurred during the thermal undulations of the sizes and shapes of the cells in the PMSM. The loss of stability of the PMSM began when these undulations caused the coalescence of two cells in the structure. The temperature dependence of the lifetime of the PMSM was found to be very weak, indicating that only a small change in surface area occurred during a critical fluctuation. The PMSM has characteristics of a pseudostable state; in particular, the PMSM seems to be at a saddle point in the nonequilibrium free energy function, F(S,L). Long-lived states with low curvature may occur in natural microstructures in part for reasons of kinetics.

6. References

1. Sato, H. and Kikuchi, R., 1976, *Acta Metall.*, **24**, 797.
2. Gschwend, K., Sato H., and Kikuchi, R., 1978, *J. Chem. Phys.*, **69**, 5006.
3. Sahni, P. S., Grest, G. S., and Safran, S. A., 1982, *Phys. Rev. Lett.*, **50**, 60.
4. Safran, S. A., Sahni, P. S., and Grest, G. A., 1983, *Phys. Rev. B* **28**, 2693.
5. Grest, G. S., and Srolovitz, D. J., 1984, *Phys. Rev. B*, **30**, 5150.
6. Huse, D. A., 1986, *Phys. Rev. B*, **34**, 7845.
7. Amar, J. G., Sullivan, F. E., and Mountain, R. D., 1988, *Phys. Rev. B*, **37**, 196.
8. Anthony, L. and Fultz, B., 1989, *J. Mater. Res.*, **4**, 1132.
9. Fultz, B., 1990, *J. Mater. Res.*, **5**, 1419.
10. Fultz, B., 1992, *J. Mater. Res.*, **7**, 946.
11. Anderson, D. M., Davis, H.T., Scriven, L. E., and Nitsche, J. C. C., 1990, *Adv. Chem. Phys.*, **77**, 337.
12. Schwarz, H. A., 1890, <u>Gesammelte Mathematische Abhandlung</u>, Springer-Verlag, Berlin.
13. Smoluchowski, R., 1951, *Phys. Rev.*, **83**, 69.
14. Turnbull, D., 1951, *Trans. AIME*, **191**, 661.
15. Cahn, J. W., 1961, *Acta Metall.*, **9**, 795.
16. Cahn, J. W., 1962, *Acta Metall.*, **10**, 179.
17. Allen, S. M. and Cahn, J. W., 1979, *Acta Metall.*, **27**, 1085.
18. Helfrich, W. and Rennschuh, H., 1990, *J. Phys. Colloq.France*, **51**, C7-189.
19. Thomas, E. L., Anderson, D. M., Henkee, C. S., Hoffman, D., 1988, *Nature*, **334**, 598.
20. Bruinsma, R., 1991, *J. Phys. II.*, **2**, 425.
21. Fultz, B., 1994, *Philos. Mag. A*, **70**, 607.
22. Fultz, B., 1987, *J. Chem. Phys.*, **87**, 1604.
23. Ouyang, H. and Fultz, B., 1989, *J. Appl. Phys.*, **66**, 4752.
24. Anthony, L., 1993, *Ph.D. Thesis*, California Institute of Technology.
25. Kikuchi, R., 1951, *Phys. Rev.*, **81**, 988.
26. Sanchez, J. M., de Fontaine, D., 1978, *Phys. Rev. B*, **17**, 2926.
27. Fultz, B. and Anthony, L., 1989, *Phil. Mag. Lett.*, **59**, 237.
28. Kikuchi, R., Mohri, T., and Fultz, B., 1992, *Mater. Res. Soc. Symp. Proc.*, **205**, 387.
29. Fultz, B., 1993, Philos. Mag. **67B**, 253.
30. Hyde, J. M., Cerezo, A., Hetherington, M. G., Miller, M. K., and Smith, G. D. W., 1992, *Surf. Sci.*, **266**, 370.
31. Miller, M. K., 1992, *Surf. Sci.*, **266**, 494.

MODELING MAGNETOSTRICTIVE MICROSTRUCTURE UNDER LOADING

Richard JAMES

Department of Aerospace Engineering and Mechanics
University of Minnesota
Minneapolis, MN 55455

David KINDERLEHRER and Ling MA

Center for Nonlinear Analysis
Carnegie Mellon University
Pittsburgh, PA 15213-3890

Abstract

Active materials, like shape memory alloys, whose behavior depends on a reversible phase transformation or an exchange of stability among variant structures, frequently develop complex and mobile microstructures. Taking as a guide magnetoelastic interactions, eg, the giant magnetoelastic material Terfenol-D, the variation of microstructural configurations and the consequent effect on the response of the material is examined. Can we optimize performance, for example, by promoting an optimal microstructure? What is an appropriate format for modeling and what new elements from nonlinear analysis must be developed to make it useful? A second and related issue is the nature of metastable configurations, as evidenced by the hysteretic behavior of the material. With the help of simulation, we explore the subtle nature of this behavior. New methods for analysis and surprises in simulation result.

Active materials like shape memory alloys, magnetostrictive materials, and ferroelectrics exploit, in their application, a reversible phase transformation or the mutation of a mobile microstructure. A complicated microstructural configuration is one signature of a nonlinear material, in particular, of martensitic materials,cf.[2,3,4,5,6,7,8,11,16,17,21,30,31,32,33]. Our focus here is the analysis of a specific material $Tb_xDy_{1-x}Fe_2$, of special interest because it displays the largest known room temperature magnetostrictive strain, [13,14,15,34]. Our first step is the introduction of an appropriate format and the determination of the possible equilibrium microstructures. In our constitutive theory, cf. [22,23,24,26], cf. also DeSimone [16,17], we adopt the large body approximation, as is appropriate to bulk magnetostrictive devices. In this approximation, one considers the minimization of the total free energy on a sequence of larger and larger bodies; the resulting sequence of minimizers can be shown, under mild conditions, to be a minimzing sequence of the total free energy with exchange energy omitted. This gives a prominent role to the location of the energetic potential wells. Energy minimizing fine phase twin systems and fine phase magnetic domain structures tend to limits of infinite fineness. Information about variant arrangement and location as well as state functions like energy, stress, and the macroscopic magnetostrictive strain, is obtained by analysis of Young measure [35,37] extrema of the variational principle.

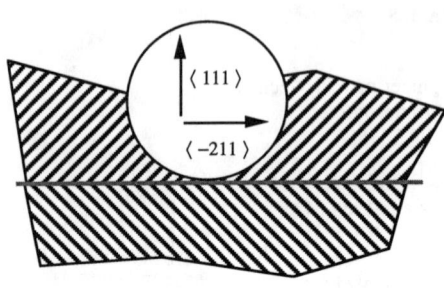

Figure 1. Schematic depiction of the microstructure in a sample of TbDyFe$_2$ illustrating the twinned dendritic herringbone structure of two sets of lamellar fine twins. The shaded midrib line represents the position of a growth twin. $\langle-211\rangle$ is the direction of the rod axis and the sample is viewed on the $\langle 0-11\rangle$ plane. Crystallographic directions are with reference to the high temperature nonmagnetic phase.

Following W. F. Brown, [9,10], the energy of a static magnetoelastic configuration is expressed in terms of

> y   deformation of a reference configuration Ω
> m   magnetization
> W   stored energy density
> H   applied field, constant in this exposition, and
> u   potential of the magnetic field vector

and is given by

$$E = \int_\Omega (W(\nabla y, m, x) - H \cdot m \det \nabla y) \, dx + \frac{1}{2} \int_{\mathbb{R}^3} |\nabla_y u|^2 \, dy \qquad (1.1)$$

subject to the constraints of Maxwell's Equation and macroscopic saturation, respectively,

$$\operatorname{div}(-\nabla_y u + m) = 0 \text{ in } \mathbb{R}^3 \text{ and } |m \det \nabla y| = M \text{ in } y(\Omega). \qquad (1.2)$$

We shall take $M = 1$ in the sequel. Note that only the body is magnetized so $m = 0$ outside $y(\Omega)$. We now focus on the properties of W which distinguish the material. A general W must satisfy the properties of frame indifference and symmetry

$$W(RFH, mR^T, x) = W(F, m, x) \text{ for } R \in SO(3) \text{ and } H \in \mathbb{H}_x,$$

where $\mathbb{H}_x$ is the symmetry group of the material at $x \in \Omega$. In our case, $\mathbb{H}_x$ will be the proper cubic group $\mathbb{H}$ of order 24 or one of its conjugacy classes.

In Terfenol-D, onset of ferromagnetism is associated with a stretch of the high temperature cubic unit cell along a main diagonal, inducing a magnetization parallel to that diagonal. Thus the energy density W for a single crystal achieves its minimum on the eight pairs of transformation strains $(U_i, \pm m_i)$ given by

$$U_i = \eta_1 \mathbf{1} + (\eta_2 - \eta_1) m_i \otimes m_i, \quad 0 < \eta_1 < \eta_2, \quad i = 1, 2, 3, 4, \tag{1.3}$$

$$m_1 = \frac{1}{\sqrt{3}}(1,1,1), \quad m_2 = \frac{1}{\sqrt{3}}(-1,1,1), \quad m_3 = \frac{1}{\sqrt{3}}(1,-1,1), \quad m_4 = \frac{1}{\sqrt{3}}(1,1,-1).$$

This leads to

$$W(RU_j, m_j R^T) = W(RU_1 H, m_1 R^T) = \min W, \quad R \in SO(3), \quad H \in \mathbb{H}, \quad j = 1, 2, 3, 4.$$

As suggested in Figure 1, the typical configuration of TbDyFe$_2$ rods is a twinned dendritic structure consisting of lamellar domains separated by grain boundaries or growth twin interfaces. The entire rod is viewed as a composite, that we take here to be one domain $\Omega^+ = \{ x \cdot m_1 > 0 \} \cap \Omega$ and a second one $\Omega^- = \{ x \cdot m_1 < 0 \} \cap \Omega$. The lower lamellar structure arises as a rotation of 180° about the m_1 axis of the original upper lattice. Denoting by R_o this rotation, we arrive at an energy density of the composite given by

$$W(F, m, x) = \begin{cases} W(F, m) & x \in \Omega^+ \\ W(FR_o, m) & x \in \Omega^- \end{cases} \tag{1.4}$$

Note that R_o is not a symmetry element of the original energy and, although holding invariant the wells of $(U_1, \pm m_1)$, gives a different set of transformation strains and magnetizations $(U'_i, \pm m'_i) = (R_o U'_i R_o, \pm m_i R_o)$.

We may now investigate the collection of twinned dendritic equilibrium structures corresponding to $H = 0$. From theory ([23,24]), we know that a minimizing sequence (y^k, m^k) giving rise to such a configuration must satisfy

$$\int_\Omega W(\nabla y^k, m^k, x) \, dx + \frac{1}{2} \int_{\mathbb{R}^3} |\nabla_y u^k|^2 \, dy \to \min W |\Omega|$$

For the Young measure $\nu = (\nu_x)_{x \in \Omega}$ generated by $(\nabla y^k, m^k)$, we then have that

$$\int_\Omega \int_{M\times S^2} W(A,\mu,x) \, dv_x(A,\mu)dx = \min W \, |\,\Omega\,|, \text{ and} \tag{1.5}$$

$$\mathrm{div}_y \, \overline{m} = 0 \quad \text{where} \quad \overline{m} = \int_{M\times S^2} \mu \, dv_x(A,\mu)$$

and $\mathbb{M}$ denotes the set of 3×3 matrices. This determines the variational condition for the support of the measure v, namely

$$\mathrm{supp}\, v_x \subset \Sigma^+ = \{(RU_i, \pm m_i R^T): R \in SO(3), i = 1,2,3,4\} \text{ for } x \in \Omega^+ \text{ and}$$

$$\mathrm{supp}\, v_x \subset \Sigma^- = \{(RU'_i, \pm m'_i R^T): R \in SO(3), i = 1,2,3,4\} \text{ for } x \in \Omega^-. \tag{1.6}$$

The twinned dendritic laminates are the Young Measures with the simple form

$$v_x = \begin{cases} \tfrac{1}{2}(1-\lambda)(\delta_{(M,m)} + \delta_{(M,-m)}) + \tfrac{1}{2}\lambda(\delta_{(N,q)} + \delta_{(N,-q)}) & x\in\Omega^+ \\ \tfrac{1}{2}(1-\lambda')(\delta_{(M',m')} + \delta_{(M',-m')}) + \tfrac{1}{2}\lambda'(\delta_{(N',q')} + \delta_{(N',-q')}) & x\in\Omega^- \end{cases}$$

where $(M,m), (N,q) \in \Sigma^+$ and $(M',m'), (N',q') \in \Sigma^-$. (1.7)

variants	twin planes	intersection of twin plane with ⟨0–11⟩
1 2	(100) twin	⟨011⟩
	(011) reciprocal	⟨100⟩
1 3	(010) twin	⟨100⟩
	(101) reciprocal	⟨–111⟩
1 4	(001) twin	⟨100⟩
	(110) reciprocal	⟨–111⟩
2 3	(001) twin	⟨100⟩
	(110) reciprocal	⟨111⟩
3 4	(–100) twin	⟨011⟩
	(01–1) reciprocal	parallel to ⟨01–1⟩
2 4	(0–10) twin	⟨100⟩
	(10–1) reciprocal	⟨111⟩

Table 1. Twinning data for the compatible variants. The third column gives the intersection of the twin plane with the (0–11) plane of observation

We confine our attention to the deformation gradients alone. It then follows from the minors relations that we are reduced to solving an algebra problem for macroscopic deformation gradients F composed of matrices M and N and F' composed of matrices M' and N' which satisfy

$$F = (1-\lambda)M + \lambda N \text{ and } F' = (1-\lambda')M' + \lambda'N', \quad 0 \le \lambda, \lambda' \le 1,$$
$$M - N = \alpha \otimes n, \quad M' - N' = \alpha' \otimes n', \text{ and} \quad (1.8)$$
$$F' - F = b \otimes m_1$$

where $(M,m), (N,q) \in \Sigma^+, (M',m'), (N',q') \in \Sigma^-$

The middle line above is the twinning equation for the individual laminates subject to the constraint that (1.5) holds. The transformation strains (1.3) determine a coherent well structure: any pair of wells admits two lamellar configurations, and hence 12 in all for each of Ω^+ and Ω^-. Combining these, there are 144 possible combinations satisfying the middle line of (1.8), but imposing the condition of coherence across the growth twin boundary, which is the last line of (1.8), reduces these to twelve. They must have $i = k, j = l$, and the fractions $\lambda = \lambda'$, although λ may be taken as a parameter. One solution corresponds to the twin, with n a (100) direction, and the other solution to the reciprocal twin, with n a (110) direction. For all of these configurations, the average magnetization $\bar{m} = 0$. Details are given in [24].

The solution to the twinning equation (1.8) may be written

$$F = (1-\lambda)U_i + \lambda R U_j \text{ and } F' = P'((1-\lambda')U'_i + \lambda'R'U'_j), \quad 0 \le \lambda, \lambda' \le 1,$$

$$RU_j = U_i(1 + a \otimes n), \quad R'U'_i = U'_j(1 + a' \otimes n'), \quad a \cdot n = a' \cdot n' = 0, \text{ and} \quad (1.9)$$

$$F' - F = b \otimes m_1, \quad i,j = 1,2,3,4.$$

The rotations R and R' depend on i,j and P' depends on i,j, and λ. To label the various fine phase laminates, we adopt the notation (ij/i'j') to denote the composite configuration arising from the wells ij and i'j' where the lamella are twins or (ij/i'j')$_r$ where the lamella ij and i'j' are reciprocal twins. Both (12/1'2') and (34/3'4') have the appearance of Figure 1, which agrees with photomicrographs of Al-Jiboory and Lord [1]. It is not difficult to ascertain the appearance of the other possible configurations on the (0,–1,1) plane and all have been seen, [2]. In particular, the solution with twin plane parallel to the (0–11) plane of observation has been seen TEM studies of samples cut on a (–2,1,1) plane, Dooley and DeGraef [18]. This fascinating study shows magnetic domain structures as well.

Since in the laboratory we do not expect to see infinitely fine lamellar structures, the fine phase laminate admits a small transition zone near the interface which enables it to achieve coherence, as depicted in Figure 1.

2. Understanding the role of microstructure in magnetostriction

A first approach to understanding the mechanism of magnetostriction is in terms of an exchange of stability among variants, or perhaps among variant systems. In response to a magnetic field H applied parallel to the (–211) rod axis, the magnetization vectors in the lamella rotate toward

H. The system decreases its energy by choosing the mechanical variant most agreeable to the magnetization, creating a strain.

We may now ask these questions:

Do all configurations have the same kinematic properties?
Are some microstructures optimal for the magnetostrictive effect?
Can we promote the formation of these microstructures?

A minimizing sequence (y^k, m^k) whose limit satisfies the conditions of (1.9) will, as noted above, satisfy the compatibility condition (1.9)$_3$ in the fine phase limit, so although in practice we may never see the transition layer, the minimizing sequence constructed for the variational principle will have

$$\int_\Omega W(\nabla y^k, m^k, x) \, dx = O(\tfrac{1}{k}). \tag{2.1}$$

On the other hand, it may be possible to find a sequence which is exactly coherent at the growth twin boundary, that does not require a transition layer, and thus obeys

$$\int_\Omega W(\nabla y^k, m^k, x) \, dx = 0. \tag{2.2}$$

The condition for this is, in lieu of (1.9)$_3$,

$$P'U'_i - U_i = b' \otimes m_1 \quad \text{and} \quad P'RU'_j - RU_j = b'' \otimes m_1. \tag{2.3}$$

We have found a simple rule to insure this which is, in the notations of (1.9), that

$$a \cdot m_1 = 0. \tag{2.4}$$

The rule is satisfied for the pairs (23/2'3'), (24/2'4'), and (34/3'4'), but no others. Exactly coherent laminates can be coarse or fine since each term of (2.3) may be satisfied separately. These systems will presumably form in the absence of any additional constraints because they demand slightly less energy or because exchange energy.

A lamellar structure

$$F(\lambda) = F_{ij}(\lambda) = (1-\lambda)U_i + \lambda RU_j = U_i(1 + \lambda a \otimes n), \quad 0 \leq \lambda \leq 1,$$

deforms a line segment of unit length in the direction of the rod axis by

$$|F(\lambda)e|, \quad e = \tfrac{1}{\sqrt{6}}(-2,1,1).$$

Comparing the expressions $\max_\lambda |F(\lambda)e|$ and $\min_\lambda |F(\lambda)e|$ for the choices of (ij) in Table 1, reveals that the (12/1'2') composite has the maximum magnetostrictive response. It does not appear on the list of exactly coherent solutions. Our analysis suggests that it is promoted by a planar growth twin interface and compressive stress applied parallel to the rod axis [25].

3. Suggestions from experiment and simulation

The exchange of stability of laminate systems when a field is applied requires that they exhibit coherence during the process. In this context, it is useful to know if an (ij) laminate is kinematically compatible with a (kl) laminate, cf. Tickle [36], thus providing a transformation path. Recent experiments suggest studying the example of the (12) and (34) laminates given by

$$F_{12} = (1-\lambda)U_1 + \lambda RU_2 = U_1(1 + \lambda a \otimes n) \text{ and} \quad (3.1)$$

$$F_{34} = (1-\lambda)U_3 + \lambda R^\dagger U_4 = U_3(1 + \lambda a^\dagger \otimes n),$$

where $n = (1,0,0)$, cf. Table 1. The deformation gradients (3.1) can be viewed as determining energy wells with Cauchy-Green strains $C_{12} = (F_{12})^T F_{12}$ and $C_{34} = (F_{34})^T F_{34}$. We can find a transformation path if it is possible to solve

$$PF_{12} - F_{34} = c \otimes p, \quad (3.2)$$

where $P \in SO(3)$ and $c, p \in \mathbb{R}^3$ may depend on λ. One criterion for this, cf. Ericksen [19,20], is the availability of a 180° rotation $Q = -1 + 2q \otimes q$ such that

$$C_{34} = QC_{12}Q \quad (3.3)$$

and in this case, q is a normal direction for the interface between the two systems, either coarse or fine phase in our Young Measure sense.

The choice $q = (0,1,0)$ has the property

$$Qm_1 = -m_3 \text{ and } Qm_2 = m_4.$$

Note incidentally that since $n \cdot q = 0$, $Qn = -n$ and the normal direction is preserved. Now

$$QU_1Q = U_3 \text{ and } QU_2Q = U_4,$$

so by choosing $R^\dagger$ appropriately, we may solve (3.3). Thus a transformation path may be

found and moreover, the interface normal is $(0,1,0)$, independent of the volume fraction λ. For further discussion of the compatibility between different systems of variants and connections with experiment, see Tickle [36].

Actual device performance under magnetic loading shows hysteretic behavior. Simulations have been employed to capture this phenomena, [27,28,29], and open the issue of the nature of metastable magnetic configurations. For a description of the experimental magnetostrictive curve we refer to Clark [14] and the references there. Metastability from the experimental, computational, and analytical points of view is an active topic of research we we do not have space to review here. We wish to remark only that the ability to produce a reliable and robust hysteresis curve by applying a general numerical method, like the conjugate gradient method, to a nonconvex functional was unanticipated and has led to some new ideas for analyzing such curves.

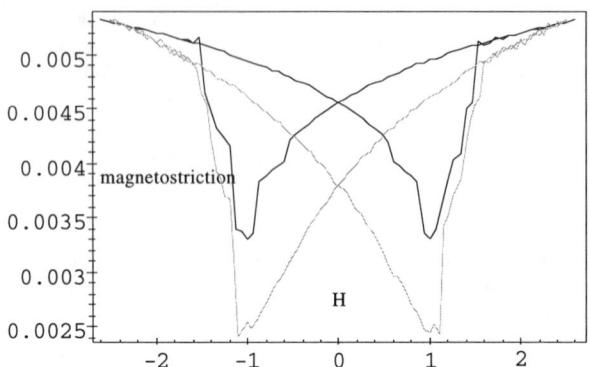

Figure 2. Computed hysteresis curves for a laminate (gray lower curve) and for a laminate with constrained growth twin interface and under compression (black upper curve).

The simulation which gives rise to Figure 2 is based on linear theory in two dimensions but includes the magnetostatic energy term. The curve corresponding to the constrained growth twin interface more closely resembles the experimental picture, suggesting a role for the growth twin in the magnetostrictive process which to date is not clearly understood.

On the other hand, the width of the hysteresis loop seems to be relatively insensitive to loading or constraints. Indeed, additional simulations under a wide variety of conditions, for example, without magnetic field energy or elasticity or single crystals with magnetic field energy, suggest that the width of the curve is extremely robust. The simulation which gives Figure 2 is the first, to our knowledge, where competition between magnetic and elastic effects is prominent.

References

[1] Al-Jiboory, M., and Lord, D.G., 1990, Study of the magnetostrictive distortion in single crystal terfenol-D by x-ray diffraction, IEEE Trans Mag 26, 2583-2585.
[2] Al-Jiboory, M., and Lord, D.G., Bi, Y.J., Abell, J. S., Hwang, A.M.H., and Teter, J.P., Magnetic domains and microstructural defects in Terfenol-D, J. Appl. Phys. (to appear).
[3] Ball, J.M. and James, R.D.   1987   Fine phase mixtures as minimizers of energy, Arch. Rat. Mech. Anal. 100, 13-52.
[4] Ball, J.M. and James, R.D.   1991   Proposed experimental tests of a theory of fine microstructure and the two well problem, Phil. Trans. Roy. Soc. Lond. A338, 389-450
[5] Ball, J.M., Chu, C., James, R.D. 1994   Metastability of Martensite, preprint.
[6] Bhattacharya, K.   1991   Wedge-like microstructure in martensite, Acta Metall Mater 39, 2431-2444.
[7] Bhattacharya, K.   1992   Self accomodation in martensite, Arch. Rat. Mech. Anal. 120, 201-244.
[8] Bhattacharya, K., Firoozye, N., James, R., and Kohn, R.   1994   Restrictions on microstructure, Proc. R. Acad. Edin., 124A, 843-878
[9] Brown, W.F.   1963   Micromagnetics, John Wiley and Sons.

[10] Brown, W.F. 1966 Magnetoelastic Interactions, Vol. 9 of Springer Tracts in Natural Philosophy (C. Truesdell, ed.), Springer-Verlag.
[11] Chipot, M., and Kinderlehrer, D. 1988 Equilibrium configurations of crystals, Arch. Rat. Mech. Anal. 103, 237-277.
[12] Chu, C., James, R.D., and Kinderlehrer, D., (to appear).
[13] Clark, A. E. 1980 Magnetostrictive rare earth - Fe2 Coumpounds, Ferromagnetic Materials, Vol 1 (Wohlfarth, E.P., ed) North Holland, 532-589.
[14] Clark, A. E. 1995 High power magnetistrictivematerials from cryogenic temperatures to 250 C, Materials for Smart Systems, (George, E. P., Takahashi, S., Trolier-McKinstry, S., Uchino, K., and Wun-Fogle, M., eds.) Mat. Res. Soc. Vol. 360,171-182
[15] Clark, A. E., Verhoven, J.D., McMasters, O.D., and Gibson, E.D. 1986 Magnetostriction of twinned [112] crystals of Tb.27Dy.73Fe2, IEEE Trans Mag. 22, 973-975.
[16] DeSimone, A. 1993 Energy minimizers for large ferromagnetic bodies, Arch. Rat. Mech. Anal. 125, 99-143.
[17] DeSimone, A.. Magnetoelastic solids: macroscopic response and microstructure evolution under applied magnetic fields and loads, J. Intel. Mat. Sys. Structures (to appear).
[18] Dooley, J. and DeGraef, M. 1995 TEM study of twinning and magnetic domains in Terfenol-D,Materials for Smart Systems, (George, E. P., Takahashi, S., Trolier-McKinstry, S., Uchino, K., and Wun-Fogle, M., eds.) Mat. Res. Soc. Vol. 360, 189-194
[19] Ericksen, J.L. 1987 Twinning of crystals I, Metastability and Incompletely Posed Problems, (S. Antman, J.L. Ericksen, D. Kinderlehrer, I. Müller,eds) IMA Vol. Math. Appl. 3, Springer, 77-96.
[20] Ericksen, J.L. 1991 On kinematic conditions of compatibility, J. Elas. 26, 65-74.
[21] James, R. and Kinderlehrer, D. 1989 Theory of diffusionless phase transitions, PDE's and continuum models of phase transitions, (Rascle, M., Serre, D., and Slemrod, M., eds.) Lecture Notes in Physics 344, Springer, 51-84.
[22] James, R. and Kinderlehrer, D. 1990 Frustration in ferromagnetic materials, Continuum Mech. and Thermodynamics 2, 215-239.
[23] James, R. and Kinderlehrer, D. 1992 Frustration and Microstructure: an example in magnetostriction, Progress in PDE, calculus of variations, applications (Bandle, et al., eds) Pitman Res. Notes Math 267, 59-81.
[24] James, R. and Kinderlehrer, D. 1993 Theory of magnetostriction with application to TbxDy1−xFe2, Phil. Mag. B, 68, 237-274
[25] James, R. and Kinderlehrer, D. 1994 Theory of Magnetostriction with application to Terfenol-D, J. Appl. Phys., 76, 7012-7014
[26] James, R.D., and Müller, S. 1994 Internal variables and fine scale oscillations in micromagnetics, Continuum Mech. Thermodyn. 6.
[27] Kinderlehrer, D. and Ma, L. 1994 Computational hysteresis in modeling magnetic systems, IEEE Trans Mag.,30.6, 4380-4382
[28] Kinderlehrer, D. and Ma., L. The hysteretic event in magnetic systems, Nonlinear analysis(to appear).
[29] Kinderlehrer, D. and Ma., L. The hysteretic event in the computation of magnetization and magnetostriction, Nonlinear PDE and their applications, (Brezis, H. and Lions, J.-L.,eds) Collège de France Sem vol XI, Pit. Res Notes Math Sci (to appear).
[30] Kinderlehrer, D. and Pedregal, P.1991 Characterizations of gradient Young measures, Arch. Rat. Mech. Anal. 115, 329-365.
[31] Kinderlehrer, D. and Pedregal, P.1994 Gradient Young measures generated by sequences in Sobolev spaces, J. Geom. Anal. 59-90.
[32] Kohn, R.V. 1991 The relaxation of a double-well energy, Cont. Mech. Thermodyn. 3, 192-236.
[33] Leo, P., Shield, T., and Bruno, O. 1993 Transient heat effects on the pseudoelastic hysteresis of shape memory wires, Acta Met. 41, 477-2485.
[34] Savage, H.T., Clark, A.E., and Spano, M.L. 1984 Strain-field relationships in rare-earth iron alloys, IEEE Trans. Mag. 20, 1449-1450.
[35] Tartar, L. 1979 Compensated compactness and applications to partial differential equations, in Nonlinear Analysis and Mechanics: Heriot-Watt Symposium IV, Pitman Research Notes in Mathematics 39, 136-212.

[36] Tickle, R. Observations of the microstructure of TbxDy1−xFe2, $x \approx 0.3$, under applied field and stress, MS. Thesis, University of Minnesota, in progress.
[37] Young, L.C. 1969 Lectures on calculus of variations and optimal control theory, W.B.Saunders.

Microlocal forces and precursor oscillations

Deborah Brandon*
Department of Mathematics
Carnegie Mellon University
Pittsburgh, PA 15215

Robert C. Rogers[†]
Department of Mathematics
Virginia Tech
Blacksburg, VA 24061-0123

Abstract

In this paper we consider a model of solid-solid phase transitions that includes elastic effects and an order parameter. Our purpose is to show that such a model can exhibit small amplitude oscillations in the strain before transition from Austenite to Martensite. The model under investigation falls within the general framework of models developed by Fried and Gurtin, but we examine the special case of a triple-well free energy with a central Austenite well flanked by two symmetric Martensite wells. (The problem is posed in one space dimension, so the terms "Austenite" and "Martensite" are simply meant to be suggestive.) The relative height of the wells and the elastic modulus of the Austenite well vary with temperature. We examine the static problem and, as we would see in a pure elasticity problem, there is a uniform Austenite solution at high temperature and highly oscillatory "twinned" Martensite solutions at low temperature. However, we show that the combination of the microlocal forces from the order parameter and a softening of the Austenite modulus cause precursor oscillations in the Austenite phase to bifurcate from the uniform mode above the transition temperature. These branches of solutions connect to Martensite oscillations of the same wavelength below the transition temperature.

---

*D. Brandon has been partially supported by the National Science Foundation under grant number DMS-9494561.

[†]R.C. Rogers has been partially supported by the National Science Foundation under grant number

1 Introduction

Many shape-memory alloys exhibit fine microstructures during structural phase transformations from a high-temperature, high-symmetry Austenite phase to a low-temperature, lower-symmetry Martensite phase. (Here the term "microstructure" is being used to describe high frequency plane-wave oscillations between two Martensitic variants.) In this paper we examine a closely related phenomenon. There have been a number of observations of small-amplitude "precursor" oscillations that take place above the temperature for transition to Martensite (See, e.g. [16, 17, 18, 19, 21, 22, 23]). These oscillations often have a characteristic "tweedy" pattern and take place on roughly the same wavelength as the eventual high-amplitude Martensite oscillations.

Why would such oscillations take place? A local instability of the homogeneous Austenite ground state is not predicted by typical elastic energy minimization problems, and refining elasticity by adding higher-order terms like capillarity would only seem to stabilize a uniform Austenite state. So far, the most successful explanations focus on material inhomogeneities (either quenched in [13] or reactive [20]) that force the material out of the Austenite well and into some oscillatory mixture of Austenite and Martensite. However, one common interpretation of the relevant experimental observations is that the oscillations take place "within the Austenite well," and there is no mixture of Austenite and Martensite states. This paper is intended to show that order parameter models can exhibit a mechanism by which oscillations occur *within* the Austenite phase.

In particular, we show that a simplified one-dimensional mathematical problem based on a model of the type considered by Fried and Gurtin [10, 11] exhibits a bifurcation picture that has branches of steady-state solutions that start as small-amplitude oscillations in the Austenite phase and become "twinned" Martensite oscillations further along the branch. We should say at the outset that the problem considered here is clearly a mathematical "cartoon" of the real physical situation: the problem is posed in one space dimension; in our global analysis the elastic part of the free energy is assumed to be piecewise quadratic. Thus, we are taking a great deal of liberty in using such terms as "Austenite," "Martensite," and "twinned." We do so both to draw attention to the analogy we wish to make and to avoid the use of less felicitous (though more accurate) terms such as "central-well," "lateral-well," and "locally symmetric."

2 The mathematical model

In this section we describe a mothematical model based on Fried and Gurtin's general formulation of elastic phase transitions characterized by an order parameter [10, 11].

We consider a bar of elastic material whose reference configuration has an axis of centroids of cross sections lying on an interval on the x-axis. We wish to describe longitudinal motions of the bar. We assume that the uniform temperature θ enters the problem only as a parameter. The state of the bar is described by the following scalar fields of reference position $x \in [0, l]$ and time $t \in [0, \infty)$.

u	longitudinal displacement
$v := \dot{u}$	longitudinal velocity
$w := u_x$	longitudinal strain
ϕ	order parameter
$h := \dot{\phi}$	order parameter rate of change
$p := \phi_x$	order parameter gradient

Here and in the following, a superposed dot indicates a partial derivative with respect to time. We do not put restrictions on the strain that would ensure that the material does not interpenetrate. Such restrictions can be difficult to treat analytically (see, e.g. [1]) and the issues raised by their treatment are largely separate from the phase transition problems considered here.

The only unusual field described above is the order parameter ϕ. The physical interpretation of this field differs for various constitutive choices. Under appropriate choices it can take the role of the "phase field" in which different phases of the material are described by different arbitrary values of the parameter (e.g., $\phi = 0$ indicates liquid and $\phi = 1$ indicates solid). This is typical in some of the work of Caginalp [7, 8] and numerous others. While the general formulation described here includes models of this type, we consider a different constitutive choice when modeling precursor oscillations. Our use of ϕ is more consistent with its interpretation as a concentration (see, e.g. [9]) or as a measure of atomic shuffles (see, e.g. [2, 12]). However, we feel that for the purposes of this paper, a specific physical identification of the order parameter is unnecessary. Our intent is to show that a generic order parameter can provide a mathematical mechanism for inducing precursors in a given constitutive regime.

To describe the dynamics of the kinematic quantities above we make use of the following fields.

ρ mass density per unit reference length
f resultant external body force per unit reference length
ψ free energy per unit reference length

The functions f and ψ are assumed to be functions of reference position and time, ρ is assumed independent of time.

Fried and Gurtin [11, 10] pose balance laws for linear momentum and microlocal forces and derive constitutive restrictions using the second law of thermodynamics. In [5] we have given a full exposition of this process applied to the present one-dimensional situation. If we incorporate the constitutive restrictions into the balance laws, we get the following form of the *balance of linear momentum*

$$\rho \ddot{u} = \hat{\psi}_w(u_x, \phi, \phi_x)_x + f. \tag{1}$$

The *balance of microlocal forces* is given by

$$b\dot{\phi} = \hat{\psi}_p(u_x, \phi, \phi_x)_x - \hat{\psi}_\phi(u_x, \phi, \phi_x) + \gamma. \tag{2}$$

where

$$b(x,t) = \hat{b}(u_x, \phi, \phi_x, \phi_t) > 0. \tag{3}$$

The function

$$\sigma(x,t) = \psi_w(u_x, \phi, \phi_x) \tag{4}$$

is identified as the stress.

For the purposes of exhibiting precursor oscillations we examine only static problems. We choose the following boundary conditions. At the left end of the bar we impose a fixed displacement boundary condition

$$u(0) = 0 \tag{5}$$

while at the right end we consider a free end

$$\sigma(l) = 0. \tag{6}$$

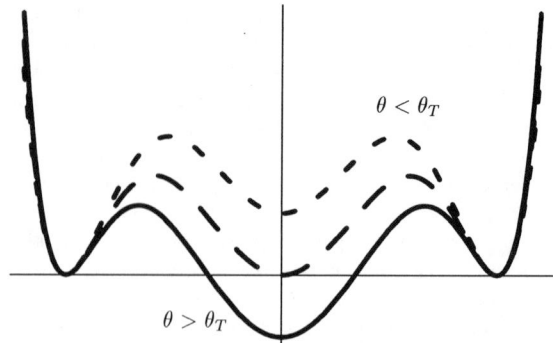

Figure 1: A generic triple-well potential G at three temperatures.

Boundary conditions for the order parameter are more problematic since we have placed such a weak physical interpretation on the variables in question. Fortunately, for the purpose of exhibiting the existence of precursors, almost any reasonable choice of linear boundary conditions suffice. For concreteness we examine Dirichlet boundary conditions in this paper.

$$\phi(0) = \phi(l) = 0. \tag{7}$$

We now consider a particular set of constitutive choices that highlights the ability of the more general model to predict precursor oscillations and the onset of microstructure. We do not make any further assumptions about the density ρ and the function $\hat{b}$ which do not enter our static calculations. Our only task is to specify a constitutive relation for the free energy ψ. We define

$$\hat{\psi}(w, \phi, p, \theta) := G(w, \theta) + \frac{\alpha}{2}|\phi|^2 - \kappa w \phi + \frac{\epsilon^2}{2}|p|^2 \tag{8}$$

where the temperature θ is used as a scalar parameter. We think of G as a symmetric triple-well potential (see Figure 1) with a central "Austenite" well flanked by two "Martensite" wells. When we do our global analysis below, we prescribe a particular piecewise quadratic function, but while we proceed, we think of the relative heights of the wells as being controlled by temperature with the Austenite well rising as temperature lowers. We think of the Martensite wells as having a relatively hard modulus that changes little with temperature while the Austenite modulus is assumed to get softer as the temperature lowers.

Thus, under this assumption, the static local balance laws with zero forcing functions ($f \equiv 0$, $\gamma \equiv 0$) reduce to

$$\bar{\sigma}(u_x, \theta)_x - \kappa \phi_x = 0, \tag{9}$$

and

$$\epsilon^2 \phi_{xx} - \alpha \phi + \kappa u_x = 0. \tag{10}$$

Which we consider with boundary conditions (5), (6), and (7).

Remark 2.1 In our constitutive model, the free energy is nonconvex in the strain, convex in the order parameter, and coupled between the strain and order parameter. This is similar to the hypothesis made in Lin and Rogers [14] and Brandon, Lin and Rogers [3] and is distinct from the assumptions made in the usual phase-field models (see [7, 8]) where the central assumption is the nonconvexity of the energy in the order parameter. We note that while Fried and Gurtin clearly have generalizations of the phase field model in mind when deriving their models, their formulations are general enough to include the type of constitutive choices made here.

3 Local Analysis in the Austenite Well

In this section we look for solutions of (9), (10), (5), (6), and (7) that lie entirely within the Austenite well. We assume that

$$\bar{\sigma}(0,\theta) = 0 \tag{11}$$

so that our nonlinear problem has the trivial solution $u \equiv \phi \equiv 0$. We now set

$$\beta(\theta) := \bar{\sigma}_w(0,\theta) = G_{ww}(0,\theta) \tag{12}$$

and examine the linearized problem about the trivial state. We assume

$$\beta(\theta) > 0, \tag{13}$$

and that

$$\frac{\partial \beta}{\partial \theta}(\theta) > 0. \tag{14}$$

The linearized problem is simply

$$\begin{aligned}
\beta(\theta)u_{xx} - \kappa\phi_x &= 0 \\
-\epsilon^2 \phi_{xx} + \alpha\phi - \kappa u_x &= 0 \\
u(0) &= 0 \\
\beta(\theta)u_x(l) - \kappa\phi(l) &= 0 \\
\phi(0) = \phi(l) &= 0
\end{aligned} \tag{15}$$

The first equation can be integrated using the free boundary condition to yield

$$u_x = \frac{\kappa}{\beta(\theta)}\phi. \tag{16}$$

This and the left-end Dirichlet condition give us

$$u(x) = \frac{\kappa}{\beta(\theta)}\int_0^x \phi(s)\,ds \tag{17}$$

Combining (16) with the balance of microlocal forces (15$_2$) and rearranging we get

$$\phi_{xx} + \left(\frac{\kappa^2 - \beta(\theta)\alpha}{\beta(\theta)\epsilon^2}\right)\phi = 0. \tag{18}$$

Of course, the linear boundary-value problem composed of (18) and the Dirichlet boundary conditions for ϕ has nontrivial, sinusoidal solutions exactly when θ_n satisfies

$$\lambda_A(\theta_n) := \frac{\kappa^2 - \beta(\theta_n)\alpha}{\beta(\theta_n)\epsilon^2} = \frac{n^2\pi^2}{l^2}, \quad n = 1, 2, 3, \ldots \tag{19}$$

A few remarks are in order,

- It follows from (14) that there is at most one solution θ_n of (19) for each n, and that the sequence θ_n is monotone decreasing.

- If there is a critical temperature $\bar{\theta}$ such that
$$\lim_{\theta \to \bar{\theta}+} \beta(\theta) = 0$$
(e.g., if β is affine: $\beta(\theta) := \bar{\beta}(\theta - \bar{\theta})$) then we have
$$\lim_{\theta \to \bar{\theta}+} \lambda_A(\theta) = \infty$$
and there exists a solution of θ_n of (19) for every n sufficiently large. (Of course, if the function $G(w, \theta)$ depends on temperature in the way suggested above, the Austenite well is far above the Martensite wells at temperatures close to $\bar{\theta}$.)

- Close to the Austenite solution, the free energy has a quadratic approximation of the form
$$\hat{\psi}(w, \phi, p) \approx \frac{1}{2}(w, \phi) \begin{pmatrix} \beta(\theta) & -\kappa \\ -\kappa & \alpha \end{pmatrix} \begin{pmatrix} w \\ \phi \end{pmatrix} + \frac{\epsilon^2}{2}|p|^2.$$
The function $\lambda_A(\cdot)$ becomes positive (and the ode (18) develops oscillations) exactly when the free energy loses local positive definiteness. In other words, the oscillations develop because the interaction of the strain and the order parameter causes the Austenite well to become a saddle point.

- If the parameters κ and α are fixed, a sufficiently small Austenite modulus $\beta(\theta)$ causes the free energy to lose *local* positive definiteness in the central well. However, if the Martensite modulus is sufficiently large or if G grows faster than quadratically, then the free energy is globally coercive in the sense that $\|w, \phi, p\|_{L^2} \to \infty$ implies $\int_0^1 \psi \to \infty$.

- The quantity ϵ/l act as a lengthscale in the problem.

- There is nothing inherently "one-dimensional" about the local analysis. A similar procedure could be performed for more realistic physical energies of deformation of three dimensional materials. (This is not true for the global analysis performed below, where ordinary differential equation techniques such as phase-plane methods are crucial.)

- The question of local stability of the bifurcating branches would be determined by higher-order nonlinearity of G. We have not done a dynamic stability analysis, but one can easily check using standard perturbation techniques that

    - if $\bar{\sigma}_{www}(0, \theta_n) < 0$ at a bifurcation point θ_n then the bifurcating branch is *subcritical*, i.e. it "pitchforks" toward the (stable) high temperature branch.
    - if $\bar{\sigma}_{www}(0, \theta_n) > 0$ at a bifurcation point θ_n the bifurcating branch is *supercritical*, i.e. it pitchforks away from the (stable) high temperature branch.

    The obvious conjecture is that all subcritical branches are unstable and the first supercritical branch is locally stable.

- In summary, we claim that the order parameter model exhibits a local bifurcation picture consistent with experimental observations of precursor oscillations if one chooses a free energy with the following properties.

- The Austenite modulus $\beta(\theta)$ is an increasing function of temperature.
- The function $\lambda_A(\theta) := \frac{\kappa^2 - \beta(\theta)\alpha}{\beta(\theta)\epsilon^2}$ becomes positive at a critical temperature θ_C substantially greater than the transition temperature θ_T at which the Austenite well rises above the Martensite well.
- There is a precursor temperature θ_P in the interval (θ_T, θ_C) such that

$$\bar{\sigma}_{www}(0,\theta) < 0 \text{ for } \theta > \theta_P$$

and

$$\bar{\sigma}_{www}(0,\theta) > 0 \text{ for } \theta < \theta_P$$

- The value of ϵ is sufficiently small so that, for l in range corresponding to realistic samples on which precursors have been observed, we have

$$l^2 \lambda_A(\theta_P) \gg 1.$$

This corresponds to very fine oscillations on the first stable branch.

4 Global Analysis

We now examine the global nonlinear problem composed of local balance laws (9) and (10) and the boundary conditions (5), (6) and (7). We describe sufficient conditions for a class of weak solutions of the problem. Our solutions will have the property that u, ϕ and ϕ_x are continuous while $w = u_x$ and ϕ_{xx} are piecewise continuous.[1]

In order to make our calculations explicit, we consider a triple-well quadratic potential. Let $w_T < w_M$, β_M, and $\beta(\theta)$ be positive scalars. We define

$$\mathcal{G}(w,\theta) := \begin{cases} \frac{\beta_M}{2}(w + w_M)^2 & -\infty < w < -w_T \\ \frac{\beta(\theta)}{2}w^2 + \bar{e}_0(\theta) & -w_T \leq w < w_T \\ \frac{\beta_M}{2}(w - w_M)^2 & w_T \leq w < \infty \end{cases} \quad (20)$$

where

$$\bar{e}_0(\theta) := \frac{1}{2}\left[\beta_M(w_M - w_T)^2 - \beta(\theta)w_T^2\right] \quad (21)$$

Here $\pm w_M$ are the "Martensite equilibrium" states, $w = 0$ is the "Austenite equilibrium" state, and $\pm w_T$ are the transition points between the Martensite wells and the Austenite well. The constants β_M and $\beta(\theta)$ are the elastic moduli in the Martensite and Austenite wells respectively. Note this potential is continuous, but that it suffers a jump in its derivative with respect to w at the transition points.

Note that the free energy now depends on the temperature θ through the parameter $\beta(\theta)$. We think of the constants w_T, w_M, β_M, α, κ, and ϵ as being fixed. If we assume that $\theta \mapsto \beta(\theta)$ is monotone increasing, then as that temperature lowers, the concavity of the central well (i.e. the Austenite modulus) decreases and the well rises. The Martensite wells and the points at which the wells are attached remain fixed.

---
[1] In fact, using arguments similar to those used in [3], one can show that all weak solutions are in this class. But since our purpose is simply to exhibit precursor oscillations (rather than to classify all solutions) we do not prove this here.

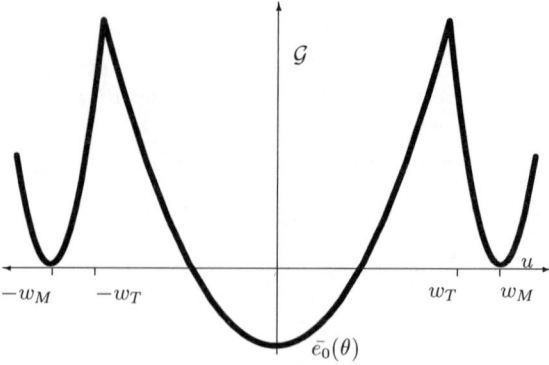

Figure 2: The triple-well potential $\mathcal{G}$.

Our solutions will be piecewise classical solutions of (9) and (10). To ensure that our functions are weak solutions we need only require the "Rankine Hugoniot" condition that the total stress be continuous at each point at which w and ϕ_{xx} undergo jump discontinuities. But in addition, we require that our weak solutions satisfy a "stability" condition. It is a standard result in the calculus of variations that strong local minimizers of the energy

$$\mathcal{E}(u,\phi) := \int_0^l \hat{\psi}(u_x, \phi, \phi_x)\,dx \qquad (22)$$

over the set of piecewise smooth functions satisfying the boundary conditions must be weak solutions of the *relaxed* version of (9):

$$(\bar{\sigma}^R(u_x, \theta) - \kappa\phi)_x = 0, \qquad (23)$$

Where $\bar{\sigma}^R$ is the derivative with respect to w of the *lower convex envelope* of $\mathcal{G}$.

We now look for solutions of the relaxed boundary-value problem. Conditions (23) and (6) imply

$$\kappa\phi(x) = \bar{\sigma}^R(u_x(x), \theta), \qquad (24)$$

for all $x \in [0, l]$. Conversely, note that if a continuous function ϕ and piecewise continuous function u_x satisfy (24) almost everywhere, then (23) and (6) are satisfied in the weak sense.

In [5], we give more details of the analysis of this problem. Here we simply describe the bifurcation diagram (figure 3) describing the solutions we obtain.

- Because the problem is linear in the Austenite well, the bifurcation curves are degenerate straight lines in the regions of pure Austenite.

- At the point where ϕ_{xx} undergoes a discontinuity ($\phi = \phi_T(\theta)$), solutions become mixtures of Austenite and Martensite. At this point, the branches always swing back toward higher temperatures. The branches then swing back toward lower temperatures and are bounded above by $\phi = \phi_M$.

- At temperatures below θ_T, the Austenite well lies above the Martensite wells and no longer is involved in the relaxed problem. The problem becomes a classic double-well problem, and the stationary points can be interpreted as Martensite "twins". Furthermore, there is no longer a lower bound on the return time of oscillating

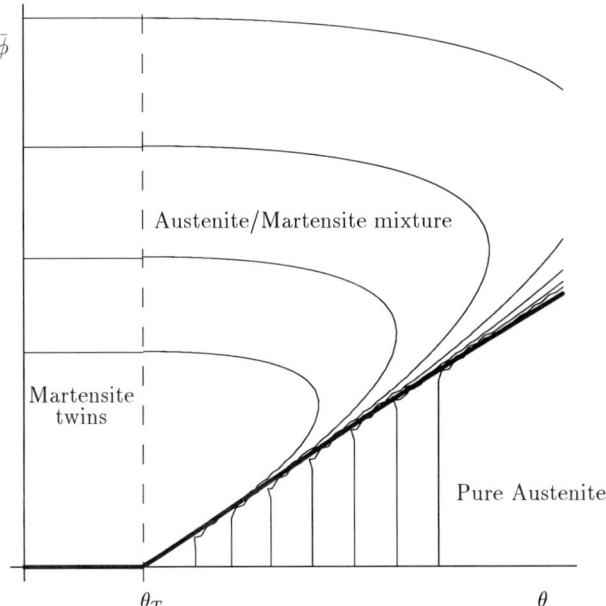

Figure 3: Bifurcation diagram showing stationary points of the order-parameter energy as a function of temperature.

about the center, and we can get stationary points of the energy with arbitrarily fine oscillations. (This is similar to the results found in [4].)

5 Conclusions

To summarize, our basic result is that a scalar order parameter that is strongly coupled to a strain with a soft modulus can cause a saddle instability in the total energy. Under appropriate constitutive assumptions, this instability causes small amplitude oscillations within the Austenite well to bifurcate from a stable uniform state.

We conclude with a few comments on future work.

- The local analysis can be extended to nonhomogeneous strains and displacements to predict stress-strain curves for the soft moduli. While the boundary conditions for the order parameter played little part in the demonstration of the existence of precursor oscillations. They may have a large effect on the predicted stress-strain laws.

- As we indicated above, it should be possible to extend the local analysis to problems in more than one space dimension. Of course, the main questions here are how many order parameters should be used and how one should couple them to the strain tensor. A number of possibilities suggest themselves.

    1. In [13], precursors were modeled using a "quenched in" scalar inhomogeneity in an energy defined for two-dimensional deformations. We could easily replace

the quenched in inhomogeneity with a "reactive" inhomogeneity generated by a scalar order parameter.

2. There are a number of well-defined proposals for constitutive functions involving "transformation strain" as an internal variable (see e.g. [6]). These might be modified to fit the framework of our order-parameter models

- The local analysis can also be extended to the dynamic case so that some predictions about the dependence of stress-strain laws on the frequency of input could be made.

- The question of local dynamic stability of the precursor oscillations needs to be addressed.

References

[1] S.S. Antman. Ordinary differential equations of nonlinear elasticity II: Existence and regularity theory for conservative boundary value problems. *Arch. Rat. Mech. Anal.*, 61:353–393, 1976.

[2] K. Bhattacharya, R.D. James, and P. Swart. Preprint. 1995.

[3] D. Brandon, T. Lin, and R.C. Rogers. Phase transitions and hysteresis in nonlocal and order-parameter models. (to appear in) *Meccanica*, submitted 1994.

[4] D. Brandon and R.C. Rogers. Nonlocal regularization of L.C. Young's tacking problem. *Applied Mathematics and Optimization*, **25**:287–301, 1992.

[5] D. Brandon and R.C. Rogers. Order parametermodels of elastic bars and precursor oscillations. IMA Preprint #1370, November 1995.

[6] O. Bruno, F. Reitich, and P. Leo. The overall elastic energy of martensitic solids, IMA Preprint. 1995. (Submitted to *J. Mech. Phys. Solids.*)

[7] G. Caginalp. An analysis of a phase field model of a free boundary. *Arch. Rat. Mech,. Anal.* 92:205–245, 1986.

[8] G. Caginalp and P.C. Fife. Dynamics of layered interfaces arising from phase boundaries. *SIAM J. Appl. Math.* 48:506–518, 1988.

[9] J.W. Cahn and A. Novick-Cohen. Evolution equations for phase separation and ordering in binary alloys. *Journal of Statistical Physics,* 76:877-909, 1994.

[10] E. Fried. "Continuum theory for coherent phase transitions incorporating an order parameter," in *Smart Structures and Materials 1993: Mathematics in Smart Structures,* H.T. Banks, Editor, Proc. SPIE 1919:328–335, 1993.

[11] E. Fried and M.E. Gurtin. Dynamic solid-solid transitions with phase characterized by an order parameter, *Physica D,* 72:287–308, 1994.

[12] R.D. James. "The stability and metastability of quartz," in *Metastability and Incompletely Posed Problems,* S. Antman, J.L. Ericksen, D. Kinderlehrer, and I. Müller, Editors. Springer-Verlag, New York, 1987.

[13] S. Kartha, J.A. Krumhansl, J.P. Sethna, and L.K. Wickham. Disorder-driven pretransitional tweed in martensitic transformations. *Phys. Rev. B,* 52:803, 1995.

[14] T. Lin and R.C. Rogers. On an order-parameter model for a binary liquid. (to appear in) *Journal of Computational Materials Science.* submitted 1994.

[15] G.A. Maugin and W. Muschik. Thermodynamics with internal variables. Part 1. General concepts, *Journal of Non-Equilibrium Thermodynamics,* 19:217–249, 1994.

[16] D. Schryvers, L.E. Tanner, and S.M. Shapiro. "Electron-microscopy and neutron-scattering studies of the premartensitic behavior in ordered Ni-Al-β_2 phase" in *Shape Memory Materials*, K. Otsuka and K. Shimizu, Editors. Materials Research Society International Symposium Proceedings, Vol. 9, 1989.

[17] D. Schryvers and L.E. Tanner. 'On the interpretation of high resolution electron microscopy images of the premartensitic microstructures in the Ni-Al-β_2 phase" *Ultramicroscopy,* 32:241–254, 1990.

[18] D. Schryvers and L.E. Tanner. Electron microscopy of stress induced martensite and pretransition microstructures in $Ni_{62.5}Al_{37.5}$. Materials Research Society Symposium Proceedings, Vol. 246, 1992.

[19] D. Schryvers. Martensitic and related transformations in Ni-Al alloys. *Journal de Physique IV,* 5:C2-225–234, 1995.

[20] S. Scmenovskaya, Y. Zhu, M. Suenaga, A. Khachaturyan, *Phys. Rev. B*, May 1993.

[21] S.M. Shapiro, B.X. Yang, Y. Noda, L.E. Tanner, and D. Schryvers. Neutron-scattering and electron-microscopy studies of the premartensitic phenomena in $NI_x Al_{100-x}$ alloys. *Physical Rev. B*, 44:9301–9313, 1991.

[22] L.E. Tanner, D. Schryvers, and S.M. Shapiro. Electron-microscopy and neutron-scattering studies of the premartensitic behavior in ordered Ni-Al-β_2 phase. *Materials Science and Engineering,* A127:205–213, 1990.

[23] L.E. Tanner, A.R. Pelton, G. VanTendeloo, D. Schryvers, and M.E. Wall. Pre-martensitic microstructures in Ni-Al ordered β_2 phase: I-effects induced by cooling. *Scripta Metallurgica et Materialia,* 24:1731–1736, 1990.

REAL-SPACE LATTICE STATICS DESCRIPTION OF A RANDOM LATTICE

J.R. Willis and T.J.Gosling

University of Cambridge
Department of Applied Mathematics and Theoretical Physics
Silver Street, Cambridge CB3 9EW, U.K.

Abstract

An approach is outlined for the analysis of stress and deformation in a random alloy of substitutional type. A description is developed in the framework of lattice statics, so that microscopic arrangements of atoms can be examined. The theory is built so that it correctly reproduces behaviour of a composite continuum, in the limit of large clusters of atom types. Methodology of proven efficiency for the analysis of randomly inhomogeneous continua is adapted to the situation of a random lattice. This includes use of a stochastic variational principle in conjunction with trial fields generated from "polarizations" introduced into a comparison lattice. One use of the theory is to estimate the energy associated with particular local patterns of atoms, described through multipoint correlation functions. As a first application, however, the theory is applied here to the estimation of the mean lattice parameter of a perfectly disordered alloy. Knowledge of the mean lattice parameter is important for assessment of the electronic performance and mechanical stability of strained-layer semiconductor heterostructures. Predicted deviations from Vegard's law are discussed in relation to published experimental data.

1. Introduction

The purpose of this work is to set up a minimal framework for describing the stress field in an alloy of random constitution. The theory to be described is a lattice theory, so that microstructure down to atomic dimensions can be addressed. It is required also to reproduce correctly known continuum behaviour, in the special case that atom types are clustered so as to form an inhomogeneous continuum, or composite material. Pure lattices of different components of the alloy will, in general, display both lattice mismatch and differences in elastic constants, and the composite continuum will as a result be in a state of self-stress, even in the absence of external loading. Furthermore, the analysis of stress resulting from external loading has to make due allowance for the elastic heterogeneity of the composite. The underlying lattice theory has to contain sufficient detail to yield both of these effects, in the limit of "large clusters".

Even the continuum theory of composite materials with random microstructure contains too much complication to permit exact analysis. There exists, however, a substantial body of theory for provision of approximate solutions. Of particular relevance here are stochastic variational formulations which provide "best estimates" given limited statistical information, in the form of low-order correlation functions. Corresponding lattice formulations can be developed. The simplest results for a composite continuum follow from a variational principle introduced by Hashin and Shtrikman [1]. Its particular simplicity is, however, lost in the corresponding formulation for a lattice; it is more convenient, in fact, to employ a stochastic version of the principle of minimum energy.

The theory to be outlined has potential application to the evolution of microstructure during materials processing. Such application requires the assessment of energy changes upon interchange of atoms in specified local environments, for substitution into kinetic equations. There is at present no (other) method of allowance for local variations in elastic response. A simpler application, however, is to the estimation of the mean lattice parameter of a random alloy. This has relevance to strained-layer heterostructures such as those employed in semiconductor devices. Here, a thin layer of alloy is deposited epitaxially on a substrate (for instance, Ge_xSi_{1-x} on Si). If the substrate is thick and the layer is too thin to allow stress relaxation through the introduction of dislocations, the layer is obliged to adopt the lattice spacing of the substrate and hence is in a state of stress, which modifies its electronic properties. The stress depends upon the mean lattice spacing that the alloy would adopt when stress-free. This is invariably estimated by invoking Vegard's law, which is nothing more than the law of mixtures. In view of the importance of stress both for the operation and mechanical stability of such devices, a more fundamental assessment of the mean lattice parameter appears to be warranted.

The plan of this article is as follows. Section 2 presents the basic lattice model of an alloy. At least in this first study, it is restricted to be of "substitutional" type, in the sense that the same basic crystal structure would be adopted by pure lattices of each component, though with different elastic constants and lattice parameters. This model is closely related to that employed by Lee [2,3] in his "discrete atom method". Here, however, a variational formulation is emphasised, in preparation for introduction, later, of a "stochastic variational principle" which will permit the development of systematic approximations applicable to a random alloy. Section 3 introduces "trial fields" derived from the introduction of "polarizations" into a uniform "comparison lattice". This is an idea borrowed directly from the continuum theory of composites [4,5]. Section 4 develops the stochastic variational principle and its use in conjunction with configuration-dependent trial fields derived from polarizations in the comparison lattice. Section 5 addresses the problem of estimating the mean lattice parameter of a random lattice, in the case that this is statistically uniform and spatially uncorrelated, admitting only nearest-neighbour interactions. Illustrative numerical results are presented in the concluding Section 6, and are compared both with a new simple alternative to Vegard's law developed in the course of the analysis and to some published experimental data. Certain detailed calculations are recorded in an Appendix.

2. The lattice model

It is supposed that the lattice is composed from atoms of several different types, with labels $A = 1, 2, \cdots$. At least at present, it is assumed that pure material of each type A would have the same crystallographic structure, and that the lattice constants of each would not be too dissimilar. In this situation, it is reasonable to treat the random lattice as having been composed by substituting atoms of the required types at sites in some reference lattice of pure material, of type 0 say, whose properties can be chosen for convenience: the reference lattice could, for example, correspond to pure material of type A, with A chosen as the major constituent of the alloy, but this is not obligatory. The substitutions will induce local distortions and also, in general, a change in mean lattice parameter.

When the atoms are displaced from the reference lattice sites by $\mathbf{u}(\mathbf{l})$, the energy in the medium is, in the harmonic approximation,

$$\mathcal{E}(\mathbf{u}) = -\sum_{\mathbf{l}} \sum_{\mathbf{l}'} \mathbf{F}(\mathbf{l}, \mathbf{l}').[\mathbf{u}(\mathbf{l}) - \mathbf{u}(\mathbf{l}')] + \tfrac{1}{2} \sum_{\mathbf{l}} \sum_{\mathbf{l}'} [\mathbf{u}(\mathbf{l}) - \mathbf{u}(\mathbf{l}')].\Psi(\mathbf{l}, \mathbf{l}').[\mathbf{u}(\mathbf{l}) - \mathbf{u}(\mathbf{l}')]. \quad (2.1)$$

This form builds in automatically insensitivity to rigid translation. The functions $\mathbf{F}$ and Ψ have to satisfy further restrictions to ensure insensitivity to small rotations. No assumption of pairwise interactions and central forces has (yet) been made: equation (2.1) is general, apart from the assumption of small displacements. The linear term is present because, at zero displacement in the reference lattice, the atoms would not be in equilibrium and so would experience a force. The symmetries

$$\mathbf{F}(\mathbf{l}, \mathbf{l}') = -\mathbf{F}(\mathbf{l}', \mathbf{l}), \quad \Psi(\mathbf{l}, \mathbf{l}') = \Psi(\mathbf{l}', \mathbf{l}) \quad (2.2)$$

may be assumed, without loss.

Equation (2.1) can be written equivalently

$$\mathcal{E}(\mathbf{u}) = -\sum_{\mathbf{l}} \mathbf{f}(\mathbf{l}).\mathbf{u}(\mathbf{l}) + \tfrac{1}{2} \sum_{\mathbf{l}} \sum_{\mathbf{l}'} \mathbf{u}(\mathbf{l}).\Phi(\mathbf{l}, \mathbf{l}').\mathbf{u}(\mathbf{l}'), \quad (2.3)$$

where

$$\mathbf{f}(\mathbf{l}) = 2 \sum_{\mathbf{l}' \neq \mathbf{l}} \mathbf{F}(\mathbf{l}, \mathbf{l}') \quad (2.4)$$

and

$$\Phi(\mathbf{l}, \mathbf{l}) = 2 \sum_{\mathbf{l}' \neq \mathbf{l}} \Psi(\mathbf{l}, \mathbf{l}'),$$
$$\Phi(\mathbf{l}, \mathbf{l}') = -2\Psi(\mathbf{l}, \mathbf{l}'), \quad \mathbf{l}' \neq \mathbf{l}. \quad (2.5)$$

Then automatically,

$$\sum_{\mathbf{l}'} \Phi(\mathbf{l}, \mathbf{l}') = 0. \quad (2.6)$$

At equilibrium under no additional forces, the displacements $\mathbf{u}(\mathbf{l})$ take the values $\mathbf{u}^e(\mathbf{l})$. These minimize $\mathcal{E}(\mathbf{u})$ and so satisfy the equations (employing matrix notation)

$$\Phi \mathbf{u}^e = \mathbf{f}, \quad \text{or} \quad \mathbf{u}^e = \mathbf{G}\mathbf{f}, \tag{2.7}$$

where

$$\mathbf{G} = \Phi^{-1}. \tag{2.8}$$

The inverse $\mathbf{G}$ of Φ exists so long as $\mathbf{u}^e$ is rendered unique by insisting that it has zero mean value and zero "first moment", just as $\mathbf{f}$ must have, to ensure overall equilibrium. Then, $\mathbf{G}$, like Φ, is a mapping of the vector space of such "self-equilibriated" fields to itself.

If additional (self-equilibriated) body-force $\mathbf{g}$ (with value $\mathbf{g}(\mathbf{l})$ at site $\mathbf{l}$) is applied, the equilibrium displacement must minimize the total energy

$$\mathcal{E}(\mathbf{u}) - \mathbf{u}^T \mathbf{g} = \tfrac{1}{2}\mathbf{u}^T \Phi \mathbf{u} - \mathbf{u}^T (\mathbf{f} + \mathbf{g}), \tag{2.9}$$

so that

$$\mathbf{u} = \mathbf{G}(\mathbf{f} + \mathbf{g}) = \mathbf{u}^e + \mathbf{G}\mathbf{g}. \tag{2.10}$$

The matrix $\mathbf{G}$ is the Green's function for the lattice.

It is convenient, in applications, to consider a lattice of infinite extent by admitting any assignment of atom types within a (large) cell Q, and then building an infinite medium by periodic replication of Q throughout space, to produce a superlattice S. Then, any lattice site $\mathbf{l} \in Q$ is replicated at all sites $\mathbf{l} + \mathbf{s}$, where $\mathbf{s} \in S$.

Equilibrium of such an infinite lattice, in the absence of externally applied force, is achieved by displacements of the form

$$\mathbf{u}(\mathbf{l}) = \mathbf{A}.\mathbf{l} + \mathbf{u}^*(\mathbf{l}), \tag{2.11}$$

where $\mathbf{A}$ is a constant 3×3 matrix and the field $\mathbf{u}^*$ is Q-periodic. For any such displacement, the energy $\mathcal{E}$ in the whole medium is infinite. At equilibrium, $\mathbf{A}$ and $\mathbf{u}^*$ minimize the mean energy density

$$\varphi(\mathbf{u}) = \frac{1}{|Q|} \sum_{\mathbf{l} \in Q} \sum_{\mathbf{l}' \in Q} \sum_{\mathbf{s}' \in S} \{ -\mathbf{F}(\mathbf{l}, \mathbf{l}' + \mathbf{s}').[\mathbf{A}.(\mathbf{l} - \mathbf{l}' - \mathbf{s}') + \mathbf{u}^*(\mathbf{l}) - \mathbf{u}^*(\mathbf{l}')]$$
$$+ \tfrac{1}{2}[\mathbf{A}.(\mathbf{l} - \mathbf{l}' - \mathbf{s}') + \mathbf{u}^*(\mathbf{l}) - \mathbf{u}^*(\mathbf{l}')].\Psi(\mathbf{l}, \mathbf{l}' + \mathbf{s}').$$
$$[\mathbf{A}.(\mathbf{l} - \mathbf{l}' - \mathbf{s}') + \mathbf{u}^*(\mathbf{l}) - \mathbf{u}^*(\mathbf{l}')] \},$$
$$\tag{2.12}$$

where $|Q|$ represents the volume occupied by Q. If this minimization is done sequentially, by first minimizing (2.12) with respect to $\mathbf{u}^*$, with $\mathbf{A}$ fixed, there results an intermediate energy density of the form

$$w(\mathbf{A}) := \frac{1}{2|Q|} (\mathbf{A} - \mathbf{A}_0) : \mathbf{C} : (\mathbf{A} - \mathbf{A}_0). \tag{2.13}$$

The matrix $\mathbf{A}_0$ defines the distortion required to take the reference lattice to the mean lattice, and $\mathbf{C}$ is the tensor of effective elastic moduli. (The tensor $\mathbf{C}$ will possess the usual symmetries, and only the symmetric part of $\mathbf{A}_0$ will be uniquely determined.)

If, now, additional, external, Q-periodic forces $\mathbf{g}^*(\mathbf{l})$ (which must have zero resultant for the overall equilibrium of the medium) are applied, these generate additional mean deformation of the lattice, and an additional Q-periodic displacement $\mathbf{v}^*(\mathbf{l})$. It is necessary now to minimize the energy function

$$\psi(\mathbf{u}) = \varphi(\mathbf{u}) - \frac{1}{|Q|} \sum_{\mathbf{l} \in Q} \mathbf{g}^*(\mathbf{l}).\mathbf{u}(\mathbf{l}). \tag{2.14}$$

Elementary calculation shows that $\mathbf{v}^*$ must satisfy the equations

$$\sum_{\mathbf{l}' \in Q} \Phi(\mathbf{l},\mathbf{l}').\mathbf{v}^*(\mathbf{l}') = \mathbf{g}^*(\mathbf{l}), \tag{2.15}$$

where, analogously to (2.5),

$$\Phi(\mathbf{l},\mathbf{l}') = 2 \sum_{\mathbf{l}'' \neq \mathbf{l}} \sum_{\mathbf{s}'} \Psi(\mathbf{l},\mathbf{l}'' + \mathbf{s}'), \ \mathbf{l} = \mathbf{l}'$$
$$= -2 \sum_{\mathbf{s}'} \Psi(\mathbf{l},\mathbf{l}'' + \mathbf{s}'), \ \mathbf{l} \neq \mathbf{l}'. \tag{2.16}$$

Equation (2.15) has solution

$$\mathbf{v}^*(\mathbf{l}) = \sum_{\mathbf{l}' \in Q} \mathbf{G}(\mathbf{l},\mathbf{l}').\mathbf{g}^*(\mathbf{l}'). \tag{2.17}$$

The Green's function $\mathbf{G}$ that appears in (2.17) is uniquely defined if $\mathbf{v}^*$ is restricted to have mean value zero over Q, to match the requirement that the external forces have zero resultant.

3. Introduction of a comparison lattice

In the continuum theory of composites, it has proven useful to introduce a uniform "comparison medium", and to observe that the stress, strain and displacement fields in the actual composite could be generated in the comparison material by the introduction of body forces derived from "polarizations" (see, for example, [1], [4]). In the lattice context, the corresponding construction introduces a lattice characterized by potentials $\mathbf{F}_0(\mathbf{l},\mathbf{l}')$ and $\Psi_0(\mathbf{l},\mathbf{l}')$ or, corresponding to these, $\mathbf{f}_0(\mathbf{l})$ and $\Phi_0(\mathbf{l},\mathbf{l}')$. Then, if $\mathbf{u}(\mathbf{l})$ is the displacement field in the actual lattice, this can be generated in the comparison lattice by the application of external forces

$$\mathbf{t}(\mathbf{l}) = \sum_{\mathbf{l}'} \Phi_0(\mathbf{l},\mathbf{l}')\mathbf{u}(\mathbf{l}') - \mathbf{f}_0(\mathbf{l}). \tag{3.1}$$

In terms of the Green's function of the comparison lattice,

$$\mathbf{u} = \mathbf{G}_0[\mathbf{f}_0 + \mathbf{t}]. \tag{3.2}$$

Since the displacement field $\mathbf{u}$ also satisfies (2.10), it follows that the force field $\mathbf{t}$ should satisfy

$$\mathbf{f} + \mathbf{g} = \Phi \mathbf{G}_0[\mathbf{f}_0 + \mathbf{t}]. \tag{3.3}$$

The object of reproducing in the comparison lattice the forces as well as the displacements in the actual lattice is served by choosing

$$\mathbf{t}(\mathbf{l}) = \mathbf{g}(\mathbf{l}) + 2 \sum_{\mathbf{l}'} \mathbf{T}(\mathbf{l},\mathbf{l}'), \tag{3.4}$$

with

$$\mathbf{T}(\mathbf{l},\mathbf{l}') = -[\Psi(\mathbf{l},\mathbf{l}') - \Psi_0(\mathbf{l},\mathbf{l}')].[\mathbf{u}(\mathbf{l}) - \mathbf{u}(\mathbf{l}')] + \mathbf{F}(\mathbf{l},\mathbf{l}') - \mathbf{F}_0(\mathbf{l},\mathbf{l}'). \tag{3.5}$$

The two-point field $\mathbf{T}$ is a close analogue of the "polarization" introduced into the continuum theory of composites by Hashin and Shtrikman [1]. Equation (3.3), with (3.4) and (3.5), defines the field $\mathbf{T}$, which will henceforth be referred to as a polarization field. Exact solution of (3.3), with (3.4) and (3.5), for $\mathbf{T}$ would provide the solution for the actual lattice. However, since the actual lattice is complicated (and will even be taken as random in the sequel), it is helpful to seek approximations, by substituting trial displacements of the form (3.2), with $\mathbf{t}$ given by (3.4), into the energy function (2.9) and seeking a minimum over some appropriately chosen subspace of functions $\mathbf{T}(\mathbf{l}, \mathbf{l}')$.

This general notion will be pursued in the context of the Q-periodic lattice introduced above. Then, the comparison lattice is taken to be infinite and uniform, so that $\mathbf{F}_0$ and Ψ_0 become functions of $\mathbf{l} - \mathbf{l}'$ only, and the two-point field $\mathbf{T}$ is taken to be Q-periodic. The displacement field generated by $\mathbf{t}$ then has the form (2.11), with

$$\mathbf{u}^* = \mathbf{G}_0[\mathbf{f}_0 + \mathbf{t}]. \tag{3.6}$$

The complete translational invariance of the comparison lattice, with the fact that Ψ_0 is an even function, ensures that $\mathbf{u}^*$ is determined independently of the mean distortion $\mathbf{A}$.

4. Characterization of the random alloy

In the sequel, the above formalism will be applied to a random alloy. For this purpose, it will be assumed specifically that the fields $\mathbf{F}(\mathbf{l}, \mathbf{l}')$ and $\Psi(\mathbf{l}, \mathbf{l}')$ depend on the atom types at sites $\mathbf{l}$ and $\mathbf{l}'$ but not on the atom types at other sites. Thus,

$$\mathbf{F}(\mathbf{l}, \mathbf{l}') = \sum_{A,B} \mathbf{F}_{AB}(\mathbf{l}, \mathbf{l}') \chi_A(\mathbf{l}) \chi_B(\mathbf{l}'), \tag{4.1}$$

$$\Psi(\mathbf{l}, \mathbf{l}') = \sum_{A,B} \Psi_{AB}(\mathbf{l}, \mathbf{l}') \chi_A(\mathbf{l}) \chi_B(\mathbf{l}'), \tag{4.2}$$

where χ_A is the characteristic function of the lattice sites occupied by atoms of type A:

$$\chi_A(\mathbf{l}) = 1 \quad \text{if atom } A \text{ occupies } \mathbf{l}$$
$$= 0 \quad \text{otherwise.} \tag{4.3}$$

With a uniform comparison lattice chosen as described above, the exact polarization field $\mathbf{T}$ should depend on the atom type at every lattice site. The simplest polarization fields that take account of the microgeometry of the actual lattice have the form

$$\mathbf{T}(\mathbf{l}, \mathbf{l}') = \sum_{A,B} \mathbf{T}_{AB}(\mathbf{l}, \mathbf{l}') \chi_A(\mathbf{l}) \chi_B(\mathbf{l}'), \tag{4.4}$$

where the functions $\mathbf{T}_{AB}(\mathbf{l}, \mathbf{l}')$ are sure. It is emphasised that the form (4.4) is not in general exact, in spite of its similarity with (4.1), because equation (3.5) introduces coupling with all sites through the presence of $\mathbf{u}$. Such trial fields do, nevertheless, take some account of the random configuration, through the presence of the factors χ_A and χ_B.

A prescription for estimating the fields $\mathbf{T}_{AB}$ is now developed, by noting the *stochastic variational principle*, that requiring the *ensemble averaged* energy, $\langle \psi(\mathbf{u}) \rangle$, to be minimized over the set of all configuration-dependent displacement fields $\mathbf{u}$, generates the solution in each realization of the lattice. Explicitly, if realizations of the lattice correspond to labels $\alpha \in \mathcal{A}$, where $\mathcal{A}$ denotes the (discrete) sample space over which probabilities $p(\alpha)$ are defined,

$$\langle \psi(\mathbf{u}) \rangle := \sum_{\alpha \in \mathcal{A}} p(\alpha) \psi(\mathbf{u}; \alpha) \tag{4.5}$$

is minimized over displacement fields $\mathbf{u}(\mathbf{l};\alpha)$ by the actual fields in each realization.

This observation leads naturally to a prescription for developing an approximate solution: substitute into (4.5) configuration-dependent displacement fields $\mathbf{u}(\mathbf{l};\alpha)$ which are given by (3.2), with $\mathbf{t}$ having some restricted dependence on the configuration α. Then, minimize with respect to the functions $\mathbf{t}(\mathbf{l};\alpha)$. In particular, in the sequel, $\mathbf{t}$ will be chosen as in (3.4), with $\mathbf{T}$ given by (4.4). The minimization is then with respect to the functions $\mathbf{T}_{AB}(\mathbf{l},\mathbf{l}')$. The general form of the functional (2.12) is:

$$\varphi(\mathbf{u};\alpha) = \frac{1}{|Q|}\Bigg\{ -\sum_{A,B}\sum_{\mathbf{l}_1 \in Q}\sum_{\mathbf{l}_2}\sum_{\mathbf{l}_3 \in Q}$$
$$\mathbf{F}_{AB}(\mathbf{l}_1 - \mathbf{l}_2).[\mathbf{G}_0(\mathbf{l}_1 - \mathbf{l}_3) - \mathbf{G}_0(\mathbf{l}_2 - \mathbf{l}_3)].\chi_A(\mathbf{l}_1)\chi_B(\mathbf{l}_2)\mathbf{t}(\mathbf{l}_3;\alpha)$$
$$-\sum_{A,B}\sum_{\mathbf{l}_1 \in Q}\sum_{\mathbf{l}_2}\mathbf{F}_{AB}(\mathbf{l}_1 - \mathbf{l}_2).\mathbf{A}.(\mathbf{l}_1 - \mathbf{l}_2)\chi_A(\mathbf{l}_1)\chi_B(\mathbf{l}_2)$$
$$+ 2\sum_{A-D}\sum_{\mathbf{l}_1 \in Q}\sum_{\mathbf{l}_2}\sum_{\mathbf{l}_3 \in Q}\sum_{\mathbf{l}_4}\sum_{\mathbf{l}_5 \in Q}\mathbf{F}_{CD}(\mathbf{l}_3 - \mathbf{l}_4).[\mathbf{G}_0(\mathbf{l}_1 - \mathbf{l}_3) - \mathbf{G}_0(\mathbf{l}_2 - \mathbf{l}_3)].\Psi_{AB}(\mathbf{l}_1 - \mathbf{l}_2).$$
$$[\mathbf{G}_0(\mathbf{l}_1 - \mathbf{l}_5) - \mathbf{G}_0(\mathbf{l}_2 - \mathbf{l}_5)].\chi_A(\mathbf{l}_1)\chi_B(\mathbf{l}_2)\chi_C(\mathbf{l}_3)\chi_D(\mathbf{l}_4)\mathbf{t}(\mathbf{l}_5;\alpha)$$
$$+ 2\sum_{A-D}\sum_{\mathbf{l}_1 \in Q}\sum_{\mathbf{l}_2}\sum_{\mathbf{l}_3 \in Q}\sum_{\mathbf{l}_4}\mathbf{F}_{CD}(\mathbf{l}_3 - \mathbf{l}_4).[\mathbf{G}_0(\mathbf{l}_1 - \mathbf{l}_3) - \mathbf{G}_0(\mathbf{l}_2 - \mathbf{l}_3)].\Psi_{AB}(\mathbf{l}_1 - \mathbf{l}_2).$$
$$\mathbf{A}.(\mathbf{l}_1 - \mathbf{l}_2)\chi_A(\mathbf{l}_1)\chi_B(\mathbf{l}_2)\chi_C(\mathbf{l}_3)\chi_D(\mathbf{l}_4)$$
$$+ \tfrac{1}{2}\sum_{A,B}\sum_{\mathbf{l}_1 \in Q}\sum_{\mathbf{l}_2}\sum_{\mathbf{l}_3 \in Q}\sum_{\mathbf{l}_4 \in Q}\{[\mathbf{G}_0(\mathbf{l}_1 - \mathbf{l}_3) - \mathbf{G}_0(\mathbf{l}_2 - \mathbf{l}_3)].\Psi_{AB}(\mathbf{l}_1 - \mathbf{l}_2).$$
$$[\mathbf{G}_0(\mathbf{l}_1 - \mathbf{l}_4) - \mathbf{G}_0(\mathbf{l}_2 - \mathbf{l}_4)]\}: \chi_A(\mathbf{l}_1)\chi_B(\mathbf{l}_2)\mathbf{t}(\mathbf{l}_3;\alpha) \otimes \mathbf{t}(\mathbf{l}_4;\alpha)$$
$$+ \sum_{A,B}\sum_{\mathbf{l}_1 \in Q}\sum_{\mathbf{l}_2}\sum_{\mathbf{l}_3 \in Q}\{[\mathbf{G}_0(\mathbf{l}_1 - \mathbf{l}_3) - \mathbf{G}_0(\mathbf{l}_2 - \mathbf{l}_3)].\Psi_{AB}(\mathbf{l}_1 - \mathbf{l}_2).$$
$$\mathbf{A}.(\mathbf{l}_1 - \mathbf{l}_2)\}.\chi_A(\mathbf{l}_1)\chi_B(\mathbf{l}_2)\mathbf{t}(\tau,\mathbf{l}_3)$$
$$+ \tfrac{1}{2}\sum_{A,B}\sum_{\mathbf{l}_1 \in Q}\sum_{\mathbf{l}_2}(\mathbf{l}_1 - \mathbf{l}_2).\mathbf{A}.\Psi_{AB}(\mathbf{l}_1 - \mathbf{l}_2).\mathbf{A}.(\mathbf{l}_1 - \mathbf{l}_2)\chi_A(\mathbf{l}_1)\chi_B(\mathbf{l}_2)\Bigg\}. \quad (4.6)$$

Summations over lattice sites not explicitly specified to be over Q are over all sites. The additional term required for $\psi(\mathbf{u};\alpha)$ (see equation (2.14)) is not given, because it is not used in the sequel.

As given, equation (4.6) applies for any choice of polarization field. The prescription adopted here is obtained by use of the form (4.4), for which the resulting $\mathbf{t}(\mathbf{l};\alpha)$ is expressed as a sum of terms, each involving the characteristic functions of two points. Ensemble averaging the resulting form for $\varphi(\mathbf{u};\alpha)$ thus involves multipoint probabilities:

$$p_A(\mathbf{l}_1) = \langle \chi_A(\mathbf{l}_1)\rangle, \quad p_{AB}(\mathbf{l}_1,\mathbf{l}_2) = \langle \chi_A(\mathbf{l}_1)\chi_B(\mathbf{l}_2)\rangle \quad (4.7)$$

and so on. Inspection of (4.6) shows that probabilities involving up to six points participate. These must either be known, or else must be plausibly estimated, but then the functional to be minimized is completely explicit. Even though the trial polarization field is not exact, it bears emphasis, perhaps, that the resulting expression always provides an upper bound for the exact mean energy. Apart from this, the procedure can be viewed as a way of closing the infinite hierarchy of equations that would result from conditional ensemble averaging of equation (3.5), which defines $\mathbf{T}$ exactly.

5. The mean lattice parameter of a statistically uniform random alloy

In order to illustrate the use of the formalism developed in previous sections we now consider its simplest possible application: calculation of the mean lattice parameter of a statistically uniform, simple cubic, two-constituent random alloy, in which forces between atoms are central and pairwise, with only nearest-neighbour interactions being significant. It will be assumed that no external forces are acting and that atom types at different lattice sites are uncorrelated. This set of approximations does, of course, make it inappropriate to compare the results obtained with experiment in anything other than a cursory fashion. However, the essential physics of the practical situation is preserved.

There are various simplifying results that follow from the above approximations. Since their development would obscure the central message of this section, the details have been consigned to an appendix. However, it is appropriate to outline some of these results now. The lattices are simple cubic, and so taking the lattice spacing in the reference lattice to be the unit of length, the reference positions may be described by vectors of the form $m_i \mathbf{e}_i$, where m_i are integers and $\{\mathbf{e}_i\}$ is a set of orthonormal basis vectors.

The statistical uniformity means that the functions $\mathbf{F}_{AB}(\mathbf{l},\mathbf{l'})$, $\Psi_{AB}(\mathbf{l},\mathbf{l'})$, and $\mathbf{T}_{AB}(\mathbf{l},\mathbf{l'})$ must, in fact, all be functions of $\mathbf{l} - \mathbf{l'}$. The assumptions that interactions are governed by pairwise interatomic potentials and that only nearest-neighbour interactions are significant place further constraints on the forms of $\mathbf{F}_{AB}(\mathbf{l})$ and $\Psi_{AB}(\mathbf{l})$. In particular, it is shown in the Appendix that under the stated approximations these functions are zero except for the cases

$$\mathbf{F}_{AB}(\pm \mathbf{e}_i) = \pm \alpha_{AB} \mathbf{e}_i \tag{5.1}$$

$$\Psi_{AB}(\pm \mathbf{e}_i) = (\alpha_{AB} + \beta_{AB})\mathbf{e}_i \otimes \mathbf{e}_i - \alpha_{AB}\mathbf{I}. \tag{5.2}$$

The constants α_{AB} and β_{AB} may be related directly to the interatomic potentials, as in (A.10). The final result of significance in the Appendix is that for a pure crystal of some material A, the two constants α_{AA} and β_{AA} are determined by the lattice parameter and stiffness of that pure material. In particular, suppose that a pure crystal of material A experiences a strain $-\epsilon_A$ when its lattice is forced into registry with the reference lattice. Moreover, let γ_A be the elastic constant C_{1111} for material A. Then the constants α_{AA} and β_{AA} are defined by the equations

$$\epsilon_A = \alpha_{AA}/\beta_{AA}, \tag{5.3}$$

$$\gamma_A = 2(\alpha_{AA} + \beta_{AA}). \tag{5.4}$$

So the parameters governing interaction of two atoms of the same material may be fully defined by reference to measurable macroscopic properties of that material. However, the parameters α_{AB} and β_{AB}, $A \neq B$ govern interaction between atoms of different types. It is not straightforward to obtain reliable values for these parameters. In order to make progress we make a further simplifying assumption. Let $\Omega_{AB}(r)$ be the potential governing interactions between atoms of types A and B (which may or may not be the same type). We assume that for some value of the parameter k the following relationship holds:

$$\Omega_{AB}(r) = k\Omega_{AA}(r) + (1-k)\Omega_{BB}(r).$$

It is clear from (A.10) that this assumption induces the relationships

$$\alpha_{AB} = k\alpha_{AA} + (1-k)\alpha_{BB} \quad \text{and} \quad \beta_{AB} = k\beta_{AA} + (1-k)\beta_{BB}. \tag{5.5}$$

We now have a scheme for defining all of the parameters that govern atomic interactions in the lattice.

It is now necessary to choose a subclass of polarizations over which to minimize the ensemble average of (4.6). We persist with the general class of fields described by (4.4), and we pursue further the similarity between (4.1) and (4.4) by choosing $\mathbf{T}_{AB}(\mathbf{l}-\mathbf{l'})$ to have precisely the same form as $\mathbf{F}_{AB}(\mathbf{l}-\mathbf{l'})$, defined in (5.1). Thus $\mathbf{T}_{AB}(\mathbf{l})$ is a 'nearest-neighbours' interaction, equal to zero except for the cases

$$\mathbf{T}_{AB}(\pm \mathbf{e}_i) = \pm \tfrac{1}{2}\theta_{AB}\mathbf{e}_i. \tag{5.6}$$

The quantity $\mathbf{t}$ in (3.4), which is derived from the polarizations $\mathbf{T}$, thus mimics the force $\mathbf{f}$ in (2.4), which is derived from the quantities $\mathbf{F}$. It should be noted, however, that although we shall impose the symmetry $\theta_{BA} = \theta_{AB}$, the choice of θ_{AB} for $A \neq B$ will be quite independent of θ_{AA} and θ_{BB}. This contrasts with the choice of α_{AB} in (5.5).

Invoking the above relations, the ensemble average of (4.6) may be written

$$\langle \varphi \rangle (\theta_{AB}, \mathbf{A}) = \frac{1}{|Q|} \{ y_{AB}\theta_{AB} + \tfrac{1}{2}z_{ABCD}\theta_{AB}\theta_{CD} + \mathbf{A} : \mathbf{H}_{AB}\theta_{AB} + (\mathbf{J}_1 + \mathbf{J}_2) : \mathbf{A}$$
$$+ \tfrac{1}{2}\mathbf{A} : \mathbf{T} : \mathbf{A} \}. \tag{5.7}$$

Using the notation
$$\Gamma_i = \{\mathbf{l} | \mathbf{l} - \mathbf{l}_i = \pm \mathbf{e}_j, \text{ for some } j\}$$
to describe the set of vectors within a unit vector of $\mathbf{l}_i$, the new quantities in (5.7) may be written

$$y_{EF} = -\sum_{A,B}\sum_{\mathbf{l}_1 \in Q}\sum_{\mathbf{l}_2}\sum_{\mathbf{l}_3 \in Q}\sum_{\mathbf{l}_4 \in \Gamma_3} \mathbf{F}_{AB}(\mathbf{l}_1 - \mathbf{l}_2).[\mathbf{G}_0(\mathbf{l}_1 - \mathbf{l}_3) - \mathbf{G}_0(\mathbf{l}_2 - \mathbf{l}_3)].(\mathbf{l}_3 - \mathbf{l}_4)$$
$$\times \langle \chi_A(\mathbf{l}_1)\chi_B(\mathbf{l}_2)\chi_E(\mathbf{l}_3)\chi_F(\mathbf{l}_4) \rangle$$
$$+ 2\sum_{A-D}\sum_{\mathbf{l}_1 \in Q}\sum_{\mathbf{l}_2}\sum_{\mathbf{l}_3 \in Q}\sum_{\mathbf{l}_4}\sum_{\mathbf{l}_5 \in Q}\sum_{\mathbf{l}_6 \in \Gamma_5} \mathbf{F}_{CD}(\mathbf{l}_3 - \mathbf{l}_4).[\mathbf{G}_0(\mathbf{l}_1 - \mathbf{l}_3) - \mathbf{G}_0(\mathbf{l}_2 - \mathbf{l}_3)].\mathbf{\Psi}_{AB}(\mathbf{l}_1 - \mathbf{l}_2).$$
$$[\mathbf{G}_0(\mathbf{l}_1 - \mathbf{l}_5) - \mathbf{G}_0(\mathbf{l}_2 - \mathbf{l}_5)].(\mathbf{l}_5 - \mathbf{l}_6)\langle \chi_A(\mathbf{l}_1)\chi_B(\mathbf{l}_2)\chi_C(\mathbf{l}_3)\chi_D(\mathbf{l}_4)\chi_E(\mathbf{l}_5)\chi_F(\mathbf{l}_6) \rangle,$$

$$\mathbf{H}_{CD} = \sum_{A,B}\sum_{\mathbf{l}_1 \in Q}\sum_{\mathbf{l}_2}\sum_{\mathbf{l}_3 \in Q}\sum_{\mathbf{l}_4 \in \Gamma_3} \{\mathbf{\Psi}_{AB}(\mathbf{l}_1 - \mathbf{l}_2).[\mathbf{G}_0(\mathbf{l}_1 - \mathbf{l}_3) - \mathbf{G}_0(\mathbf{l}_2 - \mathbf{l}_3)].(\mathbf{l}_3 - \mathbf{l}_4)\} \otimes (\mathbf{l}_1 - \mathbf{l}_2)$$
$$\times \langle \chi_A(\mathbf{l}_1)\chi_B(\mathbf{l}_2)\chi_C(\mathbf{l}_3)\chi_D(\mathbf{l}_4) \rangle,$$

$$z_{CDEF} = \sum_{A,B}\sum_{\mathbf{l}_1 \in Q}\sum_{\mathbf{l}_2}\sum_{\mathbf{l}_3 \in Q}\sum_{\mathbf{l}_4 \in \Gamma_3}\sum_{\mathbf{l}_5 \in Q}\sum_{\mathbf{l}_6 \in \Gamma_5} (\mathbf{l}_3 - \mathbf{l}_4).[\mathbf{G}_0(\mathbf{l}_1 - \mathbf{l}_3) - \mathbf{G}_0(\mathbf{l}_2 - \mathbf{l}_3)].\mathbf{\Psi}_{AB}(\mathbf{l}_1 - \mathbf{l}_2).$$
$$[\mathbf{G}_0(\mathbf{l}_1 - \mathbf{l}_5) - \mathbf{G}_0(\mathbf{l}_2 - \mathbf{l}_5)].(\mathbf{l}_5 - \mathbf{l}_6)\langle \chi_A(\mathbf{l}_1)\chi_B(\mathbf{l}_2)\chi_C(\mathbf{l}_3)\chi_D(\mathbf{l}_4)\chi_E(\mathbf{l}_5)\chi_F(\mathbf{l}_6) \rangle,$$

$$\mathbf{J}_1 = -\sum_{A,B}\sum_{\mathbf{l}_1 \in Q}\sum_{\mathbf{l}_2} \mathbf{F}_{AB}(\mathbf{l}_1 - \mathbf{l}_2) \otimes (\mathbf{l}_1 - \mathbf{l}_2)\langle \chi_A(\mathbf{l}_1)\chi_B(\mathbf{l}_2) \rangle,$$

$$\mathbf{J}_2 = 2\sum_{A-D}\sum_{\mathbf{l}_1 \in Q}\sum_{\mathbf{l}_2}\sum_{\mathbf{l}_3 \in Q}\sum_{\mathbf{l}_4} \{\mathbf{\Psi}_{AB}(\mathbf{l}_1 - \mathbf{l}_2).[\mathbf{G}_0(\mathbf{l}_1 - \mathbf{l}_3) - \mathbf{G}_0(\mathbf{l}_2 - \mathbf{l}_3)].\mathbf{F}_{CD}(\mathbf{l}_3 - \mathbf{l}_4)\} \otimes (\mathbf{l}_1 - \mathbf{l}_2)$$
$$\times \langle \chi_A(\mathbf{l}_1)\chi_B(\mathbf{l}_2)\chi_C(\mathbf{l}_3)\chi_D(\mathbf{l}_4) \rangle,$$

$$\mathbf{T} = \sum_{A,B}\sum_{\mathbf{l}_1 \in Q}\sum_{\mathbf{l}_2} (\mathbf{l}_1 - \mathbf{l}_2) \otimes \mathbf{\Psi}_{AB}(\mathbf{l}_1 - \mathbf{l}_2) \otimes (\mathbf{l}_1 - \mathbf{l}_2)\langle \chi_A(\mathbf{l}_1)\chi_B(\mathbf{l}_2) \rangle.$$

Each of the above quantities should be symmetrized in an appropriate fashion to take account of the symmetries $\theta_{AB} = \theta_{BA}$ and $A_{ij} = A_{ji}$.

As an aside, it is worth making some comments about the multi-point probabilities. Consider, for example, the two-point probability $\langle \chi_A(\mathbf{l}_1)\chi_B(\mathbf{l}_2) \rangle$. If $\mathbf{l}_1 \neq \mathbf{l}_2$ then, because of the assumption that atom types at different lattice sites are uncorrelated, this two-point probability is just the product of the probabilities of finding type A at $\mathbf{l}_1$ and type B at $\mathbf{l}_2$. Furthermore, because of the assumed statistical uniformity, these last probabilities are constant and equal to the expected volume fractions of the relevant constituent. If, however, $\mathbf{l}_1 = \mathbf{l}_2$, then the two point probability is zero if $A \neq B$ (since atoms of different types cannot co-exist at the same lattice site) and is just equal to the expected volume fraction of A if $A = B$. Denoting by x_A the volume fraction of material A, the above discussion may be summarized as follows:

$$\langle \chi_A(\mathbf{l}_1)\chi_B(\mathbf{l}_2) \rangle = (1 - \delta_{\mathbf{l}_1 \mathbf{l}_2})x_A x_B + \delta_{\mathbf{l}_1 \mathbf{l}_2} \delta_{AB} x_A,$$

where δ is the Kronecker delta, equal to one if its suffixes are equal but equal to zero otherwise.

For higher order correlation functions, note that

$$\langle \chi_{A_1}(\mathbf{l}_1)...\chi_{A_N}(\mathbf{l}_N) \rangle = \frac{\sum_{i=1}^{N-1} \delta_{\mathbf{l}_i \mathbf{l}_N} \delta_{A_i A_N}}{\sum_{j=1}^{N-1} \delta_{\mathbf{l}_j \mathbf{l}_N}} \langle \chi_{A_1}(\mathbf{l}_1)...\chi_{A_{N-1}}(\mathbf{l}_{N-1}) \rangle$$

if $\delta_{\mathbf{l}_j \mathbf{l}_N} \neq 0$ for some $j < N$ and

$$\langle \chi_{A_1}(\mathbf{l}_1)...\chi_{A_N}(\mathbf{l}_N) \rangle = \langle \chi_{A_1}(\mathbf{l}_1)...\chi_{A_{N-1}}(\mathbf{l}_{N-1}) \rangle x_{A_N}$$

if $\delta_{\mathbf{l}_j \mathbf{l}_N} = 0$ for all $j < N$. These relations become useful for evaluating the higher order correlation functions on noting that

$$\langle \chi_{A_1}(\mathbf{l}_1)...\chi_{A_N}(\mathbf{l}_N) \rangle = \left[\prod_{i=1}^{N-1} [(1 - \delta_{\mathbf{l}_i \mathbf{l}_N}) + \delta_{\mathbf{l}_i \mathbf{l}_N}] \right] \langle \chi_{A_1}(\mathbf{l}_1)...\chi_{A_N}(\mathbf{l}_N) \rangle.$$

Expansion of the product yields a series of terms each of which either contributes only when $\delta_{iN} = 0$ or contributes only when $\delta_{iN} = 1$. Each of these terms may be evaluated using the expressions given above, generating a recursive definition of each multi-point probability. It is clear that the complexity of the resulting expression increases very rapidly. In fact, the evaluation of probabilities involving five or more points at a time forms the major part of the computational workload.

Returning to the main thrust of this section, minimization of (5.7) yields the equations

$$\begin{aligned} y_{AB} + \mathbf{A} : \mathbf{H}_{AB} + z_{ABCD}\theta_{CD} &= 0 \\ \mathbf{H}_{AB}\theta_{AB} + (\mathbf{J}_1 + \mathbf{J}_2) + \mathbf{T} : \mathbf{A} &= 0. \end{aligned} \qquad (5.8)$$

These may be solved for θ_{AB} and $\mathbf{A}$. The lattice parameter, which is defined by $\mathbf{A}$, then follows. The estimate defined by (5.8) was obtained by minimizing (4.5) over the class of trial fields

$$\mathbf{u} = \mathbf{G}_0(\mathbf{f}_0 + \mathbf{t}) + \mathbf{A}.\mathbf{1}, \qquad (5.9)$$

where $\mathbf{t}$ was restricted to the form of (3.4), (4.4), and (5.6). It is instructive to consider for comparison the results obtained from two simpler classes of trial field:

$$\mathbf{u} = \mathbf{G}_0 \mathbf{f} + \mathbf{A}.\mathbf{1}, \qquad (5.10)$$

and

$$\mathbf{u} = \mathbf{A}.\mathbf{1}, \qquad (5.11)$$

with optimization only being sought with respect to the choice of **A**. The form (5.10) arises from substituting into (5.9) the **t** that results from (3.5) on neglecting in that equation the term involving **u**(l). Its use provides the following set of equations for the optimum value of **A**:

$$\mathbf{J}_1 + \mathbf{J}_2 + \mathbf{T} : \mathbf{A} = 0, \quad (5.12)$$

since in this case $\theta_{AB} = 0$. For the even simpler trial fields of (5.11), **A** is defined by

$$\mathbf{J}_1 + \mathbf{T} : \mathbf{A} = 0. \quad (5.13)$$

6. Numerical results

Consider a two-material alloy $A_x B_{1-x}$. Let the lattices of the two materials be mismatched in size by an amount f:

$$f = \frac{a_A - a_B}{a_A}.$$

Normalize the reference lattice spacing to that of material A. That is, set $a_A = 1$. Then $\epsilon_A = 0$ and $\epsilon_B = -f$. Therefore, from (5.3) and (5.4)

$$\alpha_{AA} = 0, \quad \beta_{AA} = \gamma_A/2,$$

$$\alpha_{BB} = -\left(\frac{1-3f}{1-f}\right)\frac{f\gamma_B}{2}, \quad \beta_{BB} = \left(\frac{1-3f}{1-f}\right)\frac{\gamma_B}{2}.$$

The corresponding quantities for interaction between atoms of different types are defined by (5.5). Hence all of the material parameters are set by the lattice mismatch f and the stiffness measures γ_A and γ_B. In fact, the predicted value of **A** depends not on γ_A and γ_B separately, but rather on the ratio γ_A/γ_B, which may be replaced by the ratio of any pair of appropriate elastic moduli for the two materials (e.g. by the ratio of their bulk moduli).

Now, $\epsilon = -f = \epsilon_B$ when $x = 0$ and $\epsilon = 0 = \epsilon_A$ when $x = 1$, since for an $A_x B_{1-x}$ alloy, these two cases correspond to pure crystals of, respectively, B and A. A relevant question to ask is what the value of ϵ (or equivalently the lattice parameter) should be for an alloy of the two materials.

The simplest estimate is supplied by Vegard's law:

$$\epsilon = x\epsilon_A + (1-x)\epsilon_B = -(1-x)f,$$

which is simply a rule of mixtures. The next simplest estimate is obtained by inverting (5.13), which results from using the simple class of trial fields described by (5.11). This estimate may be obtained by hand and yields the formula (accurate to first order in f)

$$\epsilon = \frac{[x^2 + 2k(x-x^2) - 1]f}{1 + (\gamma_A/\gamma_B - 1)[x^2 + 2k(x-x^2)]}. \quad (6.1)$$

It may be shown that (6.1) holds not just for nearest-neighbour interactions, but for *any* central pairwise-interaction force law. Further estimates are obtained by inverting (5.8) and (5.12), the former being our 'best' estimate.

Results are presented in Figure 1 for the case $\gamma_A/\gamma_B = 0.5$, $f = 0.05$, $k = 0.5$. In performing the calculation, the reference lattice Green's tensor was taken to have the elastic properties of material B. To simplify the calculation, the periodic cell was taken to be two-dimensional, and for illustration was taken to be 4 x 4. The straight line represents Vegard's law; the

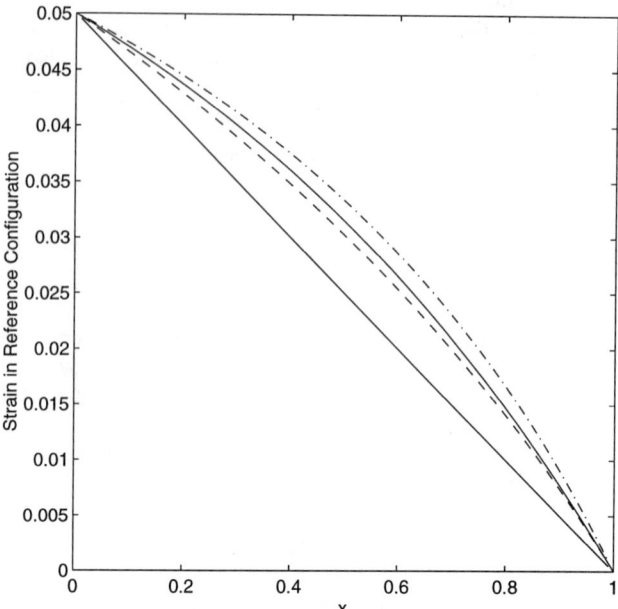

Figure 1: Deviation from Vegard's law in a random lattice illustrated by a plot of mean strain in the reference configuration as a function of composition for a two material alloy.

dot-dashed line represents our simplest estimate given by (6.1); the dashed line represents the estimate obtained by inverting (5.12); and the solid line represents our best estimate, obtained by inverting (5.8). It is interesting that the very crude approximation leading to (6.1) in fact reproduces rather closely the results of the much more sophisticated best estimate. The agreement is closer for values of γ_A/γ_B closer to 1.

Because of the approximations made to simplify the numerical calculations, one should not expect to obtain a quantitatively accurate description of the experimental results. However, it is of interest to make some comparison where possible. King [6] assessed the accuracy of Vegard's law for a range of material systems by introducing a VLF (Vegard's Law Factor). For an alloy $A_x B_{1-x}$ this is defined as follows. Let $V(x)$ be the mean atomic volume as a function of x. Let V_A and V_B be the atomic volumes of the pure materials and define

$$V_A^\star = V_B + \left.\frac{\partial V}{\partial x}\right|_{x=0}.$$

($V_A^\star$ is the atomic volume obtained at $x = 1$ by extrapolating from V_B at $x = 0$ along the initial gradient of $V(x)$.) The VLF is defined to be

$$\text{VLF} = \frac{V_A^\star - V_A}{V_A} \times 100.$$

Using our simplest estimate, (6.1), we can obtain an explicit expression for VLF:

$$\text{VLF} = 300[2k(\gamma_A/\gamma_B) - 1]f. \tag{6.2}$$

We have not undertaken a thorough comparison of the results of (6.2) and the experimental results of King. However, we do note that for Ge_xSi_{1-x}, which is a common example of the type of semiconductor system that forms part of the motivation for this study, $f = 0.040$ and $\gamma_{Ge}/\gamma_{Si} = 0.75$. Using these values we obtain from (6.2) the results

$$\text{VLF}_{GeSi} = 6(3k - 2).$$

Conversely, for Si_xGe_{1-x} we get

$$\text{VLF}_{SiGe} = -4[8(1 - k) - 3].$$

(Note that $f_{SiGe} = -f_{GeSi}$ and $k_{SiGe} = 1 - k_{GeSi}$; in both of the above equations, $k = k_{GeSi}$.) These are to be compared with King's measured values of, respectively, -6.1 and -12. By choosing $k = 0.3$, we obtain values of -6.6 and -10.4, which are in quite good agreement with the experimental values. Importantly, a single choice of the parameter k can yield good agreement with *both* VLF_{GeSi} *and* VLF_{SiGe}, which augurs well for the robustness of the model.

The value of $k = 0.3$, which results in rather good agreement with the experiments of King [6] does not seem to be an unreasonable value. Reference to (5.5) shows that it implies a potential for GeSi that is slightly closer to that of Si than that of Ge. In fact, systematic fitting with the results of King for different material systems is a possible way of postulating a value for k.

Appendix

In this appendix we derive some results that follow from the assumptions listed at the beginning of Section 5.

In view of (4.1) and (4.2), the mean energy density in (2.12) may be written as

$$|Q|\,\varphi(\mathbf{u}) = -\sum_{A,B}\sum_{\mathbf{l}_1 \in Q}\sum_{\mathbf{l}_2} \chi_A(\mathbf{l}_1)\chi_B(\mathbf{l}_2)\mathbf{F}_{AB}(\mathbf{l}_1 - \mathbf{l}_2).[\mathbf{u}(\mathbf{l}_1) - \mathbf{u}(\mathbf{l}_2)]$$
$$+ \tfrac{1}{2}\sum_{A,B}\sum_{\mathbf{l}_1 \in Q}\sum_{\mathbf{l}_2} \chi_A(\mathbf{l}_1)\chi_B(\mathbf{l}_2)[\mathbf{u}(\mathbf{l}_1) - \mathbf{u}(\mathbf{l}_2)].\Psi_{AB}(\mathbf{l}_1 - \mathbf{l}_2).[\mathbf{u}(\mathbf{l}_1) - \mathbf{u}(\mathbf{l}_2)]. \quad (A.1)$$

Now, assuming that the interaction between atom types A and B is governed by a potential $\Omega_{AB}(r)$, which is a function only of the distance r between the atoms, we can also write

$$\mathcal{E}(\mathbf{u}) = \tfrac{1}{2}\sum_{A,B}\sum_{\mathbf{l} \in Q}\sum_{\mathbf{l}'} \chi_A(\mathbf{l})\chi_B(\mathbf{l}')\Omega_{AB}(|\mathbf{l} - \mathbf{l}' + \mathbf{u}(\mathbf{l}) - \mathbf{u}(\mathbf{l}')|). \quad (A.2)$$

Expansion of (A.2) to quadratic terms in displacement and comparison with (A.1) immediately reveals that

$$\mathbf{F}_{AB}(\mathbf{l}) = -\frac{\Omega'_{AB}(|\mathbf{l}|)}{2\,|\mathbf{l}|}\mathbf{l} \quad (A.3)$$

and

$$\Psi_{AB}(\mathbf{l}) = \frac{1}{2}\left\{\left[\frac{\Omega''_{AB}(|\mathbf{l}|)}{2\,|\mathbf{l}|^2} - \frac{\Omega'_{AB}(|\mathbf{l}|)}{2\,|\mathbf{l}|^3}\right]\mathbf{l} \otimes \mathbf{l} + \frac{\Omega'_{AB}(|\mathbf{l}|)}{|\mathbf{l}|}\mathbf{I}\right\}. \quad (A.4)$$

In order to relate $\Psi_{AB}(\mathbf{l})$ and $F_{AB}(\mathbf{l})$ to the overall elastic properties of the medium, consider applying a uniform deformation

$$\mathbf{u} = \mathbf{A}.\mathbf{l} \quad (A.5)$$

to the lattice. The mean energy density, from (A.1), may then be rewritten in terms of the Green-strain

$$\mathbf{E} = \tfrac{1}{2}[(\mathbf{I}+\mathbf{A})^{\mathrm{T}}(\mathbf{I}+\mathbf{A}) - \mathbf{I}]$$

(where $\mathbf{I}$ is the identity matrix) as

$$|Q|\,\varphi(\mathbf{u}) = -\left[\sum_{A,B}\sum_{\mathbf{l}_1 \in Q}\sum_{\mathbf{l}_2} \chi_A(\mathbf{l}_1)\chi_B \mathbf{F}_{AB}(\mathbf{l}_1-\mathbf{l}_2) \otimes (\mathbf{l}_1-\mathbf{l}_2)\right] : \mathbf{E}$$

$$+\tfrac{1}{2}\mathbf{E} : \left[\sum_{A,B}\sum_{\mathbf{l}_1 \in Q}\sum_{\mathbf{l}_2} \chi_A(\mathbf{l}_1)\chi_B[(\mathbf{l}_1-\mathbf{l}_2) \otimes \Psi(\mathbf{l}_1,\mathbf{l}_2) \otimes (\mathbf{l}_1-\mathbf{l}_2) + \mathbf{F}(\mathbf{l}_1,\mathbf{l}_2) \otimes \mathbf{I} \otimes (\mathbf{l}_1-\mathbf{l}_2)]\right] : \mathbf{E}.$$

A natural definition for the effective elastic modulus tensor $\mathbf{C}$ is thus

$$\mathbf{C} = \frac{1}{|Q|}\left[\sum_{A,B}\sum_{\mathbf{l}_1 \in Q}\sum_{\mathbf{l}_2} \chi_A(\mathbf{l}_1)\chi_B[(\mathbf{l}_1-\mathbf{l}_2) \otimes \Psi(\mathbf{l}_1,\mathbf{l}_2) \otimes (\mathbf{l}_1-\mathbf{l}_2) + \mathbf{F}(\mathbf{l}_1,\mathbf{l}_2) \otimes \mathbf{I} \otimes (\mathbf{l}_1-\mathbf{l}_2)]\right], \tag{A.6}$$

appropriately symmetrized. If we have a parameterized interatomic potential Ω, then prescribing the overall elastic properties of the pure constituents of the alloy defines some of the parameters involved.

To define further such parameters note that a uniform deformation of the form (A.5) applied to a uniform material A yields energy

$$-\left[\sum_{\mathbf{l}} \mathbf{F}_{AA}(\mathbf{l}) \otimes \mathbf{l}\right]^{*} : \mathbf{A} + \frac{1}{2}\mathbf{A} : \left[\sum_{\mathbf{l}} \mathbf{l} \otimes \Psi_{AA}(\mathbf{l}) \otimes \mathbf{l}\right]^{*} : \mathbf{A},$$

where the '$\star$' superscript indicates appropriate symmetrizations. The lattice parameter(s) of the material $\mathbf{A}$ are thus defined by

$$\left[\sum_{\mathbf{l}} \mathbf{l} \otimes \Psi_{AA}(\mathbf{l}) \otimes \mathbf{l}\right]^{*} : \mathbf{A} = \left[\sum_{\mathbf{l}} \mathbf{F}_{AA}(\mathbf{l}) \otimes \mathbf{l}\right]^{*}. \tag{A.7}$$

Matching $\mathbf{A}$ to known values for the material A transforms (A.7) into a further equation that can be used to define any parameters that may appear in the interatomic potential Ω.

Now develop these relations explicitly for the nearest-neighbours, simple-cubic lattice approximation defined in the text. Normalizing lengths to the reference lattice spacing we have reference lattice sites at $m_i \mathbf{e}_i$, where m_i are integers, and where $\{\mathbf{e}_i\}$ is a set of orthonormal basis vectors. The nearest-neighbours approximation means that $\Psi_{AB}(\mathbf{l})$ and $\mathbf{F}_{AB}(\mathbf{l})$ are zero except for the following cases:

$$\mathbf{F}_{AB}(\pm \mathbf{e}_i) = \pm \alpha_{AB} \mathbf{e}_i \tag{A.8}$$

$$\Psi_{AB}(\pm \mathbf{e}_i) = (\alpha_{AB} + \beta_{AB})\mathbf{e}_i \otimes \mathbf{e}_i - \alpha_{AB}\mathbf{I} \tag{A.9}$$

where

$$\alpha_{AB} = -\tfrac{1}{2}\Omega'_{AB}(1), \quad \beta_{AB} = \tfrac{1}{2}\Omega''_{AB}(1) \tag{A.10}$$

Consider a pure crystal of material A. Since the crystal is simple cubic,

$$\mathbf{A}_A = \epsilon_A \mathbf{I}$$

where ϵ_A is known from measurements made on the pure material. Using this relation together with (A.8) and (A.9) this allows (A.7) to be simplified to the relationship

$$\epsilon_A = \alpha_{AA}/\beta_{AA}. \tag{A.11}$$

Finally, note that (A.6), (A.8), and (A.9) imply that for a pure crystal of material A, $C_{1111} = \gamma_A$, where

$$\gamma_A = 2(\alpha_{AA} + \beta_{AA}). \tag{A.12}$$

References

1. Z. Hashin and S. Shtrikman, "On some variational principles in anisotropic and nonhomogeneous elasticity," *J. Mech. Phys. Solids* **10** (1962), 335–342.

2. J. K. Lee, "Coherency strain analysis via a discrete atom method," *Scripta Met. Mat.* **32** (1995), 559–564.

3. J. K. Lee, "Shape evolution of a coherent precipitate and its interaction with an edge dislocation via a discrete atom method," *Micromechanics of Advanced Materials: A Symposium in honor of Professor James Li's 70th Birthday*, ed. S. N. G. Chu, P. K. Liaw, R. J. Arsenault, K. Sanada, K. S. Chan, W. W. Gerberich, C. C. Chau, and T. M. Kung (Warrendale, PA: The Materials, Metals and Minerals Society, 1995), 41–50.

4. J. R. Willis, "Variational and related methods for the overall properties of composites," *Advances in Applied Mechanics* **21**, ed. C. S. Yih (New York: Academic Press, 1981), 1–78.

5. J. R. Willis, "Elasticity Theory of Composites," *Mechanics of Solids, the Rodney Hill 60th Anniversary Volume*, ed. H. G. Hopkins and M. J. Sewell (Oxford: Pergamon Press, 1982), 653–686.

6. H. W. King, "Quantitative size-factors for metallic solid solutions," *J. Mat. Sci.* **1** (1966), 79–90.

MODULATED PATTERNS AND PINNING EFFECT IN

PHASE-SEPARATING ALLOYS

Akira Onuki

Department of Physics, Kyoto University, Kyoto 606, Japan

Abstract

We briefly discuss three types of long-range interactions among the composition fluctuations in binary alloys. They are derived from a Ginzburg-Landau model which takes account of the elastic effects. The first interaction arises from elastic anisotropy and is well known, while the other two interactions arise from elastic misfit and were first derived in our theory. We then show numerical results of spinodal decomposition in our model. We obtain modulated microstructures in the presence of cubic elasticity and/or anisotropic external stresses. Furthermore, when the two phases have different shear moduli, we derive a very unique elastic-misfit interaction. It makes elastic deformations anisotropic in softer regions and isotropic in harder regions in late stages, resulting in glassy states with very slow coarsening rates. In such states softer regions form a network enclosing harder regions even if the volume fraction of the softer regions is small. This aspect has been unnoticed but should be fundamental in metallurgy.

Ginzburg-Landau Apporach

Elastic effects drastically influence domain morphologies in phase-separating alloys [1-3]. However, dynamical theories for the effects have been premature yet in metallurgy. As is well known, time-dependent Ginzburg-Landau models have been widely used in physics to describe critical dynamics [4] and dynamics of first-order phase transitions in various systems [5,6]. In metallurgy, although Cahn originally presented a Ginzburg-Landau theory in which the composition and the elastic field are coupled [7], subsequent developments in this direction have been inadequate. The present author hence proposed a Ginzburg-Landau theory [8] to systematically analyze the effects under the coherent condition. We start with a Ginzburg-Landau free energy for a conserved order parameter c representing the composition of alloys and the elastic field u. They are coupled in the free energy density in the following form

$$F\{c, u\} = \int dr [f_0(c) + \frac{1}{2}(\nabla c)^2 + \alpha c(\nabla \cdot u) + f_{el}(u)] \quad (1)$$

where $f_0(c)$ is a free energy density of c and α is the coupling constant between c and the volume dilation $\nabla \cdot u$. We measure c from some reference value and $f_{el}(u)$ is the usual elastic free energy of cubic symmetry in the reference state in which $c = 0$. Due to the bilinear coupling ($\propto \alpha$) the volume changes linearly as a function of the average composition in the disordered phase in accord with the empirical Vegard law [1]. The parameter α can thus be determined experimentally [7]. In phase separation it gives rise to a difference in the lattice constants, "a lattice misfit", of the two phase structures. In our theory all the elastic interactions arise only in the presence of a lattice misfit ($\alpha \neq 0$).

Because c changes slowly in time at long wavelengths, we can obtain a closed description in terms of c by eliminating u from the mechanical equilibrium condition

$$\delta F\{c, u\}/\delta u_i = -\sum_j \partial \sigma_{ij}/\partial x_j = 0 \quad (2)$$

where σ_{ij} is the elastic stress tensor. As a new ingredient we assume that the elastic moduli linearly depend on c as

$$C_{ij} = C_{ij}^0 + c C_{ij}^1 \quad (3)$$

Then there is "an elastic misfit" between the emerging two phases in phase separation. In particular we found that the composition dependence of the shear modulus $\mu = C_{44}$ drastically alters the domain morphology in late stage phase separation. The resultant free energy consists of four parts after the elimination of u as

$$F\{c\} = \int dr \left[f(c) + \frac{1}{2}(\nabla c)^2 \right] + F_{cub} + F_{ex} + F_{em} \quad (4)$$

Here $f(c)$ is the so-called coherent free energy density [7] and will be assumed to be of the usual form $f = -\frac{1}{2}\tau c^2 + \frac{1}{4}u_0 c^4$, where $\tau \propto T_c - T$ and u_0 is a constant.

(i) Interaction due to Cubic Anisotropy

The F_{cub} may be written in terms of the elastic moduli C_{ij} with cubic symmetry. For simplicity we assume that the elastic anisotropy $\xi_a \equiv (C_{11} - C_{12} - 2C_{44})/C_{44}$ is small.

Then, to first order in ξ_a, we obtain in the Fourier space

$$F_{cub} = \frac{1}{2}\tau_a \int_{\mathbf{k}} \sum_{i\neq j} \hat{k}_i^2 \hat{k}_j^2 \, |c_{\mathbf{k}}|^2 \qquad (5)$$

where $\tau_a \cong 2\alpha^2 C_{44}\xi_a/(C_{11}+2C_{44})^2$, $\int_{\mathbf{k}} \cong (2\pi)^{-d}\int d\mathbf{k}$, and $c_{\mathbf{k}}$ is the Fourier transform of $c(\mathbf{r})$. Hereafter $\hat{k}_i = k^{-1}k_i$ represents the direction of the wave vector $\mathbf{k}$. In the real space F_{cub} is the integral of $\phi_{cub}(\mathbf{r}-\mathbf{r}')c(\mathbf{r})c(\mathbf{r}')$ over $\mathbf{r}$ and $\mathbf{r}'$ where the pairwise potential $\phi_{cub}(\mathbf{r})$ decays as r^{-d} and is angle-dependent. This bilinear contribution, arising from the cubic anisotropy, was first obtained by Cahn [9] and has subsequently been refined by other authors[1,10,11]. Its most general form in cubic crystals under external stress was presented in Ref.8.

(ii) <u>Interaction due to Anisotropic External Stress</u>

On the other hand, external anisotropic stresses can influence the domain growth only when the elastic moduli are different in the two phases. Let S_{ij} be the following symmetric traceless tensor

$$S_{ij} = A_{ij} + A_{ji} - \frac{2}{d}\delta_{ij}\sum_i A_{ii} \qquad (6)$$

where $A_{ij} = <\partial u_i/\partial x_j>$ is the spatial average of the strain tensor produced by external stresses. Then F_{ex} is of the form of a dipolar interaction [8]

$$F_{ex} = \frac{1}{2}g_{ex}\sum_{i,j}S_{ij}\int_{\mathbf{k}}\hat{k}_i\hat{k}_j|c_{\mathbf{k}}|^2 \qquad (7)$$

where $g_{ex} = \alpha\mu_1/K_{L0}$ in terms of the longitudinal elastic modulus $K_{L0} = K_0 + (4/3)\mu_0$, K_0 being the bulk modulus. The shear modulus is written as $\mu = \mu_0 + c\mu_1$. In the real space F_{ex} may be expressed in terms of a pairwise potential $\phi_{ij}(r) \propto \delta_{ij}/r^d - dx_ix_j/r^{d+2}$. In the uniaxially deformed case, (7) is of the same form as the dipolar interaction in uniaxial spin or ferroelectric systems[12,13]. In our elastic case, depending on the sign of g_{ex}, cylindrical or lamellar domain structures are eventually produced in two phase states. To get more insight let us evaluate F_{ex} for a single spheroidal precipitate whose shape is represented by $x^2/R_\parallel^2 + (y^2+z^2)/R_\perp^2 = 1$. We assume a uniaxially deformed matrix where $A_{ij} = \lambda_i\delta_{ij}$ with $\lambda_x = \lambda_\parallel$ and $\lambda_y = \lambda_z = \lambda_\perp$ and that c is discontinuous at the interface by $\Delta c = c_{in} - c_{out}$, c_{in} and c_{out} being the compositions inside and outside the domain. Then

$$F_{ex} = -2g_{ex}(\lambda_\parallel - \lambda_\perp)(\Delta c)^2(N_x - \frac{1}{3})V_s \qquad (8)$$

where N_x is the depolarization factor in electrostatics [14] and $V_s = (4\pi/3)R_\parallel R_\perp^2$ is the volume of the domain. The N_x increases from 0 to 1, as $R_\perp/R_\parallel$ increases from 0 to ∞, and is equal to 1/3 for a sphere. We can see that F_{ex} is lowered as $R_\perp/R_\parallel$ increases(or decreases) for positive (or negative) $g_{ex}(\lambda_\parallel - \lambda_\perp)$. These results are consistent with an elastic theory by Pineau [15] and a microscopic simulation by Gayda and Srolovitz [16].

(iii) <u>Interaction due to Elastic Misfit</u>

The third contribution F_{em} is a really intiriguing interaction arising from the elastic misfit. In the limit $\xi_a \to 0$ it becomes a cubic (third order) and isotropic interaction

$$F_{em} = g_{em}\int d\mathbf{r}\, c\hat{Q} \qquad (9)$$

where $g_{em} = \mu_1(\alpha/K_{L0})^2$, K_{L0} being the longitudinal elastic modulus. The $\hat{Q}$ is positive-definite as

$$\hat{Q} = \sum_{i,j}[\nabla_i\nabla_j w - \frac{1}{d}\delta_{ij}(c-\bar{c})]^2 \tag{10}$$

where $\nabla_i \equiv \partial/\partial x_i$ and w is defined by

$$\nabla^2 w = c - \bar{c} \tag{11}$$

$\bar{c}$ being the average order parameter. The elastic strain has been expanded as

$$\frac{\partial u_i}{\partial x_j} = A_{ij} - (\alpha/K_{L0})\frac{\partial^2 w}{\partial x_i \partial x_j} + \cdots \tag{12}$$

Obviously F_{em} is the composition dependent part of the shear deforrmation energy. The $\hat{Q}$ is proportional to the shear deformation energy. As a result the contributions in (9) from spatial regions with $g_{em} c > 0$ are positive, while those with $g_{em} c < 0$ are negative. Thus F_{em} is lowered if softer regions are anisotropically deformed and harder regions become isotropic. Such asymmetric shape changes occur when F_{em} overcomes the interface free energy or when the domain size exceeds the following crossover length

$$R_{em} = \sigma/|\Delta\mu|\varepsilon^2 = \sigma/|g_{em}(\Delta c)^3| \tag{13}$$

where σ is the surface tension, $\Delta\mu = \mu_1\Delta c$ is the difference in the shear moduli, and $\varepsilon = \alpha\Delta c/K_{L0}$ is the typical strain in phase separation. If the system is close to the critical point, we obtain $|\Delta c| = 2(\tau/u_0)^{1/2}$ and $R_{em} \sim u_0^{1/2}/g_{em}$ in the mean field theory. Hereafter we assume $R_{em} \gg \xi = \tau^{-1/2}$ (=the interface thickness). The reverse case $R_{em} \lesssim \xi$ may be realized sufficiently near the critical point, though the physics there has not been studied.

As in (8) we consider a single spheroidal precipitate to obtain

$$F_{em} = \frac{3}{2}g_{em}(\Delta c)^3(N_x - \frac{1}{3})^2 V_s \tag{14}$$

This term vanishes for a sphere $(N_x - \frac{1}{3})$ and changes its sign dependending on the sign of $\Delta\mu = \mu_1\Delta c$. For the softer domain case $(\Delta\mu = \mu_1(c_{in} - c_{out}) < 0)$, F_{em} decreases as the shape deviates from sphericity. This gives rise to a shape-change transition of a softer domain when its radius R exceeds a critical size of order R_{em} [17], while an isolated harder domain is stable against shape changes. However, shape changes are crucial even for harder domains if their number is more than two [18]. Its numerical evience will be given in Fig. 7 below.

It is straightforward to examine the Mullins-Sekerka instability of a growing, nearly spherical domain in a metastable matrix [19] by taking account of F_{em} [8c]. Let D, Δ, and d_0 be the diffusion constant, the supersaturation in the coherent phase diagram assumed to be much smaller than 1, and the capillary length on the order of the correlation length ξ. Assuming $\Delta c = c_{in} - c_{out} > 0$ we obtain the equation of the radius

$$\frac{\partial R}{\partial t} = \frac{D}{R}\left(-\frac{2d_0}{R} + \Delta - \frac{1}{3}g^*_{em}\right) \tag{15}$$

where $g_{em}^* = g_{em}/(\tau u_0)^{1/2} = g_{em}|\Delta c|\xi^2$ is the dimensionless strength of the elastic misfit assumed to be much smaller than 1. The critical radius R_c is equal to $2d_0/(\Delta - g_{em}^*/3)$ if the shape remains spherical. Let the radius be slightly deformed as $R+\sum_{\ell,m} \delta_{\ell m} Y_{\ell m}(\theta,\varphi)$, where $Y_{\ell m}(\theta,\varphi)$ is the spherical harmonic fnction, θ and φ being the polar coordinates. The amplitude $\delta_{\ell m}$ is governed by

$$\frac{1}{\delta_{\ell m}}\frac{\partial}{\partial t}\delta_{\ell m} = (\ell - 1)\frac{1}{R}\frac{\partial R}{\partial t} - \frac{D}{R^2}(\ell^2 + \ell - 2)\left[(2\ell+1)\frac{d_0}{R} + (\frac{2\ell}{\ell+2})g_{em}^*\right] \quad (16)$$

A similar expression in the presence of $\Delta\mu$ was derived by Leo and Sekerka [20]. In the harder domain case ($g_{em}* > 0$) the Mullis-Sekerka instability is suppressed. In the softer domain case ($g_{em}* < 0$) the mode $\ell = 2$ first becomes unstable for $R > R_2^* = 22d_0/(\Delta - 11g_{em}^*/3)$. Thus, for $R_2^* < R_c$ or for $\Delta < (11/15)|g_{em}^*|$, the critical droplets cannot be spherical. In the reverse case the crtical droplets are spherical and undrgo a shape-change transition as R grows up to R_2^*. However, we note that the Mullins-Sekerka instability has not been unambiguously observed on solid-solid interfaces [21], whereas it is ubiquitous in solidification. This is probably because of complex shape changing processes in the solid-solid case.

Remarks

Notice that the above three elastic interactions drastically change the Ising-like features of the phase transition and separation however small the coefficients τ_a, g_{ex} and g_{em} are. In phase separation the elastic energy contributions of each domain are proportional to its volume (since the strains are of order $\varepsilon = \alpha\Delta c/K_{L0}$), whereas the surface energy is proportional to its area. Therefore all the above three interactions are crucially important in various late stage phase separation processes. We believe that combinations of these three interactions can reproduce many salient features of real microstructures in binary alloys as has been shown in our simulations.[18, 22-25]. Very recently our theory has been modified to describe systems such as Fe +Al alloys, in which a nonconserved order parameter and a composition are coupled to the elastic field [26]. On the other hand, in the disordered phase, the three interactions are relevant perturbations changing the Ising-like critical behavior. We remark that the crossover due to F_{em} is highly nontrivial and will be studied in a future paper.

Simulation

We numerically solved the diffusive equation

$$\begin{aligned}\frac{\partial}{\partial t}c &= \nabla^2 \delta F/\delta c = \nabla^2[-1 - \nabla^2 + c^2]c \\ &+ \frac{1}{2}\tau_a \sum_{i\neq j}\nabla_i^2 \nabla_j^2 w + g_{ex}\sum_{i\neq j} S_{ij}\nabla_i \nabla_j c + \nabla^2(\delta F_{em}/\delta c)\end{aligned} \quad (17)$$

where $\tau = 1$ in f and $\nabla_j = \partial/\partial x_j$. The three terms on the second line arise from the three long-range interactions discussed above. The explicit form of $\nabla^2(\delta F_{em}/\delta c)$ is complicated and is not written explicitly in (17). We use the method of Oono-Puri [27] in the time integration of (17) on a 128 x 128 lattice with the periodic boundary condition. However, most of the computation times were spent to solve the Laplace equation (11). As the initial configuration, values of $c - \bar{c}$ at the sites are random numbers uniformly distributed between ± 0.3.

(i) F_{cub} and F_{ex} in Cubic Elasticity

First we explain results without F_{em} [22]. Figs.1 and 2 display domain structures after quenching at $t = 0$ only in the presence of F_{cub} at $\tau_a = 0.675$ (namely, $g_{ex} = g_{em} = 0$) for the volume fraction of one component being at $\phi = 0.5$ and 0.3, respectively. The softest directions are [01] and [10]. Domains are rectangular stripes aligned in [10] or [01], if there is no anisotropic external stress, in agreement with some experiments [1-3]. We can see rafting of domains. It occurs in cubic crystals with a lattice misfit even without elastic misfit. The lengths of shorter sides have a sharp peak at a length $R_c(t)$, while the lengths of the longer sides are broadly distributed. Similar simulations with cubic elastic anisotropy were also performed by several groups [1,16,28].

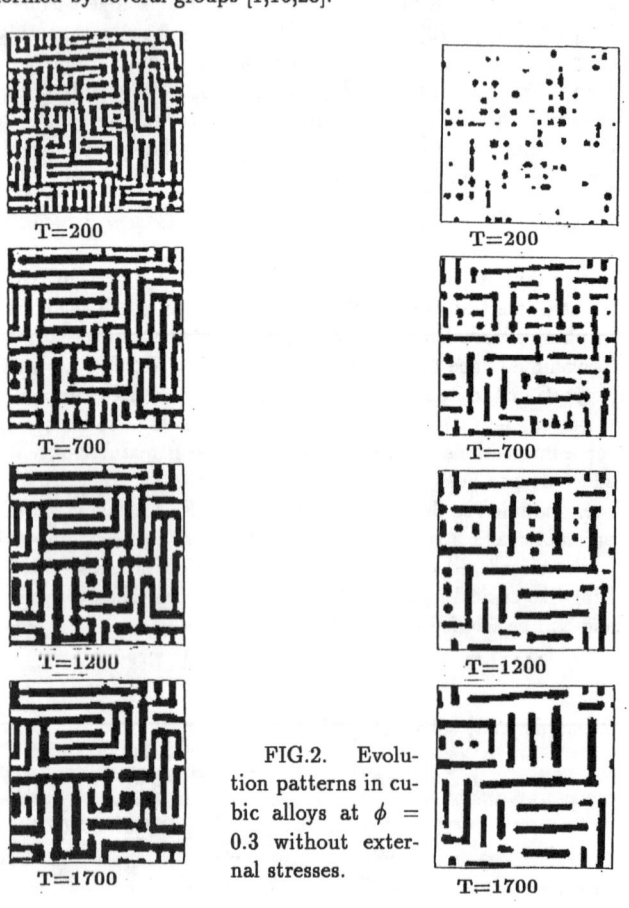

FIG.1. Evolution patterns in cubic alloys at $\phi = \frac{1}{2}$ without external stresses. The numbers below the figures are the times after the quench.

FIG.2. Evolution patterns in cubic alloys at $\phi = 0.3$ without external stresses.

Fig.3 shows patterns in the presence of F_{cub} and an uniaxial stress along [10] with $g_{ex}S_{xx} = -g_{ex}S_{yy} = -0.15$. There have been many observations of lamellar or cylindrical domain structures in phase-separating cubic alloys under uniaxial stress [29]. Fig.4 is a unique pattern produced by F_{cub} and a shear stress with $g_{ex}S_{xy} = -0.226$, which seems to have not been reported in the literature. In all these cases in Figs. 1 $\sim$ 4 the domain size R_{11} in the [11] direction grows as $R_{11} \sim t^a$ after quenching at $t = 0$. The time exponent a was found to be about 0.2, which is considerably smaller than the classical value 1/3. However, there is no pinning effect without F_{em}.

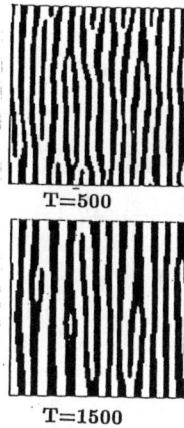

FIG.3. Lamellar patterns under uniaxial stress. The system is compressed or stretched along [10].

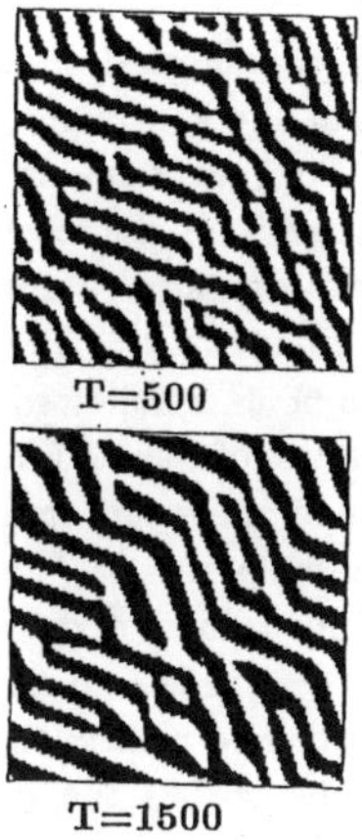

FIG.4. Patterns under shear stress. The system is softest in the two directions making angles of 21° and 69° with respect to the horizonatl axis. These patterns change into lamellar patterns for larger shear stresses.

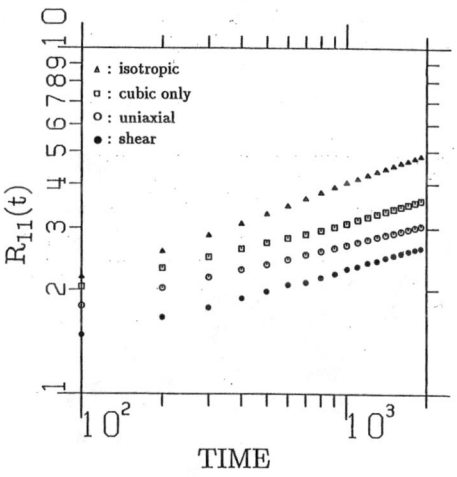

FIG.5. Coarsening of R_{11} in cubic alloys without external stress ($\square$), under a uniaxial stress ($\circ$), and under a shear stress ($\bullet$), respectively. The usual isotropic case without elasticity is also shown ($\triangle$), which has the slope of 1/3.

(ii) F_{em} in Isotropic Elasticity

Next we explain the effect of F_{em} neglecting F_{cub} and F_{ex} assuming isotropic elasticity without external stresses [18,23,25]. Fig.6 shows that a softer domain in a harder matrix undergoes a shape change transformation for three values of g_{em}. Its initial shape is slightly deformed from a circle. In Fig.7 we demonstrate that harder domains change their shapes in the presence of more than two domains. Here we prepare 2 or 4 circular domains at $t = 0$ and the subseuent time development is followed at $g_{em} = 0.05$. The initial values of c are $+1$ inside the domains and -1 outside them and small random numbers are imposed at $t = 0$. The values of ± 1 are the equilibrium values of the coherent free energy. Then there arises no appreciable change of the total area and no tendency of coagulation of the domains within the computation time

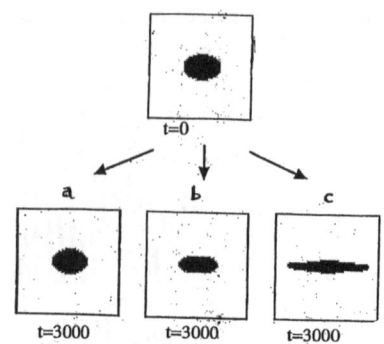

FIG. 6. Shape change of a single softer domain in a harder matrix. $g_{em} = 0.004, 0.008$, and 0.016 in a, b, and c, respectively.

($t < 1000$). However, we do not reject the possibility that the morphology of Fig.7 is still transient ultimately leading to a single domain probably in the presence of thermal noises. On the other hand, if the initial mean value outside the domains is taken to be -0.8 or -0.9 (or matastable values), the surface area increases appreciably in time. We thus believe that the shape changes in Fig. 7 are solely due F_{em}. We can see that the harder regions are isotropically deformed ($\hat{Q} \cong 0$), while the interfaces facing each other are flattened and the softer region between them is uniaxially deformed. This process occurs frequently throughout the system and can be very important in spinodal decomposition as will be illustrated in the next figures.

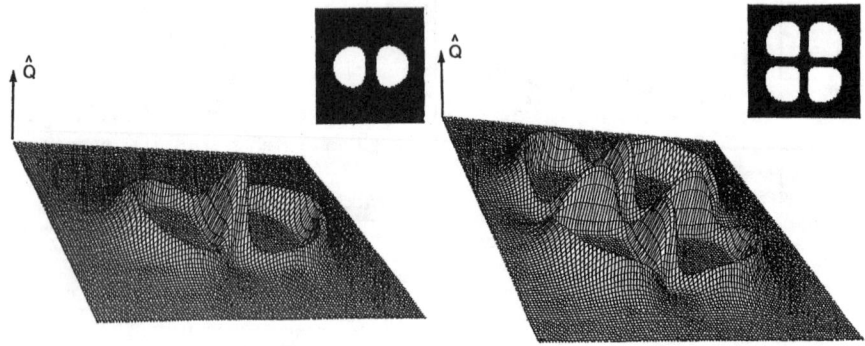

FIG.7. $\hat{Q}$ defined by (10) for two and four harder domains in a softer matrix at $t = 1000$. The profiles of the domains are also shown.

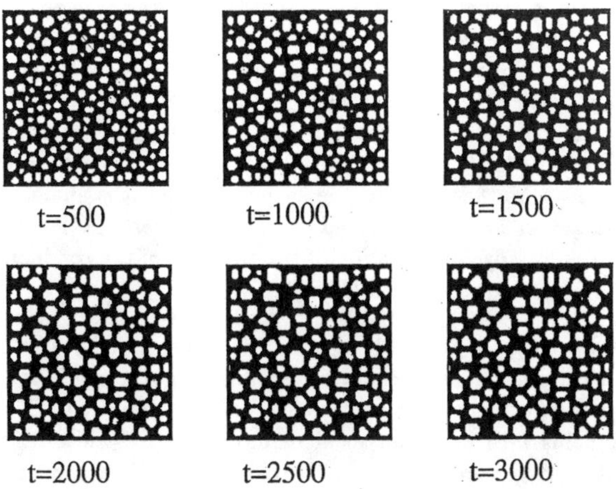

FIG.8. Evolution patterns at $\phi_s = 0.7$. The numbers below the figures are the times after quenching.

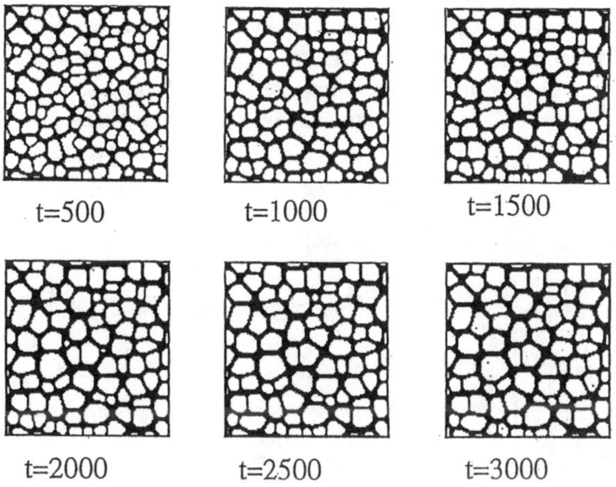

t=500　　t=1000　　t=1500

t=2000　　t=2500　　t=3000

FIG.9. Evolution pattens at $\phi_s = 0.5$.

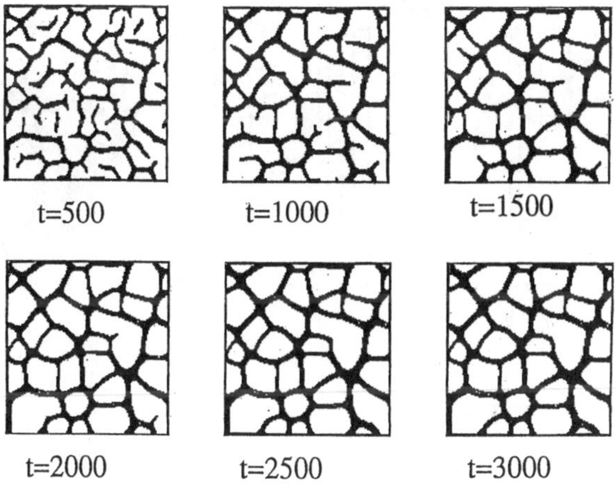

t=500　　t=1000　　t=1500

t=2000　　t=2500　　t=3000

FIG.10. Evolution patterns at $\phi_s = 0.3$.

t=500　　t=1000　　t=1500

t=2000　　t=2500　　t=3000

We remark that harder precipitates adjust their shapes to lower F_{em}. In Ref.18 it was shown that the so-called Eshelby interaction among spherical harder precipitates [3], which was first suggested by Ardell and then calculated by Eshelby, is exactly canceled to vanish due to this shape adjustmant. The adjusted shapes are ellipsoids if the domainr are far apart. (By this reason calling F_{em} as the Eshelby interaction [26] is not appropriate, so F_E and g_E in the previous notation of our work have been changed to F_{em} and g_{em} in this review.) Mathematically, the shape adjustment is analogous to redistribution of surface charges on conductors placed in vacuum. It renders the electric field in the conductors to vanish for any configuration of the positions of the conductors. This aspect seems to have not been well recognized [30].

Now we can show results of unique spinodal decomposition with elastic misfit [23]. Figs. 8-10 display the time evolution of domains at $g_{em} = 0.07$ for $\phi_s = 0.7, 0.5, 0.3$, respectively, where ϕ_s is the volume fraction of the softer component. The black regions represent softer domains with a smaller shear modulus and the white regions represent harder domains with a larger shear modulus. We can see considerable shape deformations from sphericity in Fig.8, which may well be expected from Fig.7. In Figs. 9 and 10 the softer regions form networks enclosing harder domains, which are natural configurations lowering the elastic energy. In Fig. 11 the total perimeter length of the interface regions is shown for $g_{em} = 0.05$ and 0.07. It is inversely proportional to the domain size in two dimensions. At $\phi_s = 0.7$ the time exponent of the domain size is $a \sim 0.12$. More dramatically for $\phi_s = 0.5$ and 0.3 the coarsening almost stops after a crossover time t_{em} at which $R \sim R_{em}$. The two-phase states are here driven into metastable glassy states, where anisotropically

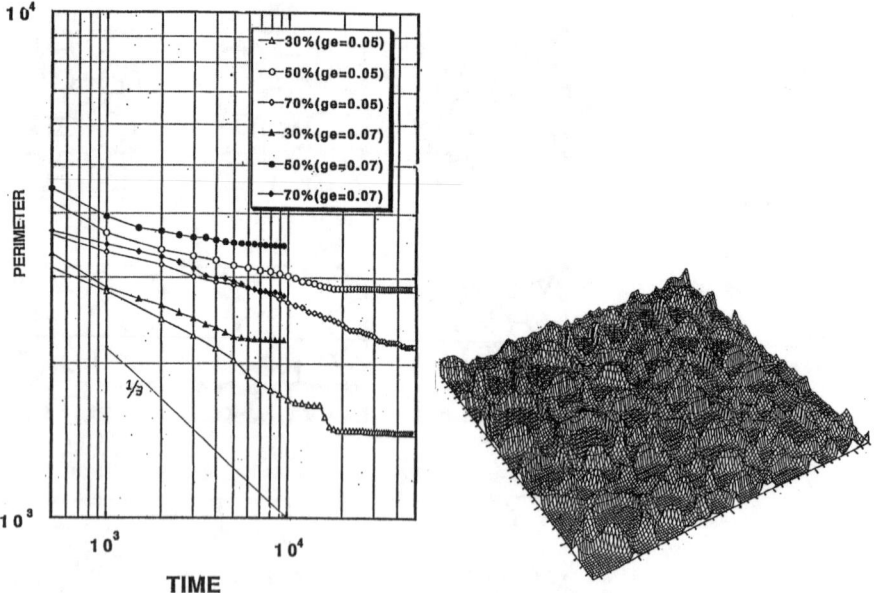

FIG.11. Perimenter length vs time.

FIG.12. The degree of anisotropic deformations $\hat{Q}$ at $t = 45000$ for $g_{em} = 0.05$.

deformed softer regions form a percolated network wrapping elastically isotropic harder domains. The elastic state is evidently illustrated in Fig.11 which shows the shear deformation energy $\hat{Q}$ in a pinned state at $t = 45000$ for $\phi_s = 0.5$.

In addiion in Ref.25 we investigated how softer domains are elongated to percolate to form a network for three values of g_{em}, $0.02, 0.05$, and 0.07, at $\phi_s = 0.2$. There, we found that the thickness of the elongated softer regions and the size of the harder regions increase with decreasing g_{em} in accord with the expectation $R \sim R_{em} \propto 1/g_{em}$. At $\phi_s = 0.2$ the softer domains are initially isolated and elongated as in Fig. 6. Then they encounter and coagulate to form a network. As a result the total perimeter length first increases due to the elongation and then it begins to decrease very slowly due to the coasening.

(iii) $\underline{F_{cub} \text{ and } F_{em} \text{ in Cubic Elasticity with Elastic Misfit}}$

Finally we display time evolutions and freezing of two phase patterns in the presence of F_{cub} and F_{em} at $\tau_a = 0.675$ and $g_{em} = 0.07$ in Figs.13-15 [23]. The pinning occured earlier in the presence of F_{cub} than in the absence of it in our simulation. The patterns obtained strikingly resemble real patterns in Ni-base cubic crystals with relatively large misfits observed by Miyazaki et al [3], in which the softer component indeed forms a network [31]. They found that the coarsening rate becomes extremely slow and the domain growth virtually stops near the critical composition. Carpenter also reported that the time exponent a of the domain size strongly depend on the composition as $1/a = 4.8, 9.3, 3.2$, respectively, for $40 : 60, 60 : 40, 80 : 20$ in Au+Pt alloys [32]. The origins of these observations might be ascribed particularly to the elastic effects arising from elastic misfit. However, comparison of our theory and experiments is still a future problem at present.

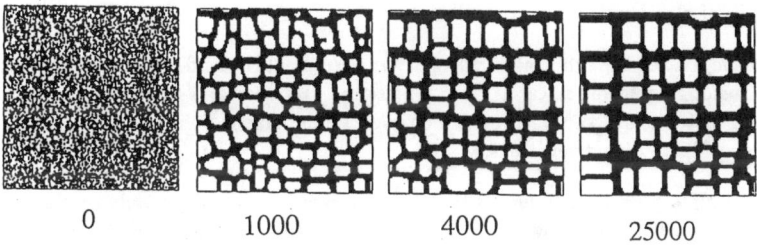

FIG.13. Evolution patterns for $\phi_s = 0.5$ in the presence of elastic misfit in a cubic solid. The numbers below the figures are the times after quenching.

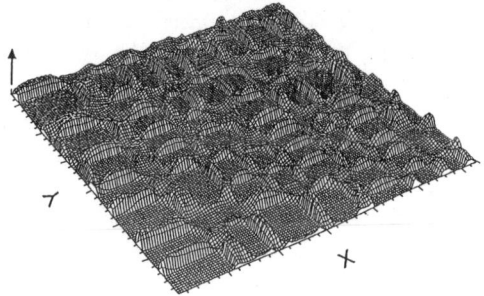

FIG.14. $\hat{Q}$ defined by (10) for $\phi_s = 0.5$ at $t = 24000$.

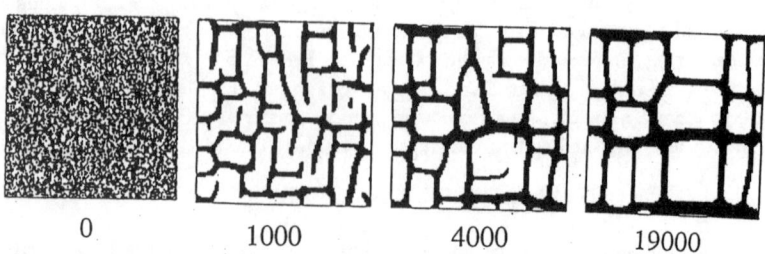

FIG.15. Evolution patterns for $\phi_s = 0.3$ at $\tau_a = 0.675$.

Concluding Remarks

(i) We have demonstrated the far-reaching efficiency of the TDGL approach in studying the pattern formation in phase-separating alloys. In our view, the effects of cubic elasticity and external stress are rather natural and obvious, while the effects of elastic misfit are highly intriguing. It goes without saying that we should perform more simulations hopefully in three dimensions. Moreover, we strongly need systematic and informative experimnts, in which the distance to the critical point, the degree of cubic anisotropy, and the degree of elastic misfit are controlled.

(ii) We have treated the simplest diffusive model. We should extend our approach to elastic systems belonging to other categories. They are (1) systems in which a nonconserved order parameter and a composition are coupled to the elastic field [26] and (2) systems undergoing Martensitic transitions.

(iii) In our theory we have assumed that the diffusive process occurs only on the spatial scale of microstructures. However, in H + metal systems, H atoms are known to diffuse extremely rapidly through samples. The H + metal systems exhibit macrosopic shape changes [33] under the stress-free boundary condition. This macroscopic instability is triggered above the usual spinodal decomposition temperature. Its time scale is short in H + metal systems but extremely long in usual alloys, although the phenomenological free energy (1) can be used both in the two cases. We have thus neglected this effect in our theory.

(iv) We mention another fascinating pattern formation in polymer gels consisting of networks swollen by solvent [34]. There, the coupling between thermodynamic instability and elasticity gives rise to a unique volume-phase transition, at which the bulk modulus tends to zero and elstic deformations can be very large.

References

1. A.G. Khachaturyan, Theory of Structural Transformations in Solids (Wiley, New York, 1983).
2. A.J. Ardell, R.B. Nicolson, and J.D. Eshelby, "On the Modulated Structure of Aged Ni-Al Alloys Acta. Metall. 14 (1966) 1295-1309 ; A. Maheshwari and A.J. Ardell, "Mophological Evolution of Coherent Misfitting Precipitates in Anisotropic Elastic Media", Phys. Rev. Lett. 70 (1993), 2305-2308.
3. T. Miyazaki and M. Doi, "Shape Bifurcation in the Coarsening of Precipitates in Elastically Constrained Systems" Mater. Sci. and Eng., A110 (1989), 175-185 ; T. Miyazaki, M. Doi and T. Kozakai, "Shape Bifurcations in the Coarsening of Precipitates in Elastically Constrained Systems, Solid State Phenomena 384 (1988)227-246.
4. P.C. Hohenberg and B.I. Halperin, "Theory of Dynamic Critical Phenomena" Rev. Mod. Phys. 49 (1977) 435-477.
5. J.D. Gunton, M. san Miguel and P.S. Sahni, "The Dynamics of First-Order Phase Transition" Phase Transitions and Critical Phenomena, vol.8, eds. C. Domb and J.L. Lebowitz (London, Academic Press, 1983), 267-466.
6. K. Binder, "Spinodal Decomposition" Materials Science and Technology, vol.5, eds.R.W. Cahn, P. Hansen and E.J.K. Kramer (Weinheim, VCH, 1991), 406-471.
7. J.W. Cahn, "On Spinodal Decomposition" Acta Metall. 9 (1961), 795-801.
8. a) A. Onuki, "Ginzburg-Landau Approach to Elastic Effects in Phase Separation of Solids" J. Phys. Soc. Jpn. 58 (1989), 3065-3068 ; b) "Long-Range Interactions through Elastic Fields in Phase-Separating Solids" ibid. 58 (1989) 3069-3072 ; c) "Interface Motion in Two-Phase Solids with Elastic Misfits" ibid. 60 (1991), 345-348.
9. J.W. Cahn, "On Spinodal Decomposition in Cubic Crystals" Acta Metall. 10 (1962), 179-183.
10. J.E. Hillard, "Spinodal Decomposition", Phase Transformations, American Society for Metals, Cleveland OH, 1970, 497-560.
11. H. Yamauchi and D.de Fontaine, "Elastic Interaction of Defect Clusters with Arbitrary Strain Fields in an Anisotropic Continuum Acta Metall. 27 (1979), 763-776.
12. T. Garel and S. Doniach, "Phase Transition with Spontaneous Modulation - the Dipolar Ising Ferromagnets", Phys. Rev. B (1982), 325-329 and references quoted therein.
13. A. L. Larkin and D. E. Khmel'nitskii, "Phase Transition in Uniaxial Ferroelectrics", Sov. Phys. JETP 29, 1123-1128.
14. L.D. Landau and E.M. Lifshitz, Electrodynamics of Continuous Media, Vol.8 (Pergamon Press, 1984).
15. A. Pineau, "Influence of Uniaxial Stress on the Morphology of Coherent Precipitates during Coarsening - Elastic Energy Consideration", Acta. Metall. 24 (1976) 559-564.
16. J. Gayda and D.J. Srolovitz, "A Monte Carlo-Finite Element Model for Strain Energy Controlled Microstructrural Evolution: "Rafting" in Superalloys", Acta Metall. 37 (1989) 641-650.
17. W. C. Johnson and J. W. Cahn, "Elastically Induced Shape Bifurcations of Inclusions" Acta Metall. 32 (1984), 1925-1933.
18. A. Onuki and H. Nishimori, "On Eshelby's Interaction in Two-Phase Solids" J. Phys. Soc. Jpn., 60 (1991) 1-4.
19. W.W. Mullins an R.F. Sekerka, "Morphological Stability of a Particle Growing by Diffusion of Heat Flow", J. Appl. Phys. 34 (1963), 323-329.
20. P.H. Leo and R.F. Sekerka, "The Effect of Elastic Fields on the Morphological Sta-

bility of a Precipitate Grown from Solid Solution", Acta Metall. 37 (1989), 3139-3149.
21. P.G. Shewmon, "Interfacial Stability in Solid-Solid Transformations", TMS AIME 233 (1965), 736-748; R.D. Doherty, Physical Metallurgy, eds. R.W. Cahn and P. Haasen (North-Holland, 1983).
22. H. Nishimori and A. Onuki, "Pattern Formation in Phase-Separating Alloys with Cubic Symmetry" Phys. Rev. B 42 (1990), 980-983.
23. A. Onuki and H. Nishimori, "Anomalous Slow Domain Growth due to a Modulus Inhomogeneity in Phase-Separating Alloys", Phys. Rev. B 43 (1991), 13649-13652.
24. H. Nishimori and A. Onuki, "Freezing of Domain Growth in Cubic Solids with Elastic Misfit" J. Phys. Soc. Jpn., 60 (1991), 1208-1211.
25. H. Nishimori and A. Onuki, "Evolution of Soft Domains in Two-Phase Alloys: Shape Changes, Surface Instability, and Network Formation", Phys. Lett. A 162 (1992), 323-326.
26. C. Sagui, A.M. Somoza and R.C. Desai, "Spinodal Decomposition in an Order-Disorder Phase Transition with Elastic Fields", Phys. Rev. E 50 (1994), 4865-4879.
27. Y. Oono and S. Puri, "Study of Phase Separation Dynamics by Use of Cell Dynamical Systems" ,Phys. Rev. A 38 (1988), 434-453.
28. Y. Wang et al, "Strain-Induced Modulated Structures in Two-Phase Cubic Alloys" Scripta METALLURGICA et MATERIALIA 25 (1991), 1969-1974 ; "Kinetics of Strain-Induced Morphological Transformation in Cubic Alloys with a Miscibility Gap" Acta. Metall. 41 (1992), 279-296.
29. J. K. Tien and S. M. Copley, "The Effect of Uniaxial Stress on the Periodic Morphology of Coherently Prime Precipitates in Nickel-Base Superalloy Crystals" Met. Trans. 2, (1971), 215-219 ; T. Miyazaki, K. Nakamura and H. Mori, "Experimental and Theoretical Investigations on Morphological Changes of γ' Precipitates in Ni-Al Single Crystals during Uniaxial Stress Annealing", J. Materials Sci. 14 (1979), 1827-1837 ; M. V. Nathal and L. J. Ebert, "Gamma Prime Shape Changes During Creep of a Nickel-Base Superalloy", Scripta METALLURGICA 17 (1983), 1151-1154.
30. Y. Enomoto and K. Kawasaki, "Computer Simulation of Ostwald Ripening with Elastic Field Interactions", Acta Matall. 37 (1989), 1399-1406 ; P.H. Leo, W.W. Mullins, R.F. Sekerka and J. Vinals, "Effects of Elasticity on Late Stage Coasening", Acta Matall. 38 (1990), 1573-1580.
31. T. Miyazaki and T. Koyama, private communication.
32. R.W. Carpenter, "Growth of Modulated Structures in Gold-Platinum Alloys", Acta. Metall. 15 (1967), 1567-1572.
33. H. Wagner and H. Horner, "Elastic Interaction and the Phase Transition in Coherent Metal-Hydrogen Systems", Adv. Phys. 23 (1974), 587-637 ; H. Zabel and H. Peisl, "Sample-Shape-Dependent Phase Transition of Hydrogen in Niobium", Phys. Rev. Lett. 42 (1979), 511-514.
34. T. Tanaka et al, "Mechanical Instability of Gels at the Phase Transition", Nature 325 (1987),796-798 ; E.S. Matsuo and T. Tanaka, "Patterns in Shrinking Gels", Nature 358 (1992), 482-485 ; K. Sekimoto, N. Suematsu and K. Kawasaki, "Spongelike Domain Structure in a Two-Dimensional Model Gel Undergoing Volume-Phase Transition", Phys. Rev. A 39 (1989), 4912 ; A. Onuki, "Theory of Phase Transition in Polymer Gels", Advances in Polymer Science 109 (Berlin-Heidelberg, Springer-Verlag, 1993), 64-121.

DISCRETE ATOM METHOD FOR MORPHOLOGICAL EVOLUTION

OF COHERENT PARTICLES

J. H. Choy, S. A. Hackney and J. K. Lee

Department of Metallurgical & Materials Engineering
Michigan Technological University
Houghton, MI 49931

Abstract

Morphological evolution of coherent second-phase particles with an arbitrary transformation strain is studied. With a purely dilatational misfit strain, a soft particle tends to have a plate-like equilibrium shape, whereas a hard particle takes on a shape of high symmetry such as a circle. In an anisotropic system with a comparable stiffness between the matrix and particle, however, the equilibrium shape depends on the degree of anisotropy, misfit strain, size, and interfacial energy. With either a tetragonal misfit strain of mixed signs or a pure shear misfit strain, a particle takes on a plate-like shape whose major axis lies along the direction containing an invariant line in accordance with the prediction of continuum elasticity theory. Interestingly, both elastic anisotropy and elastic inhomogeneity exert little influence on the preferred orientation relationship. In all, shape evolution goes through dynamic activities involving interfacial waves induced by the coherency strain.

Introduction

The microstructural evolution of an elastically stressed two-phase system is of great importance, as is intimately linked to alloy performance. An elastically-stressed, coherent particle undergoes a shape evolution fundamentally different from that of an unstressed one. The equilibrium shape of an unstressed particle is determined through minimization of the interfacial free energy. On the other hand, the morphology of a misfitting coherent particle is dictated by both the interfacial free energy and the elastic strain energy. Consequently, there has been a need for a computational technique, through which one can analyze the elastic state associated with *arbitrarily-shaped particles* whose elastic constants are *different* from those of the matrix phase.

Since Eshelby formulated the seminal inclusion method to evaluate the stress field associated with a coherent ellipsoidal inclusion (1), several theoretical works have been performed (2-4). Most of these treatments, however, are limited to simple particle geometries such as ellipsoids or rectangular parallelepipeds. With recent developments in micromechanics and fast computing equipment, there have emerged a few kinetic models appropriate for solids under stress, which are based on either a stochastic field theory (5) or on a variational principle (6). The major weakness of these models, however, is the basic assumption that both the particle and matrix are elastically homogeneous. In an effort to develop a tool to study general shape evolution of a coherent particle, a Discrete Atom method (DAM) was recently developed on the basis of a statistical approach (7,8). The method has been applied to a number of coherency strain problems (9,10), demonstrating that an elastically inhomogeneous, multi-particle system can be readily analyzed. In this work, the DAM is first reviewed to examine the morphological evolution of coherent particles with purely dilatational misfit strains. The shape evolution of coherent particles with *general stress-free transformation strains* is then addressed.

Discrete Atom Method

In the DAM, a triangular lattice is constructed following the work of Hoover et al (11). Atomic interactions are then mimicked through a parabolic potential function, $k(r-a)^2/2$, where k is spring constant, r is the interatomic distance, and a is the lattice parameter of the *stress-free* state. A precipitate phase, i.e., the domain of the inclusion, is represented by a different spring constant, k^*, and lattice parameter, a^*, from those of the matrix phase, k and a. The original Hookean nearest neighbor interaction is elastically isotropic. The lattice can be made elastically anisotropic if directional, instead of uniform, spring constants are introduced. In terms of Zener's anisotropy factor, $A = 2C_{44}/(C_{11}-C_{12})$, three lattices of A=2.33, 0.5 and 0.3 are studied to simulate a nickel-based superalloy, a niobium alloy, and partially-stabilized zirconia, respectively. In order to focus on the coherency strain effect, an isotropic interfacial energy is assumed, which is expressed in terms of the number of unlike nearest neighbor bonds for an interface atom. Here, an interface atom is defined as the precipitate atom having unlike nearest neighbor bonds. The time evolution of a particle shape is then investigated using Monte Carlo site exchange processes between matrix and precipitate atoms. Throughout this work, a plane strain condition is assumed and $k = 1.38 \times 10^{-18} J/a^2$, a specific interfacial energy, $\gamma = 2.5 \times 10^{-21} J/a$, and T = 50K are used unless otherwise specified.

Results

Coherent Particles with Dilatational Misfit Strain

Let us first consider the morphological evolution of coherent particles with a purely dilatational misfit strain of the form, $e_{ij}^T = \varepsilon \delta_{ij}$, where δ_{ij} is the Kronecker delta function. In Fig. 1, an isotropic, soft particle embedded in an infinite, isotropic matrix displays its shape transition from a circle to the equilibrium shape of an ellipse. The particle's shear modulus, μ^*, is equal to one half of the matrix shear modulus, μ, the particle radius R is $35a$, and the misfit strain ε is equal to 0.05. MCS indicates the number of Monte Carlo steps in units of one million. A striking feature is that the morphological evolution begins with interfacial waves induced by the coherency strain. The waves then develop small lobes, which then coarsen to lower density of larger lobes. Some of the larger lobes eventually disappear as the equilibrium shape is approached. During the entire sequence pictured in Fig. 1, the interfacial energy increases 88%, the strain energy diminishes 21%, and the total energy is reduced by 11%.

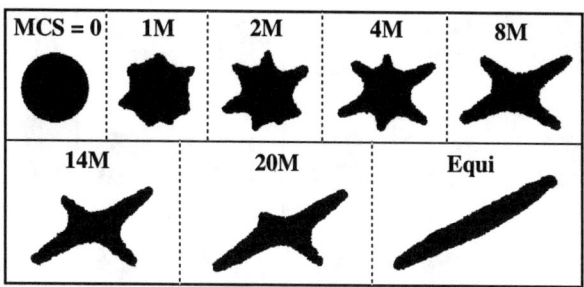

Figure 1: Morphological evolution of a soft particle with $\mu^* = \mu/2$, $R = 35a$, and $\varepsilon = 0.05$ in an isotropic matrix.

The larger the particle size, the more pronounced are the interfacial wave activities of soft particles. On the other hand, both higher stiffness and anisotropy appear to suppress the wave activities as the mean wave length diminishes. For the particles in Fig. 2, the anisotropy ratio, A, is equal to 2.33, a value similar to that of nickel. Other ancillary data are: $R = 40a$, $\varepsilon = 0.05$, and $C_{12} = C_{44} = 0.606\ k$. The soft <100> directions are marked with arrows and its Young's modulus is by a factor of 0.61 less than the one along the hard <110> directions. The top row shows the early stage of shape evolution for a soft particle ($C^*_{ijkl} = C_{ijkl}/2$), the second row for the homogeneous case ($C^*_{ijkl} = C_{ijkl}$), and the third row for a hard particle ($C^*_{ijkl} = 2C_{ijkl}$). All three cases display a symmetry-breaking transition from a radial to four-fold symmetry, consistent with the cubic anisotropy of the system. Further evolution, however, reveals that the soft particle transforms into a two-fold shape, becoming an ellipse stretched along a soft <100> direction. The four-fold form of the homogeneous case is also unstable with respect to a two-fold shape. The particle stretches, at a very sluggish rate, along a soft <100> direction, and is eventually transformed into a plate-like shape with blunt edges (8). On the other hand, the four-fold shape of the hard particle at MCS = 4M is an equilibrium morphology, as it is reproduced by the evolution process of an initially rectangular particle. Note that the soft particle displays a "concave-cuboidal" morphology in its early stage and the curvature at a corner along the hard <110> direction decreases with increases in the particle stiffness. Though two-dimensional, the overall evolution sequence observed in this simulation is consistent with many experimental observations in typical Ni-based superalloys (12,13).

Applied stress effects have been of great interest on the premise that an alloy's microstructure can be controlled (14). In Fig. 3, hard particles with effective $R = 40a$, $C^*_{ijkl} = 2C_{ijkl}$, $\varepsilon = 0.05$, and $A = 2.33$ are placed under a uniaxial stress with a magnitude equal to 4% strain along the vertical <010> direction. The top row shows the evolution process of a square particle under a compressive stress, whereas the bottom row displays shape changes under a tensile stress. The intriguing interaction behavior between the coherency strain and the applied stress polarizes the particle orientation in the opposite direction. Further, the particle under the compressive stress is shown to be on the verge of splitting. If the sense of the misfit strain or the relative stiffness of the particle is switched, the orientation is also reversed. A complete orientation relationship between an applied stress and a coherent particle also requires the inclusion of dislocation effects, some features of which are treated elsewhere (9).

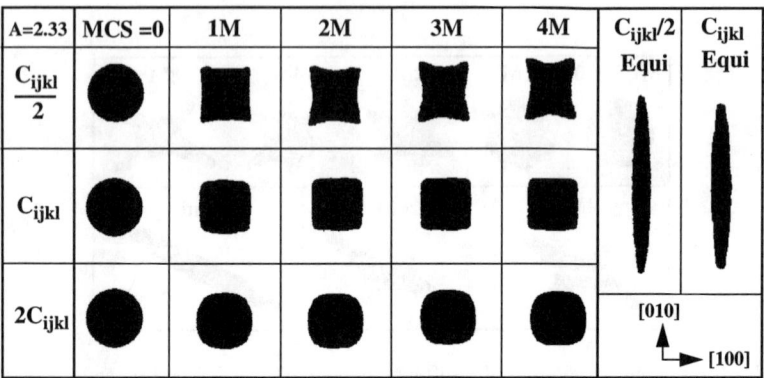

Figure 2: Morphological evolution of a soft (top row, $C^*_{ijkl} = C_{ijkl}/2$), an elastically homogeneous (middle row, $C^*_{ijkl} = C_{ijkl}$), and a hard particle (bottom row, $C^*_{ijkl} = 2C_{ijkl}$) in an anisotropic matrix with $A = 2.33$. The equilibrium shapes are an ellipse stretched along a <100> direction for the soft particle, an ellipse-like shape with blunt edges for the homogeneous case, and the four-fold shape at MCS = 4M for the hard particle.

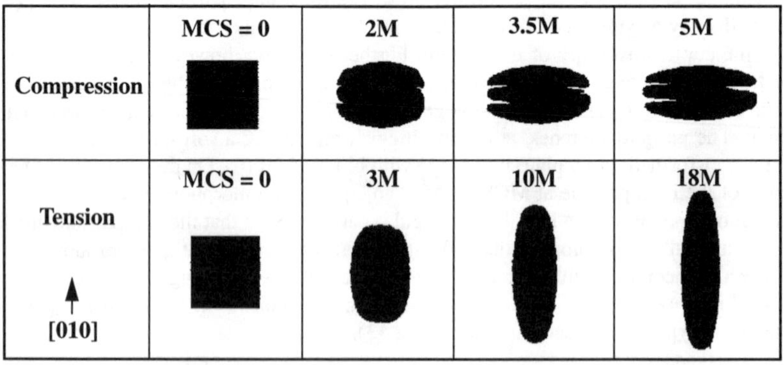

Figure 3: Effect of applied stress along the soft [010] direction on hard particles with $\varepsilon = 0.05$ and $C^*_{ijkl} = 2C_{ijkl}$ in an anisotropic matrix with $A = 2.33$. The top row shows the influence of a compressive stress, while the bottom row displays that of a tensile stress.

Coherent Particles with Tetragonal Misfit Strain

Let us consider the morphological evolution of coherent particles with a tetragonal misfit strain. In Fig. 4, a particle with R = 20a and $e_{ij}^T = 0.03(\delta_{ij} - 1.5\delta_{2i}\delta_{2j})$ is examined in an isotropic, homogeneous system. An initially circular particle passes through a two-fold rhombic-like intermediate shape at MCS = 0.2M before it reaches equilibrium. A result of a variational principle approach is also plotted in this figure. For the variational approach, the method of Voorhees et al. (6) is employed. Use of the same DAM input conditions ($C_{44} = 0.433k$, $k = 1.38 \times 10^{-18} J/a^2$ and $\gamma = 2.5 \times 10^{-21} J/a$) yields the relative energy ratio of strain to surface energy, $L = \varepsilon^2 C_{44} l/\gamma = 4.3$ with $\varepsilon = 0.03$ and $l \sim 20a$. Both equilibrium shapes of the DAM and the variational principle exhibit the preferred orientation relationship predicted by Khachaturyan (15) and Dahmen and Westmacott (16): the major axis of the particle shows an angle of about 35° ~ $\tan^{-1}(-e_{22}^T/e_{11}^T)^{1/2}$ with respect to the [010] direction.

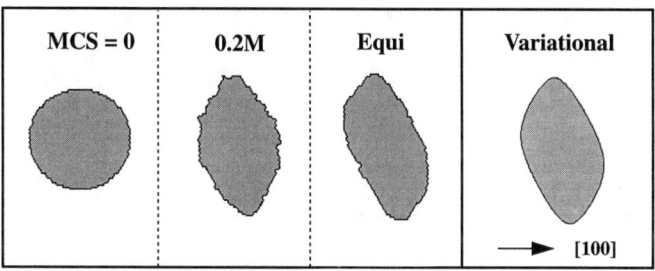

Figure 4: Shape evolution of a particle with R = 20a and $e_{ij}^T = 0.03(\delta_{ij} - 1.5\delta_{2i}\delta_{2j})$ in an elastically isotropic and homogeneous system. The equilibrium shape is also compared with the one obtained from a variational method.

In Fig. 5, another type of misfit, a Bain strain (17), is studied to examine the effects of anisotropy and inhomogeneity on a particle with R = 20a and $e_{ij}^T = \varepsilon(\delta_{ij} - 1.5\delta_{1i}\delta_{1j})$. The left two columns are for the case of A = 2.33 and ε = 0.05, while the right two columns are for the case of A = 0.3 and ε = 0.03. The case of A = 2.33 shows both soft ($C^*_{ijkl} = C_{ijkl}/2$) and homogeneous ($C^*_{ijkl} = C_{ijkl}$) particles, and a homogeneous and a hard ($C^*_{ijkl} = 2C_{ijkl}$) particle are examined for the A = 0.3 case. In all of the four examples, the preferred orientation relationship at equilibrium is again shown to be in accordance with the continuum linear elasticity prediction (16): the major axes lie along the direction of an angle, 55° ~ $\tan^{-1}(-e_{22}^T/e_{11}^T)^{1/2}$, with respect to the [010] direction. It is interesting to find that elastic anisotropy has no influence on the orientation relationship. Inhomogeneity also shows no sign of influence on the orientation, but affects the morphology in such a way that the particle aspect ratio at equilibrium tends to decrease as the particle becomes elastically stiffer.

Earlier DAM investigations (8,9) of a dilatationally misfitting precipitate showed that if a particle begins a shape transition from a highly non-equilibrium state, an evolution process of splitting followed by coalescence becomes a common feature. Splitting of a particle can be also observed for a tetragonal misfit case. In Fig. 6, a hard particle with effective R = 20a, A = 0.3, $e_{ij}^T = 0.03\delta_{2i}\delta_{2j}$, and $C^*_{ijkl} = 2C_{ijkl}$ is considered. For this case, the equilibrium shape is an ellipse with its major axis parallel to [100]. Thus, the initial rectangular shape is energetically unfavorable and undergoes splitting in the early period of evolution. The transition begins with coherency-induced interfacial waves. Wave interactions then cause the two sides to pinch off,

105

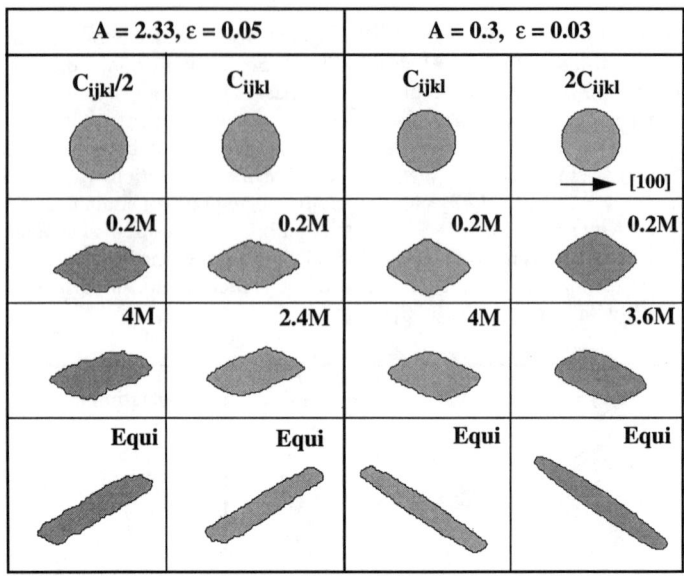

Figure 5: Shape evolution of particles with R = 20a and $e_{ij}^T = \varepsilon(\delta_{ij} - 1.5\delta_{1i}\delta_{1j})$ in an anisotropic system. The aspect ratio of the particle at equilibrium increases with a decrease in the particle stiffness, but the preferred orientation relationship is not influenced by the particle stiffness.

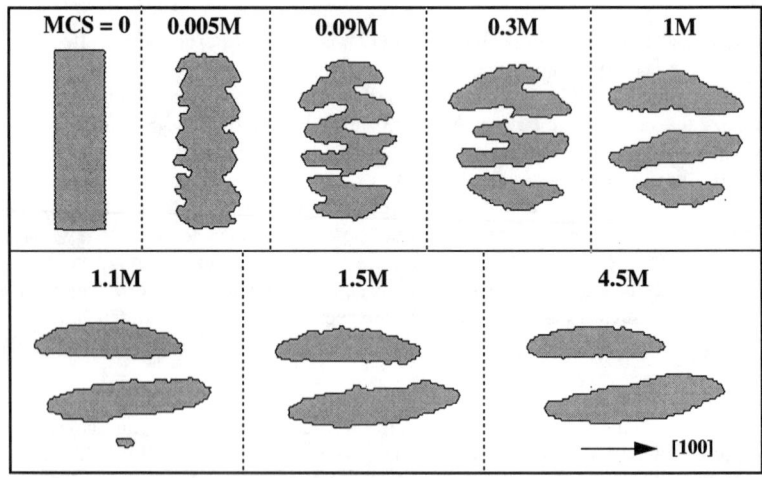

Figure 6: Shape evolution of a hard, rectangular particle with R = 20a, $C^*_{ijkl} = 2C_{ijkl}$, A = 0.3, and $e_{ij}^T = 0.03\delta_{2i}\delta_{2j}$ in an anisotropic system.

splitting the particle into three pieces at MCS = 0.3M. We note that each small particle takes on a plate-like shape, but due to the elastic repulsion between the plates, the orientation of some plates are shown to deviate from the [100] direction during the coarsening stage.

The elastic interaction between two parallel ellipses is examined in Fig. 7, where the minor axes of the two identical particles with aspect ratio equal to 0.17 and A = 0.3 are coaxial along the [010] and the misfit strain is the same for each, $e_{ij}^T = 0.03\delta_{2i}\delta_{2j}$. The interaction energy is given as the difference between the total elastic energy of the system and twice the self-strain energy of an individual ellipse, and is expressed as a fraction of twice the self-strain energy. The intercenter distance is given in units of the semi-minor axis. Note that the elastic interaction is repulsive regardless of the particle stiffness and is very strong. On the contrary, the dilatational misfit case involves an attractive interaction between two hard, elliptical particles (8).

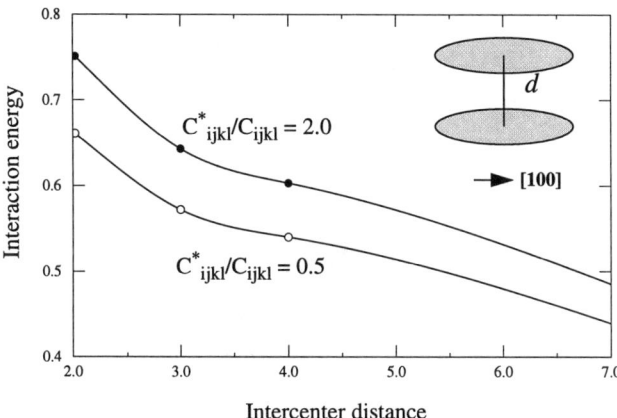

Figure 7: Elastic interaction energy of two parallel, elliptical particles with an aspect ratio equal to 0.17. The minor axes are coaxial along the [010], A = 0.3, and $e_{ij}^T = 0.03\delta_{2i}\delta_{2j}$.

Finally, Fig. 8 examines the case of misfit strains with non-zero diagonal and shear components. The misfit strain takes the form, $e_{ij}^T = \varepsilon\delta_{ij} + \eta(\delta_{1i}\delta_{2j} + \delta_{2i}\delta_{1j})$. The shape evolution of soft particles with R = 20a, C*$_{ijkl}$ = C$_{ijkl}$/2, η = 0.033, and A = 0.3 is investigated. In the left column, a soft particle with a pure shear misfit strain (ε = 0) undergoes a morphological evolution from a circle to a rectangular plate at equilibrium. The middle column represents a particle having a dilatational misfit strain whose strength is one half of that of the shear component (ε/η = 0.5). The particle in the right column has a dilatation whose magnitude is equal to that of the shear component (ε/η = 1). The preferred orientation relationship can be rationalized in terms of a tetragonal misfit strain of mixed signs, $e_{ij}^T = (\varepsilon + \eta)\delta_{ij} - 2\eta\delta_{2i}\delta_{2j}$, with [110] and [$\bar{1}$10] as the new x and y axes, respectively (i.e. as the principal axes). Applying the continuum linear elasticity theory as in the previous case, one obtains $\tan^{-1}\sqrt{(\eta-\varepsilon)/(\eta+\varepsilon)}$, the angle between the plate major axis and [$\bar{1}$10]. Thus the major axes of the equilibrium shape show <100> in the left column (ε/η = 0) and [$\bar{1}$10] in the right column (ε/η = 1), respectively. Obviously, the plate with ε/η = 0.5 lies along an orientation between [$\bar{1}$00] and [$\bar{1}$10]. The particle on the right exhibits interfacial wave activities which nearly split the particle during the evolution. A detailed studies including a tetragonal misfit strain with identical signs will be reported elsewhere.

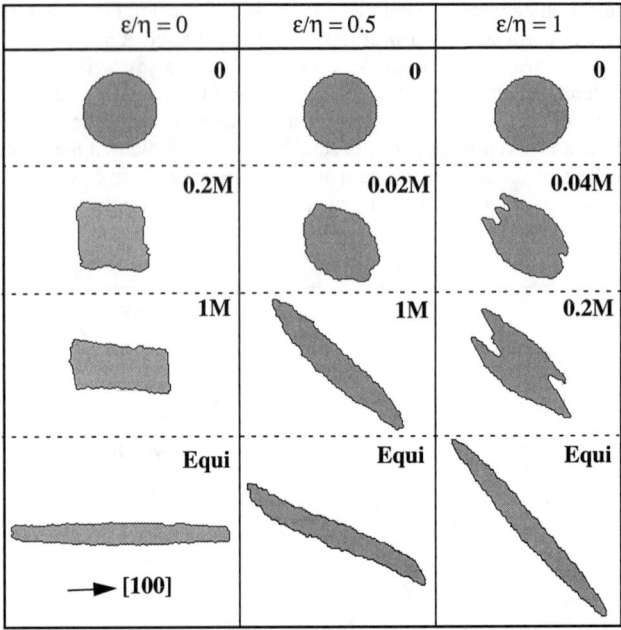

Figure 8: Shape evolution of soft particles with R = 20a, C*$_{ijkl}$ = C$_{ijkl}$/2, η = 0.033, and A = 0.3. The misfit strain is expressed as $e_{ij}^T = \varepsilon\delta_{ij} + \eta(\delta_{1i}\delta_{2j} + \delta_{2i}\delta_{1j})$. The left column shows a soft particle with a pure shear misfit strain (ε = 0). The middle column represents a particle with ε/η = 0.5, while the particle in the right column has a dilatation whose magnitude is equal to that of the shear component (ε/η = 1).

Summary

With a purely dilatational misfit strain, a soft particle tends to have a plate-like equilibrium shape, whereas a hard particle takes on a shape of high symmetry, such as a circle. In an anisotropic system with a comparable stiffness between the matrix and particle, the equilibrium shape depends on the degree of anisotropy, misfit strain, size, and interfacial energy. With a tetragonal misfit strain of mixed signs, a particle takes on a plate-like shape whose major axis lies along the direction at an angle given by $\tan^{-1}(-e_{22}^T/e_{11}^T)^{1/2}$ with respect to the [010]: this is in accordance with the prediction of continuum linear elasticity theory (15,16). Surprisingly, both elastic anisotropy and elastic inhomogeneity exert little influence on the preferred orientation relationship. For the case of a misfit strain of combined shear and dilatation, use of principal axes for the transformation strain identifies the preferred orientation. In all, shape evolution begins with interfacial wave activities induced by the coherency strain.

Acknowledgments

The authors are deeply indebted to Dr. Michael E. Thompson of Northwestern University for valuable suggestions on the manuscript. The research was supported by the U.S. Department of Energy under Grant DE-FG02-87ER45315 and by a Research Excellence Fund from the State of Michigan, for which the authors express much appreciation.

References

1. J. D. Eshelby: Prog. Solid Mech. **2**, (1961), 89.

2. A. J. Ardell, R. B. Nicholson and J. D. Eshelby: Acta Metall. **14**, (1966), 1295.

3. J. K. Lee, D. M. Barnett and H. I. Aaronson: Metall. Trans. **8A**, (1977), 963.

4. W. C. Johnson and P. W. Voorhees: J. Appl. Phys. **61**, (1987), 1610.

5. Y. Wang, L. Q. Chen and A. G. Khachaturyan: Acta Metall. **41**, (1993), 279.

6. P. W. Voorhees, G. B. McFadden and W. C. Johnson: Acta Metall. **40**, (1992), 2979.

7. J. K. Lee: Scripta Metall. **32**, (1995), 559.

8. J. K. Lee: Metall. Trans., in press.

9. J. K. Lee: in *Micromechanics of Advanced Materials*, S. N. G. Chu et al, eds., TMS, Warrendale, PA, 1995, p. 41.

10. J. K. Lee: in *Phase Transformations during the Thermal/Mechanical Processing of Steel*, E. B. Hawbolt and S. Yue, eds., The Metallurgical Society of CIM, Montreal, Canada, 1995, p. 49.

11. W. G. Hoover, W. T. Ashurt and R. J. Olness: J. Chem. Phys. **60**, (1974), 4043.

12. M. Meshkinpour and A. J. Ardell: Mater. Sci. Eng. **A185**, (1994), 153.

13. Y. S. Yoo, D. N. Yoon and M. F. Henry: Metals and Mat. **1**, (1995), 47.

14. J. K. Tien and S. M. Copley: Metall. Trans. **2**, (1971), 215.

15. A. G. Khachaturyan: *Theory of Structural Transformations in Solids*, Wiley & Sons, New York, NY, 1983, p. 248.

16. U. Dahmen and K. Westmacott: Acta Metall. **34**, (1986), 475.

17. J. M. Howe and D. A. Smith: Acta Metall. **40**, (1992), 2343.

COMPUTER SIMULATIONS OF PHASE DECOMPOSITION IN REAL ALLOY SYSTEMS BASED ON A DISCRETE TYPE DIFFUSION EQUATION

Toru MIYAZAKI and Toshiyuki KOYAMA

Department of Materials Science and Engineering
Nagoya Institute of Technology
Nagoya, 466, JAPAN

Abstract

A nonlinear diffusion equation including the information of the solute atom occupation probability at an atom site, which is named "discrete type diffusion equation", is newly proposed. The composition dependencies of atomic interchange energy and of elasticity are also taken into account so as to be applicable for the real alloy system. The two dimensional computer simulations are performed for the phase decomposition of the Fe- Mo binary alloy system by utilizing the thermodynamic data on equilibrium phase diagrams. The microstructures theoretically obtained are well coincident with experimental facts of the real alloys.

Introduction

Theoretical investigations on the diffusion-controlled phase transformation have been performed by many researchers. Since Cahn and Hilliard[1-4] proposed the nonlinear diffusion equation in the 1960's, many researchers have attempted the theoretical analysis of phase decomposition on the basis of that equation[5-7]. However, various assumptions and omissions were made in their calculations, because it was extremely difficult to get the analytical solution of the nonlinear term in the differential equation. Since the recent remarkable developments in computers have made the numerical analysis of the nonlinear diffusion equation possible, computer simulations have become very useful in understanding the dynamics of phase transformations in materials, i.e. not only in metallic alloys but also in ceramics and polymers. It is considered that the calculation of time-dependent phase transformation is essentially based upon one of the following four methods; Cahn-Hilliard nonlinear diffusion equation[8-10], TDGL(Time-Dependent Ginzburge-Landau) Model[11-15], Khachaturyan's diffusion equation[16-20] and Phase field method[21-23]. These methods are considered to be very useful for the basic understanding of the phase decomposition process, but it is undeniable that the calculations based on these methods have hitherto been carried for the virtual phase diagrams and have given only qualitative information on the phase decomposition. Such qualitative results are insufficient for quantitative understanding the phase decomposition of real alloy system. The composition and the temperature dependencies of the atomic interchange energy should be taken into account, because such dependencies are usually found in real alloy system.

In the present work, we newly propose a comprehensive nonlinear diffusion equation where the composition dependencies of atomic interchange energy and of elasticity are taken into account so as to be able to calculate for the real alloy system, and the solute atom occupation probability in atom site is also introduced. Finally we perform computer simulations of the phase decompositions in Fe-Mo binary alloys on the basis of the proposed theory.

Theoretical Basis

Outline of the Discrete Type Diffusion Equation

First of all, we define the nonlinear diffusion equation as follows;

$$\frac{\partial c(\mathbf{r},t)}{\partial t} = \frac{\partial}{\partial \mathbf{r}}\left[M\{c(\mathbf{r},t),T\}\left\{\frac{\partial \chi(\mathbf{r},t)}{\partial \mathbf{r}}\right\}\right] \quad (1)$$

where $\mathbf{r}$ is the position vector, t is time, and $\chi(\mathbf{r},t)$ and $c(\mathbf{r},t)$ are diffusion potential and the composition respectively at position $\mathbf{r}$ and at time t. $M\{c(\mathbf{r},t),T\}$ is the mobility of atom diffusion which is a function of $c(\mathbf{r},t)$ and temperature T. The difference expression of equation(1) is given by equation(2).

$$\frac{\partial c(\mathbf{r},t)}{\partial t} = M\{c(\mathbf{r},t),T\}\left\{\frac{\partial^2 \chi(\mathbf{r},t)}{\partial \mathbf{r}^2}\right\} + \left\{\frac{\partial M\{c(\mathbf{r},t),T\}}{\partial c}\right\}\left\{\frac{\partial c(\mathbf{r},t)}{\partial \mathbf{r}}\right\}\left\{\frac{\partial \chi(\mathbf{r},t)}{\partial \mathbf{r}}\right\}$$

$$= M\{c(\mathbf{r},t),T\}\left[\sum_{\mathbf{r}'}L_2(\mathbf{r}-\mathbf{r}')\chi(\mathbf{r}',t)\right] + \left\{\frac{\partial M\{c(\mathbf{r},t),T\}}{\partial \mathbf{r}}\right\}\left[\sum_{\mathbf{r}'}L_1(\mathbf{r}-\mathbf{r}')c(\mathbf{r}',t)\right]\left[\sum_{\mathbf{r}'}L_1(\mathbf{r}-\mathbf{r}')\chi(\mathbf{r}',t)\right] \quad (2)$$

Here, $L_i(\mathbf{r}-\mathbf{r}')$ is a matrix of kinetic coefficients of the i-th differential for solute atom diffusion from site $\mathbf{r}$ to $\mathbf{r}'$ during a unit time. Equation (2) is a general expression of the discrete diffusion equation. If the second term is omitted and only the first term is included, eq.(2) corresponds to Khachaturyan diffusion equation[16]. The diffusion potential $\chi(\mathbf{r}',t)$ is defined by eq.(3).

$$\chi(\mathbf{r}',t) = \left(\frac{\delta G_T}{\delta c}\right)_{\mathbf{r}=\mathbf{r}'} \quad (3)$$

G_T is the total free energy consisting of the chemical free energy G_c, the interfacial energy E_{surf} and the elastic strain energy E_{str} as follows;

$$G_T = G_c + E_{surf} + E_{str} \tag{4}$$

Thus, the chemical potential μ_c, the interfacial potential μ_{surf} and the elastic potential μ_{str} are given by the following equations, respectively:

$$\mu_c\{c(\mathbf{r'},t)\} \equiv \left(\frac{\delta G_c}{\delta c}\right)_{\mathbf{r}=\mathbf{r'}} \tag{5}$$

$$\mu_{surf}(\mathbf{r'},t) \equiv \left(\frac{\delta E_{surf}}{\delta c}\right)_{\mathbf{r}=\mathbf{r'}} \tag{6}$$

$$\mu_{str}(\mathbf{r'},t) \equiv \left(\frac{\delta E_{str}}{\delta c}\right)_{\mathbf{r}=\mathbf{r'}} \tag{7}$$

The diffusion potential is given by eq.(8).

$$\chi(\mathbf{r'},t) = \mu_c\{c(\mathbf{r'},t)\} + \mu_{surf}(\mathbf{r'},t) + \mu_{str}(\mathbf{r'},t) \tag{8}$$

The kinetic matrix of solute atom diffusion L_i in eq.(2) is defined by the usual difference expression, described in the next section. $M\{c(\mathbf{r},t),T\}$ in eq.(2) is explicitly given by equation (9) for AB binary system[4].

$$M\{c(\mathbf{r},t),T\} = [M_A(T)c(\mathbf{r},t) + M_B(T)\{1 - c(\mathbf{r},t)\}]c(\mathbf{r},t)\{1 - c(\mathbf{r},t)\} \tag{9}$$

,where $M_i(T)$ is the mobility of i- atom, which can be estimated from the self- diffusion coefficient $D_i(T)$. Since $M\{c(\mathbf{r},t),T\}$ is only a function of composition c for the case of isothermal ageing, if the diffusion potential $\chi(\mathbf{r'},t)$ at a position $\mathbf{r}$ is obtained in microstructure, $\partial c/\partial t$ is obtained from eq.(2), and consequently the time development of phase decomposition can be evaluated by repeating eq.(10).

$$c(\mathbf{r},t+\Delta t) = c(\mathbf{r},t) + \left(\frac{\partial c(\mathbf{r},t)}{\partial t}\right)_t \Delta t \tag{10}$$

It is easily understood from the above described that the precise evaluation of diffusion potential $\chi(\mathbf{r'},t)$ is the most essential and important in these calculation. The details of the estimation for $\chi(\mathbf{r'},t)$ are presented in next section.

Evaluation of Various Potentials

(a) Chemical Potential μ_c: The chemical free energy G_c is generally given by eq.(11) on the basis of the modified expression of the regular solution approximation[24]. Hereafter, $c(\mathbf{r},t)$ is described by $c(\mathbf{r})$, because t is fixed in the calculation.

$$G_c\{c(\mathbf{r})\} = \frac{1}{2}\sum_{\mathbf{r}} \Omega\{c(\mathbf{r}),T\}c^2(\mathbf{r}) + \sum_{\mathbf{r}} RT[c(\mathbf{r})\ln c(\mathbf{r}) + \{1 - c(\mathbf{r})\}\ln\{1 - c(\mathbf{r})\}] \tag{11}$$

The atomic interaction parameter $\Omega\{c(\mathbf{r}),T\}$, is given by eq.(12)[24].

$$\Omega\{c(\mathbf{r}),T\} = \sum_{j}\Omega_{j}(T)c^{j}(\mathbf{r}) \tag{12}$$

Substituting eqs.(11) and (12) into eq.(5), we get the chemical potential $\mu_c\{c(\mathbf{r'})\}$ as follows;

$$\mu_c\{c(\mathbf{r'})\} = \sum_{j}\frac{j+2}{2}\Omega_{j}(T)c^{j+1}(\mathbf{r'}) + RT[\ln c(\mathbf{r'}) - \ln\{1-c(\mathbf{r'})\}] \tag{13}$$

Since $\Omega_{j}(T)$ in eq.(12) can be obtained from the thermodynamic data of the equilibrium phase diagram, it is possible by using eq.(13) to get $\mu_c\{c(\mathbf{r'})\}$ as functions of composition c and position $\mathbf{r}$, which is directly corresponding to the real alloy phase diagram.

(b)Interfacial Potential μ_{surf}: The interfacial energy E_{surf} is given by eq.(14)[17].

$$E_{surf} = \frac{1}{2}\sum_{\mathbf{p'}}\sum_{\mathbf{p''}}\{W(\mathbf{p'}-\mathbf{p''},T) - \overline{W}(T)\}c(\mathbf{p'})c(\mathbf{p''}) \tag{14}$$

,where $\mathbf{p'}$ and $\mathbf{p''}$ are the position vectors of lattice site and $W(\mathbf{p'}-\mathbf{p''},T)$ shows the atomic interaction energy. $\overline{W}(T)$ is a three dimensional average of $W(\mathbf{p'}-\mathbf{p''},T)$. Since $W(\mathbf{p'}-\mathbf{p''},T)$ in eq.(14) and $\Omega_0(T)$ in eq.(12) are related to each other as described in eq.(15)[20], $W(\mathbf{p'}-\mathbf{p''},T)$ is evaluated by utilizing the thermodynamic data $\Omega_0(T)$ of the equilibrium phase diagram. The calculation procedure will be described in the next section.

$$\Omega_0(T) = \frac{1}{2}\sum_{\mathbf{p'}}\sum_{\mathbf{p''}}W(\mathbf{p'}-\mathbf{p''},T) \tag{15}$$

Eq.(15) represents the interfacial energy which is described in a form of Khachaturyan's expression, while in the Cahn-Hilliard's spinodal decomposition theory, the TDGL model and the Phase field method, the interfacial energy is defined by eq.(16), which corresponds to the excess free energy caused by the non-uniformity of ordering parameter(the ordering parameter corresponds to the composition c in the usual phase decomposition of disordered phase).

$$E_{surf} = \kappa\left(\frac{\partial c}{\partial r}\right)^2 \tag{16}$$

,where κ is a composition gradient energy coefficient, which is a function of $\Omega_0(T)$ and the interatomic distance[4]. Eq.(16) is well known to be approximated by omitting the higher order terms[4]. Therefore, eq.(16) is considered to be not more accurate than eq.(14), particularly eq.(16) is difficult to apply for sharp interfaces such as the order/disorder phase boundary. On the contrary, the transformation process of ordering phase can be calculated by taking the occupation probability of solute atom in lattice site into account.

The interfacial potential μ_{surf} is easily introduced by the next equation.

$$\mu_{surf}(\mathbf{p'}) = \sum_{\mathbf{p''}}\{W(\mathbf{p'}-\mathbf{p''},T) - \overline{W}(T)\}c(\mathbf{p''}) \tag{17}$$

Furthermore, by using the Fourier transform of $W(\mathbf{p'}-\mathbf{p''},T)$, the interfacial potential at position $\mathbf{r'}$, $\mu_{surf}(\mathbf{r'})$ is given by eq.(18). $V(\mathbf{h})$ and $Q(\mathbf{h})$ are given by eqs(19) and (20), respectively.

$$\mu_{surf}(\mathbf{r'}) = \sum_{\mathbf{h}}V(\mathbf{h})Q(\mathbf{h})\exp(i\mathbf{h}\beta\mathbf{r'}) \tag{18}$$

$$V(\mathbf{h}) = \sum_{\mathbf{p'-p''}} [W(\mathbf{p'-p''}, T) - \overline{W}(T)] \exp\{-i\mathbf{h}\beta(\mathbf{p'-p''})\} \quad (19)$$

$$Q(\mathbf{h}) = \left(\frac{1}{N^3}\right) \sum_{\mathbf{p''}} c(\mathbf{p''}) \exp(-i\mathbf{h}\beta\mathbf{p''}) \quad (20)$$

, where $\mathbf{h}$ is a wave number of Fourier wave, N is the maximum number of $\mathbf{h}$ which corresponds to the number of divided regions of microstructure in Fourier transformation, L is the length of calculation field in the real space and β is defined by $\beta = 2\pi/L$. The details of evaluation of eq.(19) is explained in the Section of "Calculation Method in Two Dimensions".

(c)Elastic Potential μ_{str} : The elastic strain energy for the cubic lattice crystal is given by eq.(21) based on Landau elasticity theory[13,14], where, C_{ij} is the elastic stiffness constant, e_{ij}^c is the constrained strain and η is the lattice mismatch between the A and B pure metal.

$$E_{str} = \sum_{\mathbf{r}} \begin{bmatrix} \frac{3}{2}(C_{11} + 2C_{12})\eta^2 \{c(\mathbf{r}) - c_0\}^2 - \eta(C_{11} + 2C_{12})\{c(\mathbf{r}) - c_0\}(e_{11}^c + e_{22}^c + e_{33}^c) \\ + \frac{1}{2}C_{11}(e_{11}^{c\,2} + e_{22}^{c\,2} + e_{33}^{c\,2}) + C_{12}(e_{11}^c e_{22}^c + e_{22}^c e_{33}^c + e_{33}^c e_{11}^c) + 2C_{44}(e_{12}^{c\,2} + e_{23}^{c\,2} + e_{31}^{c\,2}) \end{bmatrix} \quad (21)$$

Here, being assumed to be linearly proportional to composition, C_{ij} is expressed by eq.(22), where C_{ijkl}^X indicates the elastic stiffness constant of pure metal X(=A or B), which is a function of temperature T.

$$C_{ijkl}\{c(\mathbf{r}), T\} = C_{ijkl}^A(T)\{1 - c(\mathbf{r})\} + C_{ijkl}^B(T)c(\mathbf{r}) = C_{ijkl}^0 + \Delta C_{ijkl}^{AB}\{c(\mathbf{r}) - c_0\} \quad (22)$$

$$C_{ijkl}^0 \equiv C_{ijkl}^A(T)(1 - c_0) + C_{ijkl}^B(T)c_0 \;,\quad \Delta C_{ijkl}^{AB} \equiv C_{ijkl}^B(T) - C_{ijkl}^A(T) \quad (23)$$

Substituting eq.(22) into eq.(21), we obtain the elastic strain energy as shown in eq.(24).

$$E_{str} = \sum_{\mathbf{r}} \begin{bmatrix} (3/2)(C_{11}^0 + 2C_{12}^0)\eta^2\{c(\mathbf{r}) - c_0\}^2 - \eta(C_{11}^0 + 2C_{12}^0)\{c(\mathbf{r}) - c_0\}(e_{11}^c + e_{22}^c + e_{33}^c) \\ + (1/2)C_{11}^0(e_{11}^{c\,2} + e_{22}^{c\,2} + e_{33}^{c\,2}) + C_{12}^0(e_{11}^c e_{22}^c + e_{22}^c e_{33}^c + e_{33}^c e_{11}^c) + 2C_{44}^0(e_{12}^{c\,2} + e_{23}^{c\,2} + e_{31}^{c\,2}) \\ + (3/2)(\Delta C_{11}^{AB} + 2\Delta C_{12}^{AB})\eta^2\{c(\mathbf{r}) - c_0\}^3 - \eta(\Delta C_{11}^{AB} + 2\Delta C_{12}^{AB})\{c(\mathbf{r}) - c_0\}^2(e_{11}^c + e_{22}^c + e_{33}^c) \\ + (1/2)\Delta C_{11}^0\{c(\mathbf{r}) - c_0\}(e_{11}^{c\,2} + e_{22}^{c\,2} + e_{33}^{c\,2}) + \Delta C_{12}^0\{c(\mathbf{r}) - c_0\}(e_{11}^c e_{22}^c + e_{22}^c e_{33}^c + e_{33}^c e_{11}^c) \\ + 2\Delta C_{44}^0\{c(\mathbf{r}) - c_0\}(e_{12}^{c\,2} + e_{23}^{c\,2} + e_{31}^{c\,2}) \end{bmatrix}$$

(24)

Substituting eq.(24) into eq.(7), we obtain the elastic potential μ_{str} as in eq.(25). The Landau elastic strain energy is independently expanded with composition c and constrained strain e_{ij}^c, so that the elastic potential is easily expressed as indicated in eq.(25).

$$\mu_{str}(\mathbf{r'}) = 3(C_{11}^0 + 2C_{12}^0)\eta^2\{c(\mathbf{r'}) - c_0\} - \eta(C_{11}^0 + 2C_{12}^0)(e_{11}^c + e_{22}^c + e_{33}^c)$$
$$+ \frac{9}{2}(\Delta C_{11}^{AB} + 2\Delta C_{12}^{AB})\eta^2\{c(\mathbf{r'}) - c_0\}^2 - 2\eta(\Delta C_{11}^{AB} + 2\Delta C_{12}^{AB})\{c(\mathbf{r'}) - c_0\}(e_{11}^c + e_{22}^c + e_{33}^c) \quad (25)$$
$$+ \frac{1}{2}\Delta C_{11}^{AB}(e_{11}^{c\,2} + e_{22}^{c\,2} + e_{33}^{c\,2}) + \Delta C_{12}^{AB}(e_{11}^c e_{22}^c + e_{22}^c e_{33}^c + e_{33}^c e_{11}^c) + 2\Delta C_{44}^{AB}(e_{12}^{c\,2} + e_{23}^{c\,2} + e_{31}^{c\,2})$$

The constrained strain $e_{kl}^c(\mathbf{r'})$ in eq.(25) is given by eq.(26) on the basis of the equilibrium equation of the elasticity theory by A.G.Khachaturyan[25].

$$e_{kl}^c(\mathbf{r'}) = \sum_{\mathbf{h}} \frac{1}{2} C_{pqmn}^0 \eta_{mn} \{n_q G_{pl}(\mathbf{n})n_k + n_q G_{pk}(\mathbf{n})n_l\} Q(\mathbf{h}) \exp(i\mathbf{h}\beta\mathbf{r'}) \quad (26)$$

$$G_{il}^{-1}(\mathbf{n}) \equiv C_{ijkl}^0 n_j n_k \quad (27)$$

$\mathbf{n}$ is the unit vector of the wavenumber $\mathbf{h}$. The function $G_{il}(\mathbf{n})$ in eq.(26) is the inverse matrix of $G_{il}^{-1}(\mathbf{n})$ which is defined by eq.(27). Calculating eq.(26), we obtain the constrained strain $e_{kl}^c(\mathbf{r'})$ at arbitrary position in the microstructure, and substituting them into eq.(25) we obtain the elastic potential $\mu_{str}(\mathbf{r'})$ at arbitrary position. As above described the various potentials are obtained. Here it should be noted that all potentials are continuously given with position $\mathbf{r'}$, so that the diffusion potential can be evaluated at arbitrary position in the microstructure.

Calculation Method in Two Dimensions

Here, concrete expressions of $\chi(\mathbf{r'})$ are introduced for the two-dimensional (2-D) phase decomposition. On the chemical potential, $\Omega_j(T)$ is so obtained from the thermodynamic data of the equilibrium phase diagram that $\mu_c(\mathbf{r'})$ can be evaluated at arbitrary position $\mathbf{r'}$ by employing eq.(13). In order to evaluate the numerical value of interfacial potential $\mu_{surf}(\mathbf{r'})$, $V(\mathbf{h})$ and $Q(\mathbf{h})$ are needed. The $Q(\mathbf{h})$ can numerically be evaluated by the Fast Fourier Transformation of the composition field. $V(\mathbf{h})$ is analytically introduced as described below. Using eq.(15), we can evaluate $W(\mathbf{p'}-\mathbf{p''},T)$ from the values of $\Omega_0(T)$ under the following conditions; $W(i \geq 3)$ is assumed to be zero and the value of $\Omega_0(T)$ is assigned only to the first and second nearest neighbor atomic interchange energies $W(1)$ and $W(2)$. The ratio of $W(1)/W(2)$ is theoretically given to be $\sqrt{2}$ on the assumption that the interfacial energy density is independent of the direction[17].

$$W(1) = \frac{\Omega_0(T)}{(2+\sqrt{2})}, \quad W(2) = \frac{W(1)}{\sqrt{2}} \quad (28)$$

The assumption that $W(i \geq 3) = 0$ is considered to be rational because most of the equilibrium phase diagrams can adequately be characterized by using up to the second nearest neighbor parameters, $W(1)$ and $W(2)$[26]. Substituting eq.(28) into eq.(19) gives $V(\mathbf{h})$ shown in eq.(29).

$$V(\mathbf{h}) = \left[2W(1)\{\cos(h_x\beta d_1) + \cos(h_y\beta d_1)\} + 4W(2)\cos(h_x\beta d_1)\cos(h_y\beta d_1)\right] - 4[W(1) + W(2)] \quad (29)$$

,where h_x and h_y are the directional components of the wave vector $\mathbf{h}$ $[\mathbf{h} = (h_x, h_y)]$. The term d_1 in eq.(29) is the interatomic distance of the first nearest neighbors in the 2-D square lattice used for the present calculation. However, it should be noted that since the 2-D atomic interchange energies, $W(1)$ and $W(2)$, are evaluated from the 3-D thermodynamic data, the atomic distance of the 2-D square lattice is not always coincident with that of the actual 3-D lattice, because the atomic spacings are different between the 2-D and 3-D lattices when the lattices have the same free energy. Therefore, the d_1 in eq.(29) is the only one parameter which must be determined experimentally. The value of d_1 is estimated so as to equalize the wavelength of the modulated structure calculated in the present simulation with that of the experimental data for the same alloy system. Consequently, multiplying $V(\mathbf{h})$ with $Q(\mathbf{h})$, and inverse Fourier transforming the product, we can get the interfacial potential $\mu_{surf}(\mathbf{r'})$ at arbitrary position.

Finally, the elastic potential is dealt with. In the 2-D calculation the microstructure is so large along the c- axis that the relaxation of strain component along the c- axis ($= e_{i3}^c = e_{3i}^c$) is zero. Therefore, eq.(25) is rewritten as eq.(30) for 2-D.

$$\mu_{str}(\mathbf{r'}) = 3(C_{11}^0 + 2C_{12}^0)\eta^2\{c(\mathbf{r'}) - c_0\} - \eta(C_{11}^0 + 2C_{12}^0)(e_{11}^c + e_{22}^c)$$
$$+ \frac{9}{2}(\Delta C_{11}^{AB} + 2\Delta C_{12}^{AB})\eta^2\{c(\mathbf{r'}) - c_0\}^2 - 2\eta(\Delta C_{11}^{AB} + 2\Delta C_{12}^{AB})\{c(\mathbf{r'}) - c_0\}(e_{11}^c + e_{22}^c) \qquad (30)$$
$$+ \frac{1}{2}\Delta C_{11}^{AB}(e_{11}^{c\,2} + e_{22}^{c\,2}) + \Delta C_{12}^{AB}e_{11}^c e_{22}^c + 2\Delta C_{44}^{AB}e_{12}^{c\,2}$$

Furthermore, the explicit forms of constrained strain of eq.(26) are given by eq.(31) for the case of pure dilatation.

$$e_{11}^c(\mathbf{r'}) = \eta(C_{11}^0 + 2C_{12}^0)\sum_\mathbf{h} Z_{11}(\mathbf{n})Q(\mathbf{h})\exp(i\mathbf{h}\beta\mathbf{r'})$$
$$e_{22}^c(\mathbf{r'}) = \eta(C_{11}^0 + 2C_{12}^0)\sum_\mathbf{h} Z_{22}(\mathbf{n})Q(\mathbf{h})\exp(i\mathbf{h}\beta\mathbf{r'}) \qquad (31)$$
$$e_{12}^c(\mathbf{r'}) = e_{21}^c(\mathbf{r'}) = \eta(C_{11}^0 + 2C_{12}^0)\sum_\mathbf{h} Z_{12}(\mathbf{n})Q(\mathbf{h})\exp(i\mathbf{h}\beta\mathbf{r'})$$

$$Z_{11}(\mathbf{n}) = \frac{\left[1 + \xi n_2^2\right]n_1^2}{\left[C_{11}^0 + \xi(C_{11}^0 + C_{12}^0)n_1^2 n_2^2\right]}$$
$$Z_{22}(\mathbf{n}) = \frac{\left[1 + \xi n_1^2\right]n_2^2}{\left[C_{11}^0 + \xi(C_{11}^0 + C_{12}^0)n_1^2 n_2^2\right]} \qquad (32)$$
$$Z_{12}(\mathbf{n}) = \frac{(2 + \xi)n_1 n_2}{2\left[C_{11}^0 + \xi(C_{11}^0 + C_{12}^0)n_1^2 n_2^2\right]}$$

$$\xi = \frac{C_{11}^0 - C_{12}^0 - 2C_{44}^0}{C_{44}^0} \qquad (33)$$

Thus, the constrained strain at position $\mathbf{r'}$, $e_{ij}^c(\mathbf{r'})$, is evaluated by the inverse Fourier transformation of $Z_{ij}(\mathbf{n})Q(\mathbf{h})$ estimated from eqs.(32) and (33), and then the elastic potential at any position $\mu_{str}(\mathbf{r'})$ is given by substituting $e_{ij}^c(\mathbf{r'})$ into eq.(30)
Consequently, the diffusion potential $\chi(\mathbf{r'})$ is obtained as a sum of the three potentials.

$$\chi(\mathbf{r'}) = \mu_c\{c(\mathbf{r'})\} + \mu_{surf}(\mathbf{r'}) + \mu_{str}(\mathbf{r'}) \qquad (34)$$

Since the composition at arbitrary position $c(\mathbf{r'})$ is given, the phase decomposition process is evaluated by numerical calculation on the basis of the difference method. For 2-D simulation the difference matrix in eq.(2) is given by the following equations.

$$L_2(\mathbf{r} - \mathbf{r'}) = \begin{bmatrix} 0 & 1 & 0 \\ 1 & -4 & 1 \\ 0 & 1 & 0 \end{bmatrix} L(d_1) \qquad (35)$$

$$L_1(\mathbf{r} - \mathbf{r'}) = \begin{bmatrix} 0 & 0 & 0 \\ -1 & 0 & 1 \\ 0 & 0 & 0 \end{bmatrix} L_1(2d_1) + \begin{bmatrix} 0 & 1 & 0 \\ 0 & 0 & 0 \\ 0 & -1 & 0 \end{bmatrix} L_1(2d_1) \qquad (36)$$

Calculation Results For Fe-Mo Binary Systems

The computer simulations of phase decomposition of the Fe-Mo binary alloys are represented. **Figure 1(a)** presents the phase diagram of the Fe-Mo binary alloy system[27], where the dotted line and the chain line are the metastable coherent binodal and spinodal lines, respectively. The lines are biasymmetric against composition, because of proportional increment of elastic constants with Mo content. The chemical free energy of α-phase is given by eq. (37)[27].

$$G_c = \left[\sum_j \Omega_j(T)c^j\right]c^2 + RT[c\ln(c) + (1-c)\ln(1-c)]$$

(37)

$$\Omega_0(T) = -36490.7 \; (J/mol),$$
$$\Omega_j(T) = 0 \;, j \geq 1$$

The numerical values used in the calculation are summarized in **Table I**. The lattice mismatch of the Fe-Mo system, η is fairly large[27]. It should be noted that the elastic anisotropy factor A is greater than 1 for Fe, while $A<1$ for Mo. Therefore, in the Fe-Mo alloy system the elastically soft direction changes continuously with Mo-content from the <100> direction to <111> direction, as shown in **Fig.1(b)**. The elastically isotropic solid solution is expected to be at about Fe-60at.%Mo alloy. The computer simulations of phase decomposition are performed at solid marks inside the spinodal line. The initial composition fluctuations before ageing are given by the random numbers of the computer, whose total amplitude is approximately ±1%.

The time-development of microstructure calculated for the Fe-20at.%Mo alloy is presented in **Figure 2**, where the black parts indicate the Mo rich region. The phase decomposition is clearly recognized to progress with ageing time. Since Fe-Mo alloy system has a large lattice mismatch (η =0.083), Mo-rich zones along <100> directions, i.e. the <100> modulated structure, are produced during phase decomposition. During coarsening, competitive growth of particles is taking place as indicated by an arrow in (d), so that two lines of precipitates are combined to one line. A TEM microphotograph of Fe-20at.%Mo alloy aged at just the same condition as the computation has shown the <100> modulated structure whose wavelength is about 6 nm[27]. The interparticle spacing of calculated microstructure, about 6 nm, coincides quantitatively with that experimentally obtained.

Figure 3 shows the calculated microstructures of Fe-30at.%Mo alloy, where the periodic modulations along <100> direction are also observed. With progress of ageing the precipitates are competitively coarsened by the mechanism of dislocation climbing, so that the precipitates become more cuboid in shape and more periodic in alignment. The size distribution of precipitates become more uniform as is obviously recognized in the top of photograph of Fig.3(f). Such behaviour of precipitate coarsening originates from the elastic interaction energy between the precipitates. It has been known that the uniformity in particle size is theoretically explained well by the shape bifurcation theory of the particle coarsening proposed by us[28].

Figure 4 represents the micrographs of Fe-40at.%Mo alloy, which is approximately in the centre of miscibility gap. Since the volume fraction of precipitates becomes higher, the encounter between precipitates take place so that rod-shaped particles are temporarily produced in many places, as seen in Fig.4(c). However, on further ageing particle-splitting can be observed as indicated by arrows in Fig.4(d), which means that uniformity of particle size caused by the elastic interaction energy becomes energetically more dominant than the usual particle coarsening controlled by the interfacial energy[28]. The photographs from Figs.(d) to (f) clearly show that the microstructure is self-regulated by the elastic interaction energy so as to produce a periodic microstructure with uniform particle size.

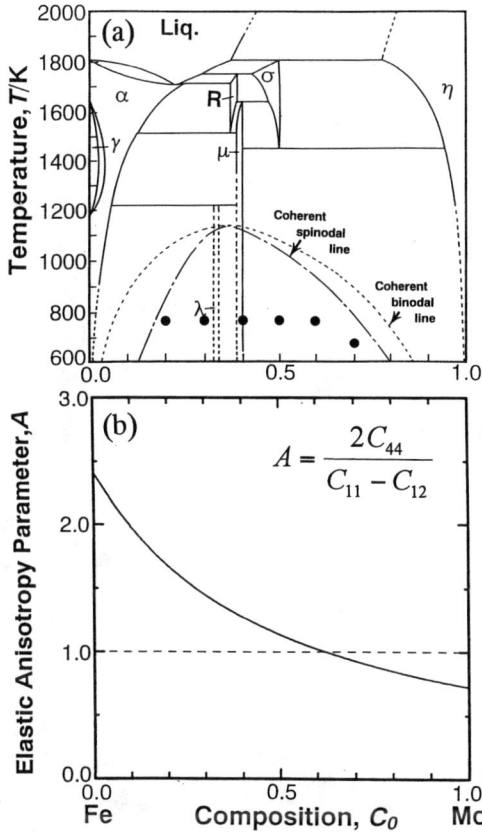

Fig.1 (a);The equilibrium phase diagram of the Fe-Mo system. A solid circle shows the chemical composition for the computer simulation. The broken and chained lines are metastable chemical binodal and spinodal lines of the α-phase, respectively. Fig.(b) shows the change in the elastic anisotropic factor A with alloy composition.

Table I The numerical values used for the calculation

Interaction parameter, Ω_0/J·mol^{-1}	-36490.7
Temperature, T/K	773, 673
Alloy composition, c_0 (Mo content)	0.2 ~ 0.7
Elastic stiffness /10^4MN·m C_{11}^{Fe}, C_{11}^{Mo} C_{12}^{Fe}, C_{12}^{Mo} C_{44}^{Fe}, C_{44}^{Mo}	23.3, 46.3 13.5, 16.1 11.8, 10.9
Lattice mismatch, η	0.083
Calculation area, L / 10^{-9} m	60
Interaction distance, d_1/10^{-10} m	2.86
Number of Fourier wave, N	128 × 128

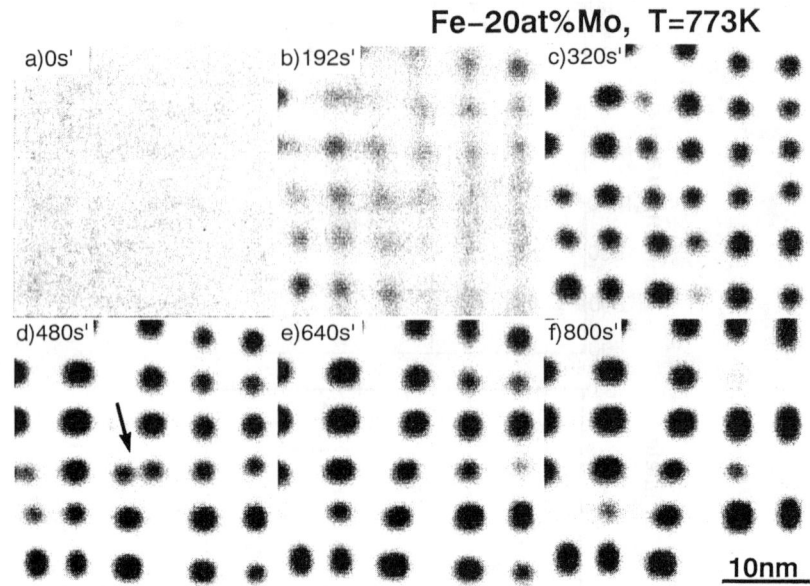

Fig.2 The time development of the phase decomposition calculated for the Fe- 20at.% Mo aged at 773 K.

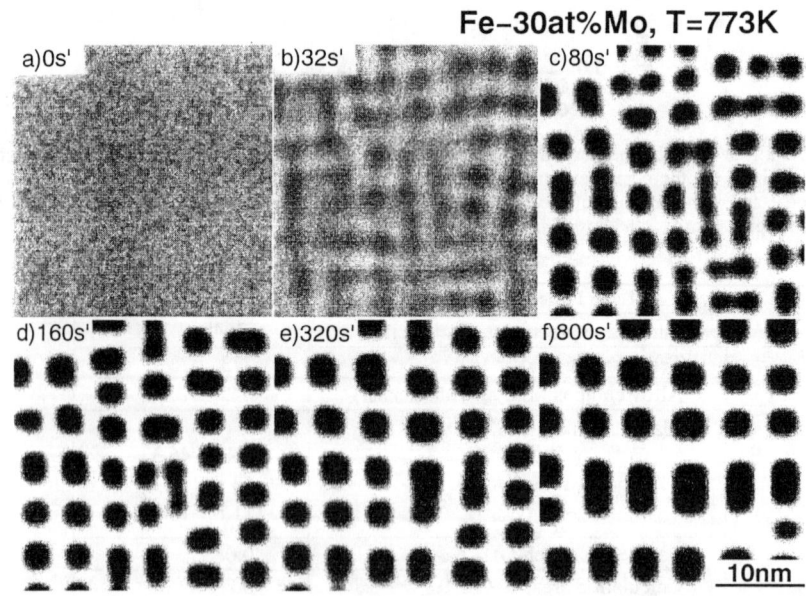

Fig.3 The time development of the phase decomposition calculated for the Fe- 30at.% Mo aged at 773 K.

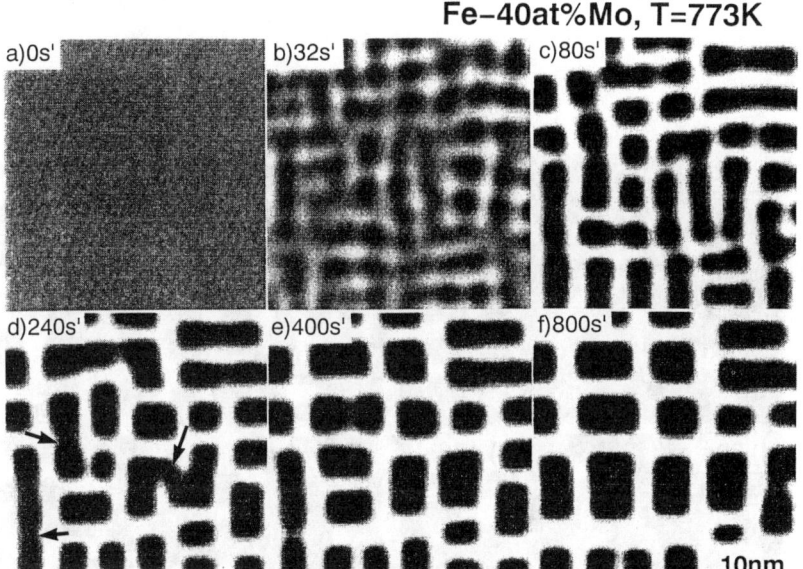

Fig.4 The time development of the phase decomposition calculated for the Fe- 40at.% Mo aged at 773 K.

Fig.5 shows the microstructural development of Fe- 50at.%Mo, - 60at.% and - 70at.%Mo alloys with ageing time. In the Fe- 50at.%Mo alloy the Mo- rich phase exceeds 50% in volume fraction(see Fig.1). Therefore, white Fe- rich precipitates are formed as the precipitate phase at the early stage of phase decomposition (see Fig.5(a)), but become connected with each other so as to form thin films wrapping the Mo- rich zones; the elastically hard Mo- rich phase is covered with the elastically soft Fe- rich phase(see Figs.(b)(c)). Such a simulation result is sure to be caused by introducing composition dependency of elastic constants into the calculations. The identical microstructures have experimentally been recognized in many real alloy systems such as Ni- base[28], Fe-base[29] alloys.

The microstructural changes with ageing for Fe- 60at.%Mo alloy are represented in Fig.5(d), (e) and (f), where a complicated microstructure can be seen. Since the supersaturated solid solution of this alloy before ageing is elastically isotropic as shown in Fig.1, the phase decomposition of Fe- 60at.%Mo alloy starts to produce the Fe- rich isotropic precipitates at the early stage of ageing. However, the elastic anisotropic factor A of Fe- rich phase becomes greater than 1 with progress of decomposition, while A becomes less than 1 for Mo- rich phase, i.e. the elastically soft direction changes continuously with progress of phase decomposition, which may induce the very complicated microstructure.

Time- development phase decomposition of Fe- 70at.%Mo alloy is represented in Fig.5(g), (h) and (i). In the present alloy the anisotropic factor A of solid solution is less than 1, so that the Fe- rich particles(white parts) nucleates along <111> direction (<11> direction in 2- D), and the shape of particle is cuboid along the <111> direction.

Figure 6 shows a comparison of microstructures calculated on the basis of (A) Khachaturyan diffusion equation and (B)that by the present discrete type diffusion equation. The right hand side

(C) in Fig.6 is the experimentally obtained TEM microstructures of Ni- Ni_4Mo[28] and Ni- Ni_3(Al,Ti)[30] alloys having large eigenstrains as same as that of Fe- Mo alloy. The cuboid precipitates are isolated and aligned along the <100> direction. Such a morphology of microstructure is perfectly reproduced in the present calculation, which comes mainly from the precise evaluation of the elastic potential and introducing the composition dependence of elastic constants. .

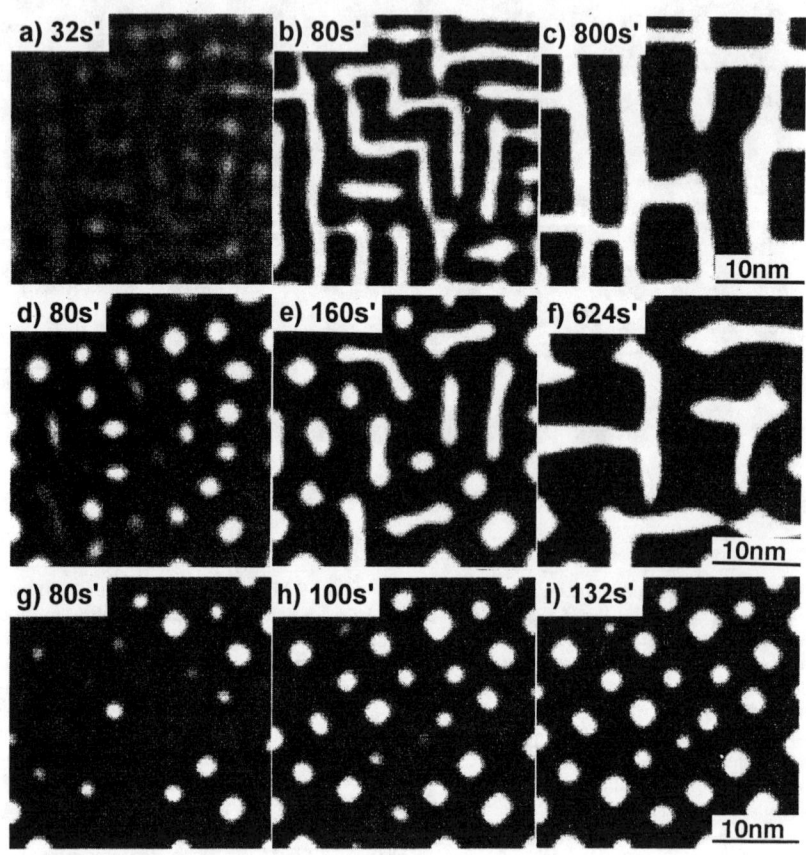

Fig.5 The time development of the phase decomposition calculated for (a) through (c) ; Fe- 50at.% Mo at 773 K, (d) through (f) ; Fe- 60at.% Mo at 773 K, and (g) through (i) ; Fe- 70at.% Mo aged at 673 K.

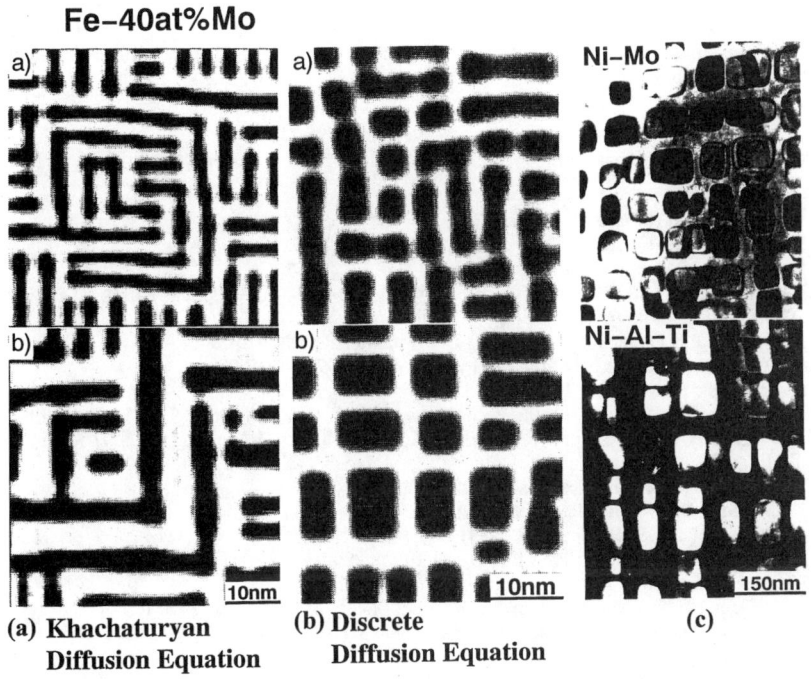

Fig.6 A comparison of microstructures calculated on the basis of Khachaturyan diffusion equation(A) and that by the present discrete type diffusion equation(B). (C) is the experimentally obtained TEM microstreuctures of Ni-Ni $_4$Mo and Ni-Ni $_3$(Al,Ti) alloys having large eigenstrains as same as that of Fe-Mo alloy.

Finally, we would like to emphasize that the thermodynamic evaluations should be supported by the experimental facts. The phenomenological equation such as the diffusion equation produces the calculation results even for unrealistic conditions, and can not judge its propriety on the application limit. The propriety of application limit can only be judged from the quantitative comparison with the experimental data of the real alloy system. It is important for us to obtain the thermodynamic calculations performed on the basis of not the virtual system but the real system.

The discrete type diffusion equation proposed here has a capability of predicting other phase transformations such as order/disorder transformations, phase decompositions in ordering alloys and the formation process of tweed structures[18], by using the degree of order s and the tetragonality of lattice η instead of composition c. The development of our calculation method is expected to open a new way of connecting the theoretical diffusion equation with the technological metallography in the real alloy systems.

Conclusions

A nonlinear discrete type diffusion equation including the solute atom occupation probability at a discrete atom site, is newly proposed. The composition dependencies of the atomic interchange energy and of the elasticity are also taken into account so as to be applicable for the real alloy system. The two dimensional computer calculations are performed for the phase decompositions of the Fe- Mo binary alloy system by using the thermodynamic data related to the equilibrium phase diagrams. The microstructures theoretically obtained are precisely coincident with experimental results of the real alloys.

References

[1] J.W.Cahn and J.E.Hilliard : J. Chem. Phys., **28**(1958), 258-67.
[2] J.W.Cahn and J.E.Hilliard : J. Chem. Phys., **31**(1959), 688-99.
[3] J.W.Cahn : Acta Metall., **10**(1962), 179-83.
[4] J.E.Hilliard, in *"Phase Transformation"*, Ed. by H.I.Aaronson, ASM, Metals Park, Ohio, (1970),497-560.
[5] L.A.Swanger,P.K.Gupta and A.R.Cooper.Jr. : Acta Metall., **18**(1970), 9-14.
[6] J.S.Langer, M.Baron and H.D.Miller : Phys.Rev.A, **11**(1975), 1417-29.
[7] T.Tsakalakos : Scripta Met., **15**(1981), 255-58.
[8] T.Miyazaki,T.Kozakai,S.Mizuno and M.Doi : Trans. JIM, **24**(1983), 246-54, 799-808.
[9] T.Miyazaki,A.Takeuchi,T.Koyama and T.Kozakai : Mater. JIM., **32**(1991), 915-20.
[10] T.Miyazaki,A.Takeuchi and T.Koyama : J. Mat. Sci., **27**(1992), 2444-448.
[11] Y.Oono and S.Puri : Phys.Rev.A, **38**(1988), 434-53.
[12] S.Puri and Y.Oono : Phys.Rev.A, **38**(1988), 1542-65.
[13] A.Onuki : J. Phys. Soci. Japan, **58**(1989), 3065-8.
[14] H.Nishimori and A.Onuki : Phys.Rev.B, **42**(1990), 980-83.
[15] S.Nambu and A.Sato : J. Am. Ceram. Soc., **76**(1993), 1978.
[16] L.- Q.Chen and A.G.Khachaturyan : Acta Metall.Mater., **39**(1991), 2533-2551.
[17] Y.Wang, L.- Q.Chen and A.G.Khachaturyan : Acta Metall.Mater., **41**(1993), 279-296.
[18] Y.Wang, L.- Q.Chen and A.G.Khachaturyan : Proc. Inter. Conf. on *"Solid → Solid Phase Trans. in Inorganic Mater."* ,(1994), 245-265.
[19] T,Koyama and T.Miyazaki : Proc. Inter. Conf. on *"Solid → Solid Phase Trans. in Inorganic Mater."* ,(1994), 365-70.
[20] T.Koyama, T.Miyazaki and A.E.Mebed : Metal. and Mater. Trans. A ,**26**(1995), 2617-23.
[21] R.Kobayashi : Physica D, **63**(1993), 410-423.
[22] W.J.Boettinger, A.A.Wheeler, B.T.Murray and G.B.McFadden : Mater.Sci. & Eng., A178(1994), 217-23.
[23] J.A.Warren and W.J.Boettinger : Acta Metall.Mater., **43**(1995), 689-703.
[24] A.P.Miodownik: "*Statics and Dynamics of Alloy Phase Transformations.* ", Plenum Press, New York, NY, (1994), 45-79.
[25] A.Khachaturyan:"*Theory of Structural Transformations in Solids*.", Wiley, New York, NY, (1983), 198-277.
[26] M.Fukaya,T.Miyazaki and T.Kozakai : J.Mater.Sci., **26**(1991), 5420-26.
[27] T.Miyazaki,S.Takagishi,H.Mori and T.Kozakai : Acta Metal., **28**(1980), 1143-53.
[28] T.Miyazaki and M.Doi : Mater.Sci. & Eng.,**A110**(1989),175-85.
[29] G.Ghosh, G.B.Olson and M.E.Fine : Proc. Inter. Conf. on *"Solid → Solid Phase Trans. in Inorganic Mater."* ,(1994), 611-6.
[30] T.Miyazaki, T.Koyama and M.Doi : Acta Metal. Mater., **42**(1994), 3417-24.

Equilibrium Particle Morphologies in Elastically Stressed Solids

M.E. Thompson and P.W. Voorhees
Department of Materials Science and Engineering
Northwestern University
Evanston, IL 60208
(June 25, 1996)

We examine the equilibrium morphologies of precipitates with a purely dilatational misfit in an elastically anisotropic medium with cubic symmetry. The case of plane-strain elasticity is reviewed and results from a fully three-dimensional calculation are presented. We find that particles with a dilatational misfit are nearly spherical at small sizes, take on cuboidal shapes with four-fold rotational symmetry at intermediate sizes and then, in the case of plane-strain, undergo a supercritical symmetry-breaking bifurcation to two-fold symmetric shapes aligned along the elastically soft directions of the crystal. The effects of the magnitude of the anisotropy ratio on the location of the bifurcation point is examined. We also find excellent agreement between the plane-strain equilibrium shapes and the mid-point cross-section of the corresponding three-dimensional equilibrium shape.

I. INTRODUCTION

The majority of theoretical work on the effects of elastic stress on the shape of a misfitting particle has been performed assuming the equilibrium shape will be that which minimizes the elastic energy within a fixed class of geometric shapes, usually ellipsoids, cubes or plates [1–6]. These types of calculations have been used to successfully predict certain properties such as the habit planes of plate-like precipitates and the existence of elastic stress induced particle shape bifurcations. They are unable, however, to predict equilibrium shapes as, once the interfacial energy is added to the problem, the chemical or diffusion potential is nonuniform in a system with such particles [7]. Thus, searching for an energy minimum within this restricted class of geometric shapes is, in essence, finding a form of constrained equilibrium. As the effects of these constraints are unclear, it is necessary to consider the equilibrium shape problem in the broader context of arbitrarily shaped particles.

Previous work on the evolution of arbitrarily shaped particles has been performed using a number of different techniques. The majority of work has concentrated on the dynamics of the single or multi-particle evolution process [8–11]. Recently, several studies have concentrated on the effects of elastic stress on the shape of an isolated, misfitting precipitate. Thompson et al. [12] used a continuum, sharp interface model to investigate equilibrium particle shapes and particle shape bifurcations in an elastically anisotropic, homogeneous system with either a dilational or tetragonal misfit strain. A discrete atom method has recently been developed by Lee [13–15] which has been used to investigate the evolution of an isolated particle in an elastically anisotropic, inhomogeneous system in the presence of an applied or edge dislocation stress field. These studies have not only shown that elastic stress can alter the equilibrium particle shape from that which minimizes the surface energy, but also the importance of not constraining the particle shape to a certain class of geometric shapes.

A major shortcoming of the above studies is that each is restricted to two-dimensional calculations. Several studies have been performed in three dimensions which examine the effects of elastic stress on coarsening kinetics [16,17] or particle shape transitions [18,19]. These studies, however, have been limited to particles which are either spherical or ellipsoidal in nature. To the best of our knowledge, there has been no previous work which has been successful in describing three-dimensional equilibrium particle morphologies in the presence of elastic stress which makes no assumptions regarding the geometric class of the particle.

To further elucidate the effects of elastic stress on particle morphology, we have determined the equilibrium shape of three-dimensional, isolated particles with a dilatational misfit in a system with the cubic anisotropic elastic constants of nickel. We have searched for particle shapes which guarantee that the diffusion potential in the system is uniform at all points, and thus do not constrain the particles to be within any class of simple geometric shapes. The results shown here are the initial results of our three-dimensional analysis and a more detailed accounting will be given in a future publication.

II. THERMODYNAMICS AND ELASTICITY

For small deviations about a *natural state*, the equilibrium state wherein the interface between the particle and matrix is flat and both phases are stress-free, Gurtin and Voorhees [20] have shown that it is reasonable to approximate the grand canonical free energy density of both matrix, α, and particle, β, as a linear function of the deviation of the diffusion potential from that state and a quadratic function of the strain. Thus, the elastic energy density, $W(\mathsf{E} - \mathsf{E}_o)$, is given by,

$$W(\mathsf{E} - \mathsf{E}_o) = \frac{1}{2}(\mathsf{E} - \mathsf{E}_o) \cdot \mathsf{C}(\mathsf{E} - \mathsf{E}_o) \tag{2.1}$$

where E is the total strain tensor, C is the elastic constant tensor, and E_o is the strain tensor of each phase relative to their natural states. Choosing stress-free α as the reference state for strain, E_o is given by

$$\mathsf{E}_o(\mathbf{x}) = \begin{cases} 0 & \mathbf{x} \text{ in } \alpha \\ \mathsf{E}^T & \mathbf{x} \text{ in } \beta \end{cases} \tag{2.2}$$

where E^T is the stress-free transformation strain or eigenstrain, and $\mathbf{x}$ is a position vector.

Under this choice for the grand canonical free energy, the lattice parameter of each phase is not a function of composition, and by neglecting the interfacial excess stress, the equilibrium shape of a particle is given by an extremum in the total energy of the system, $\mathcal{E}$, along with the requirements of a constant diffusion potential, M, constant particle volume, V^β, and mechanical equilibrium. The total energy of the system may be expressed as,

$$\mathcal{E} = \int_\Omega W(\mathsf{E} - \mathsf{E}_o)\, dV + \int_S \sigma\, dA \tag{2.3}$$

where Ω is the two-phase system which is composed of an infinite matrix of α and a single precipitate of β, S is the surface of the particle, and $\sigma(\mathbf{n})$ is the isotropic interfacial energy.

The value of the constant diffusion potential present in a system which extremizes the energy $\mathcal{E}$ is given by the value of M at the particle-matrix interface [20],

$$\Lambda M = [\![W(\mathsf{E} - \mathsf{E}_o)]\!] - [\![\mathsf{T} \cdot \mathsf{E}]\!] + \sigma K \tag{2.4}$$

where $\Lambda = \rho^\beta - \rho^\alpha$, ρ is the mass density in the designated phase, $[\![\phi]\!] = \phi^\beta - \phi^\alpha$ for a quantity ϕ at the interface, K is the mean curvature of the interface, and T is the stress as defined by linear elasticity theory. Since all of the quantities appearing in Eq. (2.4) are evaluated at the interface, then Eq. (2.4) with M chosen to be a constant can be used to derive an equation for the equilibrium shape.

In order to determine shapes which produce a constant value of the diffusion potential, as given by Eq. (2.4), we must be able to determine the elastic fields surrounding an arbitrarily shaped misfitting particle in an elastically anisotropic medium. Using the method developed by Mura [21], the entire elastic problem may be mapped onto the interface of the particle such that the elastic field at any point in the system can be represented as an integral only over the interface,

$$U_{j,k}(\mathbf{x}) = C_{ilmn} E_{mn}^T \int_S G_{ij,k}(\mathbf{x} - \mathbf{x}') n_l'\, dS' \tag{2.5}$$

where dS' denotes a surface element of the interface, G_{ij} is the elastic Green's function tensor, U_j are the components of the displacement vector measured with respect to the reference state of stress-free α, n_l is the normal to the interface pointing from β to α, and $\mathbf{x}$ and $\mathbf{x}'(S')$ locate field and interfacial points, respectively. Summation over repeated indices from 1 to 3 is assumed and the commas denote partial differentiations with respect to the noted index. The stress and strain fields in the particle and matrix may be determined from Eq. (2.5) using linear elasticity theory. In order to determine the type of solution found by this approach, an expression for the total energy of the system must be determined which does not require integrations over the entire three-dimensional space. The objective is to reduce the volume integral of Eq. (2.3) to an integral only over the interface of the particle. This can be accomplished by a modification to the technique developed by Eshelby [22] in which the total elastic energy, $\mathcal{E}_e$, of a system containing an isolated particle in an infinite matrix can be expressed as,

$$\mathcal{E}_e = \int_\Omega W\left(\mathbf{E} - \mathbf{E}_o\right) dV = \frac{1}{2} \int_S \mathbf{U}^T \cdot \mathbf{T} \mathbf{n} dS' \tag{2.6}$$

where $\mathbf{U}^T = \mathbf{U}^\beta - \mathbf{U}^\alpha$ is defined as the transformation displacement corresponding to the misfit or transformation strain. Using this expression, it is possible to determine accurately the total energy of the system and thus distinguish between energy minimizing and energy maximizing particle shapes.

As we are interested in the effects of elastic stress on the equilibrium shape of a particle, it is advantageous to choose energy and length-scales which allow the equilibrium shape of a body to be computed easily in the absence of stress. We have thus scaled the energy of the system by a characteristic interfacial energy, σl^2, where l is the characteristic size of a particle defined by the relationship $v^\beta = V^\beta / l^3$, V^β is the dimensional volume of the particle, and v^β is the dimensionless volume of the particle. Strains are scaled by ϵ and stresses are scaled by $C_{44}\epsilon$, where C_{44} is one of the three independent elastic constants in a cubic system, and ϵ is a component of the misfit strain tensor.

III. TWO-DIMENSIONAL RESULTS

The two-dimensional results discussed here are based upon anisotropic, plane-strain elasticity for a system with the elastic anisotropy ratio of Ni, $A_r = 2.5$. The precipitate has a purely dilatational misfit with the matrix. The details of the numerical method employed for these calculations are discussed in an earlier paper [12]. In all of the two-dimensional results to follow the sides of the boxes used to display the equilibrium shapes will always be oriented such that they are along the [100] and [010] directions. The results are presented in terms of a parameter, L, which is defined as $L = \epsilon^2 C_{44} l / \sigma$. L may be interpreted as the ratio of elastic to interfacial energy or, as $L \sim l$, it may also be viewed as a measure of the size of a precipitate.

The equilibrium shape at $L = 0$ is a circle as this value of L implies the absence of elastic stress in the system. If we use this shape as an initial guess for a nonzero value of L we can access the family of four-fold symmetric shapes shown in Fig. (1). Evident is the progression of particle shapes as L increases from the interfacial energy minimizing circle to square-like shapes which reflect the four-fold anisotropy of the elastic constants. The particles have regions of high curvature along the elastically hard $<110>$ directions and regions of low curvature along the $<100>$ directions, where $<>$ denotes all the symmetry-equivalent crystallographic directions in the (001) plane. For $L \gtrsim 25$ the curvature of the interface is negative near the centers of the particle sides.

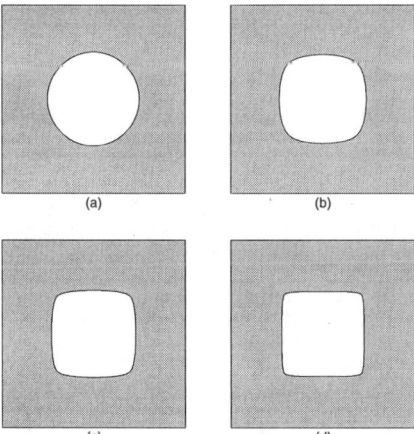

FIG. 1. Energy extremizing particle shapes for a misfitting particle with a purely dilatational misfit in a Ni matrix for various values of L: (a) 0, (b) 4, (c) 10, (d) 26.

By implementing bifurcation-tracking methods, we find a second solution family consisting of two-fold symmetric shapes at a critical value of $L = 5.6$. We can classify each family of solutions by its value of a_2^R, one component of the vector of Fourier coefficients which define a particle shape [12]. By plotting a_2^R as a function of L for both solution families, we obtain the bifurcation diagram shown in Fig. (2). As seen in the diagram, the long axis of the two-fold symmetric particles are aligned along the elastically soft [100] and [010] directions. These two-fold particle shapes no longer conform to the four-fold anisotropy of the elastic constants. While they possess the symmetry of a plate, they are definitely not plates as the curvature of the sides of the particles is positive and the corners are rounded. This reflects the influence of the interfacial energy on the equilibrium shape. As L increases, however, the equilibrium shapes appear to be approaching the elastic energy minimizing shape of an infinitely long, infinitesimally thin plate oriented along one of the elastically soft $<100>$ directions [23,24]. By computing the total energy of the energy extremizing four-fold and two-fold symmetric shapes at a given value of L, we find that four-fold shapes are energy minima for $L < 5.6$ and energy maxima for $L > 5.6$. In the bifurcation diagram, the dashed line denotes energy maximizing solutions and the solid lines indicate energy minimizing solutions.

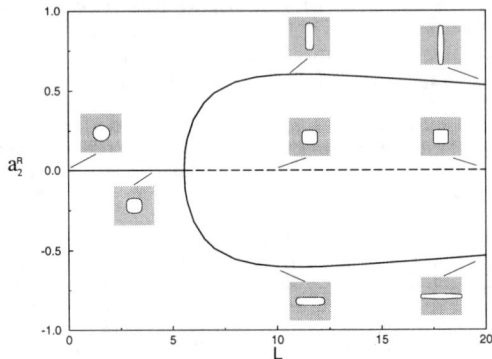

FIG. 2. Bifurcation diagram showing the supercritical bifurcation from four-fold symmetric shapes to two-fold shapes at $L = 5.6$.

A. Effects of the Anisotropy Ratio

In order to fully represent a given system using the above theory, we must specify the value of L and the three elastic constants for a cubic system denoted by C_{11}, C_{12}, and C_{44}. Appropriate scaling of the system allows us to normalize, for instance, C_{44} and thus express C_{11} and C_{12} in terms of dimensionless quantities $c_{11} = C_{11}/C_{44}$ and $c_{12} = C_{12}/C_{44}$ respectively. This reduces the parameter set of our system to three dimensionless quantites: L, c_{11}, and c_{12}. In the above results, we have chosen to express the elastic state of the system in terms of the anisotropy ratio, A_r, which in dimensionless form is defined as $A_r = 2/(c_{11} - c_{12})$. Thus, specifying A_r determines only one of the two dimensionless elastic constants. In the calculations shown in Fig. (1) and (2) we chose $A_r = 2.50$ and the following values of the dimensionless elastic constants: $c_{11} = 1.98$, $c_{12} = 1.18$, and $c_{44} = 1.00$.

To examine the dependence of the bifurcation point on the values of the elastic constants, we have selected $c_{12} = 1.30$ as an average value representative of many $Ni - X$ and $Ni_3 - X$ systems [25]. The value of c_{12} was specified as this quantity varied much less from system to system than c_{11}. Fig. (3) shows the dependence of the critical value of L at which a shape bifurcation occurs, L_c, as a function of A_r. As expected, the bifurcation point moves to increasingly higher values of L as the system becomes more isotropic, i.e. $L_c \to \infty$ as $A_r \to 1$. Larger values of A_r produce the opposite effect with $L_c \to 0$ as $A_r \to \infty$. Various values of c_{12} could have also been easily added to the figure to produce a surface representing all possible combinations of the three independent elastic constants. However, changes in

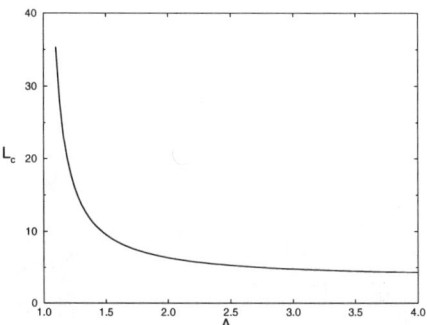

FIG. 3. L_c, the value of L at which a bifurcation occurs, versus the anisotropy ratio, A_r, for $c_{12} = 1.30$.

c_{12} on the order of 10% where found to produce changes in the value of L at the bifurcation point on the order of only 1%.

IV. THREE-DIMENSIONAL RESULTS

The development of a fully three-dimensional model to determine equilibrium morphologies required several modifications of our two-dimensional approach. The Green's function, for instance, needed in the elasticity calculation can not be cast in terms of an analytic equation for a given anisotropy factor as was used in the plane-strain model [26]. Instead, the integral representation as derived by Barnett [27] was employed and discretized such that a bilinear interpolation of the angular dependence of the Green's function yielded errors several orders of magnitude below the allowable error our elasticity calculations. The solution technique was also modified, as the Newton's method used in our two-dimensional model was found to be computationally unacceptable due to the large number of interfacial points needed to represent the interface of a three-dimensional particle. To overcome this difficulty, an equation for the diffusion potential of a kinetically-controlled interface [20] was adopted which allows the diffusion potential to be a function of the normal velocity of the interface,

$$\Lambda M = [\![W(E - E_o)]\!] - [\![T \cdot E]\!] + \sigma K - bV_n \qquad (4.1)$$

where V_n is the normal velocity and b is a kinetic coefficient. Using this expression to determine V_n, the shape of the particle is allowed to evolve toward equilibrium in a "quasi-kinetic" manner by stepping explicitly in time. This method ensures that as the shape evolves the total system energy is reduced with each time-step. The total reduction in computational time is on the order of 10^3 and the method is globally convergent. The details of this new computational method will be published in a separate paper [28]. The three-dimensional results presented here were calculated using the same system parameters as used in the two-dimensional results given in the previous section.

The spherical equilibrium shape for $L = 0$ is shown in Fig. (4). As L is increased and the elastic contribution to the total system energy is likewise increased, the particle begins to take on a cuboidal geometry with flat sides or regions of low curvature along the elastically soft $<100>$ directions and develops corners or regions of high curvature along the elastically hard $<111>$ directions. This is illustrated in Fig. (5) which shows the particle viewed from the [101] direction. The particle also develops edges along the $<110>$ directions which correspond to elastically hard directions in the $\{100\}$ plane. Viewing the particle along one of the crystal axis helps to detail the cuboidal nature of the particle, as shown in Fig. (6). The non-spherical nature of the particle shape can also be seen in a plot of the mean curvature of the interface. Fig. (7) shows the mean curvature, K, as a function of the polar angles θ and ϕ, where θ is the azimuthal angle and ϕ is the meridional angle. The cuboidal nature of the particle is reflected in the eight curvature "peaks" which correspond to the corners of the particle and the four symmetrically distributed "valleys" or regions of low curvature which represent the four

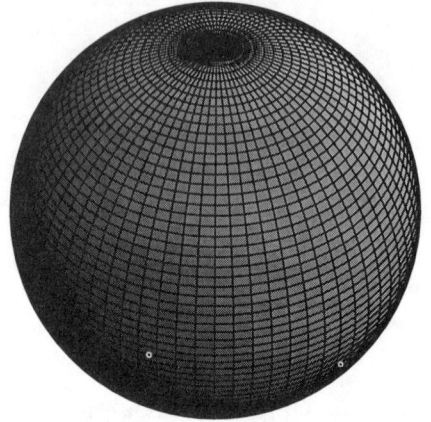

FIG. 4. Three-dimensional equilibrium shape for $L = 0$.

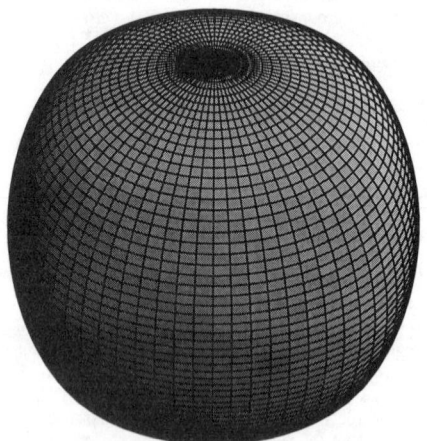

FIG. 5. Three-dimensional equilibrium shape for $L = 2$.

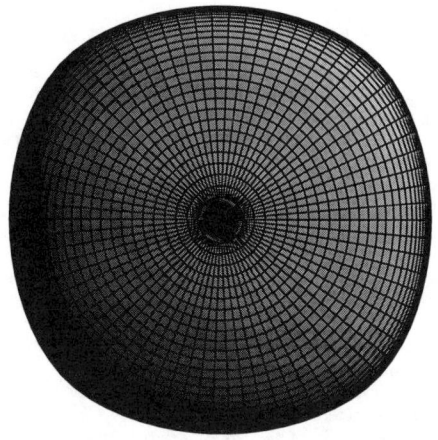

FIG. 6. $L = 2$ equilibrium shape viewed along the [001] direction.

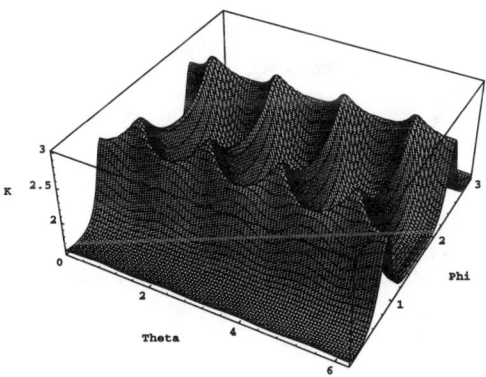

FIG. 7. Curvature of the $L = 2$ equilibrium shape as a function of the polar angles θ and ϕ.

sides of the particle. The top and bottom surfaces of the particle are denoted by the two low curvature regions located on either side of the high curvature peaks. It can be concluded from these figures that the equilibrium shape of a three-dimensional particle for $L = 2$ is certainly not spherical and posesses the geometric attributes of a cuboid with a smoothly varying curvature.

V. DISCUSSION

Using the results obtained from our two-dimensional calculations as a guide, we have begun to explore the problem of three-dimensional equilibrium particle morphologies in elastically stressed systems. We have found the existence of cuboidal equilibrium shapes in a model nickel system which corresponds to a similar solution family in two-dimensions.

A particular motivation for pursuing our work in three dimensions lies in the good agreement found between our two-dimensional particle shapes and those found in experimental studies. Fährmann et al. [29] defined a stereological curvature parameter which gives a measure of the square-like nature of an experimentally observed precipitate. They plotted the parameter as a function of the experimentally determined value of L and found surprisingly good agreement between experiment and the predictions of our theory. Considering that the equilibrium particle shapes were determined in two dimensions leads one to speculate as to why there should be such good agreement. A partial explanation can be found in Fig. (8). In this figure, we have taken the mid-point ($\phi = \pi/2$) cross-section of the three-dimensional equilibrium shape at $L = 2$ and compared that to the two-dimensional equilibrium shape at $L = 2$. As can be seen, the agreement is extremely good, with only slight deviations near the corners of each shape. As the mid-point cross-section of the three-dimensional body is the most probable projection which will be observed in a TEM micrograph, the comparison in Fig. (8) helps to explain the agreement found between our two-dimensional particle shapes and those found experimentally. A complete explanation of this phenomenon is still under consideration, but may be due to the symmetry found about the [001] axis for each cross-section. Further study of this relationship between the two and three-dimensional calculations is under way, particularly to determine if the agreement still exits at higher values of L.

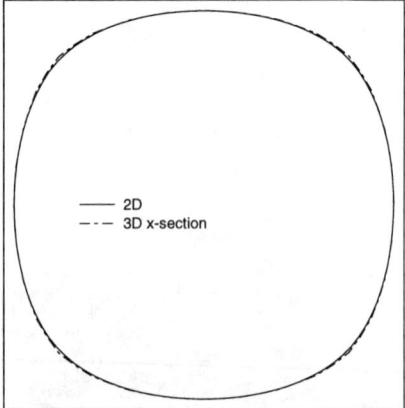

FIG. 8. Comparison of the two-dimensional equilibrium shape for $L = 2$ and the three-dimensional $L = 2$ cross-section at $\phi = \pi/2$.

VI. CONCLUSIONS

We examined the equilibrium morphologies of precipitates with a purely dilatational misfit in an elastically anisotropic medium with cubic symmetry in both a plane-strain and a fully three-dimensional model. We find that:

- the equilibrium particle morphology changes smoothly from a circle to a four-fold symmetric shape and at a critical value of L, in the case of plane-strain, bifurcates supercritically to two-fold symmetric shapes aligned along the elastically soft directions of the crystal.

- the values of the elastic constants can have a significant effect on the predicted particle size at which a shape bifurcation occurs. The experimental error involved in determining these constants should thus be considered when comparing the above theoretical predictions with experimental observations.

- there exists good agreement between the mid-point cross-section of the three-dimensional particle shapes and the corresponding two-dimensional particle morphology.

- there is also good agreement between the theoretically predicted three-dimensional particle shapes and the experimentally observed shapes for small values of L.

ACKNOWLEDGMENTS

We are grateful for the financial support of the National Science Foundation under grant DMR-9322687

---

[1] T. Miyazaki and M. Doi, Matls. Sci. and Engr. **a110**, 175 (1989).
[2] A. G. Khachaturyan, S. V. Semenovskaya, and J. W. Morris, Acta Metal. **36**, 1563 (1988).
[3] A. G. Khachaturyan, *Theory of Structural Phase Transformations in Solids* (John Wiley, New York, 1983).
[4] A. G. Khachaturyan and G. A. Shatalov, Sov. Phys. Solid State **11**, 118 (1969).
[5] I. M. Kaganova and A. L. Roitburd, Sov. Phys. JETP **67**, 1173 (1988).
[6] W. C. Johnson and P. W. Voorhees, J. Appl. Phys. **61**, 1610 (1987).
[7] M. A. Grinfeld, Phys. Earth and Planetary Interiors **50**, 99 (1988).
[8] A. Onuki, J. Phys. Soc. Japan **58**, 3069 (1989).
[9] Y. Wang, L. Q. Chen, and A. Khachaturyan, Acta Metal. et Mater. **41**, 279 (1993).
[10] P. W. Voorhees, G. B. McFadden, and W. C. Johnson, Acta Metall. **40**, 2979 (1992).
[11] H. J. Jou, P. H. Leo, and J. S. Lowengrub, J. Comp. Phys. (1995), in press.
[12] M. E. Thompson, C. S. Su, and P. W. Voorhees, Acta metall. mater. **42**, 2107 (1994).
[13] J. K. Lee, Scripta Metallurgica et Materialia **32**, 559 (1995).
[14] J. K. Lee, in *Micromechanics of Advanced Materials: A Symposium in Honor of Professor James Li's 70th Birthday*, edited by S. Chu et al. (The Minerals, Metals & Materials Society, Warrendale, Pennsylvania, 1995).
[15] J. K. Lee, Metall. Trans. A (1995), in press.
[16] W. C. Johnson, T. A. Abinandanan, and P. W. Voorhees, Acta Metall. **38**, 1349 (1990).
[17] W. Hort and W. C. Johnson, private communication.
[18] M. B. Berkenpas, W. C. Johnson, and D. E. Laughlin, J. Mater. Res. **1**, 635 (1986).
[19] W. C. Johnson, M. B. Berkenpas, and D. E. Laughlin, Acta Metall. **36**, 3149 (1988).
[20] M. E. Gurtin and P. W. Voorhees, Proc. Roy. Soc. A **440**, 323 (1993).
[21] T. Mura, *Micromechanics of Defects in Solids* (Martinus Nijhoff, the Hague, 1982).
[22] A. J. Ardell and R. B. Nicholson, Acta Metall. **14**, 1295 (1966).
[23] A. L. Roitburd, Sov. Phys. Dokaldy **16**, 305 (1971).
[24] A. G. Khachaturyan, Sov. Phys. Solid State **8**, 2163 (1967).
[25] A. J. Ardell, private communication.
[26] D. J. Bacon, D. M. Barnett, and R. O. Scattergood, Prog. Mater. Sci. **23**, 51 (1979).
[27] D. M. Barnett, Phys. stat. sol. (b) **49**, 741 (1972).
[28] M. E. Thompson and P. W. Voorhees, in preparation.
[29] M. Fährmann et al., Acta metall. mater. **43**, 1007 (1995).

ANISOTROPIC INTERFACE MOTION

Jean E. Taylor
Mathematics Department, Rutgers University, Piscataway NJ 08855, and
School of Mathematics, Institute for Advanced Study, Princeton, NJ 08540
taylor@math.rutgers.edu

Abstract

Mathematical models for various types of crystal surface motions have been made. They all reduce some energy which includes a given (usually crystalline) anisotropic surface free energy and perhaps a bulk free energy. The mechanisms which govern the motions are attachment/detachment kinetics, diffusion of atoms over surfaces, and perhaps diffusion within the bulk phases. At each instant the motions are gradient flows for the energy with respect to different inner products. Particularly important issues that the models handle are allowing for changes in topology (connectivity). The work involves proving new theorems, developing new methods of computation, and working on specific applications to problems in materials science.

This work was partially funded by the Advanced Research Projects Agency through the National Institute of Standards and Technology and by the National Science Foundation.

Introduction

In this paper I will give examples of some of the types of motions for which I have proved theorems and/or written computer programs. I will then define the surface free energy which is part of the driving force for the motions and give the derivation of the laws for the motions of these examples. The motions will then be cast as gradient flows, and a general variational approach to gradient flows will be given. Finally, the relationship to diffuse interface models will be stated.

The presentation here is brief. See my web site http://www.math.rutgers.edu/~taylor for a complete list of my publications and for .dvi or .ps files of recent preprints which describe these ideas in more detail.

Examples

Examples are given in this section; for explanation and references, see the next section.

In Figure 1 there is an example of a curve which appears to be moving by its mean curvature. The initial curve is the polyhedral curve with eight straight segments. The evolution is shown at times $2\Delta t$, $5\Delta t$, $10\Delta t$, $20\Delta t$, $40\Delta t$, $90\Delta t$, and thereafter at intervals of $50\Delta t$.

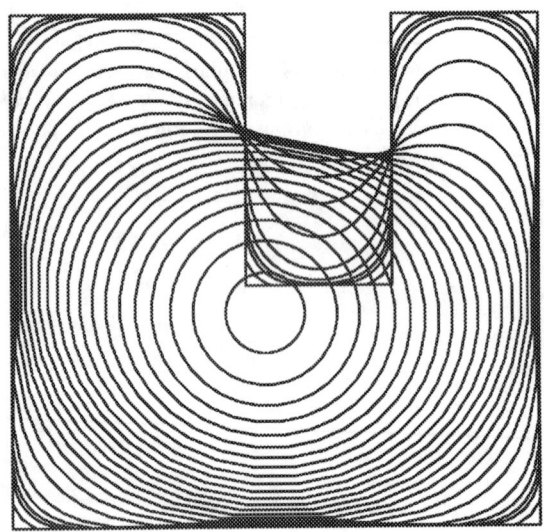

Figure 1: Approximate motion by curvature. Shown in [T3] video.

Figure 2 shows an evolution from the same initial curve, displayed at the same times, but where the evolving sets are clearly polyhedral rather than smooth. And in fact the curves in Figure 1 are also polyhedral; both are examples of motion by crystalline curvature, and illustrate how the motion can be used to approximate motion by curvature [G].

In both curves, the motion law for segment i, which is in a line with normal $\mathbf{n}_i$ and at

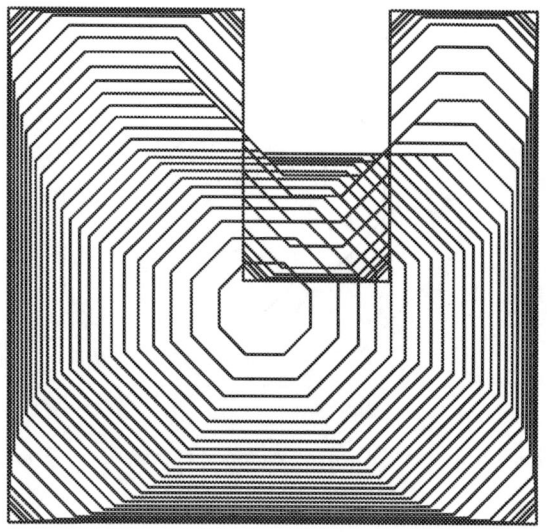

Figure 2: Motion by crystalline curvature. Shown in [T3] video.

distance $h_i(t)$ from the origin at time t and which has length $\ell_i(t)$, is

$$\frac{dh_i(t)}{dt} = -\frac{\sigma_i \Lambda_n}{\ell_i(t)};$$

here σ_i is $1, -1$, or 0 depending on the geometry of the curve around that line segment (1 if the curve is locally convex there, -1 if it is locally concave, and 0 otherwise) and Λ_n is a constant (different for the two motions). Note that usually segments keep the same normal direction $\mathbf{n}_i$ and change only the distance $h_i(t)$, but that at isolated times certain segments with the same normal come together (the intervening segment shrinks to zero length) and then those segments merge to one. The difference between the two motions is that the one of Figure 1 has more normal directions (those of a regular 32-gon) than that of Figure 2 (those of a regular octagon). Λ_n is in fact the length of a side of the regular n-gon ($n = 8$ or 32) whose sides are at distance 1 from the origin. The motion thus arises from integrating a system of coupled ordinary differential equations (since each length ℓ_i can be calculated by solving sets of linear equations in $\{h_j\}$ for the endpoints of the segment). See [T2] [T3].

Figure 3 shows a similar motion captured at one particular time, but the curves have fixed endpoints and the velocity law is now

$$\frac{dh_i(t)}{dt} = C - \frac{\sigma_i \Lambda_n}{\ell_i(t)}.$$

for some positive constant C. (Also $n = 6$.) Here it can be observed that topological changes have occurred where portions of curves of opposite orientation have collided and annihilated each other. Furthermore, when segments attached to a fixed point reach a critical length (though the motion of their adjacent segments), they pull off from the fixed points, so new segments are generated at those times. See [T2] [T4].

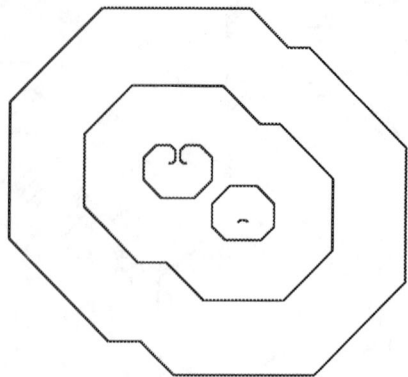

Figure 3: Snapshot of motion with fixed endpoints, extra driving force. Shown in [T4] video.

Figure 4: Final positions of motion with 2 triple junctions and 4 fixed points. Shown in [T3] video.

Figure 4 shows similar motions that involve triple junctions (here $n = 6$). New segments can be generated through the motion of the triple junctions. The precise result would be a varifold. In these computations, the program sees if it needs to insert small segments at each time step. In addition, because the insertions are found through a random variational procedure rather than explicitly, sometimes several time steps pass before the new step is put in, resulting in segments that are longer than they should be. While this clearly illustrates the nature of the program, it is a source of error; a deterministic means of moving the triple point needs to be coded. See [T2] [T3].

Figure 5 shows a snapshot of a curve undergoing motion by the system of differential equations
$$\frac{dh_i(t)}{dt} = M(\mathbf{n}_i)\left(C_i(t) - \frac{\sigma_i \Lambda(\mathbf{n}_i)}{\ell_i(t)}\right);$$
here $C_i(t)$ is the average of a temperature field over segment i which is calculated separately; it comes from an initial undercooling and then the release and diffusion of latent heat as the curve moves. $\Lambda(\mathbf{n}_i)$ now depends on the normal direction (it is the length of a side of a particular 4-fold-symmetric octagon), and $M(\mathbf{n}_i)$ is a corresponding mobility factor.

The initial curve here was a small octagon and the curve shown in the figure has over 15,000 sides; at various times, edges have been divided into two or more pieces through the insertion of small steps which then grow. The extreme sidebranching is the result of the Mullins and Sekerka-type instability amplified by numerical noise in the heat flow. See [RT]. Note the changes in topology, where sidebranches have collided.

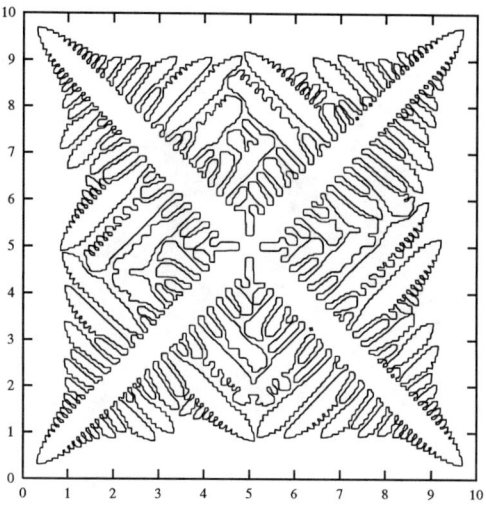

Figure 5: Motion using heat flow, average undercooling (from [RT])

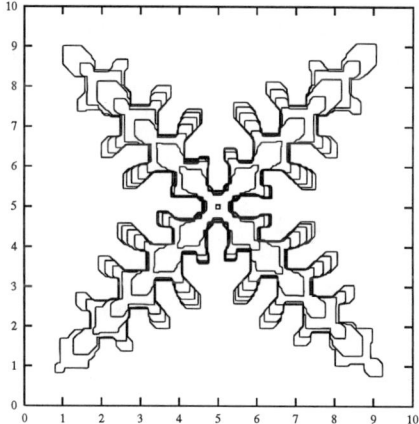

Figure 6: Motion using heat flow, maximum undercooling (from [RT])

Figure 6 uses the same system of equations as for Figure 5. It differs in that $C_i(t)$ is now not the *average* of the undercooling over the segment but the *maximum* undercooling. The conditions are also somewhat different; see [RT] for a computation which differs from Figure 6 only through the use of average undercooling.

139

Figure 7 shows a polyhedral surface; it is a snapshot of a moving surface obtained by solving a system of ordinary differential equations.

$$\frac{dh_i(t)}{dt} = M_i\left(C_i(t) - \frac{\sum_j \delta_{ij} f_{ij} \ell_{ij}(t)}{area_i(t)}\right).$$

Here $C_i(t)$ is obtained by evaluating a given function in space at a particular corner of a

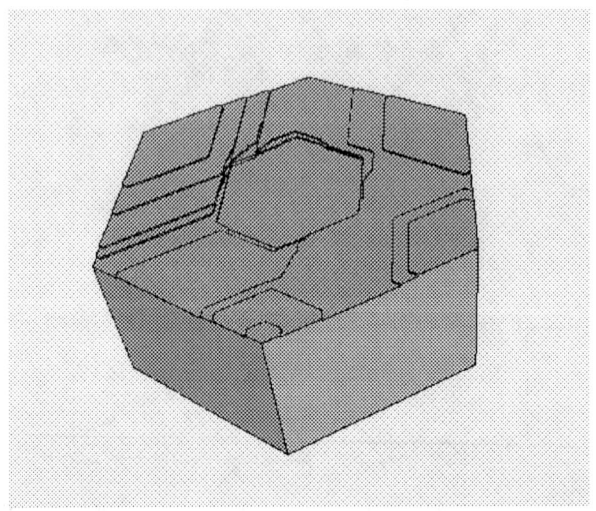

Figure 7: Example of crystalline surface motion. Shown in [T3] video.

segment, the sum $\sum$ is over all segments adjacent to segment i, $\ell_{ij}(t)$ is the length of the intersection of segments i and j, δ_{ij} is ± 1 depending on whether the intersection is locally convex or concave, f_{ij} is a numerical factor depending only on the normal directions of segments i and j, and M_i is a factor depending on the shape of the surface in a small neighborhood of segment i. In this example there has also been nucleation of new boxes at the outside corners at random times. See [T4] for more details.

In figure 8, an initial spiral moves by crystalline surface diffusion, in the process undergoing several topological changes. The equations for motion by surface diffusion are somewhat complicated; see [CRCT].

Figure 9 shows two different motions of curves, one by crystalline surface diffusion (the one that produces bulbs at the ends) and one by crystalline surface attachment limited kinetics (SALK). SALK is motion by the same set of differential equations as above

$$\frac{dh_i(t)}{dt} = C(t) - \frac{\sigma_i \Lambda_n}{\ell_i(t)}$$

(here with $n=16$), but with $C(t)$ chosen so that the net change of enclosed area is zero:

$$C(t) = \frac{\sum_i \sigma_i \Lambda(\mathbf{n}_i)}{\sum_i \ell_i(t)}.$$

Crystalline surface diffusion is more complicated to describe; see [CRCT].

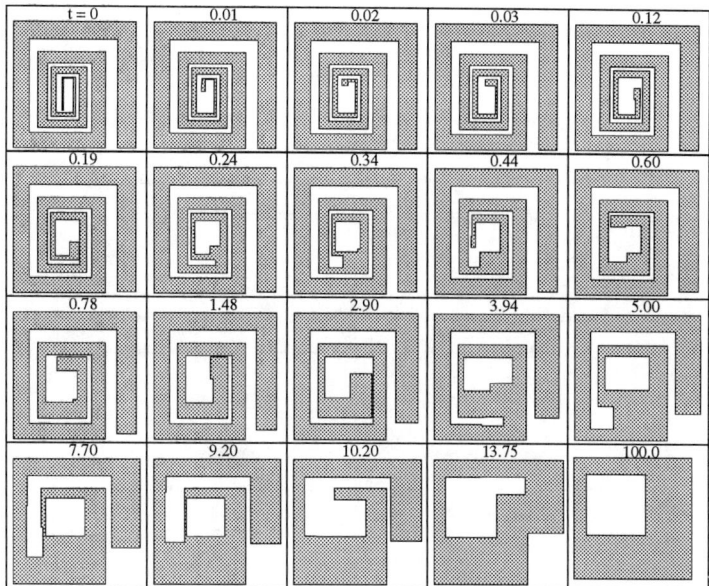

Figure 8: Successive views of an initial spiral moving by crystalline surface diffusion. From [CRCT].

Surface Free Enegy, the Wulff Shape, and Weighted Mean Curvature

Surface free energy per unit area is assumed to be a function γ from unit normal vectors in R^d (where d is the dimension of the space, usually 2 or 3) to the real numbers (usually positive). It is convenient to extend γ to all vectors, by defining $\gamma(r\mathbf{n}) = r\gamma(\mathbf{n})$ for all $r \geq 0$. If γ is then a convex function, (i.e., $\gamma(\mathbf{p}+\mathbf{q}) \leq \gamma(\mathbf{p}) + \gamma(\mathbf{q})$) and if γ is even (i.e., $\gamma(-\mathbf{n}) = \gamma(\mathbf{n})$), then γ is a *norm*. I am interested in both convex and nonconvex and in both even and noneven functions γ. The total surface free energy of a rectifiable surface S is then $\int_S \gamma(\mathbf{n}_S(x))dA$. Here $\mathbf{n}_S(x)$ denotes the oriented unit normal to S at x (which exists almost everywhere for a rectifiable surface).

The Wulff shape W_γ is defined by

$$W_\gamma = \{\mathbf{q} \in R^d : \mathbf{q} \cdot \mathbf{n} \leq \gamma(\mathbf{n}) \text{ for all } \mathbf{n}\}.$$

Thus if γ is a norm, W_γ is the unit ball of the dual norm. It is a theorem that W_γ is the unique (up to translation) body minimizing total surface free energy, among bodies having its volume, and is thus sometimes called the equilibrium crystal shape. Note that if $\gamma(\mathbf{n}) = 1$ for all unit vectors $\mathbf{n}$, then W_γ is the usual round unit ball $\{\mathbf{q} : |\mathbf{q}| \leq 1\}$.

I am particularly interested in the case where W_γ is a polyhedron. This is defined to be the *crystalline* case, although real crystals may or may not have crystalline surface free energy functions. In the crystalline case, it is sufficient to specify the values of γ solely on the finite collection of normals that are to be the normals of W_γ; this then defines a unique W, and a convex γ is then determined by the usual $\gamma(r\mathbf{n}) = r\gamma(\mathbf{n})$ for those directions, followed by linear interpolation over each region of the **n**-diagram (the decomposition of the unit sphere induced by the normal map of W; see [T0][CT0]). When a surface has its free energy determined by a crystalline γ, then portions of the surface with normal directions which

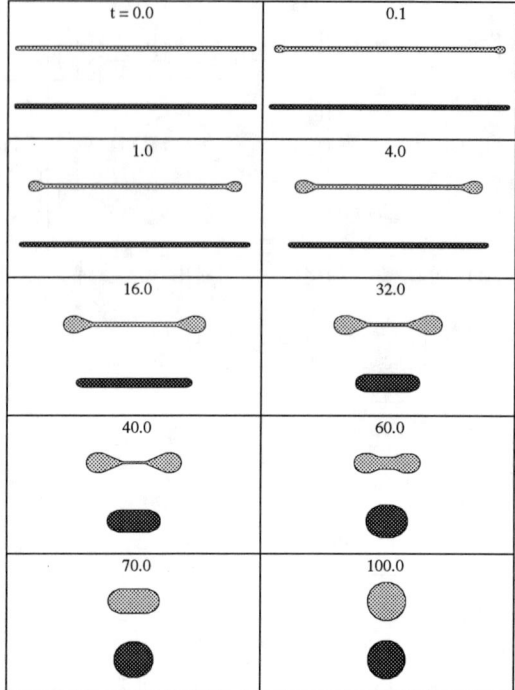

Figure 9: Successive views of motion of initial rectangle by surface diffusion (bulb development) and by SALK (remains convex). From [CRCT].

are not normals of γ are often regarded as *varifolds*, which in this case can be regarded as a way of making mathematically precise the idea of infinitesimally corrugated surfaces where the directions of the corrugations are the corners of the appropriate region or tie line of the **n**-diagram. Using varifolds is necessary for non-convex γ; for convex γ it is not necessary but can be useful, particularly in the presence of edge energies [K].

Experimentally, one does not usually measure γ as a function of **n** directly. Rather, one looks at shapes, and attempts to determine the Wulff shape up to a scale and a translation. The origin is then often determined by symmetry, and some measurement of $\gamma(\mathbf{n})$ for just one **n** is sufficient to determine the scale. This determines a unique convex γ; the Wulff shape is thus in some sense more basic than γ (though if γ is in fact nonconvex, some surfaces may be metastable to small-slope perturbations).

There is a notion of weighted mean curvature $\kappa_\gamma(x)$ at points x of smooth surfaces C governed by smooth convex (e. g., elliptic) surface free energy functions γ. It has various characterizations, which are discussed in some detail in [T1] and summarized in [CT2]. Defining $\mathcal{E}(C)$ to be the total surface free energy $\int_C \gamma(\mathbf{n}_C(x))dA$, the fundamental characterization is $-\frac{\delta \mathcal{E}(C)}{\delta V}$, the decrease in surface free energy with volume under small deformations. Its precise definition for smooth and convex γ and for a smooth surface C is as follows. The first variation for γ applied to a vector field g is defined to be

$$\delta_\gamma C(g) = \frac{d}{d\lambda} \int_{h_\lambda(C)} \gamma(\mathbf{n}_{h_\lambda(C)}(x))\, dA \Big|_{\lambda=0}$$

where $h_\lambda(x) = x + \lambda g(x)$. Thus g is the initial velocity of a deformation vector field h for the surface C. By representation theorems for bounded linear operators, one knows there exists a unique function $\kappa_\gamma(x)$ such that

$$\delta\gamma C(g) = \int_C -\kappa_\gamma(x)\mathbf{n}_C(x)\cdot g(x)\, dA.$$

This definition does not extend to the case where γ is crystalline, since the variation of energy when the surface is pushed by a vector field is no longer a linear or bounded operator on such vector fields. (Energy changes go like $\ell + h$ rather than $\sqrt{\ell^2 + h^2}$, when a line segment of length ℓ is deformed up to a height h.) The answer is to make κ_γ be nonlocal: move whole plane segments or line segments up rather than deform locally. The resulting formula for a line segment of length ℓ_i with normal $\mathbf{n}_i$ in a polygonal curve C is just

$$\kappa_\gamma(i) = -\sigma_i \frac{\Lambda(\mathbf{n}_i)}{\ell_i}$$

where σ_i is $1, -1$, or 0 depending on the geometry of S around that line segment and $\Lambda(\mathbf{n}_i)$ is the length of the part of the boundary of W_γ with normal $\mathbf{n}_i$. Note that this is the formula basic to Figures 1-6. For a plane segment, it is the formula for Figure 7:

$$\kappa_\gamma(i) = -\frac{\sum_j \delta_{ij} f_{ij} \ell_{ij}(t)}{area_i(t)}.$$

Motion by crystalline curvature, surface diffusion, and SALK for curves

Motion by crystalline curvature for curves is the motion described above for Figure 1. Both theory and computation are well-developed for crystalline curves in the plane [T2, T3, AG]. There is a general variational approach, to be described below, which has been proved to coincide with the crystalline motion for suitable curves [AT]. In this crystalline model, there are details (which have been handled successfully), such as how to do evolutions out of too-sharp corners and how to handle topological changes. The model includes not only boundaries of regions but also curves with triple junctions (modeling grain growth) and fixed boundary points. Computationally, it handles problems which are difficult to handle in other methods, such as collisions, pinching off of necks, triple junctions, etc. (Figures 1-4 were obtained from the program wmc that I wrote; wmc uses Euler's method to solve the equations and user choices of time step. Figures 5 and 6 required a considerably more sophisticated program with careful choices of time step, written by Roosen, to handle the issues of stepping and of coupling of the temperature field [R2], [RT].)

The program wmc and some of the theory has been extended to the motion law

$$\frac{dh_i(t)}{dt} = M(\mathbf{n}_i)\left(-\sigma_i \frac{\Lambda(\mathbf{n}_i)}{\ell_i(t)} + C(x,t)\right)$$

where M is a given mobility function (taking any desired positive values on the normals of W and otherwise extended linearly over the regions of the n-diagram), and C is any continuous bounded function [Y].

Theory and computation are also partly developed for motion by crystalline curvature for surfaces [T4], and again the general variational approach of [ATW] [Ca] [Y] applies in all dimensions.

Computational models for surface diffusion and SALK for curves were developed in [CRCT]. Crystalline surface diffusion is an analog to motion by the negative of the surface Laplacian of mean curvature.

For a general discussion of geometric motion in materials and a variety of ways to model it, see [TCH]. For a general discussion of surface diffusion and a variety of ways to model it, see [CT1].

Gradient Flows

Motion by weighted mean curvature, by surface diffusion, by SALK, and by a more general law incorporating both surface diffusion and attachment kinetics, can all be formulated in a variational setting, as gradient flows under various natural inner products. Doing so enables the methods of geometric measure theory to be applied in proofs, and in particular it enables one to handle singularities and topology changes. This way of looking at motions provides a unifying framework for the various motion laws, for the various possible surface energies including crystalline energies, and for sharp and diffuse interface formulations [TC1].

But beyond its usages in proofs and in conceptual understanding, it provides the underlying rationale for the crystalline approach. Since weighted mean curvature is no longer local, one needs a criterion for deciding when and how to step surfaces; gradient flow (decreasing energy as fast as possible under an appropriate notion of "distance") provides this criterion.

Assume γ is a smooth and convex function, and S a smooth surface, As before, let $\mathcal{E}(S) = \int_S \gamma(\mathbf{n}_S(x)) dA$ and $\delta_\gamma S(g)$ be the corresponding first variation of S with respect to γ operating on the vector field g, so that

$$\delta_\gamma S(g) = \int_{x \in S} -\kappa_\gamma \mathbf{n}_S(x) \cdot g(x) \, dA.$$

Now the gradient of $\mathcal{E}$ with respect to the inner product $\bullet$ is the vector field v such that

$$\delta_\gamma S(g) = v \bullet g$$

By definition, the L^2 inner product of any vector fields f and g on S is $\int_S f \cdot g \, dA$. So if $v = \kappa_\gamma \mathbf{n}$, then using this L^2 inner product we have $v \bullet g = \int_S (-\kappa_\gamma) g \cdot \mathbf{n} \, dA$, and thus $-\kappa_\gamma \mathbf{n}$ is seen to be the L^2 gradient of $\mathcal{E}$. Motion by weighted mean curvature, where the normal velocity is κ_γ, is therefore seen to be gradient flow for $\mathcal{E}$ in the L^2 inner product, and with this notion of distance, it decreases energy fastest, according to the Schwartz inequality $f \bullet g \leq f \bullet f$ if $|g| = |f|$.

But there are other inner products, and therefore other gradient flows. Surface diffusion with a scalar and constant diffusion constant D has normal velocity $-D\Delta_S \kappa_\gamma$, where Δ_S denotes the surface Laplacian. Since for scalar functions g on S

$$\delta_\gamma S(g\mathbf{n}) = -\int_S g\kappa_\gamma \, dA = -\int_S g\Delta_S^{-1}(\Delta_S \kappa_\gamma) \, dA$$

and since the H^{-1} inner product of f and g on S is (essentially by definition)

$$f \bullet g = -\int_S g\Delta_S^{-1} f \, dA,$$

by taking $f = \Delta_S \kappa_\gamma$ we see that surface diffusion is gradient flow with respect to the H^{-1} inner product. It also decreases $\mathcal{E}$ "fastest"; there is just a different definition of distance.

The motion law when interface kinetics are combined with surface diffusion turns out to be gradient flow with respect to the inner product which is the weighted sum of the L^2 and H^{-1} inner products in the case where the mobility M and the diffusivity D are constant scalars (the weights are just $1/M$ and $1/D$ respectively).

The Variational Approach to Evolving Microstructure

One way to construct mathematical evolutions modeling crystal growth with curvature effects is a limits of sequences of solutions to geometric optimization problems. The advantages of doing this include:

- There are no limitations on the geometric shapes into which crystals can evolve

- One passes naturally through times at which curvatures become unbounded, and in computations based on such a scheme it is not necessary to evaluate curvatures

- One can employ the powerful theorems of geometric measure theory in the mathematical analysis

- One can use any positive surface energy function (by convexifying it if it is not convex, and then interpreting the solution as a varifold).

The idea of the variational approach is to find a gradient flow $S(t)$ by fixing a time step Δt and and a surface S and finding the h that minimizes

$$\mathcal{E}(h(S)) - \mathcal{E}(S) + \frac{1}{2\Delta t} h \bullet h.$$

One starts with S equal to some initial surface $S(0)$ and minimizes in order to get $S(\Delta t) = h(S)$, then minimizes with $S = S(\Delta t)$ in order to get $S(2\Delta t)$, etc. One thus gets a piecewise constant flow for all time; the requirement is then to show that there is a convergent subsequence to this flow as Δt goes to zero. (This might be accomplished by proving Hölder inequalities, as was done for L^2 flow in [ATW].)

The reason this should give gradient flow is that the minimum of $v \bullet h + \frac{1}{2\Delta t} h \bullet h$ occurs at exactly $h = -v\Delta t$. But $\mathcal{E}(h(S)) - \mathcal{E}(S) \approx \delta_\gamma S(h) = v \bullet h$ when v is the gradient of $\mathcal{E}$. Therefore the quantity $\mathcal{E}(h(S)) - \mathcal{E}(S) + \frac{1}{2\Delta t} h \bullet h$ is approximately minimized at $h = -v\Delta t$, which says that you go in the opposite direction to the gradient for time Δt, as desired.

Although one cannot use first variation and vector fields for the case where γ is crystalline, the minimization still produces the right motion. For example, to get motion by crystalline curvature for a crystalline γ and a polyhedral curve S with good directions and intersections, one would minimize

$$Q = \mathcal{E}(h(S)) - \mathcal{E}(S) + \frac{1}{2\Delta t} h \bullet h$$

$$\approx \sum \sigma_i \Lambda_i h_i + \frac{1}{2\Delta t} \sum h_i^2 \ell_i$$

over all deformations $h = \{h_i\}$. Therefore we require $\partial Q / \partial h_i = 0$ for all i, which means

$$\sigma_i \Lambda_i + \frac{1}{\Delta t} h_i \ell_i = 0$$

i.e.,
$$\frac{h_i}{\Delta t} = -\frac{\sigma_i \Lambda_i}{\ell_i} = \kappa_\gamma.$$

To allow very general variations, for example through singularities, it is important to convert the surface integral to a volume integral that is approximately the same. This was done in [ATW] for the L^2 inner product in case of mobility always 1, and the proper way to do this for general mobility and additional bulk driving forces was conjectured by Roosen [R1] and proved by Yip [Y] (provided the additional driving force $C(t)$ is bounded). The H^{-1} inner product is being investigated by Chung [Ch]. Existence for the case of multiple regions (as in grain growth) has been proved by Caraballo [Ca].

Diffuse Interfaces

In the two-phase diffuse interface approach, one has an order parameter which is a real-valued function u with domain R^d, with $u \approx 1$ within one phase and $u \approx -1$ outside that phase. For any fixed positive ϵ the energy is defined to be

$$\mathcal{E}_\epsilon(u) = \int_{x \in R^d} \left(\frac{\epsilon}{2} (\Gamma(\nabla u(x)))^2 + \frac{1}{\epsilon} F(u(x)) \right) dV.$$

F is a nonnegative function having value 0 at $u = \pm 1$ and being positive elsewhere (often one takes $F(u) = (1-u^2)^2$, or perhaps $F(u) = 1 - u^2$ for $|u| \leq 1$ and $F(u) = \infty$ otherwise); it tries to force the material into one phase or the other. Γ is a given positive (except at $\mathbf{0}$) function satisfying $\Gamma(\mathbf{p}) = |\mathbf{p}|\Gamma(\mathbf{p}/|\mathbf{p}|)$; it specifies the anisotropy of the contribution of the diffuse interface to the total energy and helps determine the magnitude of that interfacial energy. There is a direct proportionality between $\Gamma(\mathbf{n})$ and the surface free energy $\gamma(\mathbf{n})$ for the limiting sharp interface as ϵ goes to zero. Specifically, one fixes a direction $\mathbf{n}$ and imposes the conditions on u that $u(x)$ depend only on $\mathbf{n} \cdot x$, with $\lim_{x \cdot \mathbf{n} \to -\infty} u(x) = 1$ and $\lim_{x \cdot \mathbf{n} \to \infty} u(x) = -1$. Let u_ϵ be the minimizer, over all such u, of the quantity $\mathcal{E}_\epsilon(u)$ per unit area perpendicular to $\mathbf{n}$. The minimum can be shown (e. g., see [TC1]) to be $\left[\sqrt{2} \int_{-1}^{1} \sqrt{F(u)}\, du \right] \Gamma(\mathbf{n})$, which is independent of ϵ. Finally, this minimum is also $\gamma(\mathbf{n})$, the surface energy per unit area of the sharp interface obtained in the $\epsilon \to 0$ limit.

If Γ is isotropic (i.e. $\Gamma(\mathbf{n}) = 1$ for every unit vector $\mathbf{n}$, then gradient flow for $\mathcal{E}_\epsilon$ in the L^2 inner product with mobility $1/\epsilon$ produces solutions to the Allen-Cahn equation

$$u_t = \Delta u - \frac{1}{\epsilon^2} F'(u),$$

and these solutions converge to motion by weighted mean curvature as ϵ goes to zero. The Cahn-Hilliard equation arises from gradient flow in the H^{-1} inner product, and the viscous Cahn-Hilliard equation arises from gradient flow in a weighted sum of these inner products. If one has the diffusivity B depend appropriately on the value of u and ϵ, so that the Cahn-Hilliard equation becomes

$$u_t = -\nabla \cdot B(u) \nabla (\Delta u - \frac{1}{\epsilon^2} F'(u)),$$

then in formal asymptotics there is convergence to motion by surface diffusion [ACE].

Many corresponding results have been proved for the case where Γ is smooth and uniformly convex [B] [BF] [OS]. In the case where Γ is nonconvex and/or nondifferentiable, we are currently determining the behavior [TC2].

Summary

Surface motion problems can be formulated in the context of optimal geometry. These motions include:

- motion by weighted mean curvature (such as might occur in an annealing polycrystal)

- motion by surface diffusion

- motion by SALK

- motion by a combination of SALK and surface diffusion

- dendritic crystal growth.

These processes are driving by surface and bulk free energy reduction but held back by mobility due to surface attachment kinetics and by diffusion on the surface and in the bulk.

By putting the problems in the context of gradient flows, and by using the crystalline approach, it is possible to tackle problems with changing geometries and with surface energies corresponding to faceted crystals, in both theory and computation.

References

[ACE] Amy Novick-Cohen, John W. Cahn, Charles M. Elliott, The Cahn-Hilliard equation: surface motion by the Laplacian of the mean curvature, preprint.

[AT] Fred Almgren and Jean E. Taylor, Flat flow is motion by crystalline curvature for curves with crystalline energies, J. Differential Geometry **42** (1995), 1-22.

[ATW] Fred Almgren, Jean Taylor and Lihe Wang, Curvature Driven Flows: A Variational Approach, SIAM Journal of Control and Optimization **31** (1993), 386-437.

[AG] S. Angenent and M. Gurtin, *Multiphase thermomechanics with interfacial structure. 2. Evolution of an isothermal interface.* Arch Rat. Mech. Anal 108 (1989) 323-391.

[B] Guy Bouchitté, Singular perturbations of variational problems arising from a two-phase transition model, Ann. Inst. H. Poincare-Anal. Non Lineaire 7 (1990), 37-65

[BF] A. C. Barroso and I. Fonseca, Anisotropic singular perturbations- the vectorial case, Proc. Royal Soc. Edin. 124A (1994), 527-571

[CRCT] Craig Carter, Andrew Roosen, John Cahn, and Jean E. Taylor, Shape evolution by surface diffusion and surface attachment limited kinetics on completely faceted surfaces, Acta Metal. Mater. **43** (1995), to appear.

[CT0] John W. Cahn and J. E. Taylor, Catalog of saddle shaped surfaces in crystals, Acta Metallurgica **34** (1986), 1-12.

[CT1] John W. Cahn and Jean E. Taylor, Surface Motion by Surface Diffusion, Acta Metall. mater. **42** (1994), 1045-1063.

[CT2] John W. Cahn and Jean E. Taylor, Thermodynamics Driving Forces and Equilibrium in Multicomponent systems with Anisotropic Surfaces, this volume.

[Ca] David Caraballo, Princeton University Mathematics Department Thesis, 1995.

[Ch] Kin Yan Chung, Princeton University Mathematics Department Thesis, in preparation.

[G] Pedro Martins Girao, *Convergence of a crystalline algorithm for the motion of a simple closed convex curve by weighted curvature*, SIAM J. Numer. Anal., 1995.

[K] M. Jeannette Kelly, Edge energy-minimizing surfaces and crystal shape, this volume

[OS] N.C. Owen and P. Sternberg, Nonconvex variational problems with anisotropic perturbations, Nonlinear Anal. 78 (1991), 705-719

[R1] Andrew Roosen, Rutgers University Mathematics Department Thesis, 1993.

[R2] Andrew Roosen, *Simulation of two-dimensional facetted crystal growth in a single diffusion field*, in Computational Crystal Growers Workshop (J. E. Taylor, ed.), Selected Lectures in Mathematics, Amer. Math. Soc. (1992) (includes video).

[RT] Andrew R. Roosen and Jean E. Taylor, *Modeling crystal growth in a diffusion field using fully faceted interfaces*, J. Computational Physics **114** (1994), 113-128.

[T0] Jean E. Taylor, Complete catalog of minimizing embedded crystalline cones, Proc. Symposia Math. **44** (1986), 379-403.

[T1] Jean E. Taylor, Mean curvature and weighted mean curvature, Acta Metall. Mater. **40** (1992), 1475-1485.

[T2] Jean E. Taylor, Motion of curves by crystalline curvature, including triple junctions and boundary points, Differential Geometry, Proceedings of Symposia in Pure Math. **51** (part 1) (1993), 417-438.

[T3] Jean E. Taylor, Motion by Crystalline Curvature, in Computing Optimal Geometries (Jean E. Taylor, ed.), Selected Lectures in Mathematics, American Mathematical Society (1991), 63-65 plus video.

[T4] Jean E. Taylor, Geometric Crystal Growth in 3D via Faceted Interfaces, in Computational Crystal Growers Workshop (Jean E. Taylor, ed.), Selected Lectures in Mathematics, American Mathematical Society (1992), 111-113 plus video 20:25-26:00.

[TC1] Jean E. Taylor and J. W. Cahn, Linking Anisotropic Sharp and Diffuse Surface Motion Laws via Gradient Flows, J. Stat. Phys. **77** (1994), 183-197.

[TC2] Jean E. Taylor and John W. Cahn, Sharp Corners and Flat Facets in Diffuse Interfaces, in preparation.

[TCH] Jean E. Taylor, John Cahn and Carol Handwerker, Geometric Models of Crystal Growth. Acta Metall. Mater. **40**(1992), 1443-1474.

[Y] NungKwan Yip, Princeton University Mathematics Department thesis, in preparation.

THERMODYNAMIC DRIVING FORCES AND EQUILIBRIUM IN MULTICOMPONENT SYSTEMS WITH ANISOTROPIC SURFACES

John W. Cahn* and Jean E. Taylor**

*Materials Science and Engineering Laboratory, NIST, Gaithersburg, MD 20899
**Mathematics Department, Rutgers University, Piscataway NJ 08855 and
School of Mathematics, Institute for Advanced Study, Princeton, NJ 08540
cahn@enh.nist.gov and taylor@math.rutgers.edu

Abstract

Problems of how surface curvature affects phase equilibria and driving forces for phase change and surface motion are formulated for surfaces with anisotropic surface free energy. Weighted mean curvature is used in a way that reveals how the free energy of curved surfaces acts like a thermodynamic driving force and thus how it interacts with bulk free energies. The relations are valid for anisotropic surfaces even when faceted. We define what we mean by driving forces, and describe how equilibria and driving forces depend on temperature, chemical potentials, and pressure when curved or facetted surfaces are present. Forms for energies that enable surface motion to be done via, say, gradient flow with respect to various inner products are obtained.

This work was partially funded by the Advanced Research Projects Agency through the National Institute of Standards and Technology and by the National Science Foundation.

Introduction

Surface free energy is an excess in a particular free energy, Ω, defined by

$$\Omega = E - TS - \sum_i \mu_i N_i. \tag{1}$$

When Ω is used as the free energy for phase change, and when the definition of mean curvature is taken as the rate of change in surface area per unit volume swept out when a curved surface is moved, the role of mean curvature in equilibria and driving forces for isotropic surfaces becomes apparent.

When the surface free energy per unit area γ is different for different normal directions **n**, a generalization of the concept of mean curvature is the weighted mean curvature, the rate of change in surface free energy per unit volume swept out. Because weighted mean curvature is a free energy change per unit volume swept out, we explore how it can be added to free energy changes such as phase change energies per unit volume for purposes of computing equilibrium conditions and driving forces. We show it is useful whether we are dealing with equilibrium or with a driving force. It applies no matter how extreme the anisotropy, and as a special case our results include familiar isotropic results.

Mean Curvature and Weighted Mean Curvature

There are a number of equivalent definitions of the mean curvature (here called κ so as not to confuse it with the enthalpy H of thermodynamics) of a surface C at a point x in C, and similarly there are equivalent definitions of weighted mean curvature. These are set out in [T1] and are reviewed here briefly. WARNING: There is a major difference of opinion as to the sign convention for the mean curvature and weighted mean curvature.[1] The mean curvature vector field is naturally defined in mathematics to point in the direction that the surface curves most (see, for example, [M]). If C is a surface with unit oriented normal $\mathbf{n}_C(x)$ at x, and if one were to write the mean curvature vector field as $\kappa(x)\mathbf{n}_C(x)$, then for a smooth convex body, naturally oriented so that $\mathbf{n}_C$ points outward, κ would be negative everywhere. Others find it unnatural to have a convex body have negative mean curvature and use the opposite sign for κ. In [T1] the FIRST convention was used, but within this paper we will abide by the SECOND convention, so that our κ and κ_γ are positive for convex bodies with surface C which are oriented so that $\mathbf{n}_C$ points outward. Thus in this paper, the mathematician's mean curvature vector field will in fact be $-\kappa \mathbf{n}_c$. (One might orient the surface inward, but then that would require additional changes such as working with the central inversion of the Wulff shape.)

Some equivalent definitions for the mean curvature of a surface C at a point x are [T1]:

(1) $\kappa = \kappa_1 + \kappa_2$, where κ_1 and κ_2 are the maximum and minimum values, over all planes perpendicular to the tangent plane to C at x, of the curvature at x of the curve which is intersection of C with that plane. The absolute value of κ_i is thus the reciprocal of the radius of curvature of the appropriate curve, and the sign of κ_i is taken to be -1 if the center of curvature is in the direction of $\mathbf{n}_C(x)$ and +1 if it is in the opposite direction. (This reversed choice of sign is precisely the result of our use of the second sign convention, as mentioned above.)

(2) $\kappa = \text{div}_C \mathbf{n}_C(x)$ (the surface divergence of the normal vector).

---

[1]There is also lack of agreement on whether to include a factor of $\frac{1}{2}$ (or more generally, the reciprocal of the dimension of the surface) in the definition. We do NOT include this extra factor.

(3) $\kappa = \frac{\delta A}{\delta V}$, where A is the area of the surface. More precisely, $\delta C(g) = \int_C \kappa \mathbf{n} \cdot g \, dA$, where g is the initial velocity of a deformation vector field h for the surface C and $\delta C(g)$ is the first variation operator applied to g (that is, $\delta C(g) = \frac{d}{d\lambda} \int_{h_\lambda(C)} 1 \, dA \big|_{\lambda=0}$ where $h_\lambda(x) = x + \lambda g(x)$).

(4) $\kappa = \mathrm{div}(-\frac{\nabla u}{|\nabla u|})$, where C is a level set of a function u defined on all space. Here the unit normal, assuming the surface is oriented in the direction of decreasing u, is the direction of the negative gradient, i.e., $\mathbf{n}_C(x) = -\frac{\nabla u(x)}{|\nabla u(x)|}$ for x in C, but the divergence is the full 3-dimensional operation, not just the surface divergence.

For weighted mean curvature κ_γ of C having (integrated) surface free energy $\int_C \gamma(\mathbf{n}_C(x)) dA$, where $\gamma(\mathbf{n})$ is a given function of unit normal directions and $\gamma(\rho\mathbf{n})$ is defined to be $\rho\gamma(\mathbf{n})$ for any $\rho \geq 0$, we have for each of the above a corresponding definition:

(1_γ) $\kappa_\gamma = \frac{\partial^2 \gamma}{\partial p_1^2}\kappa_1 + \frac{\partial^2 \gamma}{\partial p_2^2}\kappa_2$, where $\mathbf{p}_i$ is a tangent direction which together with $\mathbf{n}_C(x)$ spans the plane in which κ_i is measured. The term $\frac{\partial^2 \gamma}{\partial p_i^2}$ becomes $\gamma(\theta) + \gamma''(\theta)$ if one restricts γ to unit vectors in a circle (in the plane spanned by $\mathbf{p}_i$ and $\mathbf{n}_C$) and parametrizes the circle by angle θ.

(2_γ) $\kappa_\gamma = \mathrm{div}_C \xi(\mathbf{n}_C(x))$, where $\xi(\mathbf{n}_C(x)) = \nabla_\mathbf{p} \gamma(\mathbf{p}) \big|_{\mathbf{p}=\mathbf{n}_C(x)}$.

(3_γ) $\kappa_\gamma = \frac{\delta \mathcal{A}_\gamma}{\delta V}$, where $\mathcal{A}_\gamma = \int_C \gamma(\mathbf{n}_C(x)) dA$. More precisely, $\delta \gamma C(g) = \int_C \kappa_\gamma \mathbf{n} \cdot g \, dA$, where g is the initial velocity of a deformation vector field h for the surface C and $\delta_\gamma C(g)$ is the first variation operator for γ applied to g (that is, $\delta_\gamma C(g) = \frac{d}{d\lambda} \int_{h_\lambda(C)} \gamma(\mathbf{n}_C(x)) \, dA \big|_{\lambda=0}$ where $h_\lambda(x) = x + \lambda g(x)$).

(4_γ) $\kappa_\gamma = \mathrm{div}\left(\xi\left(-\frac{\nabla u(x)}{|\nabla u(x)|}\right)\right)$ where C is a level set of the function u as in (4) above.

These definitions all assume that both C and γ are smooth. The definition (3) has been extended to the non-smooth case for γ, which results in facets in C, by looking at deformations that move entire plane segments rather than deformations that move arbitrarily small parts of the surface. See [T1].

For an easy to understand example of weighted mean curvature for a fully faceted crystal, assume that γ is the same small quantity γ_0 for the six cube faces, and $\gamma(\mathbf{n})$ is large on all other $\mathbf{n}$. The Wulff shape (the equilibrium crystal shape) for this γ is a cube of side length $2\gamma_0$. If a surface C has a portion consisting of a horizontal rectangle with sides of length a and b and with vertical sides dropping off it all around, then varying the height of the rectangle by a distance λ results in an area change of the vertical faces only and thus a surface free energy change of $2(a+b)\lambda\gamma_0$ for a volume change of $ab\lambda$. Therefore $\kappa_\gamma = \left(\frac{2}{a} + \frac{2}{b}\right)\gamma_0$ for points in this horizontal rectangle of C. For Wulff shapes that are more general polyhedrons and for appropriate polyhedral surfaces, computing κ_γ is a trigonometry problem; the answer was already in Gibbs [G].

The important point is that the weighted mean curvature by characterization (3) is a free energy change per unit volume swept out. We next look at what kind of free energy this is.

Thermodynamics

By the first law of thermodynamics, for a single homogeneous hydrostatically stressed phase

$$dE = dQ - PdV + \sum_i \mu_i dN_i, \qquad (2)$$

and by the second,

$$dQ \leq TdS \qquad (3)$$

so

$$dE \leq TdS - PdV + \sum \mu_i dN_i, \qquad (4)$$

with equality for reversible processes or equilibrium.[2] In this paper we will be concerned with stresses that originate from surfaces of solids and which may be nonhydrostatic. We will assume that solid phases are sufficiently rigid and curvatures small enough that the elastic work of deformation can be neglected (that is, surface stresses do not induce deformation in the solid). Otherwise an elastic work term, stress times increment of strain, would have to be added to the first law.

We use the bound of equation 4 on dE and the full differential of Ω, defined in equation 1, to write

$$d\Omega \leq -SdT - PdV - \sum N_i d\mu_i. \qquad (5)$$

Thus if T, V, and μ_i are all fixed, then $d\Omega = 0$ at equilibrium and this is the condition for Ω to be minimized. Setting $\Omega_V = \Omega/V, s = S/V$, and $\rho_i = N_i/V$, this can be written

$$0 \leq -sdT - d\Omega_V - \sum \rho_i d\mu_i \qquad (6)$$

Neither G (Gibbs free energy) nor F (Helmholtz free energy) is appropriate (minimized at equilibrium) for multicomponent systems if the amount of material in a given volume is not kept fixed. All of the physics is in the first and second laws of thermodynamics and in the constitutive relations; minimizing one free energy or another is a derived concept which depends on which variables are kept fixed.

For a single homogeneous hydrostatically stressed phase at equilibrium, the thermodynamic relations are derived as follows. We start with an empty rectangular cylinder, and fill it with the phase by pulling out a piston reversibly. We assume that there are reservoirs to supply atoms as required to maintain constant chemical potential, just as we keep temperature constant from heat reservoirs. We can then use an equality in place of the inequality and integrate to obtain

$$E = TS - PV + \sum \mu_i N_i. \qquad (7)$$

We can take the full differential dE of E and subtract the formula from the first and second laws to arrive at the Gibbs-Duhem equation,

$$0 = -SdT + VdP - \sum N_i d\mu_i, \qquad (8)$$

or, in terms of quantities per unit volume,

$$0 = -sdT + dP - \sum \rho_i d\mu_i. \qquad (9)$$

---

[2] As usual in thermodynamics, E denotes the internal energy, T the temperature, S the entropy, P the pressure, V the volume, μ_i the chemical potential of the ith component, and N_i the number of moles of the ith component.

Finally, from the definition of Ω and the formula for E, we obtain

$$\Omega = -PV; \qquad \Omega_V = -P \tag{10}$$

and

$$d\Omega_V = -sdT - \sum \rho_i d\mu_i, \tag{11}$$

for a single homogeneous phase in equilibrium. This version of the Gibbs-Duhem equation holds for nonhydrostatically stressed rigid solids without having to specify what takes the place of P.[3]

Consider two fluid phases, denoted by superscripts α and β (each phase containing all the components) and separated by a planar surface C with normal $\mathbf{n}$ pointing from β into α. At equilibrium T, P, and the μ_i will be constant. We can again think of starting with an empty rectangular cylinder and creating the two phases from materials reservoirs at T, P, and the μ_i by pulling out a flat piston perpendicular to C. Now the first and second laws of thermodynamics give

$$dE \leq TdS - PdV + \sum \mu_i N_i + \gamma(\mathbf{n})dA. \tag{12}$$

Here both PdV and $\gamma(\mathbf{n})dA$ are mechanical work, with γ defined as a force per unit length along the perimeter of the surface (a mechanical definition of surface tension). We can then integrate (pull the piston out reversibly) to obtain

$$E = TS - PV + \sum \mu_i N_i + \gamma(\mathbf{n})A. \tag{13}$$

Note that $\gamma(\mathbf{n})A = E - TS - \sum \mu N + PV = \Omega + PV$, so that for fluid surfaces $\gamma(\mathbf{n})A$ is also an excess in Ω over the $-PV$ that Ω would be if there were no surface; two different physical quantities, surface tension, a force per unit length, describing the work of stretching the surface, while surface free energy per unit area, describing the work of creating the surface, have the same magnitude.

If one of the phases is solid this equality ceases to hold, because for solids the work of stretching the surface can be distinguished from the work of creating more surface. For solids this distinction becomes apparent if we define both surface area and volume in a reference configuration of no strain [C]. That they both be defined in the reference state is essential in our applications of weighted mean curvature, but it leads to such counterintuitive concepts as that an elastically bent crystalline plane has no curvature. The surface stress (termed tension by Gibbs) is now an anisotropic tensor related to the work of stretching the surface, while the surface free energy is a scalar related to creating a unit area of surface; both are functions of the orientation of the normal. For a planar surface the stress gives forces per unit length that may depend on direction in the plane. Assuming the solids to be rigid, as we do here, removes the surface stress from consideration and leaves only the surface free energy γ.

There is a large literature about how to apportion surface excesses in the extensive quantities of E, S, V and the N_i dating back to Gibbs; it is important in how γ varies with temperature, pressure, composition, etc. Gibbs couched his convention in terms of dividing surfaces in a comparison system. Some other conventions are more easily adaptable to many circumstances [C]. These arbitrary conventional choices will affect the values of some

---
[3]If the solid is not rigid, an elastic work term, stress times the increment of strain, has to be added to the first law and survives intact as an additional term in this equation.[LC]

individual excess quantities for the surface, but not those (or certain combinations such as those in Ω) that can be measured or that enter into relations among measured quantities. For planar surfaces the excess in Ω does not depend on the convention chosen. Note furthermore that for multicomponent systems $\gamma(\mathbf{n})A$ cannot be attributed to an excess in Gibbs free energy $G = E - TS + PV$. For a one component system in which the two phases differ in density, the dividing surface can always be chosen to eliminate the excess density; then the excess in Ω reduces to the excess in the Helmholtz free energy $F = E - TS$.

Extension to curved surfaces requires some additional justification. Since the pressure is no longer constant, how to define γ in terms of the excess in Ω has been extensively discussed for fluid droplets. The result is an expression for the dependence of γ on curvature. For heterophase interfaces, γ is linear in curvature with a coefficient that is of order of the thickness of the interface; the curvature dependence only becomes significant when the radii of curvature are of the same magnitude as the surface thickness (see [FW]). Such studies have not been done for anisotropic surfaces, especially ones governed by a nondifferentiable surface free energy. We assume here that for small weighted mean curvatures, we can assign an unambiguous location to the surface and that γ is independent of curvature. The net result is that the contribution of the surface to Ω is $\int_{x \in C} \gamma(\mathbf{n}_C(x))dA$. (For some possible effects when a type of curvature dependence, in the form of edge energy, is included, see [K].)

Suppose the surface C is oriented so that $\mathbf{n}_C$ points from phase β into phase α. Suppose one makes a variation δV in volume, adding or subtracting atoms from reservoirs if necessary. (δV is positive if volume is removed from α and put into β.) Let Ω_V^α denote the amount of Ω per unit volume in α and similarly for Ω_V^β. By the third characterization of κ_γ,

$$\delta\Omega = (\kappa_\gamma + \Omega_V^\beta - \Omega_V^\alpha)\delta V. \qquad (14)$$

For homogeneous and hydrostatically stressed phases $\Omega = -PV$, the jump in Ω across the interface $\Omega_V^\beta - \Omega_V^\alpha$ is $-(P^\beta - P^\alpha)$, the negative of the pressure jump. Therefore local equilibrium requires that P not be continuous across the interface and

$$-\kappa_\gamma + P^\beta - P^\alpha = 0. \qquad (15)$$

(T and μ are continuous across the interface.) For nonhydrostatically stressed solids, Ω_V remains locally well defined [LC], and we have

$$\kappa_\gamma + \Omega_V^\beta - \Omega_V^\alpha = 0 \qquad (16)$$

for local equilibrium. Since the stress or pressure might vary within a solid phase, but be defined locally, this does not become the familiar more global assertion that $\kappa_\gamma = constant$. See [LC], [JA], [AJ].

If we do not assume local equilibrium at the interface but do assume it within the individual phases, then $-\kappa_\gamma + \Delta P$ or $\kappa_\gamma + \Omega_V^\beta - \Omega_V^\alpha$ could be measures of the driving force towards equilibrium. They are equal to 0 at equilibrium and, when non-zero, represent a real unbalance of stress or pressure across the interface. See [TCH] for a survey of geometric motion and [CT] for a survey of surface diffusion driven by such forces. See [T2] for examples and more recent methods.

To extend this analysis to the case where W_γ has flat facets, all pointwise equations above should be replaced by their integrations over entire flat facets of C. An additional criterion for equilibrium is that for any way of stepping the facet into pieces (by adding between the pieces narrow strips of additional surface having directions that, in W_γ, are adjacent to the facet direction), either the integrated forces balance on each piece separately, or the forces on the pieces are such as to push the pieces back together rather than to allow them to separate. When using the driving force to do motion problems, steppings need to be done unless all ways of stepping result in velocities that push the pieces back together. See for example [CRCT].

Often the differences in pressure or Ω_V are not the desired or convenient variables; one can then use the generalized Gibbs-Duhem equations to relate curvature to other variables, such as the pressure, stress, or composition in either phase, or the temperature or chemical potentials at the interface. Consistent with the phase rule, two variables can be eliminated. The dependence of one remaining variable on another can then be determined by keeping all other remaining variables constant.

Examples

Case 1. One component, two phase (think of a droplet of phase β in matrix of phase α, with $\mathbf{n}_C$ pointing outward, so that $\kappa_\gamma > 0$), equilibrium. What follows is done for fluids, but is easily extended to rigid solids by replacing $-P$ with Ω_V.

We begin with the pressure jump related to weighted mean curvature, equation 15, and use the two Gibbs-Duhem equations, one for each phase,

$$0 = -S^\alpha dT + V^\alpha dP^\alpha - N^\alpha d\mu \tag{17}$$

$$0 = -S^\beta dT + V^\beta dP^\beta - N^\beta d\mu \tag{18}$$

to find how other thermodynamic quantities are related to curvature. Note that there is no jump in T or μ across the interface; both are constant at equilibrium. We assume that the equilibrium pressure $P_0 = P_0^\alpha = P_0^\beta = P_0(T)$ is known for plane interfaces when $\kappa_\gamma = 0$. Let $\mu_0 = \mu_0(T, P_0)$ be the value of μ at that equilibrium. Along the line of coexistence when $\kappa_\gamma = 0$, we get the Clausius-Clayperon equation by eliminating μ from the Gibbs-Duhem equations 17 and 18:

$$\frac{dP_0}{dT} = \frac{(S/N)^\beta - (S/N)^\alpha}{(V/N)^\beta - (V/N)^\alpha} = \frac{\Delta(S/N)}{\Delta(V/N)}. \tag{19}$$

(Here Δ always means we subtract the α quantity from the β quantity.) But more generally, with $P^\alpha = P^\beta - \kappa_\gamma$, we again eliminate μ (and P^α) and have

$$-(S/N)^\alpha dT + (V/N)^\alpha dP^\beta - (V/N)^\alpha d\kappa_\gamma = -(S/N)^\beta dT + (V/N)^\beta dP^\beta. \tag{20}$$

From this we obtain expressions for the effect of curvature on the pressure of either one of the phases if we fix temperature

$$\left(\frac{dP^\beta}{d\kappa_\gamma}\right)_T = -\frac{(V/N)^\alpha}{(V/N)^\beta - (V/N)^\alpha} = -\frac{1/\rho^\alpha}{1/\rho^\beta - 1/\rho^\alpha} = \frac{\rho^\beta}{\rho^\beta - \rho^\alpha} \tag{21}$$

and its effect on the temperature of coexistence if we fix the pressure of one of the phases:

$$\left(\frac{dT}{d\kappa_\gamma}\right)_{P^\beta} = -\frac{(V/N)^\alpha}{(S/N)^\alpha - (S/N)^\beta} = \frac{1}{\rho^\alpha((S/N)^\beta - (S/N)^\alpha)}. \tag{22}$$

We also obtain a generalized Clausius-Clayperon equation

$$\left(\frac{dP_0}{dT}\right)_{\kappa_\gamma} = \frac{\Delta(S/N)}{\Delta(V/N)}. \tag{23}$$

Equations such as these can be integrated from zero to nonzero curvatures, if the thermodynamic data, such as densities, are known for the phases.

We illustrate this first with the curvature dependence of P^β, holding T constant. If both phases are condensed, one can assume that the RHS of equation 21 is constant, and integrate to get[4]

$$P^\beta - P_0 = \frac{\rho^\beta}{\Delta\rho}\kappa_\gamma. \tag{24}$$

A somewhat more careful treatment might linearize compressibility of any solid or liquid - e.g., replace V^β by $V^\beta(1 - \beta^\beta P)$. (By definition, $\beta = -(1/V)(dV/dP)_T$.)

If α is a gas phase and we assume the ideal gas law $(V/N)^\alpha = RT/P^\alpha$, then equation 15 says we can replace $(V/N)^\alpha$ by $RT/(P^\beta - \kappa_\gamma)$. (Alternatively, one could use any appropriate constitutive relation for phase α.) If β is the gas phase instead, $\Delta(V/N) \approx (V/N)^\beta = (RT/P^\beta)$ (assuming ideal gas) so

$$\frac{dP^\beta}{d\kappa_\gamma} = -\frac{(V/N)^\alpha P^\beta}{RT} \tag{25}$$

i.e.

$$\frac{1}{P^\beta}dP^\beta = -\frac{(V/N)^\alpha}{RT}d\kappa_\gamma \tag{26}$$

Again assuming $\frac{(V/N)^\alpha}{RT}$ is constant, we integrate to get

$$RT \ln(P^\beta/P_0) = -(V/N)^\alpha \kappa_\gamma. \tag{27}$$

To find the effect of curvature on undercooling, we integrate equation 22. If we assume the RHS of equation 22 is constant, we can integrate from $\kappa_\gamma = 0$ (writing T_0 as the equilibrium temperature when $\kappa_\gamma = 0$) to get

$$T - T_0 = -\frac{(V/N)^\alpha}{((S/N)^\alpha - (S/N)^\beta)}\kappa_\gamma. \tag{28}$$

Using the fact that the latent heat $L = ((S/N)^\alpha - (S/N)^\beta)T_0$, we have

$$T - T_0 = -(V/N)^\alpha \frac{T_0}{L}\kappa_\gamma = -\frac{1}{\rho^\alpha}\frac{T_0}{L}\kappa_\gamma. \tag{29}$$

Suppose that one eliminates both P^α and P^β, rather than μ, between the two Gibbs-Duhem equations 17 and 18 (and again takes T to be constant). Then one gets

$$(N/V)^\alpha d\mu + d\kappa_\gamma = (N/V)^\beta d\mu \tag{30}$$

---

[4]If the two phases have exactly the same density, then $\frac{d\kappa_\gamma}{dP^\beta}$ is zero, analogous to a retrograde point on a phase diagram. If both phases have identical properties as would be true for grain boundaries and other homophase interfaces, the two Gibbs-Duhem equations are dependent, and equilibrium is impossible with κ_γ nonzero [C].

i.e.,
$$\frac{d\mu}{d\kappa_\gamma} = \frac{1}{(N/V)^\beta - (N/V)^\alpha} = \frac{1}{\Delta\rho} \qquad (31)$$
and the denominator is now dominated by the high density phase. If one assumes the RHS of equation 31 is constant, one integrates this to
$$\mu - \mu_0 = \frac{\kappa_\gamma}{\Delta\rho} \qquad (32)$$
If β is a gas, then the RHS of equation 32 is essentially $-\frac{\kappa_\gamma}{\rho^\alpha}$.

Thus the Gibbs-Duhem equations plus $\Delta P = \kappa_\gamma$ (equations 17, 18, and 15) give a large variety of relations.

Case 2. Two or more species (components), two phases which are both rigid solids in substitutional solution with incoherent phase boundary or one such solid and one fluid, equilibrium. Again when $\kappa_\gamma = 0$, the two Gibbs-Duhem equations (equation 11 within each phase) define the equilibrium surface in the phase diagram with variables T, $\Omega_V^\alpha = \Omega_V^\beta$, and concentrations $c_i^\alpha = N_i^\alpha/\sum_j N_j^\alpha$, $c_i^\beta = N_i^\beta/\sum_j N_j^\beta$ or chemical potentials. But each non-zero value of κ_γ defines another surface. We can describe the new surface relative to the old either as changes of the temperature or the chemical potentials common to both phases, or as pressures and compositions of the individual phases.

As an example, we consider the binary case with T and Ω_V^α constant. For the α phase, the Gibbs-Duhem equation (equation 11) becomes
$$d\mu_1 + \frac{c_2^\alpha}{(1 - c_2^\alpha)}d\mu_2 = 0 \qquad (33)$$
and for the β phase, with V_m^β its the molar volume (i.e., $V_m^\beta = \frac{V^\beta}{\sum_i N_i^\beta}$), equation 11 becomes
$$-V_m^\beta d\Omega_V^\beta - (1 - c_2^\beta)d\mu_1 - c_2^\beta d\mu_2 = 0. \qquad (34)$$
Substituting for $d\mu_1$, we obtain
$$-V_m^\beta d\Omega_V^\beta = \frac{(c_2^\beta - c_2^\alpha)}{1 - c_2^\alpha}d\mu_2 \qquad (35)$$
and then substituting $-d\Omega_V^\beta = d\kappa_\gamma$ we obtain
$$V_m^\beta d\kappa_\gamma = \frac{(c_2^\beta - c_2^\alpha)}{1 - c_2^\alpha}d\mu_2. \qquad (36)$$

Sometimes we wish to know how the diffusion potential [LC] $M_{12} = \mu_1 - \mu_2$ behaves with curvature. From equation 33 we have $(1 - c_2^\alpha)dM_{12} = -d\mu_2$ and so
$$V_m^\beta d\kappa_\gamma = -(c_2^\beta - c_2^\alpha)dM_{12} \qquad (37)$$
We could then continue with M_{12} in what follows rather than μ_2.

If we assume that c_2^α, c_2^β, and V_m^β are all constant, we can again integrate from $\kappa_\gamma = 0$ to obtain
$$V_m^\beta \kappa_\gamma = \frac{c_2^\beta - c_2^\alpha}{1 - c_2^\alpha}(\mu_2(\kappa_\gamma) - \mu_2(0)) \qquad (38)$$

i.e.,
$$\mu_2(\kappa_\gamma) - \mu_2(0) = \frac{1 - c_2^\alpha}{c_2^\beta - c_2^\alpha} V_m^\beta \kappa_\gamma. \tag{39}$$

When α is a dilute solution ($c_2^\alpha \ll 1$), two particular cases are often considered. If $c_2^\beta \approx 1$, then $\frac{1-c_2^\alpha}{c_2^\beta - c_2^\alpha} \approx 1$, whereas if $c_2^\beta = kc_2^\alpha < 1$, then $\frac{1-c_2^\alpha}{c_2^\beta - c_2^\alpha} \approx \frac{1}{(k-1)c_2^\alpha}$. (The latter situation is a reasonable approximation, when near an α-β phase transition that occurs in pure component 1, for two-phase equilibrium solubilities as a function of temperature, with the concentrations increasing but their ratio remaining approximately constant as the temperature is lowered.)

However, in this case where the α phase is assumed to be a very dilute solution, one can do better than to assume that c^α and c^β are constant. By Henry's law [KR],

$$\mu_2(\kappa_\gamma) - \mu_2(0) = RT \ln \frac{c_2^\alpha(\kappa_\gamma)}{c_2^\alpha(0)} \tag{40}$$

One can put this into the differential equation in place of those earlier assumptions.

<u>Graphical Method</u>

Finally, a graphical method is often used to illustrate the effect of curvature for equilibria utilizing the molar or molal Gibbs free energy. If there are originally two wells in the plot of $G/\sum N_i$ versus $c_2 = N_2/\sum N_i$ (at fixed temperature, pressure, etc.), corresponding to the stable phases α and β, then the equilibrium two-phase compositions ignoring surface free energy are found at the tangent points of the line that is tangent to both of the wells. The endpoint of the line at $c_2 = 0$ gives the chemical potential of component 1 in this two-phase region and the endpoint at $c_2 = 1$ gives the chemical potential of component 2. If, however, one includes surface free energy, and if the interface has κ_γ nonzero, then one can find the equilibrium concentrations on each side of that interface by shifting the β well by the additional pressure on β. If V_M^β is assumed constant independent of concentration and pressure, then the well is uniformly shifted up by κ_γ, corresponding to increasing the energy per mole of β by $\Delta P = \kappa_\gamma$. But again one can do better if the dependence of V_m^β on P and c_2 is known: for each individual concentration, the point on the graph should be shifted up by

$$\int_{P^\alpha}^{P^\alpha + \kappa_\gamma} V_m^\beta(c_2, P) dP, \tag{41}$$

since $\frac{\partial (G/N)}{\partial P} = V_m^\beta(c_2, P)$. The tangent points of the common tangent to the original α well and the shifted β well give the new equilibrium concentrations.

One can also interpret this another way: given a concentration c_2^α that one wishes to have for phase α, there is (if each well is uniformly convex over the range of interest) a unique concentration c_2^β and shift $\Delta P = \kappa_\gamma$ such that at equilibrium an interface with weighted mean curvature κ_γ separates a β phase with concentration c_2^β from an α phase with concentration c_2^α. Again, the endpoints of the line give the chemical potentials.

A third interpretation was used by Gibbs for finding the globally stable phase when T and the μ_i are given. In the graphical construction this specifies the line (or hyperplane for many components) that the molar Gibbs free energy surface must be tangent to. For each possible phase this defines a composition and a P. The globally stable phase is the one with the highest P, and thus with the lowest Ω_V. This highest pressure phase will displace all other phases, unless it is contained by a surface with sufficient curvature to bring it into equilibrium with phases at lower pressure. The pressure difference and the required weighted mean curvature define the critical nucleus in nucleation theory.

References

[AJ] J. Iwan D. Alexander and William C. Johnson, *Thermochemical equilibrium in solid fluid systems with curved interfaces*, J. Appl. Physics **58** (1985), 816-824.

[C] John W. Cahn, *Thermodynamics of Solid and Fluid Surfaces*, in Segregation to Interfaces, ASM Seminar Series (1978), 3-23.

[CT] John W. Cahn and Jean E. Taylor, *Surface Motion by Surface Diffusion*, Acta Metall. mater. **42** (1994), 1045-1063.

[CRCT] W. Craig Carter, Andrew R. Roosen, John W. Cahn, and Jean E. Taylor, *Shape evolution by surface diffusion and surface attachment limited kinetics on completely faceted surfaces*, Acta Metal. Mater. **43** (1995), to appear.

[FW] Matthew P. A. Fisher and Michael Wortis, *Curvature correction to the surface tension of fluid drops: Landau theory and a scaling hypothesis*, Phys. Rev. B **29** (1984) 6252-6260.

[G] J. Willard Gibbs, *The Collected Works of J. W. Gibbs, Vol. 1*, Longmans, Green and Co., New York, 1928.

[JA] William C. Johnson and J. Iwan D. Alexander, *Interfacial conditions for thermomechanical equilibrium in two-phase crystals*, J. Appl. Physics **59** (1986), 2735-2746.

[LC] Francis Larché and John W. Cahn, *The Interactions of Composition and Stress in Crystalline Solids*, Acta Met. **33** (1985), 331- 367; reprinted in Journal of Research of NBS **89** (1984), 467-500.

[K] M. Jeannette Kelley, *Edge-energy-minimizing surfaces and crystal shape*, this volume.

[KR] I. M. Klotz and R. M. Rosenberg, *Chemical Thermodynamics*, Benjamin/Cummings, Menlo Park, 3rd Edition, 1972, Chapt. 18.

[M] Frank Morgan, *Riemannian Geometry: A Beginner's Guide*, Jones and Bartlett, Boston, 1993.

[T1] Jean E. Taylor, *Mean curvature and weighted mean curvature*, Acta Metall. Mater. **40** (1992), 1475-1485.

[T2] Jean E. Taylor, *Anisotropic Interface Motion*, this volume.

[TCH] Jean E. Taylor, John W. Cahn and Carol A. Handwerker, *Geometric Models of Crystal Growth*, Acta Metall. Mater. **40** (1992), 1443-1474.

ANALYSIS OF ANISOTROPIC CHARACTERISTICS OF DISLOCATION LOOP MORPHOLOGY

Donald Schwendeman[*] and Krishna Rajan[†]
[*]Department of Mathematical Sciences and [†]Department of Materials Science and Engineering
Rensselaer Polytechnic Institute
Troy, NY

ABSTRACT

In this paper we outline a mathematical formulation of the curvature-dependent motion of planar curves and use this formulation to simulate the shrinkage of dislocation loops. It is shown that the anisotropic characteristics associated with dislocation line tension can be modeled by including a dependence on the normal direction of the dislocation curve to the assumed normal velocity function. Examples of the simulation results are provided and are shown to qualitatively capture anisotropic features of dislocation morphology of closed loops.

INTRODUCTION

Curvature dependent flow is a fundamental tenet in a variety of materials science problems. In this paper we explore one such example; namely curvature induced motion of dislocations involving purely conservative conditions. In considering dislocation motion, one has a practical materials problem which can be modeled mathematically in a relatively simple way as the dislocation itself can be treated as simply as a line. With no other parameters associated with defining the interfacial curvature such as phase changes across that line this lends itself well to explore two-dimensional computational procedures involving curve evolution. While there exists numerous studies on the mathematical and computational issues associated with shape evolution particularly as applied to crystal growth [1], there has been little discussed in terms of its applications to dislocation problems. The purpose of this communication is to present the initial stages of work on examining the dynamics of dislocation loop shrinkage.

The anisotropic characteristics associated with dislocation loops has been discussed extensively in the literature (see, for example, Kirchner [2] for an excellent review). As noted by Shintani [3] the differential equation which determines the equilibrium shape of a dislocation loop is an equilibrium equation between the line tension and the Peach-Koehler force due to an external stress, and is also equivalent to the result of an application of the Wulff construction to the dislocation problem. Most studies involving the calculation of the shape of dislocation loops have considered equilibrium conditions and not the evolution of the shape of the dislocation. In the following discussion we provide a mathematical formulation that can be used to describe the curve evolution. While we limit ourselves to curvature dependent flow as the governing physics of the problem, it is pointed out that the results at this stage of the work still provide an useful insight into the role of anistropic characteristics on the dynamics of dislocation loops.

MATHEMATICAL FORMULATION

The mathematical framework and computational method used to study the motion of dislocation loops is similar to that discussed in Schwendeman [4] which follows the formulation of geometrical shock dynamics discussed in Witham [5]. Similar formulations can be found elsewhere (see for example Yokoyama and Sekerka [6]) but these have not been applied to the evolution of dislocation loops.

Let the family of curves $\alpha(x, y) = t$ where t is time define the successive positions (x, y) of the dislocation loops or "fronts". Orthogonal projections of these curves or "rays" are defined by the family of curves $\beta(x, y) =$ constant. The two families of curves form an orthogonal "front-ray" network as shown in Figure 1. The metric for the orthogonal coordinate system (α, β) is $ds^2 = U^2 d\alpha^2 + A^2 d\beta^2$, where $U(\alpha, \beta)$ is the normal velocity of the front and $A(\alpha, \beta)$ is related to the arc length along the front. A straightforward geometric argument leads to the equations

$$\frac{\partial \theta}{\partial \alpha} = -\frac{1}{A}\frac{\partial U}{\partial \beta} \tag{1}$$

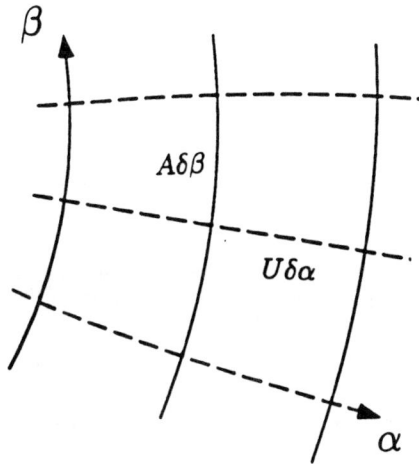

Figure 1: Schematic of front- ray network

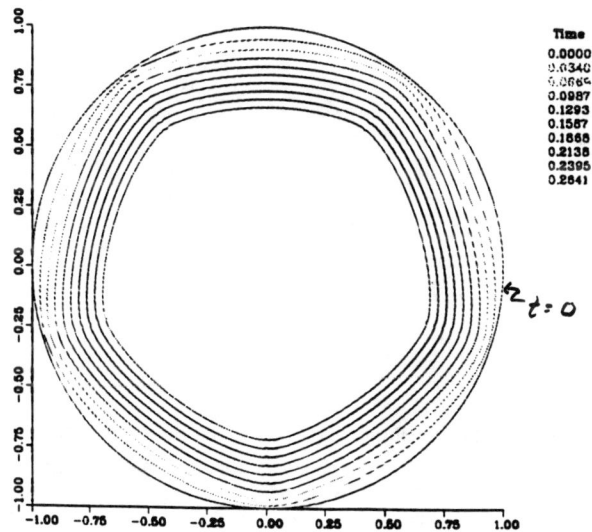

Figure 2: Shrinkage of circular loop: n=5, δ = 0.8

$$\frac{\partial \theta}{\partial \beta} = \frac{1}{U}\frac{\partial A}{\partial \alpha} \qquad (2)$$

where $\theta(\alpha,\beta)$ is the angle between a ray and some fixed reference direction.

The dynamics of the dislocation loop motion is determined by an assumed dependence of the normal velocity U on the geometry of the front. It is assumed that

$$U = U(\theta,\kappa) \qquad (3)$$

where $\kappa(\alpha,\beta)$ is the curvature of the front and is given by

$$\kappa = \left.\frac{d\theta}{ds}\right|_{front} = \frac{1}{A}\frac{\partial \theta}{\partial \beta} \qquad (4)$$

The dependence on θ is included in (3) to model the anisotropic motion and makes the formulation here slightly different than that in [4] . Using (3) and (4) in (1) and (2) leads to the following system of evolution equations

$$\left. \begin{aligned} \frac{\partial \theta}{\partial \alpha} &= -U_\theta \kappa - \frac{U_\kappa}{A}\frac{\partial \kappa}{\partial \beta} \\ \frac{\partial A}{\partial \alpha} &= AU\kappa \\ \kappa &= \frac{1}{A}\frac{\partial \theta}{\partial \beta} \end{aligned} \right\} \qquad (5)$$

Initial conditions for the equations in (5) are given by θ , A and κ along the initial curve at $\alpha = t = 0$ and periodic boundary conditions are used assuming an evolution of a closed dislocation loop curve. Once (5) is solved, the successive positions of the dislocation loops can be obtained by an integration along rays as is discussed in [4].

In order to model the motion of a dislocation loop, we take

$$U = -\kappa T(\theta) \qquad (6)$$

where $T(\theta) = 1 + \delta \sin(n(\theta - \theta_o))$ is an anisotropy function. In this simple form, δ measures the degree of anisotropy and n and θ_O give the n fold symmetry and angular orientation of the anisotropy. This type of treatment of crystalline anisotropy has been used for instance by Sethian and Strain [7] in their analysis of dendritic growth using a level set approach. We shall return later in the paper to comment further on this anisotropy function.

To handle the motion of general initial dislocation loop curves a numerical treatment of the equations in (5) is needed. The one used here is similar to the one presented in [4]. It is a second order accurate finite difference formulation of (5) that uses an implicit time marching in α. A numerical integration along rays determines a discrete set of points that give the successive positions of the dislocation loops.

RESULTS AND DISCUSSION

Using the mathematical formulation and computation method indicated above, we have conducted simulations of loop shape evolution for a variety of parameters as defined in the crystalline anisotropy function. Figure 2 shows the shrinkage of a circular loop in the presence of a five-fold symmetry (n=5). Figure 3 shows that a six sided prismatic loop will evolve into a four sided loop with n=4 but in the absence of any anisotropy (T=1; i.e.. δ=0) the same loop will evolve into a circular loop. Figure 4 shows a series of calculations with varying values of our starting reference orientation (θ_O) for an initially elliptical dislocation loop. While they all the example calculations in Figure 4 result in approximately four-sided loops (n=4) it is clear that the line lengths in different directions vary.

Qualitatively it is apparent that variations in n and θ_O as experimented with in these present calculations each affect different aspects of the shape of an anisotropic dislocation loop while the value of δ represents the magnitude of the anisotropic effect. While anisotropy in dislocation line tension is clearly one source of anisotropic effects in loop morphology, it is important to recall that it is separate from the line energy contributions. It is interesting to note that Kirchner has emphasized this point and he notes that calculations of loop morphology which have size dependent shapes need to account for the variation in line energy as the loop size changes along with the anisotropy in line tension. As he notes, the energy of each piece of the dislocation loop depends on the shape of the rest of the loop. Hence the energy has to be minimized by adjustment of the shape of the dislocation loop. In fact Kirchner has noted that the work to date on loop morphology does not take into account all the factors associated loop shape simultaneously. As he states, the difficulty of construction of loop shapes depends both on the global configuration as well as the core terms which contribute to the anisotropy in the line tension formulations.

Although we have not explicitly included elasticity equations into our analysis, the anisotropy expression used in this study appears to provide in a phenomenological manner at least the role of crystalline anisotropy calculating loop morphology. The next stage of this work is to incorporate these stress effects explicitly.

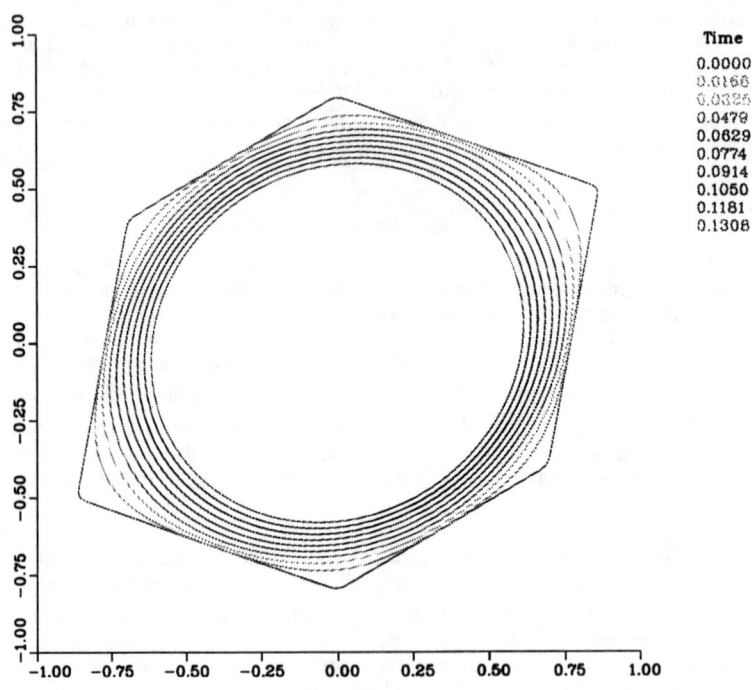

Figure 3 a: Shrinkage of prismatic loop ; n=6, δ=0 (no anisotropy)

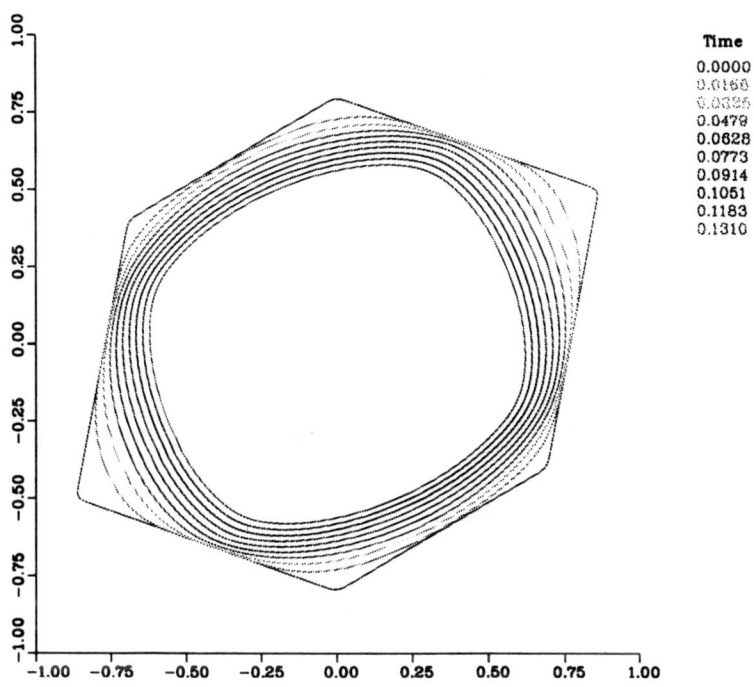

Figure 3 b: Shrinkage of prismatic loop ; n=6, δ=0.6

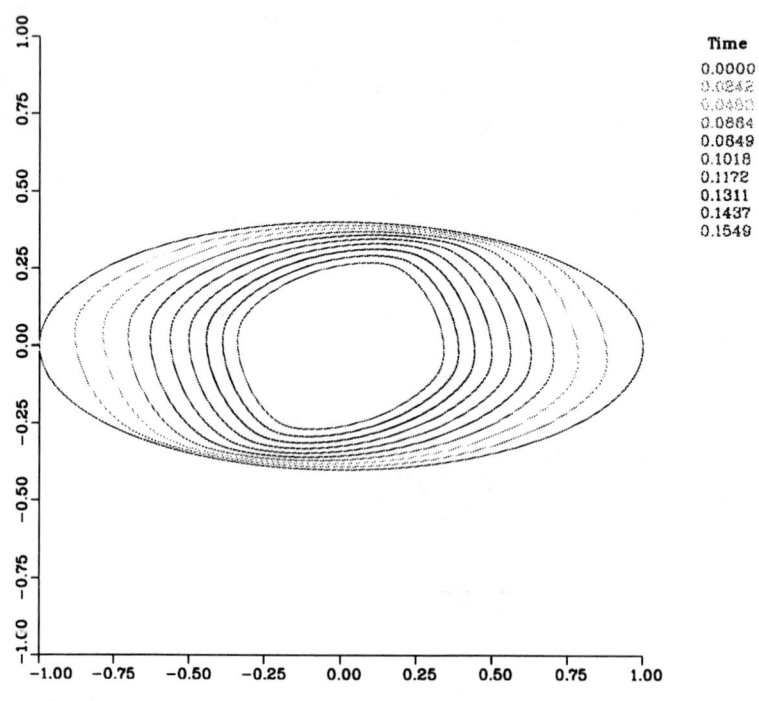

Figure 4a: Shrinkage of elliptical loop ; n=4, δ=0.6, θ_O= 0

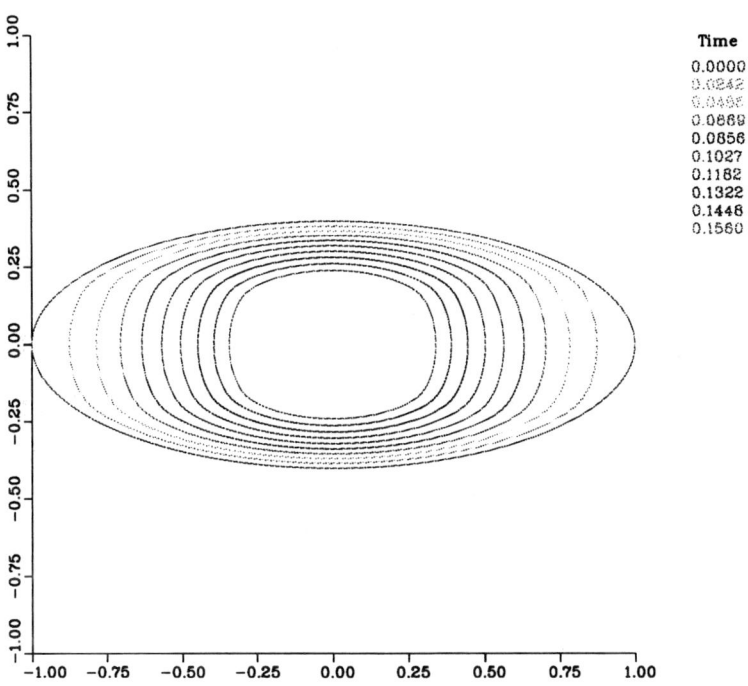

Figure 4b: Shrinkage of elliptical loop ; n=4, δ=0.6, θ_O= -π/8

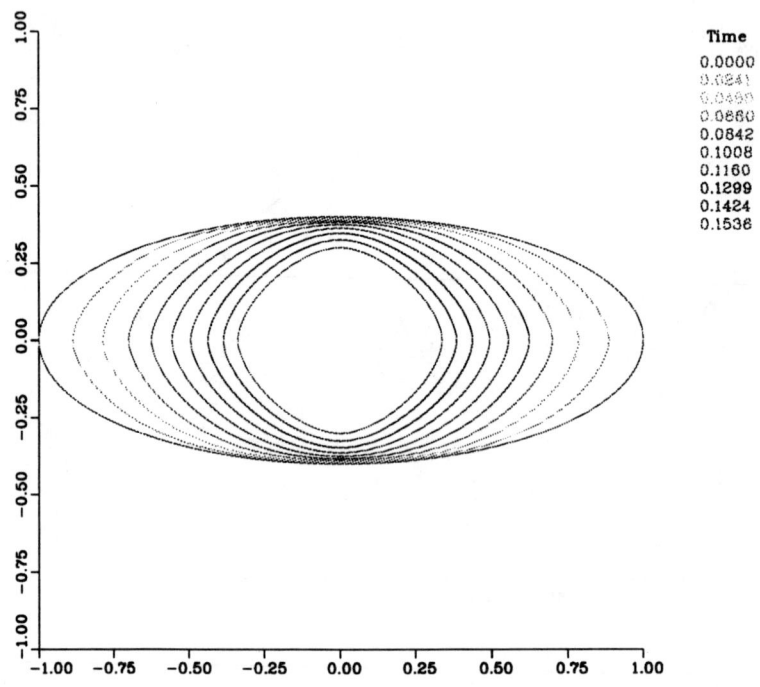

Figure 4c: Shrinkage of elliptical loop ; n=4, δ=0.6, θ_O= π/8

CONCLUSIONS

A mathematical formulation of curvature dependent motion of the morphological evolution of dislocation loops taking into account anisotropy has been described. While the simulation results are only for the local law of motion by mean curvature, some suggestions are provided on the physical interpretation of the results in terms of the anisotropy of local elasticity effects.

REFERENCES

1. J. E.Taylor, J.W.Cahn and C.A. Handwerker, "Geometric Models of Crystal Growth" Acta Metall. et. Mater. **40** 1443-1474 (1992)

2. H.O.K.Kirchner, "The Concept of Line Tension: Theory and Experiments" in Dislocations 1984 Comptes rendus du Colloque International du C.N.R.S. Dislocations - Aussois , France , March 8-17 , 1984 , CNRS Publications , Paris, 53-71 (1984)

3. K.Shitani, "Equilibrium Shapes of Dislocation Shear Loops in Lithium Niobate and Lithium Tantalate Under an External Stress in the Line Tension Approximation" Philosophical Magazine **67** 361-368 (1993)

4. D.W.Schwendeman, "A Front Dynamics Approach to Curvature Dependent Flow" SIAM Journal of Applied Mathematics (in press)

5. G.B.Whitham, "A New Approach to Problems of Shock Dynamics: Part I: Two dimensional problems" Journal of Fluid Mechanics **2** 145-171 (1957)

6. U. Yokoyama and R.F.Sekerka, "A Numerical Study of the Effect of Anisotropic Surface Tension and Intefrace Kinetics on Pattern Formation During the Growth of Two-dimensional Crystals" J. Crystal Growth **125** 389-403 (1992)

7. J.A.Sethian and J.Strain, "Crystal Growth and Dendritic Solidification" Journal of Computational Physics **98** 231-253 (1992)

ON THE PHYSICS OF SELF-ORGANIZING MICROSTRUCTURAL EVOLUTION

J.S. Kirkaldy

Institute for Materials Research
McMaster University
Hamilton, Ontario, Canada L8S 4L7

Abstract

Anderson has argued that to be self-organizing, a pattern-forming dissipative system must proceed via one or more broken symmetries into a new order parameter space and thereby exhibit *autonomy* with respect to the initial and boundary conditions, together with a property of *rigidity* which removes the ensuing degeneracies. The microstructures of condensed materials possess a high degree of formative autonomy, and because such patterns often exhibit uniqueness, must possess rigidity. Phenomenologies such as the phase field construction and dissipation stationarity are examined and compared as to their degeneracy-removing abilities and their foundations in the variational frictional mechanics of Hamilton as constrained within thermodynamics. It is demonstrated that for a class of non-linear irreversible binary materials phenomena, including migration of antiphase boundaries (APB's), global rigidity must be reflected atomistically within an analogue of Onsager Reciprocity, its validation residing ultimately in time-reversal and logical 4-group invariances. It is argued within specific models that coherency strain effects and the phase field paradigm as currently formulated do not always provide sufficient constraints for a complete specification of the dynamical steady or non-steady state configurations. Dissipation stationarity in order parameter space pertaining to nonlinear steady or quasi-steady systems where the heat bath is near equipartition is, on the other hand, sufficient when applicable, and represents a natural extension of the rigid patterning or uniqueness provided in multivariate linear systems by the principle of minimum entropy production based upon time reversal invariance (Onsager Reciprocity). It is concluded that diffusion-induced grain boundary migration, spinodal decomposition and discontinuous precipitation in binary alloys are all stabilized by orthogonal vacancy fluxes correlated according to the aforementioned dynamical time reversal and syntactic 4-group symmetries. Degeneracy-removing phenomenologies must reflect such atomistic constraints.

Introduction

Many modern physicists have come to believe that the complete understanding of spontaneous pattern formation or self-organization is one of the keys to attaining a correct theory of development in vital matter. After all, the evolution of both the biological species and the individual involves a sequence of spontaneous mesoscopic broken symmetries mediated by topological defects, otherwise known as genes. Broken symmetries automatically define internal degrees of freedom or degeneracies whose autonomous extension in phase space can usually be identified with one or more order parameters, e.g., the scales of allotriomorphs. One should consider in this light the pictured non-vital dynamical systems (Fig. 1) provided with a continuous or slowly depleting supply of available free energy by supersaturation and responding via a broken symmetry.

Since spontaneous patterns often tend to uniqueness, understanding requires that we specify correctly how the degeneracies are to be removed. Phillip Anderson, the condensed state physicist, coined the essential properties of such self-organization as *autonomy* and *rigidity* (1). The latter term is particularly apt since it could conceivably be inclusive of the microscopic solvability constraint, coherency strain effects, of phase field and dissipation principles, and equivalently in some cases of correlated motions of quasi-particles or topological defects, which Friedel (2) and Kawasaki (3) regard as essential to the maintenance of the rigidity property (Fig. 1). Quasi-particles are deemed to reside

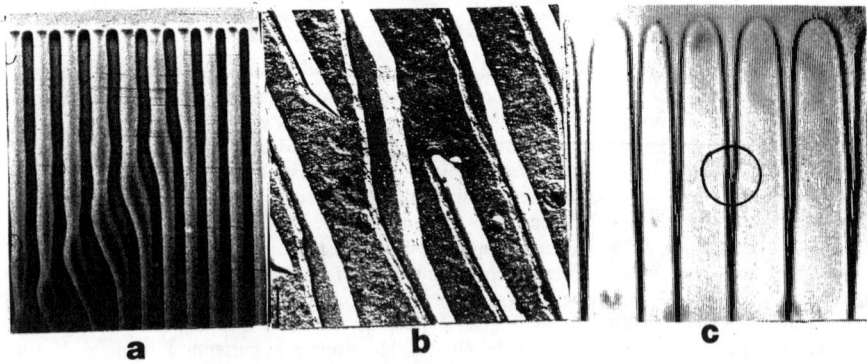

Fig. 1   Examples of pattern stabilizing topological defects. a) Forced velocity eutectic with frontal defect moving to the right to increase spacing. b) Migrating lamellar faults in isothermal eutectoid offering mechanism for spacing change. c) Forced velocity cells exhibiting groove wall kinks which upon annihilation at the roots accommodate widening of the thinnest finger.

in an ergodic coarse-grained irreversible phase space in which the path probability or path entropy serves as the stabilizing potential. The present contribution focusses upon systems whose quasiparticles, in contrast to Fig. 1, are elementary atomic combinations.

There are a number of reputable scientists who do not accept the concept of dynamical degeneracy. Such readers are directed to Fig. 2 which illustrates within a rigorously valid theory the fact that planar two-phase ternary diffusion paths on the Gibbs isotherm, which are numerical solutions of simultaneous transcendentals, are not necessarily unique for invariant boundary conditions (4, 5), and therefore define a rather profound digital selection problem. This is superposed upon the wavenumber selection problem which accompanies the symmetry-breaking implied by the local supersaturation appearing within all three paths. In this contribution, I address in turn five questions pertaining to the rigidity property and the relationship between its microscopic and macroscopic representations.

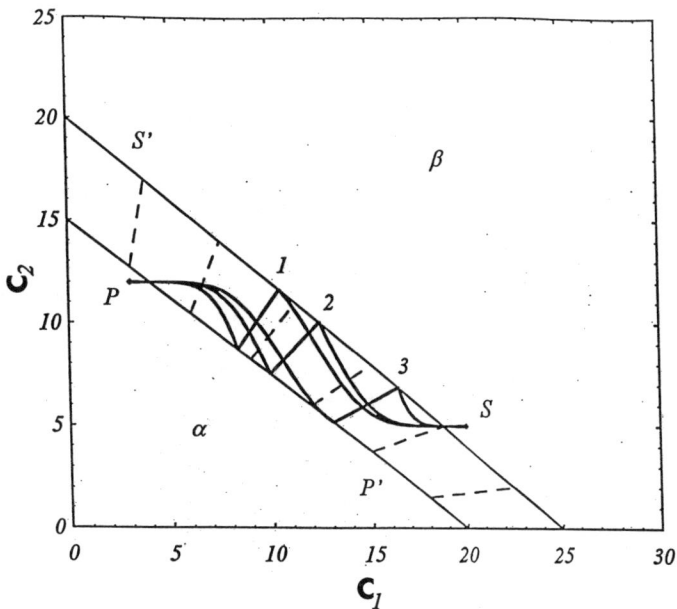

Fig. 2 Multiple ternary diffusion paths on the Gibbs isotherm corresponding to invariant constituent parameters and boundary conditions (5).

Can a Gibbsian Constraint like Coherency Strain Remove Dynamical Degeneracies?

Coherency strain is often an essential element of the early stages of spinodal decomposition (6) where strain-relaxing misfit dislocations (7–10) may be unavailable. The main degeneracy has to do with wavenumber selection which is only moderately influenced by coherency strain. Cahn (6) invoked selection to the fastest growing wavenumber without reference to coherency strain, so here the answer is in the negative. This is an independent, dynamical principle generic with a dissipation principle.

Lamellar discontinuous precipitation (DP) bifurcates from, and often dominates multidisperse precipitation (11, 12) in dilute supersaturated binary solutions (Fig. 3). It involves a multivariate degeneracy whose order parameters may initially be specified as the lamellar spacing and the four interface chemical potentials $\mu_{A,B}^{\alpha}$ and $\mu_{A,B}^{\alpha'}$. There is one differential Gibbs-Duhem constraint, in principle integrable from the triple points, to each side of the interface (13, 14) leaving three parameters unspecified. In dilute α' solutions knowledge of the solvent difference $\mu_A^{\alpha'} - \mu_A^{\alpha}$ is essential to formulation of the interface chemical reaction which lies in series with the solid diffusion field. Now coherency strain can modify both solvent chemical potentials but it cannot by itself remove the degeneracy which links the curvature-dependent free concentration difference and the spacing. All of this assumes the unlikely simplification that the interface mobility can be independently specified. The answer to the question is again, in the negative. A theory of chemically induced grain boundary migration (CIGM) based solely on coherency strain necessarily falters on the same grounds. A number of experimental investigations have emphasized this intelligence (15, 16). On the other hand, dissipation optimality has led to semi-quantitative predictions for both DP and CIGM (17, 18; see also penultimate section).

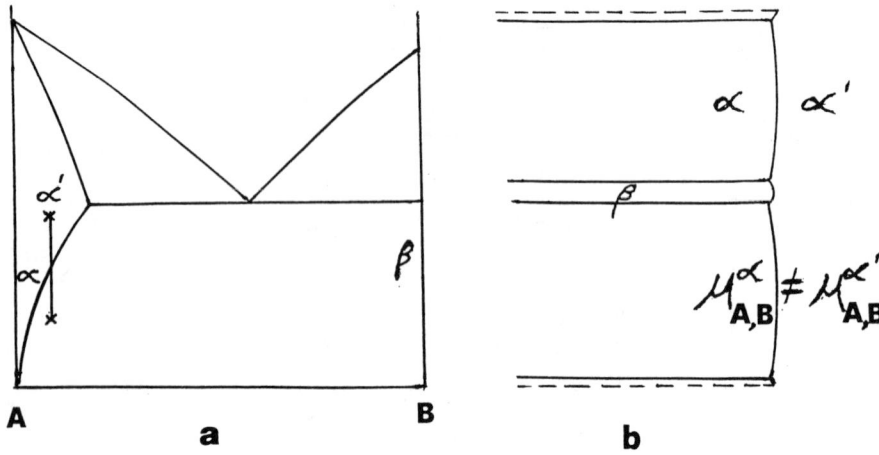

Fig. 3   Representation of discontinuous precipitation. a) Schematic phase diagram. b) Lamellar configuration.

The controlling involvement of coherency strain in liquid film migration has often been proposed, particularly by Yoon, Cahn and coworkers (19–21). Binary (Mo-Ni) and ternary (Mo-Ni-Fe) transformations of pre-sintered solid-liquid mixtures have been heat-treated to demonstrate LFM and the observations have been argued circumstantially to support the coherency strain theory. Unfortunately, the three-dimensional experiments are analyzed from the point-of-view of a one-dimensional theory, overlooking the obvious metallographic fact (Fig. 4) that a dendritic or Mullins and Sekerka instability (22) occurs with lateral liquid diffusion driven by the emergence of regions of high local curvature which attract the Ni solute. Furthermore, in the analysis of the ternary experiments, which are claimed to be definitive, Yoon, Cahn and coworkers (19–21) failed to take into account the fact that the order of the degeneracy has increased. This is because different tielines can be selected on the two sides of the film. The null in the migration as a function of iron content which is reported is most probably a constitutional effect. It is to be appreciated that with unrestricted lateral diffusion driven by capillarity, coherency strain cannot alone define the liquid *gradient* at the advancing front so a kinetic indeterminacy remains. Furthermore, with the normally large solid diffusion coefficients near the melting point abetted by an efficient vacancy source at the interface, the diffusion length will tend to be large with concomitant easy nucleation and climb of misfit dislocations which together with a correlative high vacancy density will relax the coherency strain (8–10). The investigations of Fournelle et al. (23–25) have also discounted a dominant effect of coherency strains. We conclude in general that this Gibbsian constraint is insufficient for removing dynamical degeneracies. Recognition of this shortfall has led to more elaborate theories incorporating gradient energies and the phase field paradigm (16, 26).

<u>Under What Circumstances Can Phase Field and Dissipation Optimal Procedures Act as Rigidity Agents?</u>

Here we access Hamilton's Principle for frictional systems as a general principle of rigidity (27, 28). This can be written

$$\delta \int_{t_1}^{t_2} (T - V + W)\, dt = \delta \int_{t_1}^{t_2} (2T + W)\, dt = 0 \qquad (1)$$

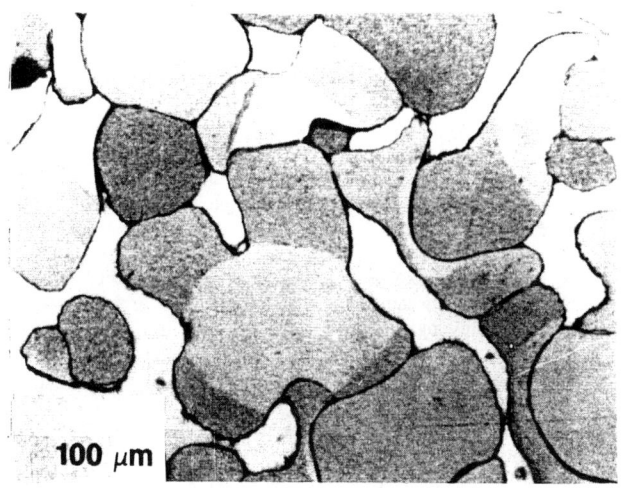

Fig. 4   Microstructure of Mo-10 Ni solid-liquid alloy quenched from 1480° to 1400°C and held for 2 hours. Lighter regions represent original solid particles, many of which have sprouted dendritic protuberances (19).

where T is the total kinetic energy while the cumulative work W−V is divided into that stored as available potential energy and the internal friction kinetic energy transferred to the heat bath, W. Assuming V = V(x, y, z) and equipartitioning in the heat bath, that trajectories for t_1 and $t_2 \to \infty$ imply $\delta W = 0$ (2nd Law), then from the mechanical equivalent of heat, $\delta T \sim \delta W$, with T + V conserved, one obtains the right-hand side of Eq. (1) and from this the three principles

$$\delta \int_{t_1}^{t_2} T\, dt = 0; \quad \delta \int_{t_1}^{t_2} V\, dt = 0; \quad \delta \int_{t_1}^{t_2} 2W\, dt = 0 \quad (2)$$

where $2W/\Theta(°K)$ is the path entropy. The virtual variations of mechanics which normally violate the kinematic constraints are translated to variations in order parameter space with t_1 and t_2 fixed. The last form in Eqs. (2), which is equivalent to Onsager's Principle of Maximum Path Probability or Entropy (29), implies stationarity in the dissipation at the steady or quasisteady state at near *equipartition* (28). Ultimately, this comes about because temperature can equalize at chemical and/or mechanical disequilibrium, but not *vice-versa*. More generally, a nonsteady combination from Eq. (2) implies

$$\delta \int_{t_1}^{t_2} \Delta F\, dt = 0; \quad \Delta F = \int_v \Delta f\, dv \quad (3)$$

which bears a relationship to the phase field postulate (3)

$$\frac{\partial \eta}{\partial t} = -M_\eta \frac{\delta \Delta F}{\delta \eta} \quad (4)$$

for non-conserved order parameter η, and

$$\frac{\partial c}{\partial t} = \nabla M_c \nabla \left[\frac{\delta \Delta F}{\delta c} \right] \quad (5)$$

for conserved order parameter c (3, 6), M_η and M_c being appropriate kinetic parameters. Eqs. (4) and (5) are not derivable from Eq. (3) but they are consistent with it subject to certain restrictions on the M's, which may in fact *require* spatial dependencies. This adds to the plausibility of Eqs. (4) and (5) but does not guarantee their correctness in all applications. On the other hand, the dissipation stationarity principle at the steady state

$$\delta(d_i S/dt) = 0; \quad \text{max or min} \tag{6}$$

is deriveable with the strict caveat that variations are to be made in order parameter spaces which are not explicit in the time. We will later argue that Eqs. (4) and (5) require revisions for valid application to the binary solute trapping problem.

The application of Eq. (4) to the relaxation of curved antiphase boundaries (APB's) is quite well-established so we will now penetrate more deeply into the physics of this apparently rudimentary kinetic configuration via the following question.

How Does Onsager Coupling Enter into the Nonlinear Phase Field Formulation of APB Migration?

In 1960, Kikuchi (30) formulated the homogeneous second order, order-disorder reaction in Onsager's linear irreversible thermodynamics derivable from Eq. (3)

$$\frac{\partial \eta_1}{\partial t} = -L_{11} \frac{\delta \Delta F}{\delta \eta_1} - L_{12} \frac{\delta \Delta F}{\delta \eta_2} \tag{7}$$

and

$$\frac{\partial \eta_2}{\partial t} = -L_{21} \frac{\delta \Delta F}{\delta \eta_1} - L_{22} \frac{\delta \Delta F}{\delta \eta_2} \tag{8}$$

proving via microscopic reversibility that $L_{12} = L_{21}$, that the short range order parameter η_2 equals the long range η_1 for $T_c >> T$, η_1 vanishes at $T = T_c$ and η_2 persists above that temperature (clustering). Microscopic reversibility here serves as the rigidity agent eliminating the virtual degeneracy involved in introducing two tentatively independent order parameters. Significantly, this is equivalent to a quasi-steady or drifting minimum in the dissipation rate (30, 31). Our question pertains to the non-linear APB problem framed within Eq. 4 where a single hybrid parameter is invoked. Lifshitz (32) has defined this locally as

$$\eta = X^\alpha - X^\beta \tag{9}$$

where the concentrations are adjacent sublattice values. This amounts to retention of two order parameters. The question then arises as to how Onsager coupling appears in the non-linear formulation.

Krzanowski and Allen (33) have argued within a Bragg-Williams model that this formulation implies segregation to an equilibrium APB profile. Proceeding in a similar fashion, but with some corrections, my collaborator Maugis (34) has found a continuum representation of a symmetric B2 alloy in which the excess biplane free energy density is

$$\Delta f(\eta, c) = \kappa \left[\left(\frac{d\eta}{dx}\right)^2 - \left(\frac{dc}{dx}\right)^2 + 2(c - \frac{1}{2}) \frac{d\eta}{dx} - 2\eta \frac{dc}{dx} \right] + \Delta f^0 \tag{10}$$

(Δf^0 = homogeneous part) and the equilibrium Euler equations corresponding to $\delta \Delta F/\delta \eta = 0$ are

$$\frac{\partial \Delta f^0}{\partial \eta} - 2\kappa \left[\frac{d^2\eta}{dx^2} + 2\frac{dc}{dx} \right] = 0 \tag{11}$$

and

$$\frac{\partial \Delta f^0}{\partial c} + 2\kappa \left[\frac{d^2c}{dx^2} + 2\frac{d\eta}{dx} \right] = 0 \qquad (12)$$

Here, Maugis has defined in terms of mole-fractions X, $c = 1/2(X^\alpha + X^\beta)$ and $\eta = 1/2(X^\alpha - X^\beta)$ so the maximum ranges are both $\pm 1/2$.

Figure 5 shows the results of a numerical computation at $T = 0.8T_c$ for a matched APB pair based

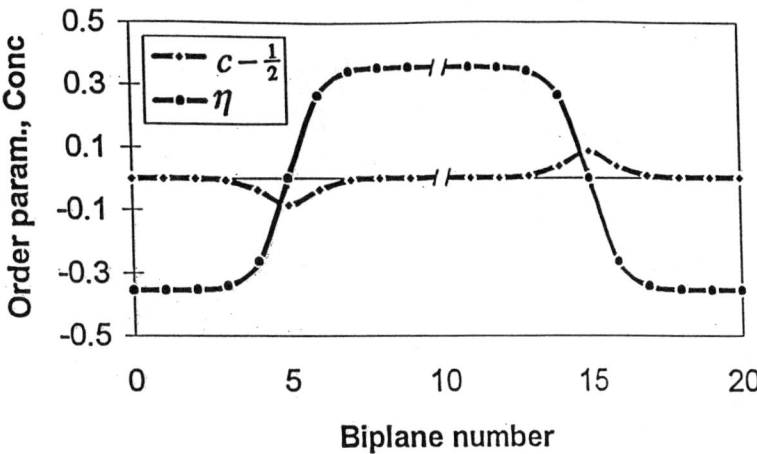

Fig. 5 Equilibrium order parameter and concentration distributions pertaining to a shear-generated pair of antiphase boundaries (34).

on the discrete representation from which Eqs. (10), (11) and (12) were derived. The relative amount of the segregation is by no means negligible.

This result gains greater significance where one considers a closed APB such as a slowly relaxing sphere formed by <111> shear in a B2 lattice, ignoring anisotropy. If we assume here, as with the Allen-Cahn solution (35) in the absence of segregation (Fig. 6), that both the η and c profiles in motion retain close to equilibrium representations then such collapse must be accompanied by correlated peripheral desegregation to opposite hemispheres. Without generalizing Eqs. (11) and (12) to the three-dimensional, time-dependent forms, we aver in the following that the necessary coupling between radial and peripheral motions can be expressed in terms of microscopically reversible trajectories.

In another contribution (36), we undertook a theoretical estimation of M_η in Eq. (4) for a relaxing sphere on the basis of an inward vacancy flux driven by capillarity with η profiles according to Allen and Cahn (Fig. 6). The unit mechanism of this Kirkendall-like process is flipping of dominant B/A dipoles on the inside (say) into dominant A/B dipoles on the outside via a 6-step response to the vacancy flow from the outside (Fig. 7, upper tier) in which the net lateral flow is zero. There is on the other hand, an equally probable local sequence for inward flow of vacancies whose net effect (lower tier) is to translate B/A dipoles and vacancies laterally. Clearly, if the region is A or B - rich the latter motions will achieve a net lateral transport of A or B in a direct probabilistic relationship to the rate of dipole flipping. Also, each trajectory set considered locally will have an exactly reversed set at a diametrically opposed position so that the global manifold satisfies microscopic reversibility. It is significant that this vectorial vacancy cross effect is analogous to the earliest

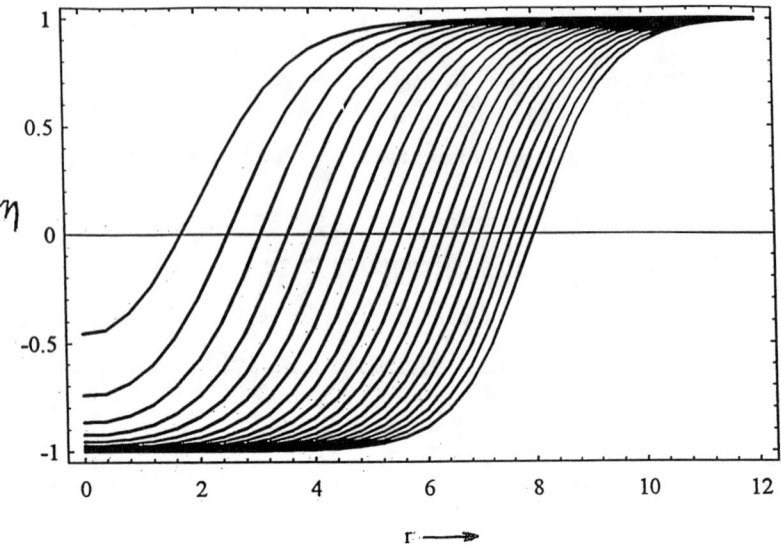

Fig. 6   Computational order parameter solution of Allan-Cahn Eq. (4) for a collapsing antiphase boundary sphere.

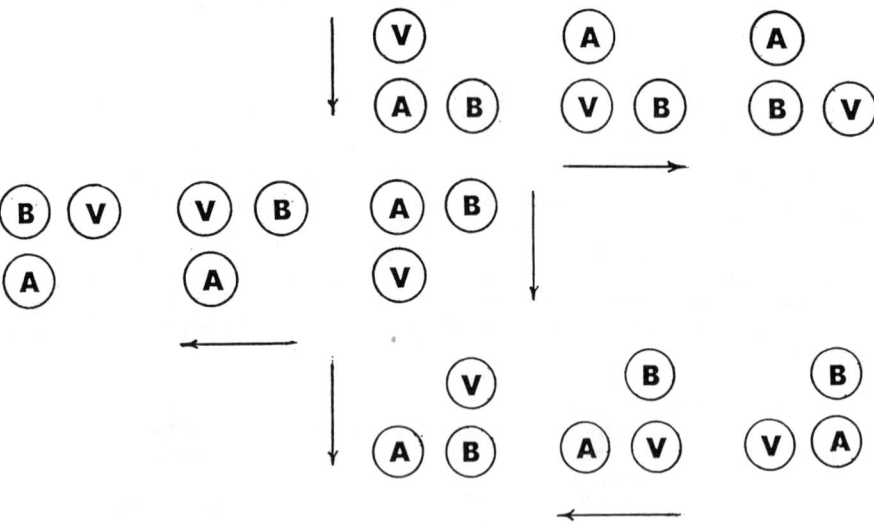

Fig. 7   Illustration of APB motion as in Fig. 6 by flipping of B/A dipoles to A/B via inward motion of vacancies V (upper sequence).   The equally probable lower sequence contributes only to lateral flow of vacancies.

example of reciprocity analyzed by Onsager (37) which pertained to orthogonal heat flows in anisotropic crystals. Note that we have chosen the term Onsager *coupling* rather than *reciprocity* since no constant symmetric macroscopic L coefficients are evident in the non-linear case. The macroscopic coupling is to be fully accommodated in principle by simultaneous solution of the three-dimensional coupled kinetic equations corresponding to Eqs. (11) and (12), with the understanding that the two mobilities have been rigorously evaluated in physics. This difficult modification remains to be investigated. Our structure gives a different viewpoint on the solute drag effect which was considered by Krzanowski and Allen as acting in the radial direction only (33).

Although the idea of a continuum or phase field boundary as with the early stages of a spinodal reaction holds up fairly well approaching $T = T_c$ in this simple APB case, its quantitative relevance to sharp boundary problems has not been firmly established even though some of the dendrite *patterns* are very impressive (38). Kobayashi, however, makes clear that these are *simulations* which make no attempt towards realism in the constitutive parameters. Our next question pertains to the phase field treatment of the solute trapping problem.

Is the Phase Field Methodology Adequate to the Multiple Degeneracies of the Solute Trapping Problem?

The solute trapping problem is generic with the DP and CIGM problems, for the degeneracies due to deviations from local equilibrium at the higher forced velocities are assignable to *a priori* unknown values of the four interface chemical potentials. Wheeler et al. (39, 40) ignored this precise statement of the problem, recognized originally by Langer and Sekerka (13) in a spinodal-based early approach to an analogous problem. They have proceeded on the basis of steady-state transformed versions of Eqs. (4) and (5) assigning a calibrated phantom double-welled potential to phase field Eq. (4) representing a continuum phase boundary and seeking an asymptotic solution wherein the boundary approaches a realistic width and surface tension. While this offers an appropriate simulated decrease in the effective partition coefficient with increasing velocity it appears to have no relevance to the main point of the problem which for a metastable solid (vanishing diffusion kinetics) has to do with *trapping of liquid structure* to produce an amorphous solid. We accordingly must answer in the negative to the question.

In contrast, an approach which invokes Gibbs-Duhem equations to remove two of the degenerate potentials and a saddle-point stationarity in the dissipation function to specify the other two (14) leads to the approximate explicit trapping formula for relative depression of the liquid interface composition.

$$\frac{\Delta X_L}{X_0} = -\frac{\rho v X_0}{L_B RT_m} \qquad (13)$$

as a ratio of the flux entering the solid to the microscopically reversible kinetic theory flux within the boundary in terms of the alloy composition X_0, the density ρ, the forced velocity v, and the solute across-interface mobility, L_B. In the full trapping limit the v in this accords in magnitude with the Mullins and Sekerka criterion for absolute stability of a planar interface (22). Furthermore, the deviation from equilibrium of the solvent chemical potential at the trapping limit

$$\mu_A^S - \mu_{A\,eq}^S \simeq RT_m \qquad (14)$$

Since empirically for metals the latent heat (41)

$$L \simeq 1.2\,RT_m \qquad (15)$$

we can conclude that the maximal solid lattice energy trappable from the amorphous state is approximately equal to the latent heat, an intuitively convincing result. This is stored available energy which could be later released and partially restored in the process of inverse creep (42), an ordering and work-producing process describable in part by Eq. (4).

A possible modification to the phase field structure would be to replace the phantom field η by the real non-conservative free volume field η_v satisfying form (4). The observations of Di Nardo and Bilgram (43) are salutary to such a phenomenological formulation for they have reported a pre-ordered frontal layer of about 1 μm thickness in the solidification of xenon, salol and cyclohexane. These authors also report a rich correlative literature which must not be further ignored by phase field specialists.

The entry of the time-reversible flux within this complex phase boundary process through phenomenology suggests that Onsager coupling as a microscopic basis for the rigidity property may offer some generality, particularly as regards patterning in vacancy-moderated binary solids.

How General is Microscopic Reversibility as a Remover of Dynamical Degeneracies?

We are reminded here of the procedure whereby (7) and (8) were coupled by Onsager Reciprocity to yield a unique description of the ordering process and its equivalence to a principle of dissipation minimality. All expressions (2) through (5) bear the condition of equipartition of the heat bath. Eq. 6 being deriveable is the strongest expression, actually generating a macroscopic equivalence by explicitly making reference to a time reversible flux (Eq. 13). Eqs. (4), (5), (7) and (8) evidently require the addition of microscopic reversibility to achieve uniqueness of solution specification.

All of this suggests that the mechanism of stabilization of CIGM must also involve this symmetry condition, and since it must also involve binary diffusion by a vacancy mechanism, an orthogonal vacancy flow interaction according to Fig. 7 would appear to be appropriate. Here we envision a circulation of vacancies out of the grain boundary and into the trailing phase to accommodate the creep of the excess atoms to the surface. The circulation closure by partial cross-boundary flow accommodates the cross-effect while delivering a sufficient number of excess vacancies to the boundary front for sustained motion. A source of excess vacancies resides in slip bands behind the interface.

Variants upon this class of models should pertain to discontinuous precipitation, which stabilizes at or near a state of maximum dissipation (17) and spinodal decomposition which, ignoring critical fluctuations, is conjectured to stabilize dynamically at a dominant fastest growing three-dimensional wavenumber. In this latter case for a symmetric spinodal decomposition growth in amplitude and wavenumber must from very early in the process be coupled by orthogonal vacancy fluxes. The countercurrent ripening mass fluxes presumably follow short-circuit paths defined by vacancy-rich regions predicated in turn and in part upon the orientation and relaxation of coherency strains. Such *essential* Onsager-coupled diffusion, driven by non-uniform incipient surface tension in solid state arrays, seems not to have received much attention, presumably because the conception challenges the adequacy of Eq. (5), which in any case is only consistent with (1) or (3) in the apparently irrelevant steady or quasi-steady state.

As background for the foregoing, which may appear controversial in some circles, it should be recognized that Cahn and Mullins long ago objected to the validity of Onsager Reciprocity in reference to solid state diffusion, and to dissipation optimality in relation to degeneracy in order parameter space, the latter on the grounds that it could not be exact in the trivial cases of non-degenerate, nonlinear systems (44). Mullins (45) ultimately moderated his views on Onsager coupling, but neither author has subsequently admitted that their expressed view on dissipation optimality was a *non-sequitur*. It can be surmised that had they done so, the emphasis and content of this symposium would have been much different.

The reader who is not directly attuned to the controversy should review Kikuchi's 1960 *tour-de-force* (30) and contemplate that the rather complete empiricism and theoretical closure on the Kirkendall effect provided by Manning (46) and many others (47, 48) within a vacancy mechanism strongly confirms indirect Onsager Reciprocity in the binary solid state. Furthermore, the direct evidence in the case of ternary alloys is highly salutary to the symmetry principle (48). As regards dissipation optimality we claim validity only for virtual variations between generalized mechanical trajectories which define an order parameter space. The experimental record is also salutary (49) with maximality attending commonly supersaturated systems and minimality attending non-uniform boundary conditions.

Discussion

During the past three decades, a common terminology has tended to permeate particle physics, critical point and phase transformation theory. For example, the terms *spontaneous symmetry breaking, hierarchies, scaling* and *renormalization* are common to all three sub-disciplines and indicative of a sense of self-organization or logical structure. The particle physicist, Stephen Weinberg, remarks that rigidity or logical connectedness is sustained in part through internal symmetry principles (50) and we have seen how time reversal invariance or microscopic reversibility plays a part in removing degeneracies following broken symmetries in both linear and non-linear pattern-forming systems. We find all of this as supportive of the idea that there exists an explicit mapping between formal logic and spontaneous pattern formation. We have indeed demonstrated in a 1985 Physical Review article (51) that phase state motion on an entropy production rate saddle surface in a bivariate or trivariate order parameter space modelling the stabilization process of an actual steady state phase transformation can be mapped to both the syntax and inferential syllogisms of logic. Furthermore, we have recently discovered and reported (52) that every overall pattern and the topological defects or quasiparticles which contribute to their construction (Fig. 1) possess the same syntactical 4-group symmetries which pertain to well-formed binary propositions in Boolean formal logic. The three 4-group symmetry operations of reflection, reflection at right angles and 180° rotation can all be matched to the configurations of Fig. 7, and their dynamical time reversed forms provide further evidence that the property of rigidity is properly interpreted as *logical rigidity*. The requirement for validation of a stable dissipation stationarity in equipartition or heat bath equilibrium is evidently equivalent in general to universality of 4-group and time reversal symmetry in the finest dispersions of the substrate quasiparticles (e.g., Fig. 7). The detailed physics of patterning thus extends into the digital foundations of mathematics rather than to the continuum representations which are the main focus of the mathematicians participating in this symposium. Materials scientists who persist in ignoring the microscopic and mesoscopic physics in favour of mathematics of the continuum will ultimately be seen to have rejected an assured path towards a complete theoretical quantification of their discipline.

References

1. P.W. Anderson, "Some General Thoughts About Broken Symmetry", in Symmetries and Broken Symmetries, ed. N. Boccara (Paris: IDSET, 1981), 11−20.

2. J. Friedel, "Concluding Remarks", in Symmetries and Broken Symmetries, ed. N. Boccara (Paris: IDSET, 1981), 197-209.

3. K. Kawasaki, "Ordering Kinetics in Phase Transitions", in Proceedings of the 1986 Summer School on Statistical Mechanics, ed. C.-K. Hu (Taipei: Institute of Physics, Academic Sinica, 1987), 171-185.

4. D.E. Coates, "Interface Stability During Isothermal Ternary Phase Transformations", Ph.D. Dissertation, (Hamilton, Canada: McMaster University, 1970).

5. P. Maugis et al., "Multiple Interface Velocity Solutions for Ternary Biphase Infinite Diffusion Couples", Acta Materialia, submitted (1995).

6. J.W. Cahn, "On Spinodal Decomposition", Acta Met., 9 (1961), 795-801.

7. J.S. Kirkaldy, discussion to F.C. Larché and J.W. Cahn, "Equilibrium and Diffusion in Stressed Solid Solutions with Defects", in Solute-Defect Interaction, Theory and Experiment, eds. S. Saimoto, G.R. Purdy and G.V. Kidson (Toronto: Pergamon Press, 1986), 1-27.

8. A.M. Beers and E.J. Mittemeijer, "Dislocation Wall Formation During Interdiffusion in Thin Bimetallic Films", Thin Solid Films, 48 (1978), 367-376.

9. J.H. van der Merwe and W.A. Jesser, "An Exactly Solvable Model for Calculating Misfit and Thickness in Epitaxial Superlattices", J. Appl. Phys., 63 (1988), 1509-1517.

10. J.P. Hirth and Xiaoxin Feng, "Critical Layer Thickness for Misfit Dislocation Stability in Multilayer Structures", J. Appl. Phys., (1990), 3343-3349.

11. M.S. Sulonen, "On the Driving Force of Discontinuous Precipitation and Dissolution", Acta Met., 12 (1964), 748-753.

12. J. Shapiro and J.S. Kirkaldy, "The Kinetics of Discontinuous Precipitation in Copper-Indium Alloys", Acta Met., 16 (1968), 1239-1252.

13. J.S. Langer and R.F. Sekerka, Acta Met., 23 (1975), 1225-1237.

14. J.S. Kirkaldy, "Stable Configuration of Chemical Potentials at a Planar Binary Alloy Solidification Interface", Scripta Met., 21 (1987), 953-958.

15. M. Hillert, "On the Driving Force for Diffusion Induced Grain Boundary Migration", Scripta Met., 17 (1983), 237-240.

16. C.Y. Ma et al., "On the Kinetic Behaviour and Driving Force of Diffusion Induced Grain Boundary Migration", Acta Met. et Mat., 43 (1995), 3113-3124.

17. I.G. Solarzano and G.R. Purdy, "Interlamellar Spacing in Discontinuous Precipitation", Met. Trans. A, 15 (1984), 1055-1063.

18. J.S. Kirkaldy and G.R. Purdy, "Chemically Induced Grain Boundary Migration as a Free Boundary Problem", Scripta Met. et Mat., 25 (1991), 901-904.

19. D.N. Yoon et al., "Coherency Strain Induced Migration of Liquid Films Through Solids", in Interface Migration and Control of Microstructure, eds. C.S. Pande et al. (Metals Park: ASM, 1986), 19-31.

20. Y.-J. Baik and D.N. Yoon, "Chemically Induced Migration of Liquid Films and Grain Boundaries in Mo-Ni-(Fe) Alloy", Acta Met., 34 (1986), 2093-2044.

21. C.A. Handwerker et al., "The Effect of Coherency Strain on Alloy Formation: Migration of Liquid Films", in Diffusion in Solids: Recent Developments, eds. M.A. Dayananda and G.E. Murch (Warrendale, PA: The Metallurgical Society, 1985), 275-292.

22. W.W. Mullins and R.F. Sekerka, "Stability of a Planar Interface During Solidification of a Dilute Binary Alloy", J. Appl. Phys., 35 (1964), 444-451.

23. M. Kuo and R.A. Fournelle, "Diffusion Induced Grain Boundary Migration (DIGM) and Liquid Film Migration in an Al-2.07 wt% Cu Alloy", Acta Met. et Mat., 39 (1991), 2835-2845.

24. R.A. Fournelle, "On the Thermodynamic Driving Force for Diffusion-Induced Grain Boundary Migration, Discontinuous Precipitation and Liquid Film Migration in Binary Alloys", Mat. Sci. Eng., A138 (1991), 133-145.

25. C.-Y. Ma et al., "Is Coherency Strain Energy the Driving Force for Diffusion Induced Grain Boundary Migration", Zeits für Metallkunde, (Munich: Carl Hanser Verlag, 1992), 633-637.

26. O. Penrose, "A Mathematical Model for Diffusion-Induced Grain Boundary Motion", this volume.

27. E.T. Whittaker, A Treatise on the Analytical Dynamics of Particles and Rigid Bodies (Cambridge at the University Press, 1952) 214-287.

28. J.S. Kirkaldy, "Equivalence of Phase Field and Dissipation Optimal Procedures in Predicting Phase-Antiphase Boundary Motion", Scripta Met. et Mat., 29 (1993) 1275-1278.

29. L. Onsager, "Reciprocal Relations in Irreversible Processes II", Phys. Rev., 38 (1931) 2265-2279.

30. R. Kikuchi, "Irreversible Cooperative Phenomena", Ann. of Phys., 10 (1960) 127-151.

31. S.R. de Groot, Thermodynamics of Irreversible Processes, (Amsterdam: North-Holland, 1952) 195-207.

32. I.M. Lifshitz, "Kinetics of Ordering During Second-Order Phase Transitions", Soviet Physics JETP, 15(1962) 939-942.

33. J.E. Krzanowski and S.M. Allen, "Segregation of Static and Migrating Diffuse Antiphase Boundaries", Surface Science, 144 (1984) 153-175.

34. P. Maugis, "From a Discrete to a Continuum Model for Static Antiphase Boundaries", Phys. Rev. B., submitted (1995).

35. S.M. Allen and J.W. Cahn, "A Microscopic Theory of Antiphase Boundary Motion and its Application to Antiphase Domain Coarsening", Acta Met., 27 (1979) 1085-1095.

36. J.S. Kirkaldy, "Atomistics of the Mobility in the Allen-Cahn Antiphase Boundary", Scripta Materialia, submitted (1995).

37. L. Onsager, "Reciprocal Relations in Irreversible Processes I", Phys. Rev., 37 (1931), 405-426.

38. R. Kobayashi, "A Numerical Approach to Three-Dimensional Dendritic Solidification", Experimental Mathematics, 3 (1994), 59-81.

39. A.A. Wheeler, W.J. Boettinger and G.B. McFadden, "Phase-Field Model for Isothermal Phase Transitions in Binary Alloys", Phys. Rev. A, 45 (1992), 7424-7439.

40. A.A. Wheeler, W.J. Boettinger and G.B. McFadden, "Phase Field Model of Solute Trapping During Solidification", Phys. Rev. E, 47 (1993), 1893-1909.

41. Collier's Encyclopedia, 11 (1966), 757.

42. J.S. Kirkaldy, "Negative Creep and the Exciton Gas-Electron-Hole Drop Spinodal in Semiconductors", in Micromechanics of Advanced Materials, ed. S.N.G. Chu et al. (Warrendale, PA: TMS, 1995), 221-223.

43. S. DiNardo and J.H. Bilgram, "Fluctuations During Freezing and Melting at the Solid-Liquid Interfaces of Xenon", Phys. Rev. B, 51 (1995), 8012-8017.

44. J.W. Cahn and W.W. Mullins, discussions to J.S. Kirkaldy, "Theory of Diffusional Growth in Solid-Solid Transformations", in Decomposition of Austenite by Diffusional Processes, eds. V.F. Zackay and H.I. Aaronson (New York: Interscience Publishers, 1962), 39-130.

45. R.F. Sekerka and W.W. Mullins, "Proof of the Symmetry of the Transport Matrix for Diffusion and Heat Flow in Fluid Systems", J. Chem. Phys., 73 (1980), 1413-1421.

46.   J.R. Manning, Diffusion Kinetics for Atoms in Crystals (Princeton NJ: D. Van Nostrand, 1968).

47.   J.S. Kirkaldy, "Clarification and Generalization of Corrections to the Darken Diffusion Equations", Scripta Met., 21 (1987), 33-38.

48.   J.S. Kirkaldy and D.J. Young, Diffusion in the Condensed State (London: The Institute of Metals, 1987).

49.   J.S. Kirkaldy, "Spontaneous Evolution of Spatiotemporal Patterns in Materials", Rep. Prog. Phys., 55 (1992), 723-795.

50.   S. Weinberg, Dreams of a Final Theory, (New York: Vintage Books, 1992).

51.   J.S. Kirkaldy, "Pattern Formation, Logistics and Maximum Path Probability", Phys. Rev. A, 31 (1995), 3376-3390.

52.   J.S. Kirkaldy, "The Symmetry 4-Group and Function of Quasiparticles in Pattern Formation", Scripta Met. et Mat., 33 (1995), 259-265.

THE GEOMETRICAL PHASE TRANSFORMATIONS

DURING EVOLUTION IN FINITE MEDIA

V. Ya. Shur, E.L. Rumyantsev, and S.D. Makarov
Ural State University,
620083, Ekaterinburg, Russia.

Abstract

The paper deals with the theoretical and experimental investigations of the phase evolution kinetics in real systems. Computer simulations of phase kinetics in finite media and for spatially nonuniform nucleation were used as ideal experiments. As a result the method of mathematical treatment within which one can obtain the kinetic parameters concealed in integral data was elaborated.

The research described in this publication was made possible in part by Grant No.NMVOOO from the International Science Foundation and by Grant No.93-02-2451 from the Russian Foundation of Fundamental Research.

Introduction

Theoretical and experimental investigation of the transformation kinetics is very important problem of general and applied physics. Any kinetic process of phase evolution, which consists of arising and subsequent growth of isolated volumes of new phase (reconstruction of heterophase structure) can be described in terms of such general kinetic parameters as nucleation probability and phase growth rate. The local methods in principal can give directly the complete information about evolution of the geometry/morphology of the momentary phase patterns during transformation. The problem is that the application of commonly used techniques such as optical or electron microscopy (TEM and SEM) and so on encounters usually with many technical problems, especially for investigation of the fast processes. That is why the universal integral methods are more popular. In this case usually obtained responses are proportional to momentary value of the fraction of new/growing phase or its derivative on time. The kinetic parameters can be obtained from experimental data only by using the mathematical treatment based on an adequate theoretical approach.

The classical approach developed in the works of Kolmogorov [1] and Avrami [2] is used commonly for obtaining the kinetic parameters from integral data. But application of K-A formulas is under question for phase evolution in real finite media in which the average distance between arising nuclei is comparable with the system sizes.

The goal of present work is to propose the modification of K-A formulas which allows to extract the kinetic parameters from integral data measured in real systems and to test proposed mathematical treatment by computer simulations and model experiments.

Classical approach

The theoretical description of the phase transformation is usually based on Kolmogorov-Avrami (K-A) statistical theory [1, 2]. Two limiting situations are considered usually [1, 3]: α model - in which the nucleation process exists during whole phase transformation (the nucleation probability per volume α usually remains constant) and β model when all nuclei involved in the process are arising instantaneously at the very beginning with the density β per volume. In both cases the phase growth rate v is usually taken to be constant for transformation in stable conditions (constant value of driving force). The main equations used for the description of the time dependence of the fraction of the volume occupied by old phase run as follows:

$q(t) = \exp[-c \, \alpha \, v^n \, t^{n+1}/(n+1)]$ for α model   (1)

$q(t) = \exp(-c \, \beta \, v^n \, t^n)$ for β model   (2)

where: n - the dimensionality of considered problem (initially K-A formula was written for 3D case (n = 3), but it can be used for 2D and 1D problems also), c - the shape constant.

The K-A theory is applicable for the description of the phase evolution which satisfies two main principles. First, the average distance between arising nuclei must be much less then the media sizes (it means that the edges effects are ignored). Second, the spatial distribution of the nucleation sites must be random.

In real situations, e.g. while fitting the transient current data (j(t) ~ dq/dt) by K-A expressions the noninteger values of the dimensionality n are usually obtained [4-6]. Moreover the decaying part of the current pulses can not be fitted by the formula. It indicates that in this case the main principles are violated. We propose the approach which allows to overcome these difficulties by modification of K-A formula. As a result the description of the phase transformation phenomena in finite media and for nonuniform spatial distribution of nucleation sites becomes possible.

Phase transformations in finite media

In real finite objects one must take into account the fact that at the moment, when the separate volumes of the growing phase touch the boundaries of the media, the change of their shape constants occurs. It is easy to show that the influence of media boundaries can be taken into account through the time variation of the shape constant averaged over the whole media at given moment [3, 7, 8]. As a result the following formula can be written for the time dependence of fraction of the old phase volume q(t) for β model

$$q(t) = \exp[-c \beta v^n t^n (1 - t/t_m)] \qquad (3)$$

where t_m accounts for the interaction of growing domains with boundaries of the switched volume.

This formula is meaningless for $t/t_m > 1$. So there are no problems for the description of phase evolution in finite media by the Equation (3) if the switching time t_s (needed for the transformation of almost whole media) answers the requirement. But it must be pointed out that there are several types of real situations for which this condition is violated. It was shown that the final part of the kinetic process must be described by the same formula with the reduced integer value of growth dimensionality. In other words a geometrical transformation occurs. As a result for 2D growth in rectangular shaped volume with area $S = A L^2$ (A - anisotropy of sizes) q(t) time dependence is different for two stages of switching process:

$$q(t) = \exp[-(t/to_1)^2 (1 + t/t_m)], \qquad \text{for } 0 < t < t_c$$
$$q(t) = \exp[-(t/to_2)], \qquad \text{for } t_c < t < t_s \qquad (4)$$

where

$t_c = L/v$, the time of catastrophe, L - the average shortest size of the media,

$to_1 = (C_1 N0/A)^{-1/2} t_c$, $to_2 = (C_2 N0/A)^{-1} t_c$,

C_1, C_2 - the shape constants of first and second stages, N0 - the number of initial nuclei.

This situation can be realized only in "anisotropic case" (Figure 1).

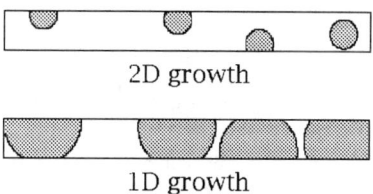

2D growth

1D growth

Figure 1 - The scheme demonstrating the geometrical transformation for β model in finite media in "anisotropic case".

The anisotropy can be induced not only by nonequivalent geometrical sizes of media (e.g. rectangular shape), but also by the anisotropy of the growth rate. For β model the crossover time characterizing the "catastrophic change" of the dimensionality corresponds to the moment when all growing nuclei have reached two opposite boundaries of the media.

For $\beta+\alpha$ model the situation is more complicated and the catastrophe is smeared. The shape of the current pulse depends on the ratio between the densities of arising and remnant nuclei r:

$$r = \beta^{-1} \int_0^{t_s} \alpha \, dt \qquad (5)$$

The formula describing the transformation for $\beta+\alpha$ model in the case, when β process prevails (for r < 1), runs as follows:

$$q(t) = \exp[-(t/t_{o1})^2 (1 + t/t_{m1})], \quad \text{for } 0 < t < t_c$$
$$q(t) = \exp[-(t/t_{o2}) (1 + t/t_{m2})], \quad \text{for } t_c < t < t_s \qquad (6)$$

where $t_{m1}^{-1} = t_m^{-1} + t_{m2}^{-1}$
$t_{m2}^{-1} = (dN/dt) \, N_0^{-1}$

So, as it is seen, in both cases (for β and $\beta+\alpha$ model) the growth dimensionality is changed drastically at $t = t_c$. For pure β model this effect is more pronounced as compare with $\beta+\alpha$ model due to existence in the latter case of additional term in the second stage (for $t > t_c$) in Equation (6) as a result of the influence of arising nuclei.

Computer simulation

In our investigations we use the simulations as an ideal experiments for testing the validity of proposed formula [3, 9]. Let us consider the phase transformation in thin square plate (with area $S = l^2$). This kinetic process can be considered as two-dimensional (2D) because the time of phase growth through the plate is always essentially shorter the time of complete transformation. Let us take into account that during any experiment there exists limited time interval between consecutive measurements Δt. Thus for any value of the anisotropy of the rate of sideways phase growth $A = v_x/v_y$ one can consider the process as the isotropic movement of the phase boundaries occurring on the discrete matrix $L_x \times L_y$, where $L_x = l/(v_x \Delta t)$ and $L_y = l/(v_y \Delta t)$.

Thus we performed the computer simulation of phase transformation kinetics on the 2D matrix. For concrete problem it is necessary to determine the matrix sizes, the number of initial nuclei (for β process), the step increasing of nuclei number for time interval Δt (for α process) and the angular dependence of sideways growth rate of separate nuclei. The coordinates of arising nuclei were chosen randomly. For every step the fraction of the area occupied by new phase q_i or its variation on time $\Delta q/\Delta t = (q_{i+1} - q_i)/\Delta t$ was calculated and averaged over the great number of independent computer experiments/realizations. The time dependence of these two quantities is proportional to the data obtained in most integral experiments, e.g. switching charge during polarization reversal in ferroelectrics is proportional to q(t) and switching current j(t) ~ dq/dt.

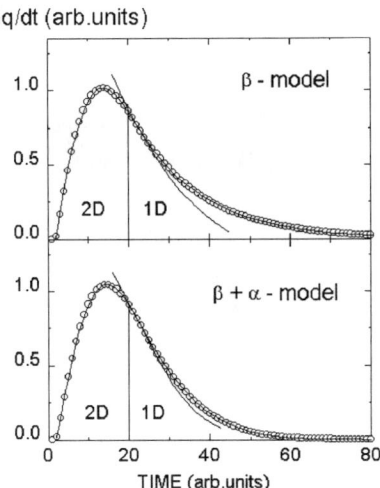

Figure 2 - The time derivative of new phase volume for β and β+α - processes in anisotropic case obtained by computer simulation (matrix size 1000 × 20). The points are the results of computer simulations, the solid curve - fitting by Equations (4) and (6).

Varying of matrix sizes and angular dependence of growth rate allow to consider the influence of media finiteness. For simulation of inhomogeneous nucleation we divide the whole matrix into regions with different values of nucleation probability.

The cluster and correlation analysis of simulated momentary heterophase patterns was carried out also. These results are of principal importance for interpretation of light, neutron and so on scattering experiments.

Typical results of computer simulation of the time derivative of new phase volume for β model in anisotropic case are shown in Figure 2. The distinct change of the shape is due to the geometrical catastrophe/crossover from two-dimensional (2D) growth to 1D one. The kinetic parameters extracted from the best fit results demonstrate very good agreement with the values of nucleation probability and growth rate taken in computer experiments. It must be stressed that proposed statistical approach (so as K-A formula) can be used only for the great ensemble of independently transformed phase volumes.

The phase transformations under spatially nonhomogeneous nucleation can be investigated by computer simulation in the similar manner. It is common knowledge that for various real systems the surface and bulk nucleation probabilities strongly differ. The results of computer simulations in limiting case when nucleation occurs only on the surface demonstrate the geometrical transformations also (Figure 3). In this case the process can be divided in three stages with the following sequence of geometrical transformations (catastrophes): $\beta(2D) \rightarrow \beta(1D) \rightarrow \alpha(2D)$.

The proposed approach can be used even in the case when initial state consists of both phases (heterophase initial structure). The time dependence of dq/dt and corresponding

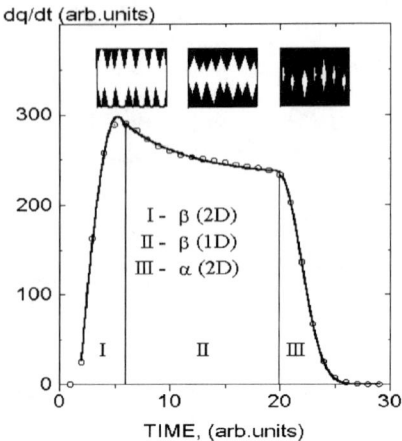

Figure 3 - Computer simulation of switching process with spatially nonuniform nucleation (β model, nucleation near sample's edges only, anisotropic growth rate). The points - computer experiment, the solid curve - theory.

geometrical transformations strongly depend on the geometry of the initial structure [3, 8]. It was shown by computer experiments that proposed formulas allow to extract the kinetic parameters from best fit results with good accuracy in all cases discussed above. The successful test of our approximations makes it possible to use the mathematical treatment based on modified formulas for obtaining kinetic parameters from real experimental data.

Experimental results

We used as an example of phase transformation process the evolution of ferroelectric domain structure during polarization reversal in electric field (in bulk single crystals and thin films) and growth of crystallites during annealing of amorphous films [10]. It must be stressed that ferroelectrics are very appropriate objects for the investigation of phase transformation kinetics. Firstly, due to the simplicity of managing the oversaturation degree by the variation of external electric field. Secondly, the usage of optical methods of *in situ* registration with high spatial and time resolutions allowed to obtain full statistically reliable information about phase transformation kinetics. Ferroelectric single crystals with optically distinguished domains were used as model objects for simultaneous investigation by local and integral experimental methods. The comparison of sequence of momentary patterns of domain structure with the time dependence of switching current was held in single crystalline thin plates of lead germanate and bismuth titanate [8, 9]. The good correlation between change of the geometry of domain patterns and t_c defined from current data was demonstrated.

The typical analysis of switching current in thin epitaxial ferroelectric film is demonstrated on Figure 4.

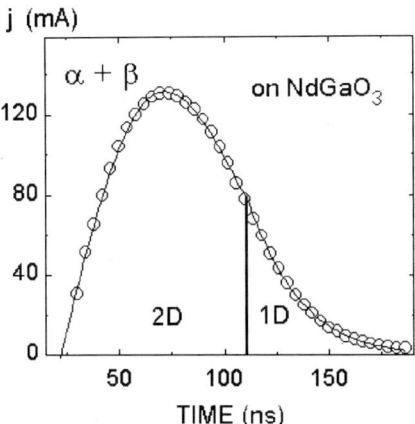

Figure 4 - The switching current for epitaxial film of PZT (thickness 400 nm, switching voltage 6.4 V). Experimental points were fitted by Equation (6).

In this case the kinetics of domain structure can not be investigated by direct methods due to superfast switching and superthin thickness. At the same time through our mathematical treatment we predict two-stage kinetics with geometrical transformation from 2D to 1D process.

It is possible to analyze in the same manner the time dependence of total scattered light intensity for crystallization of amorphous films during annealing (including Rapid Thermal Annealing) [11, 12] (Figure 5).

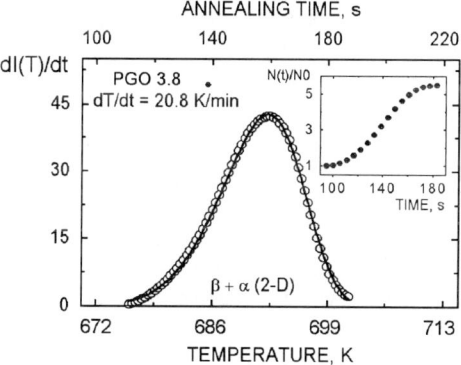

Figure 5 - Annealing time dependence of derivative of total scattered light intensity per time (crystallization rate) for constant heating rate in lead germanate film. Experimental points fitted by Equation 4. In input - time dependence of the crystallite density.

We use the generalization of proposed formulas for time dependent growth under annealing with constant heating rate [13]. From the best fit values of $dI(T)/dt \sim dq/dt$ the time dependence of the number of arising crystallites was extracted (Figure 5, in input).

Conclusion

It was shown that using of computer simulations as ideal experiments allows to test the used approach qualitatively and quantitatively. The successful testing of proposed formulas makes it possible to determine from integral data various scenarios realized during phase transformations in real media. The mathematical treatment of the set of data obtained for controlled change of experimental conditions gives the dependence of the main kinetic parameters describing the phase evolution. The proposed method is especially important for investigations of fast evolution in strongly nonequilibrium systems, which are almost inaccessible for direct observation.

References

1. A. N. Kolmogorov, "A Statistical Theory of Metal Crystallization", Izv. Acad. Nauk USSR, Ser. Math., 3 (1937), 355-359.
2. M. Avrami, "Kinetics of Phase Change. I. General Theory", J.Chem.Phys., 7 (1939), 1103-1112.
3. V. Ya. Shur et al., "How to Extract Information about Domain Kinetics in Thin Ferroelectric Films from Switching Transient Current Data", Integrated Ferroelectrics, 5 (1994), 293-301.
4. Y. Ishibashi and Y. Takagi, "Note of Ferroelectric Domain Switching", J.Phys.Soc.Jap., 31 (1971), 506-510.
5. K. Dimmler et al., "Switching Kinetics of Ferroelectric Thin-Film Memories", J. Appl. Phys., 61 (1985), 5467-5470.
6. J. F. Scott et al., "Switching Kinetics of Lead Zirconate Titanate Submicron Thin-Film Memories", J.Appl.Phys., 64 (1988), 787-792.
7. V. Ya. Shur et al., "Switching in Ferroelectric Thin Films: How to Extract Information about Domain Kinetics from Traditional Current Data", Proceedings of the Ninth IEEE ISAF, (1994), 669-673.
8. V. Ya. Shur, E. L. Rumyantsev and S. D. Makarov, "Geometrical Transformation of the Ferroelectric Domain Structure in Electric Field", Ferroelectrics, 172 (1995), 361-372.
9. V. Ya. Shur et al., "Switching Kinetics in Epitaxial PZT Thin Films", Microelectronics Engineering, 29 (1995), 153-157.
10. V. Ya. Shur and E. L. Rumyantsev, "Crystal Growth and Domain Structure Evolution", Ferroelectrics, 142 (1993), 1-7.
11. V. Ya. Shur et al., "Crystallization Kinetics of Amorphous Ferroelectric Films", Ferroelectrics, (1995), in press.
12. V. Ya. Shur et al., "Elastic Light Scattering as a Probe for Detail in situ Investigations of Domain and Phase Evolution", Ferroelectrics, 169 (1995), 63 -73.
13. V. Ya. Shur et al., "Transient Current During Switching in Increasing Electric Field as a Basis for a New Testing Method", Integrated Ferroelectrics, 10 (1995), 223-230.

Phase-Field Modeling of Crystal Growth: New Asymptotics and Computational Limits

Alain Karma and Wouter-Jan Rappel

Physics Department, Northeastern University,
Boston, Massachusetts 02115

Abstract

The phase-field method has recently emerged as a powerful computational tool to model crystal growth. However, so far, the range of the method has remained severely restricted because of computational constraints imposed by the interface thickness. We discuss the results of a new asymptotic analysis of the sharp interface limit of the phase-field model of the crystallization of a pure material from its melt which incorporates the variation of the temperature field across the interface. This analysis yields a Gibbs-Thomson condition which remains valid for a thicker interface and, consequently, greatly expands the computational range of the phase-field method. It makes it possible to model for the first time quantitatively dendritic growth at a much smaller undercooling in 2D and 3D, and with arbitrary interface kinetics. We report numerical tests of the method in 2D without kinetics. Simulations yield dendrites with tip velocities and tip shapes which agree within a few percents with 'exact' numerical Green's function solutions of the steady-state growth problem.

Table 1: List of symbols and abbreviations used in the text

ψ	phase field
u	($= c_p[T - T_M]/L$) dimensionless temperature field
W	phase-field interface thickness
τ	relaxation time of phase-field
λ	coupling constant between ψ and u
L	latent heat of melting
T_M	melting temperature of a pure substance
c_p	specific heat at constant pressure
d_0	($= \gamma_0 c_p T_M^2/L$) capillary length
Δ	($= c_p[T_M - T_0]/L$) dimensionless undercooling
T_0	initial temperature of the melt
θ	polar angle with respect to the crystal axis
Θ	($= \arctan \partial_y \psi / \partial_x \psi$) phase field angular variable
$\gamma(\theta)$	($= \gamma_0 f_S(\theta)$) anisotropic surface energy
$f_S(\theta)$	anisotropy form of $\gamma(\theta)$ and phase-field gradient energy
ϵ	anisotropy strength ($f_S(\theta) = 1 + \epsilon \cos 4\theta$)
$f_\psi(\theta)$	anisotropy form of phase-field kinetics
$\beta(\theta)$	anisotropic interface kinetic coefficient
κ	interface curvature
V_n	normal velocity of the interface
V_{tip}	steady-state dendrite tip velocity
U_{tip}	($= V_{\text{tip}} d_0 / D$) dimensionless tip velocity
U_{tip}^{GF}	U_{tip} corresponding to the solution of Eqn. 15
R_{tip}	($= 1/\kappa_{\text{tip}}$) steady-state dendrite tip radius
ρ_{tip}	($= R_{\text{tip}}/d_0$) dimensionless tip radius

Introduction

The phase-field methodology [1] originates from phenomenological continuum models of second order phase transitions [2]. Its main appeal is that it avoids to track a sharp boundary by introducing one or more phase fields to distinguish between distinct phases. Some of the earlier versions of the phase-field model were introduced to model the crystallization of a pure material from its melt [1, 3, 4]. Since then, the method has been extended to model the isothermal solidification of binary alloys [5] and eutectic growth [6, 7, 8]. The latter models combine elements of the earlier models and the Cahn-Hilliard equation [9]. A host of computations have been performed of dendritic growth in pure materials [3, 10, 11, 12] and alloys [13], directional solidification [14], eutectic growth [6, 7], coarsening phenomena using the Cahn-Hilliard equation [15] and an alloy phase-field model [16], and grain growth using a model that introduces several phase fields to distinguish between grain orientations [17].

Here, we are interested in the application of the phase-field method to model dendritic growth. The operating state of the dendrite tip is known to be sensitively controlled by a subtle balance between surface tension, diffusion, and crystalline anisotropy [18, 19]. It is a prime example of a delicate dynamical system where small quantitative details are important and where obtaining quantitative predictions by dynamical simulation has remained extremely difficult.

For a pure material, the dynamics of the phase-field is traditionally coupled to that of the dimensionless temperature field, $u \equiv (T - T_M)/(L/c_p)$ (see Table 1 for definitions), in such a way that the equations for the two fields reduce to the standard sharp-interface equations in the limit of vanishingly small interface thickness W. However, in practice, phase-field simulations are only computationally feasible for finite W because several grid points are necessary to resolve the interface region. This is problematic because the standard analysis of the sharp interface limit of the phase-field model (carried out by many authors for various model versions [1, 4, 10, 20, 21, 22]) is only strictly valid for vanishingly small W.

Recently, we have performed an asymptotic analysis of the sharp interface limit of the phase-field model of the solidification of a pure material [23]. Our analysis is based on including the variation of the diffusion field u across the interface thickness, which is assumed to be constant in previous analyses [1, 4, 10, 20, 21, 22]. This analysis yields a Gibbs-Thomson condition of the standard form: $u = -d_0 \kappa - \beta V_n$, with the same capillary length d_0 but with a different kinetic coefficient β. This condition has two new key properties: (i) it remains valid for a much smaller ratio d_0/W, which dramatically increases the computational efficiency of the method, and (ii) β can be made to vanish, or arbitrarily small, by an appropriate choice of computational parameters. This second property provides the freedom to model growth with or without interface kinetics. This is important from a modeling standpoint because the kinetic-free case (i.e. $\beta = 0$) has remained notoriously difficult to model with numerical methods that track a sharp-boundary [24, 25, 26, 27].

The mathematical details of this new asymptotic analysis are quite technical and require a lengthy exposition which will be given elsewhere [23]. In this paper, we illustrate the numerical application of our results to model dendritic growth in two dimensions. This paper is organized as follows. We first present the equations of the phase-field model. We then give the results of the analysis of the sharp-interface limit which includes temperature variations. Next, we present the results of numerical simulations of dendritic growth in 2D and numerical tests. Finally, we give concluding remarks on the prospect of the phase-field method.

Model

We consider the phase-field model defined by the equations

$$\tau f_\psi(\Theta) \frac{\partial \psi}{\partial t} = -\frac{\delta \mathcal{F}(\psi, u)}{\delta \psi} \tag{1}$$

$$\frac{\partial u}{\partial t} = D \nabla^2 u + \frac{1}{2} \frac{\partial p(\psi)}{\partial t} \tag{2}$$

where,

$$\mathcal{F}(\psi, u) = \int d\mathbf{r} \left[\frac{W^2}{2} [f_S(\Theta)]^2 |\nabla \psi|^2 + f(\psi, \lambda u) \right] \tag{3}$$

and,

$$f(\psi, \lambda u) = -\frac{\psi^2}{2} + \frac{\psi^4}{4} + \lambda u \left[\psi - \frac{2}{3}\psi^3 + \frac{1}{5}\psi^5 \right] \tag{4}$$

After functional differentiation, Eqn. 1 becomes

$$\tau f_\psi(\Theta) \frac{\partial \psi}{\partial t} = \left(\psi - \lambda u (1 - \psi^2) \right) (1 - \psi^2) \tag{5}$$

$$+ W^2 \left[\vec{\nabla} \cdot (f_S(\Theta)^2 \vec{\nabla} \psi) - \frac{\partial}{\partial x} \left(f_S(\Theta) f_S'(\Theta) \frac{\partial \psi}{\partial y} \right) \right.$$
$$\left. + \frac{\partial}{\partial y} \left(f(\Theta) f_S'(\Theta) \frac{\partial \psi}{\partial x} \right) \right]$$

The free energy function $f(\psi, \lambda u)$ has two minima at $\psi = +1$ and $\psi = -1$ which correspond to the solid and liquid phase, respectively. This choice of $f(\psi, \lambda u)$ has the advantage that it keeps the minima at the same locations, independent of the value of u, and retains a double-well free-energy landscape for values of $|\lambda u|$ much larger than unity. Since it is desirable in some computations to choose λ much larger than unity, such large values of $|\lambda u|$ can arise away from the interface where $u \to -\Delta$. The angular variable $\Theta \equiv \arctan(\partial_y \psi / \partial_x \psi)$ is the angle that the directions normal to the contours of constant ψ make with the x-axis. The function $p(\psi)$ describes the production of latent heat.

This model is directly related to previous models. It reduces to the earlier versions of the phase-field model introduced in Refs. [1, 4] with $p(\psi) = \psi$ (which is the simplest way to incorporate latent heat diffusion) and with the term in square brackets on the right-hand-side of Eqn. 4 replaced by ψ. It includes anisotropy as in Refs. [10, 11, 12, 21]. Finally, for $p(\psi) = 15(\psi - 2\psi^3/3 + \psi^5/5)/8$, it reduces to one particular case of the entropy formulation of Ref. [22], used in the computations of Ref. [12].

Sharp-interface limit

In the limit where the interface is curved on a scale which is large compared to W, and the u-field varies across the interface thickness, Eqns. 1 and 2 can be shown [23] to reduce to the Stefan problem:

$$\frac{\partial u}{\partial t} = D \nabla^2 u \tag{6}$$

$$V_n = D \left(\left. \frac{\partial u}{\partial n} \right|_S - \left. \frac{\partial u}{\partial n} \right|_L \right) \tag{7}$$

with the velocity-dependent Gibbs-Thomson condition of the standard form

$$u = -d_0 \left[f_s(\theta) + f_s''(\theta) \right] \kappa - \beta(\theta) V_n \tag{8}$$

This condition corresponds to an anisotropic surface energy $\gamma(\theta) = \gamma_o f_S(\theta)$, with

$$d_0 = \frac{I}{J} \frac{W}{\lambda}, \tag{9}$$

and an anisotropic kinetic coefficient defined by

$$\beta(\theta) = \frac{\tau}{\lambda W} \frac{I}{J} \frac{f_\psi(\theta)}{f_S(\theta)} \left[1 - \frac{\lambda W^2}{2 D \tau} \frac{f_S(\theta)^2}{f_\psi(\theta)} \frac{K + JF}{I} \right] \tag{10}$$

The positive constants F, I, J, and K, are defined by the integrals:

$$F = \int_0^{+\infty} dx \left[1 + p(\psi_0(x)) \right] \tag{11}$$

$$I = \int_{-\infty}^{+\infty} dx \left[\frac{d\psi_0(x)}{dx} \right]^2 \tag{12}$$

$$J = -\int_{-\infty}^{+\infty} dx \, (1 - \psi_0^2(x))^2 \left[\frac{d\psi_0(x)}{dx} \right] \tag{13}$$

$$K = \int_{-\infty}^{+\infty} dx \, (1 - \psi_0^2(x))^2 \left[\frac{d\psi_0(x)}{dx} \right] \int_0^x dx' \, p(\psi_0(x')) \tag{14}$$

where $\psi_0(x) \equiv -\tanh(x/\sqrt{2})$. The numerical values of the various constants are $I = 2\sqrt{2}/3$, $J = 16/15$, $K = 0.13604$ and $F = \sqrt{2}\ln 2$ for $p(\psi) = \psi$, and $K = 0.22359$ and $F = 0.49412$ for $p(\psi) = 15(\psi - 2\psi^3/3 + \psi^5/5)/8$.

In the limit $W \to 0$, the term in square bracket on the right-hand-side of Eqn. 10 reduces to unity, and $\beta(\theta)$ reduces to the 'leading order' expression for the kinetic coefficient derived in Refs [1, 4, 20], without anisotropy, and Refs. [21, 22] with anisotropy. For a given set of computational parameters, this leading order kinetic coefficient is only valid if $\lambda W^2/D\tau \ll 1$, or, equivalently, if

$$d_0 \gg W \times \frac{W^2}{D\tau}$$

Since computational constraints make it difficult to perform simulations with the ratio $W^2/D\tau$ much smaller than unity, this condition requires the capillary length to be at least comparable or larger than the interface thickness. The main advantage of our new asymptotic limit is that it removes this stringent computational constraint and makes it possible to perform computations with the capillary length smaller than the interface thickness.

Numerical Simulations of Dendritic Growth

The theory of the operating state of the dendrite tip is well established in 2D. Predictions of the dendrite tip velocity, V_{tip}, and the steady-state interface shape can be obtained numerically to arbitrary precision by solving the steady-state growth problem. This problem consists of looking for time-independent steady-state solutions of Eqns. 6 through 8 in a frame moving at velocity V_{tip} (obtained by letting $\partial u/\partial t = -V_{\text{tip}}\partial u/\partial x$ in Eqn. 6 and $V_n = V_{\text{tip}}\cos(\theta)$ in Eqns. 7 and 8, where x is the growth axis.) The steady-state equations can be recasted into a single integral equation by the standard Green's function method:

$$\Delta - d_0\left[f_s(\theta) + f_s''(\theta)\right]\kappa - \beta(\theta)V_{\text{tip}}\cos(\theta)$$
$$= \int_{-\infty}^{+\infty} \frac{dx'}{2\pi l}\exp\left[\frac{y(x) - y(x')}{l}\right]K_0\left[\frac{|r - r'|}{l}\right] \quad (15)$$

where $l \equiv 2D/V_{\text{tip}}$ is the diffusion length. Eqn. 15 was written down originally by Nash and Glicksman [28] and has subsequently been solved numerically by several groups [29, 30, 31] following the theoretical insights gained from the study of local growth models [18, 19]. It admits a discrete spectrum of solutions if a finite amount of anisotropy, in either the surface energy or the interface kinetics, is present. Only the fastest growing solution in this discrete set is stable and corresponds to the dynamically selected operating state of the dendrite tip [18, 19].

Here, we focus on modeling growth without interface kinetics and with an anisotropic surface energy with an underlying cubic symmetry:

$$f_S(\theta) = 1 + \epsilon \cos 4\theta \quad (16)$$

Following Eqn. 10, we make $\beta(\theta)$ vanish by choosing

$$f_\psi(\theta) = [f_S(\theta)]^2 \quad (17)$$

and

$$\lambda = \frac{I}{K + JF}\frac{2D\tau}{W^2} \quad (18)$$

By rescaling length by d_0, and time by D/d_0^2, the free-boundary problem defined by Eqns. 6-8 can be trivially shown to only depend on Δ and ϵ when $\beta(\theta) = 0$. This is important from a modeling standpoint because it provides the freedom to perform simulations with any values of d_0 and D, provided, of course, that they are numerically resolved and independent of computational parameters. The simulation results can always be related to a given material, with specific values of D and d_0, by a trivial rescaling. Accordingly, it is convenient to define the dimensionless tip velocity, U_{tip}, and tip radius, ρ_{tip}, defined by

$$U_{\text{tip}} = \frac{d_0}{D} V_{\text{tip}} = F_V(\Delta; \epsilon) \tag{19}$$

$$\rho_{\text{tip}} = \frac{R_{\text{tip}}}{d_0} = F_R(\Delta; \epsilon) \tag{20}$$

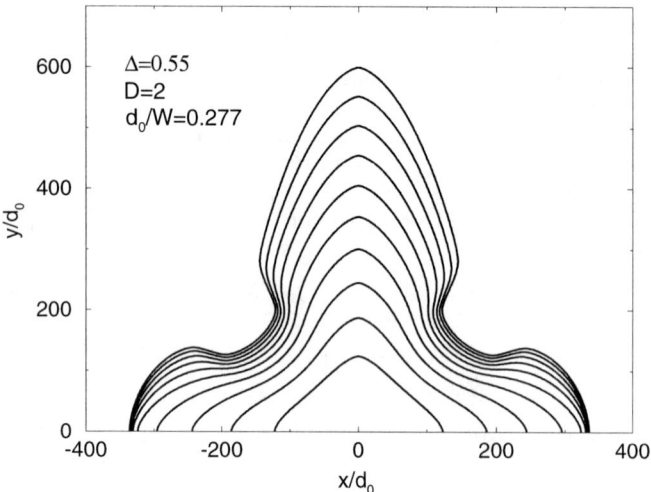

Figure 1: Sequence of interface shapes ($\phi = 0$) shown every 5,000 iterations. Note that sidebranching is absent. The addition of noise produces sidebranching.

Results reported here are for
$$p(\psi) = \psi \tag{21}$$
We have also verified by simulation that the entropy formulation $p(\psi) = 15(\psi - 2\psi^3/3 + \psi^5/5)/8$ leads to accurate results. However, specifically for dendritic growth, we have found that a smaller Δx is needed to obtain a numerically resolved tip velocity with this choice of $p(\psi)$. This is because the u-field profile in this case is steeper in the interface region. Thus, although the entropy formulation may seem more appealing than the ad-hoc incorporation of latent heat, because it is thermodynamically consistent, it turns out to be computationally less advantageous in this particular application. Note that, for the purpose of quantitative modeling of crystal growth, the particular choice of the

function $p(\psi)$ is irrelevant in the sharp-interface limit. For this reason, computational efficiency is generally more important than thermodynamic consistency in choosing a particular form of $p(\psi)$.

Eqns. 2 and 5 were discretized using standard second order finite difference formulae, except $\nabla^2 \psi$ which was discretized using a nine-point formula with nearest and next nearest neighbors which minimizes the grid anisotropy. The ψ and u fields were time-stepped using, respectively, a first order Euler scheme and a second order implicit Crank-Nicholson scheme. For smaller values of D, we also used a second version of the code where u is time-stepped with an Euler scheme. Simulations of dendritic growth were performed on two dimensional lattices of rectangular sizes $N_x \times N_y$ and constant grid spacing Δx in both directions. Simulations were seeded with a small quarter disk of solid at one corner of the lattice and a spatially uniform undercooling $u = -\Delta$. N_x was chosen sufficiently large to outlast dynamical transients and for the tip velocity to reach a steady-state value V_{tip}.

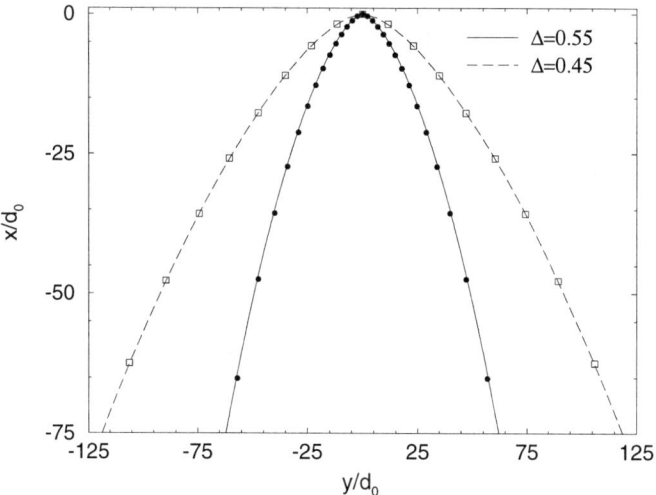

Figure 2: Comparison of steady-state tip shapes calculated by phase-field simulations (lines) and the numerical solution of Eqn. 15 (symbols). The two interfaces correspond to: $\Delta = 0.55$, $d_0/W = 0.277$ (solid line and circles), and $\Delta = 0.45$, $d_0/W = 0.185$ (dashed line and squares). For clarity, only one out of every four symbols along the interface is shown for the Green's function results.

All simulations were performed for the fixed values $W = 1$ and $\tau = 1$. The grid spacing was selected by performing simulations with decreasing Δx in steps of 0.1 until V_{tip} did not change in value by more than two percent. The study of the convergence of our results as a function of computational parameters was carried out by increasing d_0 at the fixed value of W defined above. This was done by decreasing D and concomitantly decreasing λ via Eqn. 18. This is equivalent to increasing d_0 which is inversely proportional to λ (Eqn. 9).

Table 2: Comparison of steady-state tip velocities calculated by phase-field simulations with $p(\psi) = \psi$, U_{tip}, and calculated by solving Eqn. 15 numerically with a Newton-Raphson iteration scheme, U_{tip}^{GF}. T_{CPU} denotes the CPU time in hours for simulations on a DEC Alpha 3000-700 workstation. T_{CPU} on one processor of the CRAY C90 is roughly 5 times smaller. The model parameters are: $W = 1$, $\tau = 1$, $\Delta x = 0.4$, $\Delta t = 0.016$, $\epsilon = 0.05$, and λ defined by Eqn. 18.

Δ	D	d_0/W	U_{tip}	$\tilde{U}_{\text{tip}}^{GF}$	% Error	N_x	N_y	T_{CPU}
0.55	2	0.277	0.0168	0.0170	1	1300	200	12
0.55	3	0.185	0.0175	0.0170	3	900	200	3
0.55	4	0.139	0.0174	0.0170	2	600	150	0.5
0.50	3	0.185	0.01005	0.00985	2	1500	200	10
0.45	3	0.185	0.00557	0.00545	2	1500	300	40
0.45	4	0.139	0.00540	0.00545	1	1300	250	15
0.40	5	0.111	0.00300	0.00290	3	1700	350	25

In order to benchmark our simulation results, we solved numerically the steady-state growth problem defined by Eqn. 15 using the algorithm of Ref. [29]. The input parameters of these calculations were chosen to correspond exactly to those of the phase-field computations; namely $\beta(\theta) = 0$, d_0 defined by Eqn. 9, and $f_S(\theta)$ defined by Eqn. 16. A comparison of the dimensionless steady-state tip velocities obtained by phase-field simulations and Green's function calculations is shown in Table 2. A comparison of interface shapes for two different undercoolings is shown in Fig. 2.

It can be seen that the quantitative agreement is remarkably good over the whole range of d_0 and Δ investigated here. Table 2 shows that accurate simulations are still possible at a very small d_0/W ratio with an enormous gain in computational efficiency. This gain can be estimated by noting that the CPU time scales at intermediate Δ as $\sim N_x N_y \rho_{\text{tip}}^2/D\Delta t$ where ρ_{tip}^2/D is the transient time necessary to reach a steady-state dendrite. Since, N_x and N_y can be increased proportionally to $\rho_{\text{tip}}/\Delta x$ and $\rho_{\text{tip}} \sim d_0$, this time scales roughly as $(d_0/W)^4$. An argument based on the diffusion length yields the same scaling at small Δ. It can be estimated that three-dimensional simulations of dendritic growth, which take full advantage of the cubic anisotropy to minimize the computations, should require about one to two hundred times more CPU time. Hence, quantitative three dimensional simulations are directly accessible for the smaller d_0/W ratios. We are presently carrying out such computations and hope to present results in the near future.

Conclusions

In summary, we have presented an analysis of the sharp-interface limit of the phase-field model of a pure material which includes temperature variations across the interface thickness. This analysis extends the phase-field method in two important ways. Firstly, it allows for more efficient computations with a smaller capillary length to interface thickness ratio. This, in turn, renders quantitative three-dimensional simulations directly accessible for the first time, as well as the exploration of a much wider range of growth conditions with the same computational resources. Secondly, it makes it possible to study the important limit of small or zero kinetic coefficient, which had previously been thought to be unreachable by this method.

We have demonstrated the applicability of our results in 2D by performing direct quantitative tests of dendrite velocity and shape selection. Simulations yield tip velocities which are very accurate over a wide range of undercoolings.

This new insight into the phase-field method should find a wide range of applications to other solidification problems. The most immediate application, beside 3D dendritic growth, is directional solidification. We have carried out a similar asymptotic analysis of a phase-field model of this problem, with a frozen temperature field, and found that it also yields a Gibbs-Thomson condition which remains valid for a thicker interface [32]. There is a host of interesting pattern formation questions related to the dynamics of dendrite arrays and spacing selection which become accessible to quantitative modeling using this approach.

From a broader perspective, the results of the present paper provide a new outlook on using the phase-field method, and a new direction for future studies. Namely, they show that there is much to be gained by performing the analysis of the sharp-interface-limit at 'higher order' in an expansion which involves the interface thickness [1]. This is because the leading order answer, valid for vanishingly small thickness, turns out to be computationally too stringent, and too restrictive for quantitative modeling even on current supercomputers. In view of this, it seems desirable to extend our analysis to a wide range of phase-field models describing the solidification of alloys and other thermodynamically driven phase transformations.

Acknowledgments

We thank Herbert Levine for providing the Green's function steady-state code of Ref. [29] used to test our phase-field simulations. This research was supported by US DOE grant No DE-FG02-92ER45471 and benefited from supercomputer time allocation at the National Energy Research Supercomputer Center.

References

[1] J. S. Langer, in *Directions in Condensed Matter* (World Scientific, Singapore, 1986), p. 164.

[2] B. I. Halperin, P. C. Hohenberg, and S-K. Ma, Phys. Rev. B**10**, 139 (1974).

[3] G. J. Fix, in *Free Boundary Problems: Theory and Applications, Vol. II*, edited by A. Fasano and M. Primicerio (Piman, Boston, 1983), p. 580.

[4] J. B. Collins and H. Levine, Phys. Rev. B **31**, 6119 (1985).

[5] A.A. Wheeler, W.J. Boettinger, and G.B. McFadden, Phys. Rev. A**45**, 7424 (1992); Phys. Rev. E**47**, 1893 (1993).

[6] A. Karma, Phys. Rev. E**49**, 2245 (1994).

[7] K.R. Elder, F. Drolet, J.M. Kosterlitz, M. Grant, Phys. Rev. Lett. **72**, 677 (1994).

[8] A.A. Wheeler, G.B. McFadden, and W.J. Boettinger, to appear in the *Proceedings of the Royal Society of London*.

---

[1] As will be discussed at length in Ref. [23], one of the critical small parameter in this expansion turns out to be the dimensionless interface Peclet number WV_n/D.

[9] J.W. Cahn and J.E. Hilliard, J.Chem. Phys. **28**, 258 (1958).

[10] R. Kobayashi, Physica D**63**, 410 (1993).

[11] A. A. Wheeler, B. T. Murray, and R. Schaefer, Physica D**66**, 243 (1993).

[12] S-L. Wang and R. F. Sekerka, Phys. Rev. E (in print).

[13] J.A. Warren, and W.J. Boettinger, Acta Mett. et Mater., **43**, 689 (1995).

[14] B. Grossmann, K.R. Elder, M.Grant, and J.M. Kosterlitz, Phys. Rev. Lett. **71**, 3323 (1993).

[15] A. Chakrabarti, R. Toral, and J.D. Gunton, Phys. Rev. E**47**, 3025 (1993). R. Toral, A. Chakrabarti, and J.D. Gunton, Physica A**213**, 41 (1995).

[16] J.A. Warren and B.T. Murray, *Simulations of Ostwald Ripening: a Phase-Field Model*, submitted to Modelling and Simulation in MS&E.

[17] L-Q. Chen and W. Yang, Phys. Rev. B**50**, 15752 (1994).

[18] J.S. Langer, in *Chance and Matter*, Lectures on the Theory of Pattern Formation, Les Houches, Session XLVI, edited by J. Souletie, J. Vannimenus, and R. Stora (North Holland, Amsterdam, 1987), p. 629-711.

[19] D. Kessler, J. Koplik, and H. Levine, Adv. Phys. **37**, 255 (1988).

[20] G. Caginalp, Phys. Rev. A **39**, 5887 (1989).

[21] G.B. McFadden, A.A. Wheeler, R.J. Braun, and S.R. Coriell, and R.F. Sekerka, Phys. Rev. E **48**, 2016 (1993).

[22] S-L. Wang, R.F. Sekerka, A.A. Wheeler, B.T. Murray, S.R. Coriell, R.J. Braun, and G.B. McFadden, Physica D**69**, 189 (1993).

[23] A. Karma and W-J. Rappel (to be published).

[24] J.A. Sethian and J. Strain, J. Comp. Phys. **98**, 2313 (1992).

[25] R. Almgren, J. Comput. Phys. **106**, 337 (1993).

[26] T. Ihle and H. Muller-Krumbhaar, Phys. Rev. E **49**, 2972 (1994).

[27] A.R. Roosen and J.E. Taylor, J. Comput. Phys. **114**, 113 (1994).

[28] G.E. Nash and M.E. Glicksman, Acta Metall. **22**, 1283 (1974).

[29] D.A. Kessler and H. Levine, Phys. Rev. B **33**, 7687 (1986).

[30] D.I. Meiron, Phys. Rev. A **33**, 2704 (1986).

[31] M. Ben Amar and B. Moussallam, Physica D**25**, 155 (1987).

[32] P. Kopczynski, A. Karma, and W-J. Rappel (to be published).

Phase Field Instabilities
and
Adaptive Mesh Refinement

Robert F. Almgren
The University of Chicago
Department of Mathematics
Chicago, IL 60637

Ann S. Almgren
Lawrence Livermore National Laboratory
Livermore, CA 94550

November 17, 1995

We show that the phase field model, used for numerical simulation of solidification, can possess unphysical stable intermediate states for certain values of the parameters. We demonstrate that these states can be seen in simple computations in one dimension; in two dimensions we use an adaptive mesh refinement algorithm to exhibit appearance of the intermediate states in realistic geometries of dendritic growth.

The Phase-Field Model

We study solutions of the one-dimensional phase-field model, which we write using the notation and scalings of [1]:

$$\tau \phi_t = f(\phi, u) + \epsilon^2 \phi_{xx}$$
$$e_t = D u_{xx} \tag{1}$$
$$e = u - \tfrac{1}{2} p(\phi)$$

Here $\phi(x,t)$ is a phase variable, taking the value 1 in solid and -1 in liquid, and $u(x,t)$ is temperature scaled by the latent heat of fusion divided by the heat capacity, relative to the equilibrium temperature of a planar interface. The additional variable $e(x,t)$ is the internal energy density, which depends on ϕ via the specific heat function $p(\phi)$, with $p(\pm 1) = \pm 1$. The parameter D is the thermal diffusivity; τ is a relaxation time for the phase field, and the length scale ϵ is the thickness of the phase interface.

We write the nonlinear source term f in the form

$$f(\phi, u) = -g'(\phi) - \tfrac{1}{2} \lambda \, u \, p'(\phi) \tag{2}$$

where λ is a coupling constant. We shall take the specific functional choices

$$g(\phi) = \tfrac{1}{4}(1 - \phi^2)^2, \qquad p(\phi) = \tfrac{15}{8}\left(\phi - \tfrac{2}{3}\phi^3 + \tfrac{1}{5}\phi^5\right). \tag{3}$$

The function $g(\phi)$ is a double-well potential, with smooth wells at $\phi = \pm 1$. Quintic p is the simplest form which preserves the two wells at ± 1 for general u (see [2] and below).

By virtue of the structure (2) for f, the overall dynamics has a Lyapunov function, the negative entropy [2]

$$-S = \int_V \left(-s(\phi, u) + \tfrac{1}{2}\epsilon^2 |\nabla \phi|^2\right) dV \geq 0 \tag{4}$$

where the specific negative entropy is

$$-s(\phi, u) = g(\phi) + \tfrac{1}{2}\lambda u^2. \tag{5}$$

In fact, the dynamics (1) is a gradient flow for this entropy, since

$$f(\phi, u) = \frac{d}{d\phi} s\bigl(\phi, e + \tfrac{1}{2} p(\phi)\bigr), \tag{6}$$

and thus, if $\phi(x,t)$ and $u(x,t)$ solve (1), then $-S$ is positive and monotonically decreasing in time. It is not necessary to write the dynamics in this gradient form, and we will show that existence of a global Lyapunov function does not prevent the solution from exhibiting highly oscillatory behavior.

For small ϵ, solutions of (1) approximate solutions of the one-dimensional sharp-interface Stefan problem with $u = -\beta V$ on the interface. The kinetic coefficient β is given in terms of the parameters $\lambda, \epsilon, \tau, D$ of the diffuse model by extracting a solvability condition for solutions propagating at constant speed V, and expanding for small V at fixed ϵ; surprisingly, it is possible to obtain $\beta = 0$ at leading order [1]. In one dimension, where surface tension is not important, only D and β have physical significance, and thus there are two degrees of freedom in the choice of parameters in (1).

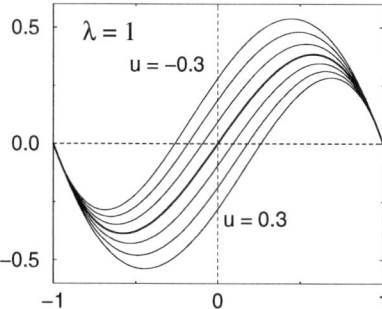 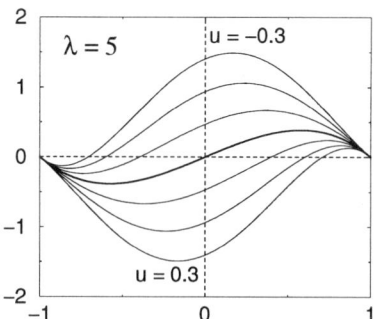

Figure 1: Source term $f(\phi, u)$ as a function of ϕ with u held fixed, taking values $u = -0.3$ to 0.3 in steps of 0.1 ($u = 0$ emphasized), for $\lambda = 1$ and $\lambda = 5$. For all λ and all u, $f(\phi, u)$ is the gradient of the double-well potential $s(\phi, u)$.

Intermediate Stable States

The purpose of this note is primarily to point out a subtlety in the meaning of the term "double-well potential" as applied to solutions of (1).

The functional forms above were constructed by requiring that the dynamics of the ordinary differential equation (ODE) $\phi_t = f(\phi, u)$ for u fixed, have exactly two stable fixed points at $\phi = \pm 1$; this requires that $f(\pm 1, u) = 0$ and $f_\phi(\pm 1, u) < 0$ for any value of u. There is one intermediate stationary point ϕ_0 at which $f(\phi_0, u) = 0$, but that point is unstable since $f_\phi > 0$ there. Figure 1 shows $f(\phi, u)$ for several values of u for $\lambda = 1$ and $\lambda = 5$; the desired structure is clearly preserved. Equivalently, the specific entropy $s(\phi, u)$ has local minima only at $\phi = \pm 1$ for any λ, considered as a function of ϕ with fixed u.

A different criterion emerges if we consider the stability of uniform states for the full time-dependent dynamics of (1). Stability against long-wavelength perturbations is determined by the behavior of the ODE obtained by suppressing the spatial derivative terms, and for this dynamics, the internal energy e is constant rather than u. Thus, stability is determined by the behavior of $f(\phi, e + \frac{1}{2}p(\phi))$ as ϕ varies. As illustrated in Figure 2, for λ large and e in a neighborhood of 0, this dynamics can develop an third stable state for a value $\phi = \phi_0$ intermediate between -1 and 1. This inversion happens for $\lambda > \lambda_* = -4g''(0)/p'(0)^2 = 256/225 \approx 1.14$ and for e in a neighborhood of zero which widens as λ increases. Equivalently, the negative entropy $-s(\phi, e + \frac{1}{2}p(\phi))$ develops an third local minimum as a function of ϕ with e constant.

If the effects of surface tension and kinetics are small, then $u \approx 0$ on the interface, and since ϕ passes through zero there, values of e around zero are exactly the range of interest.

Steady-State Solutions

We look for solutions of the system (1) which are steady-state profiles propagating to the right at speed $V \geq 0$. Making the substitution $\partial_t \mapsto -V\partial_x$, we obtain two second-order ODEs, of which one may immediately be integrated to obtain the third-order system

$$\epsilon^2 \phi_{xx} + V\tau\phi_x + f(\phi, u) = 0$$
$$Du_x + Vu - \tfrac{1}{2}Vp(\phi) = Ve_\infty,$$

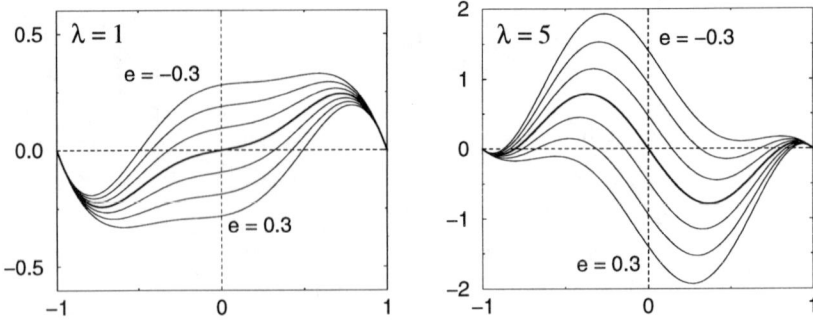

Figure 2: Source term $f(\phi, e + \frac{1}{2}p(\phi))$ as a function of ϕ with e held fixed, taking values $e = -0.3$ to 0.3 in steps of 0.1 ($e = 0$ emphasized), for $\lambda = 1$ and $\lambda = 5$. For large values of λ and for e near zero, an third intermediate stable state appears.

in which the constant of integration e_∞ is the common value of e in any constant state. We consider waves joining two constant states at $x \to \pm\infty$.

Any constant state (ϕ_0, u_0) must satisfy

$$f(\phi_0, u_0) = 0 \quad \text{and} \quad u_0 - \tfrac{1}{2}p(\phi_0) = e_\infty. \tag{7}$$

In addition, we require that the state (ϕ_0, u_0) be long-wave stable for the full time-dependent dynamics; that is, that

$$\frac{d}{d\phi} f\bigl(\phi, e_\infty + \tfrac{1}{2}p(\phi)\bigr)\Big|_{(\phi_0, u_0)} = f_\phi(\phi_0, u_0) + \tfrac{1}{2}p'(\phi_0)\, f_u(\phi_0, u_0) \;<\; 0. \tag{8}$$

With our choices above, this condition is always satisfied at $\phi_0 = \pm 1$. For large λ, it can also be satisfied at intermediate values of ϕ.

We now consider the asymptotic behavior of solutions to the ODE, as they approach the candidate constant state. Looking for a perturbed solution with space dependence $\exp(kx)$, we obtain the cubic dispersion relation

$$D\epsilon^2 k^3 + V(\epsilon^2 + \tau D)\, k^2 + \bigl(\tau V^2 + D f_\phi(\phi_0, u_0)\bigr) k + V\bigl(f_\phi(\phi_0, u_0) + \tfrac{1}{2}p'(\phi_0) f_u(\phi_0, u_0)\bigr) = 0.$$

Denoting the three roots by k_1, k_2, k_3, we have

$$k_1 k_2 k_3 = -\frac{V}{D\epsilon^2}\bigl(f_\phi(\phi_0, u_0) + \tfrac{1}{2}p'(\phi_0) f_u(\phi_0, u_0)\bigr) > 0$$

$$k_1 + k_2 + k_3 = -\frac{V}{D\epsilon^2}(\epsilon^2 + \tau D) < 0$$

by virtue of the long-wave stability condition. These two conditions require that the system have one real positive root, and two possibly complex roots with negative real parts. (This conclusion does not depend on using the entropy form (2) for f.)

Thus the system globally has three spatially growing modes: one at $x = +\infty$ and two at $x = -\infty$, and approach to constant states at $x = \pm\infty$ represents three boundary conditions. Although our system is third-order, its x-independence in effect removes one degree of freedom, so that bounded solutions can exist only for special values of the parameter

V. The important point is that this eigenvalue structure is qualitatively the same for any stable intermediate state as for the natural endpoints.

Suppose we specify a right state (ϕ_R, u_R); we want this state to represent a uniformly supercooled liquid, so $\phi_R = -1$, $u_R = -Y$, where Y with $0 \leq Y \leq 1$ is the undercooling. The constant of integration is then $e_\infty = -Y + \frac{1}{2}$. If $\lambda < \lambda_*$, the only candidate left state is $\phi_L = 1$, $u_L = u_R + 1 \approx -\beta V$. We may then solve for V and u_R together; or if $\beta = 0$ then $u_L = 0$ and steady waves are possible only if $Y = 1$.

If $\lambda > \lambda_*$, then for values of e_∞ in a suitable range, there is an additional candidate left state (ϕ_L, u_L) obtained by solving the equilibrium conditions (7,8). In addition, increase of global entropy requires that $s(\phi_L, u_L) > s(\phi_R, u_R)$, more restrictive than the long-wave stability condition. For example, for $Y = \frac{1}{2}$, $e_\infty = 0$, and $-s(\phi, e+\frac{1}{2}p(\phi)) = g(\phi)+\frac{1}{8}\lambda p(\phi)^2$; the intermediate state $\phi_0 = 0$ is a global minimum if $\lambda > 8g(0) = 2$. Under these conditions, there can be a steady wave joining the right state $(-1, -Y)$ to an unphysical intermediate left state (ϕ_0, u_0).

One-Dimensional Computations

To show that the steady-state solutions discussed above can be observed in practice and to exhibit their behavior, we show some one-dimensional time-dependent computations. We take $\epsilon = 0.1$ and $D = 1$, and take τ so $\beta = 0$ for a chosen value of λ. The computational method is second-order semi-implicit on the coupled system; we take extremely fine discretization steps $\Delta x = 0.01 = \epsilon/10$ and $\Delta t = 0.001 \approx \tau/20$.

We take initial data to be step functions in ϕ and in u; smooth profiles inhibit the transition to intermediate waves. The sharp-interface limit problem then has an exact solution which is a constant-velocity wave for $Y = 1$, and a similarity solution with scaling $x - x_0 \sim t^{1/2}$ for $Y < 1$.

We illustrate two runs with $Y = 0.7$, for $\lambda = 5$ and $\lambda = 10$, with $\tau = 0.019$ and 0.037 respectively. For $\lambda = 5$, the intermediate state is $\phi_0 = 0.311$ and for $\lambda = 10$, $\phi_0 = 0.255$; both of these are global minima of $-s$. However, for $\lambda = 5$ (Figure 3), the physical interface is stable, while for $\lambda = 10$ (Figure 4) the unphysical intermediate state appears. The Lyapunov functional $-S$ steadily decreases in both cases.

More detailed investigation shows that the intermediate state appears quite sharply at a critical value of λ, not explained by any of the requirements above. Further, intermediate waves appear to be unstable for $Y \leq \frac{1}{2}$. We do not yet understand either property.

Two-Dimensional Computations

To show that the above considerations can be a problem in higher-dimensional geometries of interest, we preview some computations from a forthcoming work [3]. Using an adaptive mesh refinement algorithm [4], we have repeated a test case chosen by Wheeler et al. [5].

This computation uses a phase field model of the same form as in (1), with a smooth double-well $g(\phi)$ and a quintic $p(\phi)$. In more than one dimension, surface tension and its anisotropy become important; we incorporate this in a standard way.

In the notation of (1), the parameters are $\epsilon = 0.005$, $\lambda = 16$, $\tau = 0.002$, $D = 1$. The spatial domain is 1.92×0.96, and the undercooling $Y = 0.5$. In all cases the coarsest spatial grid is 48×24, corresponding to a grid cell size $\Delta x_{coarse} = 0.04$, and we vary the local grid cell size in the neighborhood of the interface from $\Delta x_{fine} = 0.01$ down to 0.0025 by varying the number of levels of grid refinement and the refinement ratio at each level. Our local

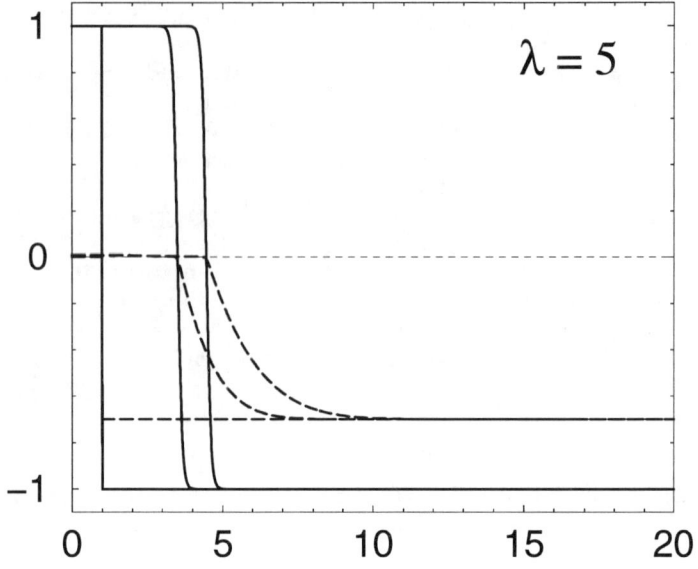

Figure 3: $Y = 0.7$, $\lambda = 5$: ϕ (solid) and u (dashed) for $t = 0, 2, 4$.

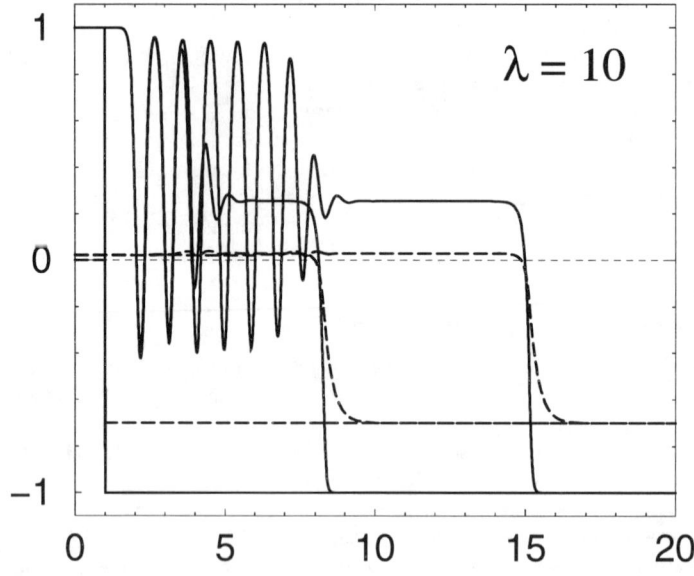

Figure 4: $Y = 0.7$, $\lambda = 10$: ϕ (solid) and u (dashed) for $t = 0, 2, 4$.

mesh refinement algorithm enables us not only to perform the computations efficiently, but also to examine the global solution structure in detail.

The initial data for ϕ represented a small solid seed in the lower left corner of the domain, and the initial data for u was uniformly equal to the undercooling $u = -Y = -0.5$ throughout most of the domain, smoothly rising to $u = 0$ inside the initial solid seed. The surface tension had four-fold anisotropy of strength $\eta = 0.01$, and thus dendritic fingers grow parallel to the coordinate axes (we have not yet carefully assessed the effect of grid anisotropy). All pictures shown are at time $t = 0.08$, although the computations proceeded until the dendrite slowed as it hit the right wall of the box at $t \approx 0.4$. The measured tip velocities agree with [5] to the two decimal places reported there.

For the coarsest grid, $\Delta x = 0.01$ in Figure 5, the interface looks roughly as it ought to. The phase field ϕ rises smoothly from its liquid value to its solid value over a zone of width a few times ϵ; there are about 5 grid points between the contours $\phi = 0.1$ and $\phi = 0.9$. Nonetheless, this computation is far from resolved for the given value of ϵ.

When Δx is reduced by a factor of 2 (Figure 6), the structure of the interface changes: it broadens out and a "plateau" with an intermediate value of ϕ appears. When Δx is again reduced by a factor of 2 (Figure 7), a real secondary maximum appears with $\phi \approx 0.224$. This intermediate value of ϕ has no significance in the context of the sharp-interface model; it is purely an artifact of the finite value of ϵ.

We have performed the same computation with one more halving of Δx; the structure does not change, indicating that Figure 7 shows, at least qualitatively, the true solution of the partial differential equation for this value of ϵ. We believe that the appearance of this secondary maximum is due to the reasons outlined above, though modified by the two-dimensional geometry. Indeed, the maximum disappears at later times as this computation progresses and the thermal fields become less steep.

Acknowledgments

This work was supported by the National Science Foundation CAREER program under award DMS-9502059, by the NSF MRSEC Program under award DMR-9400379, and by the AMS Program of the Department of of Energy Office of Mathematics, Information and Computational Science under contract No. W-7405-Eng-48.

References

[1] A. Karma and W.-J. Rappel. Phase-field method for computationally efficient modeling of solidification with arbitrary interface kinetics, 1995. Preprint.

[2] S.-L. Wang, R. F. Sekerka, A. A. Wheeler, B. T. Murray, S. R. Coriell, R. J. Braun, and G. B. McFadden. Thermodynamically-consistent phase-field models for solidification. *Physica D*, 69:189–200, 1993.

[3] A. S. Almgren, R. F. Almgren, and C. Rendleman. Adaptive mesh refinement for phase field computations, 1996. In preparation.

[4] A. S. Almgren, J. B. Bell, P. Colella, and L. H. Howell. An adaptive projection method for the incompressible Euler equations. In *Proceedings of the AIAA 11th Computational Fluid Dynamics Conference*, 1993.

[5] A. A. Wheeler, B. T. Murray, and R. J. Schaefer. Computation of dendrites using a phase field model. *Physica D*, 66:243–262, 1993.

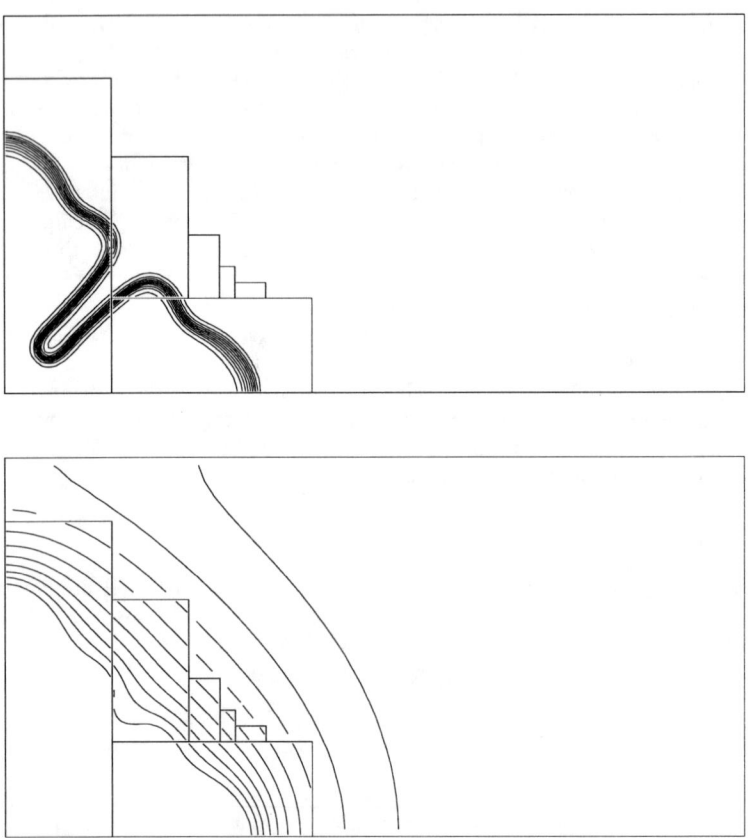

Figure 5: Contours of phase field and temperature for local $\Delta x = 0.01 = 2\epsilon$. The rectangles are the boundaries of subgrids, refined by a factor of 4 relative to the outer coarse grid. The interface is coherent, though underresolved.

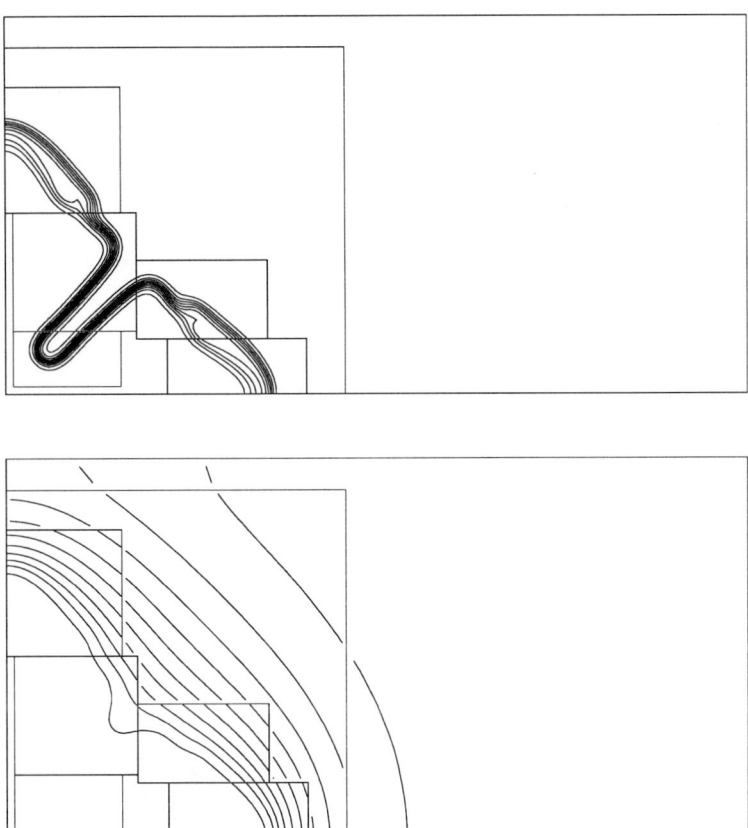

Figure 6: Phase field and temperature for local $\Delta x = 0.005 = \epsilon$. The square shape is the boundary of a grid refined by a factor of 2 relative to the outer grid; the inner rectangles are refined by an additional factor of 4. The interface begins to broaden and the intermediate state begins to appear.

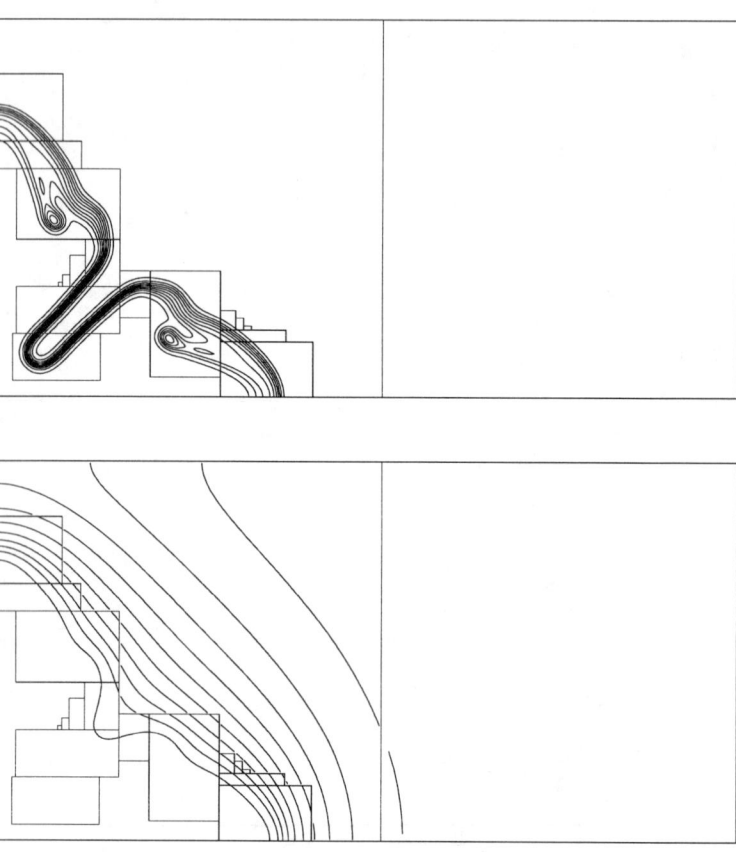

Figure 7: Phase field and temperature for local $\Delta x = 0.0025 = \frac{1}{2}\epsilon$. Grid refinement by two successive factors of 4. The little "island" is a local maximum of ϕ with a value intermediate between solid and liquid. Further refinement does not qualitatively change this structure.

MICROSTRUCTURE EVOLUTION AND GRAIN GROWTH KINETICS IN A TWO-PHASE SOLID WITH QUADRIJUNCTIONS

D. Fan and L. Q. Chen

Department of Materials Science and Engineering
The Pennsylvania State University
University Park, Pennsylvania 16802

Abstract

The stability and evolution of quadrijunctions in a two-phase solid with conserved volume fractions were studied in two dimensions by numerically solving a set of coupled Allen-Cahn and Cahn-Hilliard equations. It was shown that quadrijunctions can be stable in a two-phase solid if the grain boundary energies and the interphase boundary energy satisfy certain conditions, consistent with previous thermodynamic analyses. It was revealed that the grain topology and topological transformations in a microstructure with only quadrijunctions are dramatically different from those in a single-phase system. After a short transient, both the shape and size distributions of each phase are shown to be independent of time, whereas the average grain radius for each phase was found to increase with time as $t^{1/3}$, implying the long-range diffusion controlled grain growth in a two-phase solid with stable quadrijunctions and with conserved volume fractions.

Mathematics of Microstructure Evolution
Edited by L. Q. Chen, B. Fultz, J. W. Cahn,
J. R. Manning, J. E. Morral, and J. A. Simmons
The Minerals, Metals & Materials Society, 1996

Introduction

In a two-phase (α+β) polycrystalline material, there are three kinds of interfaces: grain boundaries in α (α/α); grain boundaries in β (β/β); and interphase boundaries between α and β (α/β). Even in the simplest possible case of isotropic grain boundary energies and isotropic interphase boundary energy, the kinetics of interface motion in such systems is quite complicated as different processes, such as grain growth and Ostwald ripening, often occur simultaneously during sintering or processing. Important practical examples include the ZrO_2-Al_2O_3 two-phase particulate composite in ceramics [1,2] and the two-phase (α + β) titanium alloys in metallic systems [3]. Second-phase precipitates were often observed at grain boundaries even in so-called single-phase materials and they may undergo Ostwald ripening during grain growth. In fact, pores may be treated as a second phase during sintering.

Grain growth is a process in which the average grain size of a single-phase polycrystalline material increases with time, driven by the reduction in the total grain boundary energy. The driving force for a given boundary to move is the difference in the chemical potentials of an atom on the opposite sides of the grain boundary, which may be written as

$$\Delta\mu = \sigma_{gb}\Omega(H_1 + H_2) \quad (1)$$

,where $\Delta\mu$ is the chemical potential change for an atom going from one side of the grain boundary to another, σ_{gb} is the grain boundary energy, Ω is the atomic volume, and $(H_1 + H_2)$ is the local mean curvature. The velocity of boundary migration is determined by the rate at which atoms jump across the boundary,

$$V = B\Delta\mu \quad (2)$$

,where V is the grain boundary migration velocity and B is the boundary mobility. The typical diffusion distance for atoms involved in a grain boundary migration is, therefore, on the order of the boundary width which, for a pure material, is about two or three lattice parameters.

On the other hand, Ostwald ripening usually refers to a process during which large second-phase particles grow while small particles dissolve in a matrix, resulting in a reduction in the total interfacial energy between the precipitates and matrix. The driving force is the difference in the chemical potentials of atoms in large and small second phase particles, which results in a difference in the compositions of solute atoms in the matrix immediately outside the second phase particles,

$$\Delta\mu = \sigma_{\alpha\beta}\Omega(1/r_1 - 1/r_2) \quad (3)$$

, where $\Delta\mu$ is the chemical potential difference of a solute atom in two particles with radius r_1 and r_2, respectively, $\sigma_{\alpha\beta}$ is the interfacial energy between the precipitate (α) and matrix (β), and Ω is the atomic volume of a solute atom. In order for a large particle to grow, however, solute atoms outside a small particle have to diffuse through the matrix to regions near the large particle. Therefore, the typical diffusion distance for atoms involved in the Ostwald ripening process is on the order of the separation distance between second-phase particles.

Recently, Cahn [4] performed a thermodynamic analysis for the stability of microstructures in a two-dimensional two-phase solid in which the volume fractions of the two phases are not conserved, i.e., the two phases have the same composition, and hence long-range diffusion and Ostwald ripening are not involved. In his theory, the microstructural stability and features were analyzed based on the energetic ratios of grain boundary energies to the interphase boundary energy: $R_\alpha = \sigma_{\alpha\alpha} / \sigma_{\alpha\beta}$ and $R\beta = \sigma_{\beta\beta} / \sigma_{\alpha\beta}$, where the $\sigma_{\alpha\alpha}$ is the grain boundary energy in the α phase, $\sigma_{\beta\beta}$ is the grain boundary energy in the β phase, and $\sigma_{\alpha\beta}$ is the interphase boundary energy between α and β phases. For $0 \leq R_\alpha \leq \sqrt{3}$, the trijunctions $\alpha\alpha\alpha$ are stable and when $R_\alpha > \sqrt{3}$ the $\alpha\alpha\alpha$ trijunctions are unstable with respect to the nucleation of β grains at these trijunctions [4]. Similarly, the trijunctions $\beta\beta\beta$ are stable for $0 \leq R\beta \leq \sqrt{3}$ and are unstable with respect to the nucleation of α grains if $R\beta > \sqrt{3}$. The trijunctions $\alpha\alpha\beta$ and $\alpha\beta\beta$ are stable under the conditions of $0 \leq R_\alpha < 2$ and $0 \leq R\beta < 2$, respectively. More interestingly, he found that the quadrijunctions $\alpha\beta\alpha\beta$ will become stable if the condition $R_\alpha^2 + R_\beta^2 \geq 4$ is satisfied [4].

Following Cahn's work, Holm *et al* performed Monte Carlo simulations on the same system, i.e., a two-phase solid in which the volume fractions are not conserved, based on the Q-states Potts model [5]. They showed quadrijunctions can indeed be stable within a certain range of the values for R_α and $R\beta$ as predicted by Cahn's thermodynamic analysis. More surprisingly, based on the Monte-Carlo simulations, they predicted that the grain growth in a system with only quadrijunctions may be frozen [5]. However, a recent study by Cahn and Van Vleck using a discrete diffuse-interface model showed that quadrijunctions of ordered domains, in which the volume fractions of ordered domains are not conserved, can not stop the coarsening process [6].

The main objective of this paper is to investigate the stability and evolution kinetics of quadrijunctions in a two-phase solid in which the volume fractions are CONSERVED, using a continuum diffuse-interface grain growth model, i.e., by numerically solving a set of coupled continuum Allen-Cahn and Cahn-Hilliard equations. This model is similar to the phase-field model for solidification [7-8]. We have applied this model to grain growth in single-phase systems [9-11] and to coupled grain growth and Ostwald ripening in two-phase solids [12]. It should be emphasized that most two-phase solids in real applications belong to the case of conserved volume fractions.

Diffuse-Interface Description of a Two-Phase Microstructure

Within the diffuse-interface context, we describe an arbitrary two-phase polycrystalline microstructure using a set of continuous field variables [9-11],

$$\eta_1^\alpha(r), \eta_2^\alpha(r),..., \eta_p^\alpha(r), \eta_1^\beta(r), \eta_2^\beta(r),..., \eta_q^\beta(r), C(r),$$

where η_i^α (i = 1, ..., p) and η_j^β (j = 1,..., q) are called orientation field variables with each representing grains of a given crystallographic orientation of a given phase. Those variables assume continuous values ranging from -1.0 to 1.0. For example, a value of 1.0 for $\eta_i^\alpha(r)$, with the values for the rest of the orientation field variables zero at r, means that the material at position r belongs to an α-phase grain with the crystallographic orientation labeled as i. At the grain boundary region, orientation variables will have absolute values intermediate between 0.0 and 1.0. $C(r)$ is the composition field which takes the value of C_α within an α grain and C_β within a β grain.

Diffuse-Interface Description of the Energetics of a Two-Phase Solid

In the diffuse-interface model [13], the total free energy of a two-phase system can be written as

$$F = \int \left[f_o\left(C(r); \eta_1^\alpha(r), \eta_2^\alpha(r),..., \eta_p^\alpha(r); \eta_1^\beta(r), \eta_2^\beta(r),..., \eta_q^\beta(r)\right) \right. \\ \left. +(\kappa_C/2)(\nabla C(r))^2 + (1/2)\sum_{i=1}^{p} \kappa_i^\alpha \left(\nabla \eta_i^\alpha(r)\right)^2 + (1/2)\sum_{i=1}^{q} \kappa_i^\beta \left(\nabla \eta_i^\beta(r)\right)^2 \right] d^3r \quad (4)$$

, where f_o is local free energy density, κ_C, κ_i^α and κ_i^β are the gradient energy coefficients for the composition field and orientation fields, and p and q represent the number of orientation field variables for the α and β phases.

The energy of a planar grain boundary, $\sigma_{\alpha\alpha}$, between an α-grain of orientation 1 and another α-grain of orientation 2 can be calculated as :

$$\sigma_{\alpha\alpha} = \int_{-\infty}^{+\infty} \left[\Delta f + (\kappa_C/2)(dC/dx)^2 + \left(\kappa_1^\alpha/2\right)\left(d\eta_1^\alpha/dx\right)^2 + \left(\kappa_2^\alpha/2\right)\left(d\eta_2^\alpha/dx\right)^2 \right] dx \quad (5)$$

in which

$$\Delta f = f_o\left(\eta_1^\alpha, \eta_2^\alpha, C\right) - f_o\left(\eta_{1,e}^\alpha, \eta_{2,e}^\alpha, C_\alpha\right) - (C - C_\alpha)(\partial f_o/\partial C)_{\eta_{1,e}^\alpha, \eta_{2,e}^\alpha, C_\alpha} \quad (6)$$

where $f_o\left(\eta_{1,e}^\alpha, \eta_{2,e}^\alpha, C_\alpha\right)$ represents the free energy density minimized with respect to η_1^α and η_2^α at a fixed composition C_α.

Similarly, the interphase boundary energy between an α-grain with orientation 1 and a β-grain with orientation 1 is given by

$$\sigma_{\alpha\beta} = \int_{-\infty}^{+\infty} \left[\Delta f + (\kappa_C/2)(dC/dx)^2 + \left(\kappa_1^\alpha/2\right)\left(d\eta_1^\alpha/dx\right)^2 + \left(\kappa_1^\beta/2\right)\left(d\eta_1^\beta/dx\right)^2 \right] dx \quad (7)$$

where

$$\Delta f = f_o\left(\eta_1^\alpha, \eta_1^\beta, C\right) - f_o\left(\eta_{1,e}^\alpha, \eta_{1,e}^\beta, C_\alpha\right) - (C - C_\alpha)(\partial f_o/\partial C)_{\eta_{1,e}^\alpha, \eta_{1,e}^\beta, C_\alpha} \quad (8)$$

The Coupled Allen-Cahn and Cahn-Hilliard Equations

Based on the diffuse-interface description of the microstructure and energetics, the kinetics of microstructural evolution of a two-phase system can be described by the spatial and temporal evolution of orientation and composition field variables by numerically solving the coupled Allen-Cahn [14] and Cahn-Hilliard [15] equations,

$$d\eta_i^\alpha(r,t)/dt = -L_i^\alpha \left(\delta F/\delta \eta_i^\alpha(r,t)\right), \quad i = 1, 2, ..., p, \tag{9a}$$

$$d\eta_i^\beta(r,t)/dt = -L_i^\beta \left(\delta F/\delta \eta_i^\beta(r,t)\right), \quad i = 1, 2, ..., q, \tag{9b}$$

$$dC(r,t)/dt = \nabla\{L_C \nabla[\delta F/\delta C(r,t)]\} \tag{9c}$$

where L_i^α, L_i^β and L_C are kinetic coefficients related to grain boundary mobilities and atomic diffusion coefficients, which may be functions of local orientation and composition field variables, t is time, and F is the total free energy given in equation (4).

Temporal Evolution of a Two-Phase Microstructure With Quadrijunctions

In order to study the stability of a two-phase microstructure with only quadrijunctions, we construct the following relatively simple free energy density function,

$$f_o = f(C) + \sum_{i=1}^{p} f\left(C, \eta_j^\alpha\right) + \sum_{i=1}^{q} f\left(C, \eta_j^\beta\right) + \sum_\alpha \sum_\beta \sum_{i=1}^{p} \sum_{j=1}^{q} f\left(\eta_i^\alpha, \eta_j^\beta\right) \tag{10}$$

in which

$$f(C) = -(A/2)(C - C_m)^2 + (B/4)(C - C_m)^4 + (D_\alpha/4)(C - C_\alpha)^4 + (D_\beta/4)(C - C_\beta)^4$$

$$f\left(C, \eta_j^\alpha\right) = -(\gamma_\alpha/2)(C - C_\beta)^2 \left(\eta_i^\alpha\right)^2 + (\delta_\alpha/4)\left(\eta_i^\alpha\right)^4$$

$$f\left(C, \eta_j^\beta\right) = -(\gamma_\beta/2)(C - C_\alpha)^2 \left(\eta_i^\beta\right)^2 + (\delta_\beta/4)\left(\eta_i^\beta\right)^4$$

$$f\left(\eta_i^\alpha, \eta_j^\beta\right) = \left(\varepsilon_{ij}^{\alpha\beta}/2\right)\left(\eta_i^\alpha\right)^2\left(\eta_j^\beta\right)^2$$

where $\eta_i^\alpha, \eta_i^\beta$ are orientation field variables, C is the local composition, C_α and C_β are the equilibrium compositions of α and β phases, $C_m = (C_\alpha + C_\beta)/2$, A, B, D_α, D_β, γ_α, γ_β, δ_α, δ_β, and $\varepsilon_{ij}^{\alpha\beta}$ are phenomenological parameters. It can be shown that f_o has p degenerate minima with equal depth located at $\left(\eta_1^\alpha, \eta_2^\alpha, ..., \eta_p^\alpha\right) = (1,0,...,0), (0,1,...,0), ..., (0,0,...,1)$ in p-dimension orientation space at the equilibrium composition C_α, and has q degenerate minima located at $(\eta_1^\beta, \eta_2^\beta, ... \eta_q^\beta) = (1,0,...,0), (0,1,...,0), ..., (0,0,...,1)$ at C_β. This requirement ensures that each point in space can only belong to one crystallographic orientation of a given phase.

The parameters in the free energy density function and the gradient energy coefficients are chosen in such a way that grain boundary energies and interphase boundary energy satisfy the so-called "double" wetting condition, $R_\alpha^2 + R_\beta^2 \geq 4$, in which no trijunctions are stable and the only stable interfaces are the α/β interphase boundaries [4]. In the following computer simulation, the following numerical values are employed: $C_\alpha = 0.05$, $C_\beta = 0.95$, $C_m = 0.5$, A=1.0, B=4.94, Da = Db = 6.09, $\gamma_\alpha = \gamma_\beta = 2.47$, $\delta_\alpha = \delta_\beta = 1.0$, and $\varepsilon_{ij}^{\alpha\beta} = 7.0$. Using these values, R_α and R_β are found to be $R_\alpha = R_\beta = 2.1$. We assumed that $L_\eta^\alpha = L_\eta^\beta = 1.0$ and $L_C = 0.5$.

The coupled Allen-Cahn and Cahn-Hilliard equations are numerically solved using the simple explicit Euler technique in two-dimensions with 256×256 grid points and with periodic boundary conditions applied along both directions. The time step and grid size are chosen to be 0.1 and 2.0, respectively. The initial condition is generated by assigning small random values to all the orientation field variable and the overall average composition to the composition variable at all positions.

The microstructural evolution of a two-phase solid with 50% volume fractions for both phases is shown in Fig. 1. It can be seen that, as expected, the microstructures are comprised only of quadrijunctions αβαβ and α/β interphase boundaries during the microstructural evolution. The angles of the quadrijunctions vary within a certain range, i.e., there is not a thermodynamically fixed angle for quadrijunctions. The above observations seem to be consistent with Cahn's thermodynamic predictions even though his analysis was made on a system in which the volume fractions are nonconserved [4].

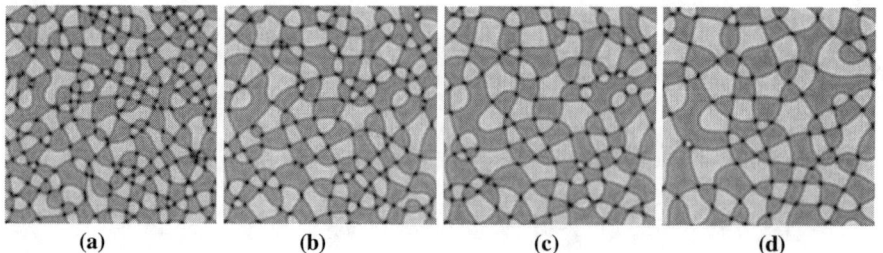

(a) (b) (c) (d)

Fig. 1 The microstructural evolution in all quadrijunction microstructures ($R_\alpha = R_\beta = 2.1$). The volume fraction of α phase is 50%. (a) t=5000; (b) t=10000; (c) t=30000; (d) t=50000.

Topological Transformations

The topological transformations during grain growth can be directly obtained from the temporal microstructures generated from the computer simulations. It is observed that two-sided grains (grains surrounded by two quadrijunctions) and three-sided grains (grains surrounded by three quadrijunctions) can directly undergo vanishing during grain growth of a two-phase solid with only quadrijunctions. The vanishing of a two-sided grain results in the disappearance of two quadrijunctions and the appearance of a new quadrijunction while the two former grains neighboring to the vanishing grain lose one quadrijunction each. The vanishing of three quadrijunction grains leads to the formation of a hexajunction (a junction at which six grains, with three of each phase, meet). However, a hexajunction is highly unstable in a two-phase solid and it quickly splits into two new quadrijunctions. As a result, each of the two adjacent grains of the other phase loses one interface and one quadrijunction; one of the same phase grains gains one interface and one quadrijunction; and other grains remain unchanged in terms of the topology.

The time dependence of topological distributions is shown in Fig. 2. It can be seen that the shapes of these distributions do not change with time, i.e., the topological distribution is time-invariant in a two-phase microstructure with only quadrijunctions, similar to the grain growth in single phase systems [11]. However, in a single-phase system with trijunctions, 2-sided grains never appear whereas in Fig. 2 there are quite a few two-sided grains whose occurrence is due to the fact that grain growth in a two-phase solid with conserved volume fractions is controlled by long-range diffusion, not just by the mean curvature as in single-phase systems. As a result, the average number of grain edges in a two-dimensional two-phase solid will not be 6 per grain, which comes from the requirement of space filling and balance of surface tension for a single-phase system. Actually, the average number of grain edges varies with the energetic conditions, i.e., values of R_α and R_β, as well as the volume fraction. The average number of grain edges obtained from the microstructures in Fig. 1 is about 5.92 per grain.

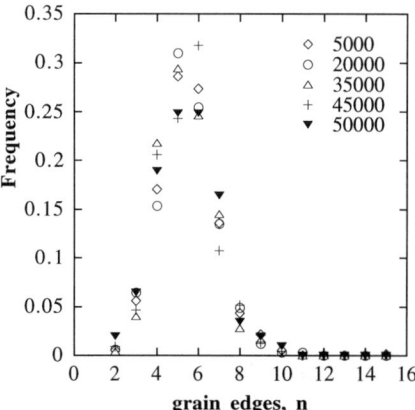

Fig. 2. The time dependency of topological distributions of α phase in all quadrijunction microstructures. The volume fraction of α phase is 50%.

Time-Dependence of Grain Size and Size Distributions

The time dependencies of grain size distributions for α and β phases corresponding to the microstructures in Fig. 1 are shown in Fig. 3 and 4 respectively. It can be seen that the distributions for α and β phase are almost identical and the peaks of size distributions for both phases occur at average size ($\log_{10}(R/<R>) = 0.0$) position. The shape of size distributions is independent of time for both phases, indicating that this system has reached the dynamic steady state after 5000 time steps. The average grain size for the two phases are shown in Fig. 5 and 6, respectively. In contrast to the freezing of grain growth predicted in the Potts model simulations for a two-phase solid with non-conserved volume fractions [5], it is shown that the average grain sizes of the two phases increase with time and follows the power growth law $R_t^m - R_0^m = kt$ with m = 3, indicating the long-range diffusion controlled coarsening (Fig. 5 and 6).

Conclusions

The stability and evolution of quadrijunctions in a two-dimensional two-phase solid with conserved volume fractions were investigated by computer simulations based on a diffuse-interface field model. It is shown that under certain thermodynamic conditions quadrijunctions can be

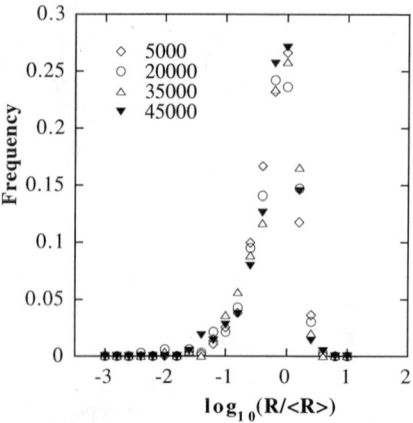

Fig. 3. The time dependency of grain size distributions of α phase in all quadrijunction microstructures. The volume fraction of α phase is 50%.

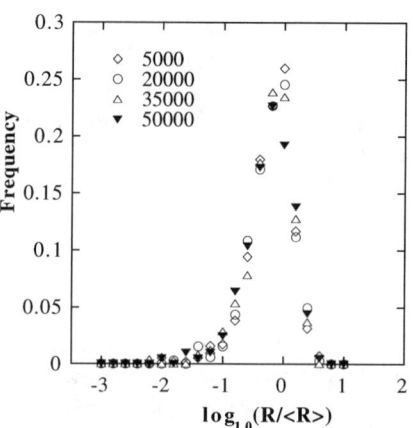

Fig. 4. The time dependency of grain size distributions of β phase in all quadrijunction microstructures. The volume fraction of β phase is 50%.

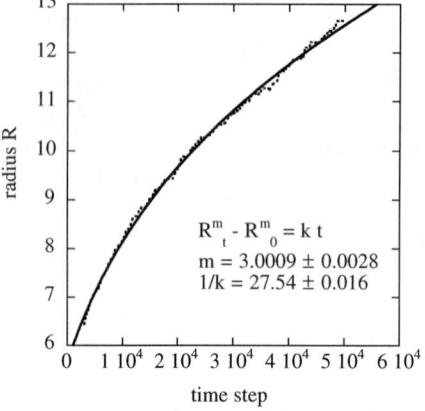

Fig. 5. The time dependency of the average grain size of α phase in all quadrijunction microstructures. The volume fraction of α phase is 50%. The dashed line is simulation data. The solid line is a fitting into equation $R^m_t - R^m_0 = k\,t$.

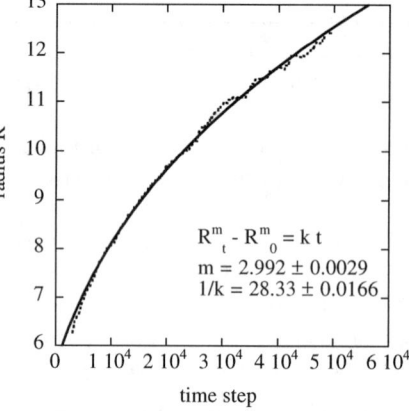

Fig. 6. The time dependency of the average grain size of β phase in all quadrijunction microstructures. The volume fraction of β phase is 50%. The dashed line is simulation data. The solid line is a fitting into equation $R^m_t - R^m_0 = k\,t$.

stable in a two-phase solid. The kinetic simulations show that 2-sided grains can be stable in a two-phase solid with conserved volume fractions and the average number of grain edges is less than six. The topological transformations in a two-phase microstructure with only quadrijunctions is dramatically different from those in a single-phase system or a two-phase with only trijunctions. It was shown that in the steady state, both the shape and size distributions of each phase are invariant with time, whereas the average grain radius for each phase with time was found to follow $t^{1/3}$ law, implying the long-range diffusion controlled grain growth in a two-phase solid with stable quadrijunctions and with conserved volume fractions.

Acknowledgments: The work is supported by the National Science Foundation under the grant number DMR 93-1898 and the simulations were performed at the Pittsburgh Supercomputing Center.

References

1. F. F. Lange and M. M. Hirlinger, "Grain Growth in Two-Phase Ceramics -Al_2O_3 Inclusions in ZrO_2", J. Am. Ceram. Soc., 70, 827 (1987).
2. K. B. Alexander, P. F. Becher, S. B. Waters, and A. Bleier, "Grain Growth Kinetics in Alumina-Zirconia (CeZTA) Composites", J. Am. Ceram. Soc. **77**, 939 (1994), and references therein.
3. S. Ankem and H. Margolin, "Grain Growth Relationships in Two-Phase Titanium Alloys", Proc. Fifth Int. Conf. on Titanium, Munich, 1984. Published by Deutsche Gesellsschaft Fur Metallkunde E. V., F.R.G., Vol. 3, 1705 (1985).
4. J. W. Cahn, "Stability, Microstructural Evolution, grain Growth, and Coarsening in a Two-Dimensional Two-Phase Microstructure," Acta Metall. Mater., 39, 2189-99 (1991).
5. E. A. Holm, D. J. Srolovits, J. W. Cahn, "Microstructural Evolution in Two-Dimensional Two-Phase Polycrystals", Acta metall. mater. **41**, 1119 (1993).
6. J. W. Cahn and E. S. Van Vleck, "Quadrijunctions Do Not Stop Two-Dimensional grain Growth," submitted to Scripta Metall., 1995.
7. J. A. Warren and W. J. Boettinger, " Prediction of dendritic growth and microsegregation patterns in a binary alloy using the phase-field method," Acta Metall. Mater., 43, 689 (1995).
8. G. Caginalp and E. Socolovsky, " Phase field computations of single-needle crystals, crystal growth, and motion by mean curvature," SIAM J. Sci. Comput., 15, 106(1994).
9. L. Q. Chen, "A Novel Computer Simulation Technique for Modeling Grain Growth", Scr. metall.et Mater., 32, 115 (1995).
10. L. Q. Chen and W. Yang, "Computer Simulation of the Domain Dynamics of a Quenched System with a Large Number of Nonconserved Order Parameters: Grain Growth Kinetics", Phys. Rev. B, **50**, 15 752 (1994).
11. Danan Fan and L. Q. Chen, "Computer Simulation of Grain Growth Using a Continuum Field Model," submitted to Acta Metall. Mater., 1995.
12. L. Q. Chen and Danan Fan, "A Computer Simulation Model for Coupled Grain Growth and Ostwald Ripening - Application to Al_2O_3 - ZrO_2 Two-phase Systems," accepted in J. Am. Ceram. Soc., 1995.
13. J. W. Cahn and J. E. Hilliard, "Free Energy of a Nonuniform System. I. Interfacial Free Energy", J. Chem. Phys. **28**, 258 (1958).
14. S. M. Allen and J. W. Cahn, "A Microscopic Theory for Antiphase Domain Boundary Motion and Its Application to Antiphase Domain Coarsening", Acta metall. **27**, 1085 (1979).
15. J. W. Cahn, "On spinodal decomposition," Acta metall, 9, 795-801(1961).

Anisotropic Interfaces and Ordering in fcc Alloys: A Multiple-Order-Parameter Continuum Theory

R. J. Braun[*], J. W. Cahn, J. Hagedorn,
G. B. McFadden, and A. A. Wheeler[†]
National Institute of Standards and Technology
Gaithersburg, MD 20899
USA

Abstract

A multiple-order-parameter theory of ordering on a binary face-centered-cubic (fcc) crystal lattice is developed, and adapted to provide a continuum formulation that incorporates the underlying symmetries of the fcc crystal in both the bulk and gradient-energy terms of the free energy. The theory is used to compute the orientation dependence of the structure and energy of interphase and antiphase boundaries. The structure of these interfaces compares favorably with previous lattice calculations by Kikuchi and Cahn. Anisotropy is a natural consequence of the lattice calculation and the multiple-order-parameter continuum formulation presented here.

---

[*]Permanent Address: Department of Mathematical Sciences, University of Delaware, Newark, DE 19716, USA
[†]Permanent address: Faculty of Mathematical Studies, University of Southampton, Highfield, Southampton SO17 1BJ, UK.

Mathematics of Microstructure Evolution
Edited by L. Q. Chen, B. Fultz, J. W. Cahn,
J. R. Manning, J. E. Morral, and J. A. Simmons
The Minerals, Metals & Materials Society, 1996

1 Introduction

In this contribution we summarize recent work on a continuum mean-field theory of anisotropic interfaces in ordered fcc crystals; this work is described in detail in Ref. [1].

In many continuum theories of phase change, such as spinodal decomposition and ordering reactions, the interface between the two phases is regarded as being diffuse rather than sharp, with a finite thickness over which properties vary smoothly from one set of bulk values to another. Such diffuse interface theories arise naturally in the equilibrium descriptions of critical phenomena [2], where the interface thickness scales with the correlation length and diverges as the critical temperature is approached. Recently this approach has been extended to other phase changes, such as solidification, with significant computational advantages for describing such complex behavior as dendritic growth.

Crystalline anisotropy is an important component of many of these phase changes. There have been a variety of methods for introducing the anisotropy into the theory. Some of these arise from physical principles, while others are *ad hoc* and can be the origin of inconsistencies. Some types of anisotropy are associated with the interfaces, while others, such as anisotropic elastic effects, are associated with the volumes, even when they arise from coherency constraints at interfaces [3].

One method for treating diffuse interfaces is to use a free energy functional for the system in terms of continuum parameters that are spatially varying. The functional is written as the integral of the sum of two kinds of terms; bulk energies that are multiple well functions of these parameters that have minima, or common tangents, at values that characterize the adjacent phases, and gradient energies that are functions, commonly the square, of the gradients of the order parameters. Both terms contribute to the energy in the transition regions that separate bulk phases. Such gradient energy models date back to work by Rayleigh [4] and Van der Waals [5], and are useful in a variety of contexts.

One such use is for dynamical calculations in which evolution equations for the system are derived by using variational arguments on these free energy functionals. When there is a single nonconserved scalar order parameter, the usual form of the resulting equation is the Cahn-Allen equation [6, 7] describing antiphase boundary motion between domains of an ordered phase, as described in more detail below. For a single conserved parameter, such as the composition, the result is the Cahn-Hilliard equation [3, 8] used to describe the spinodal decomposition of a binary alloy. Phase-field descriptions of the solidification of binary alloys have been developed recently by Wheeler *et al.* [9, 10], Caginalp and Xie [11], and Warren and Boettinger [12]; they can be viewed as combining elements of the Cahn-Allen and the Cahn-Hilliard models. In many cases gradient energy models can be viewed as mean-field approximations to models that provide atomic-level descriptions; examples of such mean field theories are given by the Landau theory of a second-order phase transition [13] or density functional theories (see, e.g., [14, 15]).

Phase-field models can be particularly useful for numerical computation in situations in which the interphase boundary is expected to be geometrically complicated. For example, during the dendritic growth of a pure material into its undercooled melt, the generation and propagation of sidebranches along the primary branch of the dendrite leads to a wide range of length and time scales, and phase-field models have been employed to provide numerical tests of theories of tip selection [16, 17, 18]. During dendritic growth an important role is played by the crystalline anisotropy of the growing solid phase, which selects the growth direction of the dendrite, the symmetry of the sidebranch structure, and the velocity of the dendrite tip.

For a scalar order parameter, formulations that use the square of a gradient are inherently isotropic. For a phase-field model to be of practical importance, it must include a description of the anisotropic nature of the crystal, through such effects as the dependence of the solid-liquid surface free energy (sometimes called the surface tension) on the interface orientation with respect to the crystal lattice, and also the variation of the rate of attachment kinetics with interface orientation. One way that anisotropic effects have been modeled is to allow the phenomenological phase-field parameters (such as the gradient energy coefficient and the mobility coefficient for temporal relaxation) to depend on the direction of the gradient of the phase field parameter [19]. It is possible to re-express the gradient energy term as the square of a function of the gradient which is homogeneous of degree one [20]; these two forms for the anisotropy are equivalent. This approach is an example where the anisotropy is introduced in an *ad hoc* manner, and its connection with the underlying crystalline anisotropy is indirect. Still, it successfully generalizes the original isotropic phase-field methodology to allow the computation of dendrites with the proper qualitative behavior, and leads to the proper anisotropic version of the Gibbs-Thompson equation at the crystal-melt interface in the sharp interface limit [21, 22]. With this type of formulation, anisotropies with general symmetries can be treated using a single order parameter; for example, two-dimensional dendrites with six-fold symmetry can be computed using finite differences on a rectangular mesh [12, 23].

Anisotropy occurs naturally in discrete lattice calculations of interfaces (see, e.g., [24, 25]). The symmetry of a crystal imposes symmetry constraints on the properties of both bulk and interfacial properties. For example, it is well known that for bulk diffusion processes the second rank conductivity tensor is constrained to be isotropic for a cubic material. In contrast, surface energies are generally not tensor properties, and even cubic crystals are expected to have anisotropic interfacial properties. The symmetry constraints for surfaces between two crystalline phases depends on both their symmetries and their relative positioning.

Anisotropy that is inherent in a lattice description can be lost when lattice models are approximated by continuum descriptions that employ gradient energy terms. For a single order parameter [26], gradient energy coefficients become matrices or second rank tensors. For cubic crystals this results in an isotropic tensor, and the gradient energy reduces to the square of the gradient, leading to an isotropic surface energy. For crystals of lower symmetry a quadratic gradient energy term can be made isotropic by a rescaling of the spatial coordinates, and so the computed anisotropy in the physical system is restricted to a simple (elliptical) form.

Ordering reactions in crystalline solids provide a number of simple types of interfaces with anisotropies that are constrained by symmetry arguments. We distinguish two general types: interphase boundaries (IPBs) between two different ordered phases or between an ordered and a disordered phase, and antiphase boundaries (APBs) between two domains of ordered variants of the same phase. The elucidation of the symmetry properties in terms of Shubnikov groups is particularly simple, if the inversion centers of both abutting phases fall on a common lattice [27]. In that case the dependence of the interface properties on the direction of the interface normal are constrained by the intersection group of the two abutting domains.

Since the ordering reaction generally produces a phase with symmetry properties that form a subgroup of those of the disordered phase, the symmetry properties for the IPB are those of the ordered phase. For the APB, the symmetry depends on the relative positioning of the two domains; symmetries that are no longer common to both domains are lost. For particular orderings in two cubic lattices, bcc and fcc, to give cubic ordered phases, general symmetry arguments [28, 29] show that the APB between two cubic ordering domains has cubic symmetry for ordering to the $B2$ and $DO3$ structures in bcc and tetragonal symmetry for ordering to the $L1_2$ structure in fcc. (The prototypes for these structures are respectively, CsCl (or β brass), BiF_3 (or Fe_3Al) and Cu_3Au.) Thus the different orderings studied in these two cubic lattices illustrate another distinct source of the interfacial anisotropy in lattice models.

Kikuchi and Cahn have performed discrete calculations for both bcc [28] and fcc lattices [29] for a limited set of orientations as a function of temperature and composition. In the case of bcc, the lattice calculations were compared with isotropic continuum calculations.

In this paper we report interfacial anisotropy between parallel cubic phases or domains using a continuum formulation with multiple order parameters that preserves the underlying anisotropies of the crystalline lattice. We will focus on binary alloy fcc crystals in the context of the order-disorder transition that give the tetragonal symmetry. We derive the continuum gradient energy model and use it to obtain for all orientations the orientation dependence of the structure and free energy of static planar antiphase and interphase boundaries. We compare the results obtained with those obtained for a limited set of orientations by Cahn and Kikuchi [30] using cluster variation methods and by Landman *et al.* [31, 32, 33] using molecular dynamics simulations of solidification. We note the facetting and wetting transitions in these interfaces. Interfacial energies can be used to compute equilibrium microstructural shapes, such as single particles in a matrix (the Wulff shapes) and domain structures, which for Cu_3Au have long been known to show a high degree of anisotropy [34]. We will ignore elastic effects due to mismatching lattice parameters between the phases. These are important only for the coherent coexistence of large-enough particles of a phase with a different lattice parameter in a matrix or in domain structures of noncubic phases. Depending on the mismatch parameter, the size where elastic effects become important can be as small as nanometers.

2 Formulation

The ordering of fcc to the $L1_2$ structure (with space group $Pm3m$; prototype Cu_3Au) gives four interpenetrating primitive cubic lattices, as seen in Figure 1. This ordering transition cannot be a higher-order transition [35]. At the transition temperature there is coexistence between the ordered and disordered cubic phase and hence there is an IPB between them. The symmetry of the orientational properties of the IPB is also $m3m$; there is no reason for it to be isotropic. APBs are interfaces between any two of the four ordered domains that are shifted by a vector of type $(1/2) < 110 >$ relative to one another. Such shifts between two cubic domains break the symmetries among the three cube axes; although the axes of the two domains are parallel, all are shifted, and two of them cease to be four-fold axes of the two-domain system. Only in one cube direction a four-fold axis through the cube corners in one domain is retained, and continues as a four-fold axis through the face center in the other domain. The four-fold axes of the other two cube directions continue as two-fold axes in the other domain, and are lost as four-fold axes. The shift also means that no three-fold axis threads both domains. This leads to interfaces with a lowered tetragonal symmetry ($4/mmm$) in their orientational properties. The anisotropy due to this tetragonality was found to be severe enough to make some orientations unstable [29]; the interfacial energy is a nonconvex function of orientation. Furthermore there are structural changes in the interface with increasing temperature, that are described as an infinite set of interfacial phase transitions [36]. A proper continuum model should preserve the tetragonal anisotropy of the APB energy due to the $(1/2) < 110 >$ shift. The continuum model developed in this paper will give the tetragonal symmetry for the APB and at the same time will yield cubic anisotropy in the IPB.

The model we consider for order-disorder transitions on an fcc crystal lattice is de-

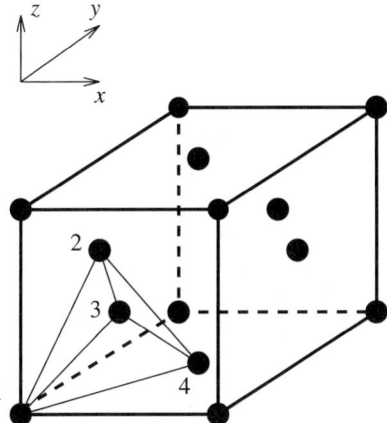

Figure 1: A schematic diagram of an fcc lattice. There are four distinguished sites corresponding to a corner and one each for the faces intersecting at that corner.

scribed geometrically by four inter-penetrating simple cubic sublattices defined by the lattice points labeled 1, 2, 3, and 4, respectively (see Fig. 1). For the disordered fcc structure all sublattices are crystallographically equivalent. In Figure 1, the site 1 is chosen to be the corner of the fcc cube, sites 2, 3 and 4 correspond to the face centers of the planes $x = 0$, $y = 0$, and $z = 0$, respectively. The overall state of the crystal is then assumed to be completely described by the four parameters ρ_1, ρ_2, ρ_3, and ρ_4 representing the atomic fraction of A on each sublattice.

It is convenient to introduce four new parameters W, X, Y, and Z in place of the parameters ρ_1, ρ_2, ρ_3, and ρ_4; they are defined by

$$W = \frac{1}{4}(\rho_1 + \rho_2 + \rho_3 + \rho_4), \tag{1a}$$

$$X = \frac{1}{4}(\rho_1 + \rho_2 - \rho_3 - \rho_4), \tag{1b}$$

$$Y = \frac{1}{4}(\rho_1 - \rho_2 + \rho_3 - \rho_4), \tag{1c}$$

$$Z = \frac{1}{4}(\rho_1 - \rho_2 - \rho_3 + \rho_4). \tag{1d}$$

The parameter W represents the atomic fraction of the system as a whole, and X, Y, and Z are non-conserved order parameters that can vary between plus and minus one half. In this model a disordered state is represented by $\rho_1 = \rho_2 = \rho_3 = \rho_4 = W$, which implies that $X = Y = Z = 0$.

We next consider a thermodynamic description of the crystal for the case of an isothermal system, based on a generalized free energy functional assumed to have the form

$$\mathcal{F} = \int_V \left\{ f(W, X_1, X_2, X_3) + a_{jk}\frac{\partial W}{\partial x_j}\frac{\partial W}{\partial x_k} + b_{jlkm}\frac{\partial X_j}{\partial x_l}\frac{\partial X_k}{\partial x_m} \right\} dV. \tag{2}$$

For notational convenience we have suppressed the dependence of the free energy on temperature, and the suffix notation is given by $X_1 = X$, $X_2 = Y$, $X_3 = Z$, and $(x_1, x_2, x_3) = (x, y, z)$; repeated indices are summed. The term $f(W, X_1, X_2, X_3)$ is the bulk Helmholtz free energy density, and the remaining terms are gradient energy contributions.

The symmetries associated with the fcc crystal structure restrict the possible forms of f, a_{jk}, and b_{jlkm}. In the following development, we will discard the a_{ij} term, since symmetry considerations require that these terms lead only to isotropic contributions. We will also take f to be a low order polynomial in X_1, X_2, X_3, and will assume that the coefficients b_{jlkm} are independent of W.

Imposing the required fcc symmetries implies that the energy f can be written in the form [1]

$$f(W, X, Y, Z) = a_0 + a_2(X^2 + Y^2 + Z^2) + a_3 XYZ + a_{41}(X^4 + Y^4 + Z^4)$$
$$+ a_{42}(X^2Y^2 + X^2Z^2 + Y^2Z^2) + a_5 XYZ(X^2 + Y^2 + Z^2) + a_{61}(X^6 + Y^6 + Z^6) \quad (3)$$
$$+ a_{62}\left\{X^4(Y^2 + Z^2) + Y^4(X^2 + Z^2) + Z^4(X^2 + Y^2)\right\} + a_{63}X^2Y^2Z^2$$

through terms of degree six, where the coefficients are generally functions of W (and temperature). We note that symmetry permits the cubic term XYZ in the free energy, which plays a fundamental role in the subsequent analysis [1]. The order parameters do not transform in the manner of the components of a vector under simple rotations; for example, a rotation of the crystal by $\pi/4$ about the z axis effectively interchanges the roles of ρ_2 and ρ_3 while leaving ρ_1 and ρ_4 unchanged. This implies that X and Y are interchanged with no change in sign, whereas such a rotation would be expected to produce a sign change in one of the components of a vector quantity.

A free energy of this type can provide correct qualitative descriptions of the disordered fcc phase (vanishing order parameters) and two of the commonly-observed ordered phases that occur in fcc systems [37, 38]. In a temperature-composition phase diagram, the $L1_0$ ordered phase has a congruent temperature located at or near the composition $W = 1/2$; this phase is described by a single non-zero order parameter. The $L1_2$ ordered phases have congruent temperatures located at compositions at or near $W = 1/4$ (for the stoichiometry A_3B) and $W = 3/4$ (for AB_3); for the $L1_2$ phase all three order parameters are non-zero and are equal in absolute value. A phase diagram with the correct qualitative features can be obtained with this free energy by prescribing an appropriate temperature and composition dependence to the coefficients a_0, a_2, and a_3; the higher-degree coefficients may be taken to be constants. The fcc-$L1_0$ phase transition is first order if a_{41} and a_{61} are negative and positive, respectively, and the fcc-$L1_2$ transition is first order if the coefficient a_3 is non-zero [39]. More generally, one would choose the coefficients in the free energy to obtain agreement with a given phase diagram or other measured properties of the alloy, such as heat capacity or equilibrium pair correlations [40, 41, 42], or first principle quantum mechanical calculations [43].

Here we focus on the fcc-$L1_2$ phase transition, and simplify the analysis by truncating the free energy at fourth degree and assuming that the fourth-degree terms are positive definite. Such a truncation would not provide a correct description of the fcc-$L1_0$ transition at $W = 1/2$; instead, one obtains the multicritical point of second-order transitions originally found by Nix and Shockley [44]. In future work we plan to investigate the sixth-degree theory; however, we may note that it is possible to choose the free energy coefficients in the sixth-degree theory in such a way that the fcc-$L1_0$ transition remains first order, while in the expression for the bulk energy of the $L1_2$ phase the fifth and sixth degree terms are absent. Preliminary studies indicate that the results obtained here for the fcc-$L1_2$ transition are in qualitative agreement with a more realistic sixth-degree model. In the present work we will describe phase boundaries by prescribing the coefficients a_2, a_3, a_{41} and a_{42} themselves; without loss of generality we will set $a_0 = 0$. Since our principal concern is with the surface tension anisotropy introduced by the multiple order parameters, we will also simplify the model by ultimately assuming that the composition W is uniform throughout the system. Except at a congruent point, the equilibrium compositions for a two-phase system would be expected to differ; assuming a uniform composition generally is equivalent to working along T_0 curves [45] in the phase diagram that describe states of equal free energies.

By invoking the symmetry of the fcc crystal we find that the gradient energy term can be written in the simple form [1]

$$b_{jlkm}\frac{\partial X_j}{\partial x_l}\frac{\partial X_k}{\partial x_m} = \frac{A}{2}\left(X_x^2 + Y_y^2 + Z_z^2\right) + \frac{B}{2}\left(X_y^2 + X_z^2 + Y_x^2 + Y_z^2 + Z_x^2 + Z_y^2\right), \quad (4)$$

where A and B are independent constants. A heuristic motivation of the form of the gradient energy term can be based on near-neighbor interactions on the fcc lattice [1, 47]. The discrete free energy is assumed to consist of contributions from pointwise energies and from nearest- and second nearest-neighbor interactions. The second-nearest-neighbor interactions involve points on the same primitive cubic sublattices. In the continuum limit the second nearest-neighbor interactions then contribute the gradient term

$$\mathcal{F}^{(2)} = \frac{\tilde{B}}{8}\int_V \left(|\nabla \rho_1|^2 + |\nabla \rho_2|^2 + |\nabla \rho_3|^2 + |\nabla \rho_4|^2\right) dV, \quad (5)$$

where $\tilde{B}$ is proportional to the second nearest-neighbor interaction energy. Inserting the definitions of the atomic fractions in terms of the order parameters gives

$$\mathcal{F}^{(2)} = \frac{\tilde{B}}{2}\int_V \left\{|\nabla X|^2 + |\nabla Y|^2 + |\nabla Z|^2\right\} dV. \quad (6)$$

The nearest-neighbor interactions are less straightforward, since they involve coupling between sublattices. In the continuum limit, collecting all the terms gives the contribution

$$\mathcal{F}^{(1)} = \frac{\tilde{A}}{2}\int_V \left\{X_x^2 + Y_y^2 + Z_z^2\right\} dV. \quad (7)$$

We note that $\tilde{A} > 0$ and $\tilde{B} > 0$ follow from the specific choice of interaction energies; these choices correspond to repulsive nearest-neighbor and attractive second nearest-neighbor interactions. In terms of the interaction energies $\tilde{A}$ and $\tilde{B}$, the gradient energy coefficients are given by $A = \tilde{A} + \tilde{B}$ and $B = \tilde{B}$.

The system free energy thus has the form

$$\mathcal{F} = \int \left\{f(X,Y,Z) + \frac{A}{2}\left(X_x^2 + Y_y^2 + Z_z^2\right) + \frac{B}{2}\left(X_y^2 + X_z^2 + Y_x^2 + Y_z^2 + Z_x^2 + Z_y^2\right)\right\} dV.$$

This form of gradient energy contribution has also been studied by Lai [46], who was concerned with scaling laws for domain growth via Langevin dynamics; our concern in this paper is the structure and anistropy of steady-state interfaces.

Evaluating the functional derivatives to minimize $\mathcal{F}$ gives steady-state equations we shall study:

$$0 = AX_{xx} + BX_{yy} + BX_{zz} - f_X, \quad (8a)$$

$$0 = BY_{xx} + AY_{yy} + BY_{zz} - f_Y, \quad (8b)$$

$$0 = BZ_{xx} + BZ_{yy} + AZ_{zz} - f_Z. \quad (8c)$$

3 Single phase bulk states

Equilibrium states [1] in which the order parameters X, Y, and Z are constant require that the free-energy of the system is stationary. The trivial solution $X = Y = Z = 0$ represents the disordered or fcc phase, and is stable providing $a_2 > 0$. The variable a_2 plays the role of temperature, with $a_2 = 0$ corresponding to the limit of metastability of the fcc phase.

There are solutions for which $X \neq 0$ and $Y = Z = 0$, which exist only when a_2 and a_{41} have opposite sign. In this situation the order parameters satisfy $\rho_1 = \rho_2$ and $\rho_3 = \rho_4$, representing a layered structure consisting of planes with different copper concentrations alternating in the x-direction (c.f. Fig. 1). We refer to this type of solution as the CuAu or $L1_0$ phase. Because of symmetry there are analogous solutions with $Y \neq 0$ and $X = Z = 0$, and with $Z \neq 0$ and $X = Y = 0$, that also represent the $L1_0$ phase with the layering occurring in the y and z directions. Symmetry allows changing the sign of any two of the order parameters, which allows a total of six equivalent variants of the $L1_0$ phase.

There are solutions of the form $X = Y = Z(\neq 0)$. For these solutions, $\rho_2 = \rho_3 = \rho_4 \neq \rho_1$, so that site 1 is distinguished from the other sites. We refer to this case as the $L1_2$ phase. The symmetries that correspond to changing the signs of two of the variables imply that there are related solutions of the form $Y = Z = -X$, and so forth, resulting in four variants on each branch of $L1_2$ phase. There are also "mixed-mode" solutions that are never stable, but play a role in determining the stability of the bulk $L1_0$ and $L1_2$ phases [1].

4 Interphase boundaries

In this section we consider those one-dimensional solutions to the governing equations (8) which represent a stationary, planar, interfacial region separating an ordered $L1_2$ bulk phase from a disordered bulk phase at the same composition (usually a congruent point on a phase diagram, but in this case it is a point on the T_0 curve, since we fix W to be the same value in both phases). The order parameters vary only in a direction parallel to the unit normal to the interface, denoted by $\hat{\mathbf{n}} = (n_x, n_y, n_z)$, and so

$$X(\mathbf{x}) = \hat{X}(\hat{\mathbf{n}} \cdot \mathbf{x}), \quad Y(\mathbf{x}) = \hat{Y}(\hat{\mathbf{n}} \cdot \mathbf{x}), \quad Z(\mathbf{x}) = \hat{Z}(\hat{\mathbf{n}} \cdot \mathbf{x}). \tag{9}$$

The governing equations then reduce to the following system of nonlinear ordinary differential equations

$$\xi_x^2 \hat{X}_{\zeta\zeta} = f_X(\hat{X}, \hat{Y}, \hat{Z}), \tag{10a}$$

$$\xi_y^2 \hat{Y}_{\zeta\zeta} = f_Y(\hat{X}, \hat{Y}, \hat{Z}), \tag{10b}$$

$$\xi_z^2 \hat{Z}_{\zeta\zeta} = f_Z(\hat{X}, \hat{Y}, \hat{Z}), \tag{10c}$$

where $\zeta = \hat{\mathbf{n}} \cdot \mathbf{x}$, and

$$\xi_x^2 = An_x^2 + Bn_y^2 + Bn_z^2, \quad \xi_y^2 = Bn_x^2 + An_y^2 + Bn_z^2, \quad \xi_z^2 = Bn_x^2 + Bn_y^2 + An_z^2. \tag{11}$$

Assuming that the bulk disordered phase exists for $\zeta \to -\infty$ requires that

$$\hat{X}(\zeta), \hat{Y}(\zeta), \hat{Z}(\zeta) \to 0, \text{ as } \zeta \to -\infty. \tag{12a}$$

We further require $a_2 > 0$ so that the bulk disordered phase is stable. Assuming that the the ordered $L1_2$ phase exists for $\zeta \to \infty$ requires that

$$\hat{X}(\zeta), \hat{Y}(\zeta), \hat{Z}(\zeta) \to \chi_1, \text{ as } \zeta \to +\infty. \tag{12b}$$

where χ_1 represents the common value of the order parameter in the stable $L1_2$ bulk phase. The equations (10) have a first integral

$$\frac{\xi_x^2}{2}[\hat{X}_\zeta]^2 + \frac{\xi_y^2}{2}[\hat{Y}_\zeta]^2 + \frac{\xi_z^2}{2}[\hat{Z}_\zeta]^2 - f(\hat{X}, \hat{Y}, \hat{Z}) = 0 \tag{13}$$

where the constant of integration vanishes in view of the far-field boundary conditions $f(0,0,0) = 0$. We note that coexistence of the $L1_2$ and disordered phases as stable equilibrium determines a unique value of a_2, representing the transition temperature.

The interfacial energy γ is given by the excess free energy per unit area, which may be written as

$$\gamma = \int_{-\infty}^{\infty} \left\{ \frac{\xi_x^2}{2}[\hat{X}_\zeta]^2 + \frac{\xi_y^2}{2}[\hat{Y}_\zeta]^2 + \frac{\xi_z^2}{2}[\hat{Z}_\zeta]^2 + f(\hat{X}, \hat{Y}, \hat{Z}) \right\} d\zeta. \tag{14}$$

4.1 Numerical solutions

In the computations of interphase boundaries presented below we set $a_3 = -12$, $a_{41} = a_{42} = 1$, which determines the critical value $a_2 = a_2^c = 2$. For the case $A = B$, the solutions are independent of the interface normal $\hat{\mathbf{n}}$, which results in an isotropic surface energy [1]. Anisotropic surface energies are obtained for $A \neq B$; here we consider the case $B/A = 0.005$. To ensure the accuracy of our numerical solutions we have used two methods to compute solutions of the nonlinear system of ordinary differential equations (10), with the boundary conditions (12). In the first approach we employed a central finite difference approximation to the spatial derivatives and solved the resulting nonlinear algebraic equations by Newton's method, allied with a continuation method to provide suitable initial guesses. The second approach used the software package COLNEW [48], which solves nonlinear boundary value problem using adaptive meshing on a variable number of subintervals.

4.2 The orientation dependence

We next discuss the dependence of the interfaces on their orientation. Because of the underlying $m\bar{3}m$ symmetries of the $L1_2$ and the disordered fcc, only those orientations whose normals $\hat{\mathbf{n}}$ are subtended by the spherical triangle whose vertices are given by the intersection of the unit sphere and radial vectors in the directions of $\hat{\mathbf{n}}$, [100], [110] and [111] need to be considered. The $m\bar{3}m$ symmetry dictates that scalar properties, such as the surface energy, will have extrema at $<111>$ and $<100>$, an extremum or saddle at $<110>$, and no gradients normal to the sides of the spherical triangle.

As an example, we consider an interface oriented in the [100] direction ($n_x = 1$ and $n_y = n_z = 0$). There are one-dimensional solutions with $\hat{Y}(\zeta) = \hat{Z}(\zeta)$. As can be seen clearly from the corresponding plot of the sublattice atomic fractions shown in Fig. 2, there is a transition region within the interfacial layer where there are just two distinct pairs of sublattices, each sublattice of the pair having the same atomic fraction of Au. This region is therefore occupied by the $L1_0$ phase, although this is not a stable bulk phase.

This interesting layered structure of the interface was first found by Kikuchi and Cahn [29] in their cluster variation simulation of the copper-gold alloy; recently, more refined calculations have been carried out by Finel et al. [36, 49]. In those models, the composition

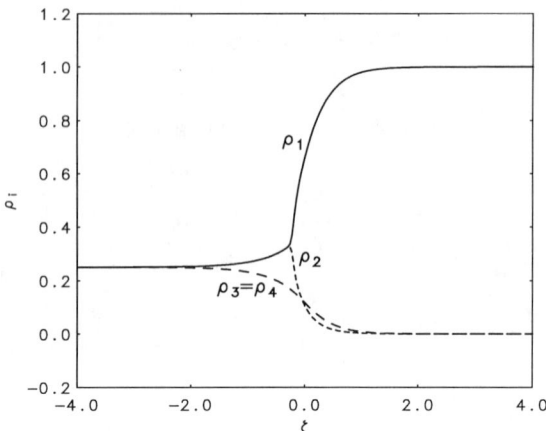

Figure 2: The sublattice atomic fractions ρ_i for the [100] orientation.

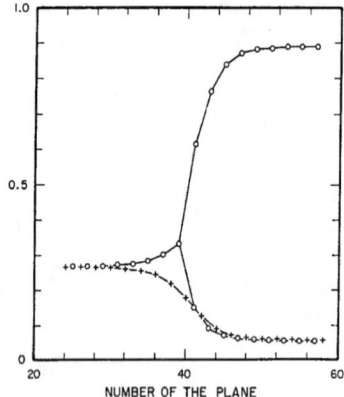

Figure 3: The interfacial structure predicted by Kikuchi and Cahn (1979) for the [100] IPB; the curves represent the various occupation densities of the lattice as in Figure 2.

was allowed to vary across the interface, unlike our simpler model. In Fig. 3 we reproduce the corresponding result of Kikuchi and Cahn [29] (their figure 11b), which shows the occupation densities of the four sublattices of the (spatially discrete) cluster variation method plotted against distance through the interfacial layer. Comparison of Fig. 2 and Fig. 3 shows the striking qualitative agreement between our work and theirs. While our result is not directly analogous with theirs, our more restricted model appears to retain the essence of the interfacial layering. It is interesting to note that layering in interfaces between phases has been observed in the molecular dynamics computations by Landman et al. [31, 32, 33], though in that context the layering is strictly onto a lattice arrangement for a pure material (see also Cahn and Kikuchi [30]).

4.3 Interfacial energy

From our numerical solutions we have evaluated the interfacial energy as a function of interface orientation [1]. Figure 4 shows contours of the interfacial energy on the unit

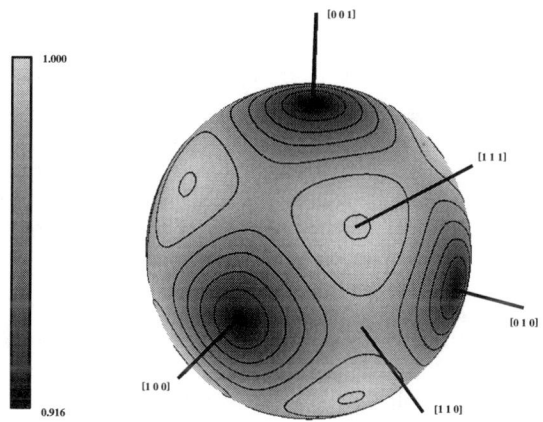

Figure 4: The variation of surface energy on the unit sphere; contours on the sphere represent orientations with common values of γ.

sphere for $B/A = 0.005$. The interfacial energy has a global maximum in the [111] direction and a global minimum in the [100] direction; in the [110] direction, there is a saddle point. This behavior of the interfacial energy shows the expected $m\bar{3}m$ symmetry. Our findings are consistent with Kikuchi and Cahn's finding of the ranking $\gamma_{100} < \gamma_{110}$ [29]. It is natural to ask whether the equilibrium shapes will develop missing orientations; a two-dimensional ordered particle with interfacial energy of the form $\gamma_0 + \gamma_4 \cos 4\theta$ would not develop corners with the level of anisotropy we have found at $B/A = 0.005$. In order to study the three dimensional case, we adopt the vector thermodynamics formalism of Hoffman and Cahn [50, 51] and use the "ξ-vector" to determine the equilibrium shapes given the anisotropy of interfacial energy. For $B/A = 0.005$, the equilibrium shape determined from the interfacial energy shown in Figure 4 does not develop missing orientations. While the anisotropy of the interfacial energy is relatively weak, the equilibrium shape has nevertheless developed a rounded cuboidal shape with no angular edges, even for the highest anisotropies studied.

5 Antiphase boundaries

An antiphase boundary separates two variants of the $L1_2$ phase that necessarily share the same free energy, and can be expected to exist for a range of values of a_2. This is in contrast to interphase boundaries that only exist when $a_2 = a_2^c$; that is, where the free energy of the $L1_2$ and disordered fcc bulk phases are equal.

5.1 The orientation dependence

Antiphase boundaries separate two ordered ($L1_2$) domains that are shifted relative to one another by a $(1/2) < 110 >$ vector. Because this shift breaks the cubic symmetry, it matters which vector is chosen from this set of vectors. Take $(1/2)[101]$ as an example; the y direction is then the distinguished direction and remains as the common 4-fold axis that threads both domains. For a $(1/2)[110]$ shift the z direction becomes the tetragonal axis. These cases are related by an appropriate rotation. For symmetry arguments and in the presentation of orientation dependencies we will always take the z axis as the unique 4-fold axis.

For a $(1/2)[110]$ shift the atomic fractions of the two ordered domains are given respectively by $\rho_2 = \rho_3 = \rho_4 \neq \rho_1$ for one and $\rho_1 = \rho_2 = \rho_3 \neq \rho_4$ for the other, and the domains are differentiated by the distinguished atomic fractions on sublattices 1 and 4. The shift changes the sign of two of the order parameters; $X = Y = Z$ and $-X = -Y = Z$, respectively. Without loss of generality, we will consider this type of shift in our calculations of the structure of antiphase boundaries for all the orientations.

The resultant symmetry of the orientational properties of the APB is $4/mmm$, a thirding subgroup of $m\bar{3}m$. For this group of order 16, the spherical triangle has corners at the [001] (a 4-fold axis), at the [100] (no longer a 4-fold axis) and at the [110] two-fold axes, and encompasses three of the $m\bar{3}m$ triangles. This symmetry dictates that scalar properties, such as the surface energy, will have an extremum at $< 001 >$, extrema or saddles at $< 100 >$ and $< 110 >$, and no gradients normal to the sides of this larger triangle.

Most of the $\{110\}$ mirrors that were important in the IPB calculations have been lost. The only special orientations for the computations are the [hhk] (but not the [khk]) that lie in the $(1\bar{1}0)$ mirror plane for which $X = Y$. In particular, for the [111] orientation, this is the only symmetry that applies. All orientations other than the [hhk], including [100], are general orientations with distinct values for X, Y, and Z.

The dimensionless governing equations are given by Eq. (10), but with boundary conditions

$$\hat{X}(\zeta), \hat{Y}(\zeta) \to -\chi_1, \hat{Z}(\zeta) \to +\chi_1 \text{ as } \zeta \to -\infty, \tag{15}$$

and

$$\hat{X}(\zeta), \hat{Y}(\zeta), \hat{Z}(\zeta) \to +\chi_1, \text{ as } \zeta \to +\infty, \tag{16}$$

where χ_1 is the common value of the order parameters in the bulk states.

As an example, we consider the [100] APB. In contrast to the [100] IPB considered above, the imposed boundary conditions Eq. (15) and Eq. (16) for the [100] APB do not allow the symmetry $\hat{Y}(\zeta) = \hat{Z}(\zeta)$, but instead allow solutions in which $\hat{X}(\zeta)$ and $\hat{Y}(\zeta)$ are

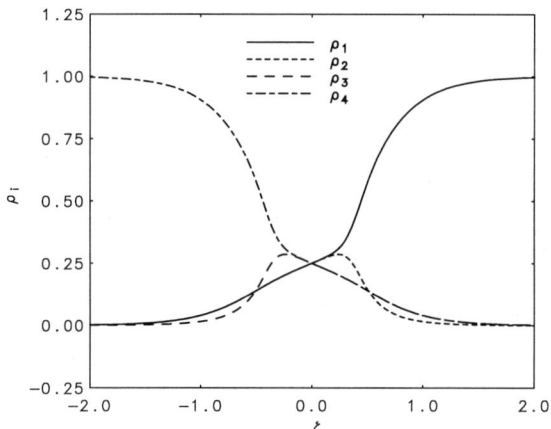

Figure 5: The occupation densities in the [100] APB.

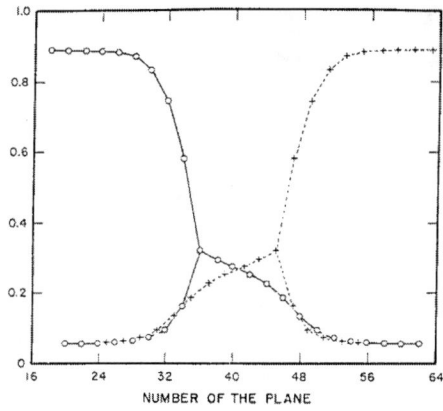

Figure 6: The interfacial structure predicted by Kikuchi and Cahn (1979) for the [100] APB.

odd and $\hat{Z}(\zeta)$ is even. The governing equations (10) may then be solved on a semi-infinite interval with appropriate symmetry conditions at the endpoint. Numerical solutions of the governing equations for the [100] APB for different values of B/A and a_2 show somewhat more complicated interfacial structures than that found for the [100] IPB [1]: depending on the values of B/A and a_2, the [100] APB can show the formation of an intervening layer composed of either $L1_0$ or disordered phase, or, in intermediate cases, combinations of the two.

As $B/A \to 0$ for fixed values of $a_2 \neq a_2^c$, an intervening $L1_0$ layer is observed that is analogous to the [100] IPB layering by $L1_0$. This is illustrated in Fig. 5 for $a_2 = 1.9$ and $B/A = 0.005$. They may be compared with the APB results of Kikuchi and Cahn [29] (their figure 11a), reproduced in Fig. 6; again the similarity is striking. Similar results have been obtained by the improved cluster variation calculations of Finel et al. [36].

5.2 Interfacial energy

The orientation dependence of the interfacial energy of the APB solutions is illustrated in Fig. 7. The interfacial energy is now a function of two parameters, the "temperature" a_2 and the ratio of the gradient energy coefficients B/A. Because of the broken symmetry due to the $1 \leftrightarrow 4$ shift, the APB interfacial energy has $4/mmm$ tetragonal symmetry as shown in Fig. 7. The [001] APB has low energy, and the [100] and [110] APBs have high energies [1].

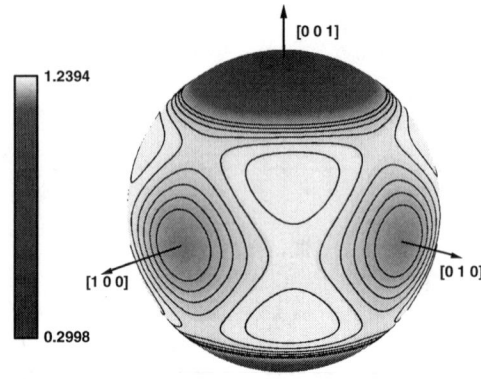

Figure 7: APB interfacial energy variation for $a_2 = 1.99$ and $B/A = 0.005$.

The location and number of extrema in the energy depend on the parameters. As a_2 approaches a_2^c, the maxima of the interfacial energy tend to the $<111>$ directions and the anisotropy diminishes. This behavior is consistent with the energy of the APB approaching the behavior of two separated IPB as a_2^c is approached [29]. As a_2 decreases from a_2^c, the maxima in the interfacial energy tends to the $<110>$ directions and the anisotropy becomes stronger. At some distinguished value of a_2, there is a bifurcation of the maxima at the $<110>$ directions into two maxima and saddle, while the saddles at $<100>$ become minima and two saddles each of which move toward [001] and [00$\bar{1}$] along the $<h0k>$.

In the CVM calculations of Kikuchi and Cahn [29], the interfacial energy of some of the orientations vanished as the critical point was approached because they only took into account nearest-neighbor interactions in their tetrahedral approximation. Because our continuum formulation models both first and second nearest-neighbor interactions, our model has nonzero interfacial energy for all orientations.

The anisotropic interfacial energies were used to compute the ξ-vectors in the manner of Cahn and Hoffman [50, 51]. We refer to the analogous shape computed from the APB surface energy as a ξ-surface, and refer to the convex inner portion of the possibly non-convex ξ-surface as the Wulff shape. One octant from the resulting ξ surfaces is shown in Fig. 8. For $a_2 = 1.99$ and $B/A = 0.005$ the anisotropy is strong enough to cause "ears" to form; this implies the the ξ-surfaces for these parameters have edges that are almost circular. The Wulff shapes correspond to the appropriately reflected ξ surface with the ears removed; in these cases the Wulff shape resembles a slightly bulging almost circular coin. Such Wulff shapes could be expected on the basis of the tetragonal symmetry caused by the different behaviors of the $1 \leftrightarrow 4$ interchange in the [001] direction vs. the [100] and [110] directions. Congruent rotated shapes are obtained for the other interchanges. The formation of the edges can be considered to develop parametrically with decreasing B/A for fixed a_2. For a given a_2, the edges may first appear at either $< 0kl >$ or $< hhl >$ directions for decreasing B/A. The edges then widen over a very small range of B/A until a complete ear rings the ξ surface. This picture of domain structures composed of large flat segments of low energy surfaces in the three cube orientations together with sharp corners to smoothly curved segments is readily seen in the micrographs of the classic study of Fisher and Marcinkowski [34]. These shapes confirm the conclusions of Kikuchi and Cahn [29]. The competition of these faces in the three-dimensional ordering of an $L1_2$ alloy can now be computed using our model; we are currently studying this situation.

Figure 8: One-eighth of the ξ surface for $a_2 = 1.99$ and $B/A = 0.005$.

6 Discussion

The model developed in this paper is an attempt to overcome the ad hoc approach employed by phase-field models to represent anisotropic interfaces. Here we have focussed on the case of an fcc lattice and have developed a model which is intimately related to the lattice and is formulated in terms of physically based order parameters. Our model employs an energy functional in which the gradient energy terms are simple square terms and results in a simple continuum description of an interface. This model provides a natural development of phase-field models. By conducting both asymptotic analysis and numerical solutions of the resulting system of nonlinear ordinary differential equations that represent stationary interfaces we have been able to analyse many interesting features of both IPBs and APBs, and have exhibited phenomena that have only previously been observed in much more complicated numerical simulations. We anticipate that the approach adopted here can be adapted to other crystalline structures.

One way to incorporate anisotropic interfacial properties in a single-order-parameter diffuse interface theory is to allow the gradient energy coefficient and the mobility coefficient to depend on the spatial gradient of the order parameter; in this way the surface energy σ and the kinetic coefficient μ can be assigned a given anisotropy. While this approach allows a great deal of flexibility, it is also somewhat *ad hoc*. Another approach to introducing anisotropy is through generalized gradient energy terms that include higher-order derivatives; this approach can also be difficult to justify on theoretical grounds. The use of lattice models, or continuum models based on an underlying lattice such as we have have considered here, have the advantage that the anisotropy is introduced in a natural way, and correctly incorporates the crystal symmetries that are present.

We obtain good comparison with previous discrete models [29] for the spatial variation of the occupation of the lattice for IPBs and APBs. Our model is able to easily calculate one-dimensional interfacial profiles for general orientation. This allows us to determine the parametric variation of the interfacial energy of both IPBs and APBs, and via the Cahn-Hoffman ξ-vector, to determine equilibrium shapes or ξ-surfaces. We find that our model successfully incorporates both the cubic anisotropy of the IPBs and the tetrahedral anisotropy of the APBs in fcc crystals.[1]

Although we have described the model in terms of the ordering of a binary alloy, the resulting model can also be interpreted in terms of solidification if we associate the liquid phase of a pure material with the disordered state of the alloy (a "lattice gas" liquid), and identify the solid phase of the material with a particular ordered state of the lattice that undergoes a first-order transition in passing from the ordered to the disordered state. This description then provides an alternate description of phase change allowing the anisotropy of the interface to arise in a natural way from the underlying crystal.

Our approach has focussed on the role played by the three order parameters X, Y, and Z that appear in the fcc model in determining the anisotropy of interphase boundaries and antiphase boundaries. We have not examined the role played by the overall composition variable W, which we have taken to be constant. This might be expected to give useful results for the description of antiphase boundaries that separate equivalent bulk domains having the same composition. For interphase boundaries, however, requirements of thermodynamic equilibrium generally require distinct values of W in each phase. Before thermodynamic equilibrium is attained, however, our model might be appropriate to the

---

[1] We have computed surface energies based on one-dimensional solutions that correspond to stationary, if not minimal, surface energies. We have not examined the possibility that the actual minimum energy solutions have lateral variations along the interface, and are described by two or three-dimensional solutions (see, for example, [52]).

early stages of order-disorder transitions in systems that are initially of uniform composition, when the time scales for compositional diffusion are long compared to the times scales required for ordering to take place. Extension of our model to include concentration variation and to compute interfacial properties at specific locations on the phase diagram is currently underway. In particular, the determination of the equilibrium phase diagram determines the coefficients of the bulk free energy, and matching the measured surface energy from experiment would then determine the gradient energy coefficients. We are also studying the motion of APBs in an effort to understand the facetting into (001) and (hk0) faces in three dimensions.

Acknowledgements

The authors are grateful for helpful discussions with W.C. Carter, S.R. Coriell, K. Elder, M.E. Gurtin, R.V. Kohn, D.J. Muraki, R.F. Sekerka, and particularly with W.J. Boettinger. We also thank H.E. Rushmeier for help with visualization of data. JWC and GBM were partially supported by the Applied and Computational Mathematics Program of ARPA. AAW and GBM were partially supported by a NATO collaborative research grant.

References

[1] R. J. Braun, J. W. Cahn, G. B. McFadden, and A. A. Wheeler, "Anisotropy of interfaces in an ordered alloy: A multiple-order parameter model," submitted to *Phil. Trans. Roy. Soc. (London)*, Ser. A (1995); also National Institute of Standards and Technology report NISTIR 5641 (April, 1995).

[2] H. E. Stanley, *Introduction to Phase Transitions and Critical Phenomena*, (Oxford: Oxford University Press, 1971).

[3] J. W. Cahn, "On spinodal decomposition," *Acta Metall.* 9 (1961) 795–801.

[4] Lord Rayleigh 1892 "On the theory of surface forces.–II. Compressible fluids," *Phil. Mag.* 33 (1892) 209–220.

[5] J. D. van der Waals, *Verhandel. Konink. Akad. Weten. Amsterdam, Sect.* 1 (1893). For a translation, see J. S. Rowlinson, *J. Stat. Phys.* 20 (1979) 197–244.

[6] J. W. Cahn and S. M. Allen, "A microscopic theory for domain wall motion and its experimental verification in Fe-Al alloy domain growth kinetics," *J. Phys. (Paris) Colloque C7*, (1977) C7-51–C7-54.

[7] S. M. Allen and J. W. Cahn, "A microscopic theory for antiphase boundary motion and its application to antiphase domain coarsening," *Acta metall. mater.* 27 (1979) 1085–1095.

[8] J. E. Hilliard, "Spinodal Decomposition," in *Phase Transformations*, ed. H. I. Aaronson (Metals Park, Ohio: American Society of Metals, 1970).

[9] A. A. Wheeler, W. J. Boettinger, and G. B. McFadden, "Phase-field model for isothermal phase transitions in binary alloys," *Phys. Rev. A* 45 (1992) 7424–7439.

[10] A. A. Wheeler, W. J. Boettinger, and G. B. McFadden, "Phase-field model of solute trapping during solidification," *Phys. Rev.* E 47 (1993) 1893-1909.

[11] G. Caginalp and W. Xie, "Phase-field and sharp-interface alloy models," *Phys. Rev.* E 48 (1993) 1897-1909.

[12] J. A. Warren and W. J. Boettinger, "Prediction of dendritic growth and microsegregation patterns in a binary alloy using the phase-field method," *Acta metall. mater.* 43 (1995) 689-703.

[13] L. D. Landau, On the theory of phase transitions (1937). Reprinted in *Men of Physics: L. D. Landau*, Vol. 2: Thermodynamics, Plasma Physics, and Quantum mechanics. ed. D. ter Haar (London: Pergamon, 1969) pp. 61-84.

[14] R. Evans, "The nature of the liquid-vapor interface and other topics in the statistical mechanics of non-uniform, classical fluid," *Adv. Phys.* 28 (1979) 143-200.

[15] D. W. Oxtoby, "Crystallization of liquids: a density functional approach," in *Liquids, Freezing and Glass Transition* ed. J. P. Hansen, D. Levesque, and J. Zinn-Justin, Les Houches, Session LI, 1989 (Amsterdam: Elsevier Science Publisher, 1989), pp. 147-191.

[16] D. A. Kessler, J. Koplik, and H. Levine, "Pattern selection in fingered growth phenomena," *Advances in Physics* 37 (1988) 255-339.

[17] M. E. Glicksman and S. P. Marsh, "The Dendrite," in the *Handbook of Crystal Growth*, ed. D. T. J. Hurle, (Amsterdam: North Holland, 1993) pp. 1077-1122.

[18] A. A. Wheeler, B. T. Murray, and R. J. Schaefer, "Computation of dendrites using a phase field model," *Physica* D 66 (1993) 243-262.

[19] R. Kobayashi, "Modeling and numerical simulations of dendritic crystal growth," *Physica* D 63 (1993) 410-423.

[20] J. E. Taylor and J. W. Cahn, "Linking anisotropic sharp and diffuse surface motion laws via gradient flows," *J. Stat. Phys.* 77 (1994) 183-197.

[21] G. B. McFadden, A. A. Wheeler, R. J. Braun, S. R. Coriell, and R. F. Sekerka, "Phase-field models for anisotropic interfaces," *Phys. Rev.* E 48 (1993) 2016-2024.

[22] A. A. Wheeler and G. B. McFadden, "A ξ-vector formulation of anisotropic phase-field models: 3-D asymptotics," *Euro. J. Appl. Math.* (1995) in press.

[23] B. T. Murray, W. J. Boettinger, G. B. McFadden, and A. A. Wheeler, "Computation of dendritic solidification using a phase-field model," in *Heat transfer in melting, solidification and crystal growth*, ed. I. S. Habib and S. Thynell, (New York: The American Society of Mechanical Engineers, 1994) pp. 67-76.

[24] C. Herring, "Some theorems on the free energies of crystal surfaces," *Phys. Rev.* 82 (1951) 87-93.

[25] F. C. Frank, "Geometrical thermodynamics of surfaces," Chapter 1, *Metal Surfaces* (American Society of Metals, 1962) pp. 1-15.

[26] J. W. Cahn and J. E. Hilliard, "Free energy of a nonuniform system. I. Interfacial free energy," *J. Chem. Phys.* 28 (1958) 258–267.

[27] G. Kalonji and J. W. Cahn, "Symmetry constraints on the orientation dependence of interfacial properties: The group of the Wulff plot," *J. Phys. Coll. C6* 43 (1982) C6-25–C6-30.

[28] R. Kikuchi and J. W. Cahn, "Theory of domain walls in ordered structures–II. Pair approximation for nonzero temperatures," *J. Phys. Chem. Solids* 23 (1962) 137–151.

[29] R. Kikuchi and J. W. Cahn, "Theory of interphase and antiphase boundaries in fcc alloys," *Acta Metall.* 27 (1979) 1337-1353.

[30] J. W. Cahn and R. Kikuchi, "Transition layer in a lattice-gas model of a solid-liquid interface," *Phys. Rev. B* 31 (1985) 4300–4304.

[31] U. Landman, C. L. Cleveland, and C. S. Brown, "On the dynamics of epitaxial phase transformations," in *Ordering in Two Dimensions*, ed. S. K. Sinha (Amsterdam: Elsevier-North Holland, 1980) pp. 335–338.

[32] U. Landman, C. L. Cleveland, C. S. Brown, and R. N. Barnett, R. N. 1981 "On the dynamics of epitaxial phase transitions," in *Nonlinear Phenomena at Phase Transitions and Instabilities* ed. T. Riste, (New York: Plenum, 1981) pp. 379–389.

[33] C. L. Cleveland, U. Landman, and R. N. Barnett, "Molecular dynamics of a laser-annealing experiment," *Phys. Rev. Lett.* 49 (1982) 790–793.

[34] R. M. Fisher and M. J. Marcinkowski, "Direct observation of antiphase boundaries in the $AuCu_3$ superlattice," *Phil. Mag.* 6 (1961) 1385–1405.

[35] L. D. Landau and L. Lifshitz, *Statistical Physics*, 3rd ed., Part 1 (New York: Pergamon, 1980).

[36] A. Finel, V. Mazauric, and F. Ducastelle, "Theoretical Study of Antiphase Boundaries in fcc Alloys," *Phys. Rev. Lett.* 65 (1990) 1016–1019.

[37] L. Ansara, B. Sundman, and P. Wilemin, "Thermodynamic modeling of ordered phases in the Ni-Al system," *Acta metall. mater.* 36 (1988) 977–982.

[38] N. Dupin, "Contribution à l'évaluation thermodynamique des alliages polyconstitués à base de nickel" (Ph.D. Thesis, Laboratoire de Thermodynamique et de Physico-Chimie Métallurgiques de Grenoble, Institut National Polytechnique de Grenoble, 1995).

[39] R. J. Braun, J. W. Cahn, G. B. McFadden, and A. A. Wheeler, Unpublished research (1995).

[40] P. C. Clapp and S. C. Moss, "Correlation Functions of Disordered Binary Alloys - I," *Phys. Rev.* 142 (1966) 418–427.

[41] P. C. Clapp and S. C. Moss, "Correlation Functions of Disordered Binary Alloys - II," *Phys. Rev.* 171 (1968) 754–763.

[42] S. C. Moss, and P. C. Clapp, "Correlation Functions of Disordered Binary Alloys - III," *Phys. Rev.* 171 (1968) 764–777.

[43] D. de Fontaine, "Cluster approach to order-disorder transformations in alloys," *Solid State Phys.* 47 (1994) 33-176.

[44] F. C. Nix and W. Shockley, "Order-disorder transformations in alloys," *Rev. Mod. Phys.* 10 (1938) 1–71.

[45] J. W. Cahn, "Thermodynamics of Solidification," in ASM Seminar Series on *Solidification* (Metals Park, Ohio, 1971) pp. 23–58.

[46] Z. W. Lai, "Theory of ordering dynamics for Cu_3Au," *Phys. Rev.* B 41 (1990) 9239–9256.

[47] J. W. Cahn and A. Novick-Cohen, "Evolution Equations for Phase Separation and Ordering in Binary Alloys," *J. Stat. Phys.* 76 (1994) 877–909.

[48] G. Bader and U. Ascher, "A New Basis Implementation for a Mixed Order Boundary Value ODE Solver," *SIAM J. Sci. Stat. Comput.* 8 (1987) 483–500.

[49] A. Finel, "Thermodynamical properties of antiphases in fcc ordered alloys," In *Ordering and Disordering in Alloys*, ed. A. Yavari (New York: Elsevier Applied Science, 1992).

[50] D. W. Hoffman and J. W. Cahn, "A vector thermodynamics for anisotropic surfaces I. Fundamentals and application to plane surface junctions," *Surface Science* 31 (1972) 368–388.

[51] J. W. Cahn and D. W. Hoffman, "A vector thermodynamics for anisotropic surfaces II. Curved and facetted surfaces," *Acta Metall. Mater.* 22 (1974) 1205–1214.

[52] Frank Morgan, "Lowersemicontinuity of energy of clusters," preprint, Department of Mathematics, Williams College (1995).

Diffusional Phase Transformations on the Atomic Scale : Experiment and Modelling

J.M. Hyde,[1] A. Cerezo,[2] M.K. Miller,[3] R.P. Setna,[2] G.D.W. Smith[2]

*[1]AEA Technology,
Harwell,
Didcot, Oxon OX11 0RA, U.K.*

*[2]Department of Materials,
University of Oxford,
Parks Road, Oxford OX1 3PH, U.K.*

*[3]Metals and Ceramics Division,
Oak Ridge National Laboratory,
Oak Ridge, TN 37831-6376, U.S.A.*

Abstract

In the early stages of phase transformations, microstructures are generated with the dimensions of only a few atomic spacings. Investigation of structures on such a fine scale poses severe difficulties, not only for experimental studies, but also for conventional theories based on continuum models. In this paper we report on a combination of atomic-scale microanalysis using the position-sensitive atom probe and atomistic simulations using the dynamic Ising model. Three phase transformations have been studied : spinodal decomposition in Fe-Cr alloys, nucleation and growth in dilute Cu-Co alloys and finally a conditional spinodal reaction in Ti-Al. The model provided a good quantitative match to the kinetics of spinodal decomposition observed in Fe-45%Cr, and nucleation and growth in Cu-1%Co. It also predicts the development of coupled ordering and phase separation in Ti-15%Al.

1. Introduction

Many technologically important alloys used in service are thermodynamically metastable or unstable against some form of phase separation at the atomic scale. These alloys include duplex stainless steels used in power generating systems [1], copper-cobalt and iron-chromium multilayers whose giant magnetoresistance is being used for high density magnetic read heads, and low density titanium-aluminium based alloys used in the aerospace industry [2-4]. Conventional theories for phase separation have been developed using continuum approximations in which the local alloy concentration can be explicitly defined. However, in each of the alloy systems mentioned above, concentration modulations on a sub-nanometre scale generated during the early stages of phase separation have a dramatic effect on the physical properties of the alloy. Recent advances in microscopy now enable the examination of materials with near atomic resolution. Results show that structures on this fine scale cannot be adequately described by continuum approximations, and alternative atomistic microstructural models must be developed and compared with three-dimensional experimental information.

In this paper an atomistic model based on the dynamic Ising model is described. With suitable choice of interaction parameters, the model can predict spinodal decomposition, nucleation and growth and also a conditional spinodal reaction. To facilitate a detailed comparison with experimental results, binary model alloys based on technologically relevant materials have been chosen for study. The model predictions are compared with experimental results obtained from the position-sensitive atom probe [5, 6]. In this paper all concentrations are defined in atomic %.

2. The Position-Sensitive Atom Probe

The position sensitive atom probe is an instrument used for ultra-high resolution 3D microscopy. A specimen in the form of a sharp needle is held in a vacuum chamber and subjected to a high positive potential. If the potential is increased sufficiently, the field at the tip apex will be high enough to field evaporate individual atoms from the specimen surface. These ions are projected approximately radially towards a detector giving a desorption image with a magnification of the order of a million times. Time of flight mass spectrometry is used to determine the chemical identity of each ion. The position of impact on the detector is used to determine the approximate position of field evaporation. Thus an atomic map is built up of the distribution of elements on the specimen surface. Continued removal of material allows a full three-dimensional reconstruction of the atomic positions and thus local variations in chemical composition. The atomic scale structure of the ferrite phase of a duplex stainless steel in which spinodal decomposition has generated a two phase microstructure with a wavelength less than 10 nm is shown in Fig. 1.

3. The Dynamic Ising Model

The driving force for phase separation is the reduction of free energy. At a microscopic level this is achieved by the stochastic process of diffusion of atoms through the microstructure. The dynamic Ising model can be used to simulate this process. A Hamiltonian is defined in terms of bond energies either between pairs of atoms or over small clusters of atoms and summed over all possible interactions. The change in enthalpy (ΔH) on swapping a nearest neighbour pair of atoms is then calculated and the probability of acceptance determined. If only nearest neighbour interactions are considered, the change in energy, ΔH, will be a multiple of the interaction parameter, $\varepsilon = E_{AA} + E_{BB} - 2E_{AB}$, where E_{AA}, E_{BB} and E_{AB} are the energies associated with a A-A, B-B and A-B nearest neighbour interactions respectively. Provided that the ratio of the probability of accepting a swap with energy change ΔH to a swap with energy change $-\Delta H$ is $e^{-\Delta H / kT}$, then the free energy of the system will tend towards equilibrium. The simplest acceptance criteria is given by the Metropolis algorithm [7], in which the new configuration (x') is always accepted if the enthalpy decreases and accepted with probability $P = e^{-\Delta H / kT}$ if the enthalpy increases (equation 1).

$$P(x \to x') = \begin{cases} e^{-\Delta H / kT} & \Delta H > 0 \\ 1 & \Delta H \leq 0 \end{cases} \quad (1)$$

The absence of a pre-exponential factor to take into consideration the attempt frequency and activation energy shows that this algorithm does not simulate exact microscopic behaviour. However, provided that this factor is independent of local atomic configuration, and is the same for both species, it simply scales the time and so can be ignored. Moreover, inclusion of a pre-exponential factor would necessitate many millions of steps in which no swaps were accepted and unnecessarily increase the required computing power. Direct swapping of nearest neighbour atoms is unphysical in that phase separation is more accurately represented by vacancy diffusion. However, the two methods produce similar results provided that the vacancy migration attempt frequency is independent of local atomic configuration [8]. In this paper, only direct swapping of atoms has been considered because introducing a vacancy to the simulations also necessitates introducing unknown 'vacancy-atom' interactions.

The precise choice of interaction energies will determine the mode of phase transformation. If like bonds are more favourable than unlike bonds, the alloy will tend to phase separate at temperatures and concentrations within the miscibility gap (Fig. 2). As the alloy temperature is increased, the probability of accepting an 'unfavourable' swap increases and so the width of the miscibility gap decreases. At high temperatures the alloy tends towards a random solid solution. If unlike bonds are favoured, ordering will occur. In many alloys, interactions between atoms further apart are important. The choice of interaction energies for each of the three phase transformations reviewed in this paper will be discussed in the relevant sections. The phase diagram will depend on the choice of interaction parameters, but can easily be determined by simulation and matched to the alloy phase diagram.

Time in the model is defined in terms of Monte Carlo steps (MCS). If there are N atoms in the simulation then one MCS corresponds to N attempted swaps. Time in the model is directly proportional to real time assuming that the vacancy concentration in the real alloy is constant. The constant of proportionality is most easily determined by matching microstructural parameters such as the scale between experimental data and model data.

4. Spinodal Decomposition in Fe-Cr alloys

4.1 Introduction

In a spinodal alloy there is no thermodynamic barrier to phase separation [9]. Any random fluctuations in local composition reduce the free energy and will therefore be amplified. Spinodal decomposition in the ferrite phase of duplex stainless steels increases the alloy hardness and can lead to an increase in the ductile-to-brittle transition temperature. Since these alloys are used for heavy engineering applications such as the cooling pipes in pressure vessel reactors, it is important to be able to predict the development of phase separation and, from that, the resulting change in physical properties. During decomposition the atoms in the ferrite unmix to form a fine scale microstructure consisting of Fe-rich and Cr-enriched veins. The Fe-Cr is an ideal system to study because it is a model alloy for the ferrite phase of duplex stainless steels and because its phase diagram is similar in shape to the dynamic Ising model miscibility gap as shown in Fig. 2 [10-13]. The only competing phase transformation is the sluggish formation of sigma phase which can be ignored. Moreover the lattice mismatch is small and so it is not necessary to consider the effect of lattice strain in the computer model.

4.2 Experimental Alloys

An Fe-45%Cr alloy was made from high purity elemental Fe (>99.99% purity) and Cr (>99.996% purity) by arc-melting in a dry argon atmosphere. Wires of diameter 0.25 mm were fabricated from the master ingot by swaging and wire drawing operations. The wires were then solution treated in argon (0.4 atm) for 2 h at 1273K, water quenched and isothermally aged at 773 K for various times up to 500 h and water quenched. Each specimen was analysed in a position-sensitive atom probe (PoSAP).

The change in microhardness as a function of ageing time was determined from the ends of the wires with the use of a Shimadzu hardness tester with a 200 g load. An increase in hardness from 210 to 470 VHN was observed clearly demonstrating that a phase transformation was occurring [14]. Fig. 3 shows reconstructions of the Cr-enriched regions from PoSAP analyses. As thermal ageing proceeds the scale of the Cr-enriched and Fe-rich regions increases. After 500 h of ageing at 773 K the mean width of the domains had increased to approximately 5 nm.

4.3 Simulations

On a bcc lattice, the six second nearest neighbours are not much further away than the eight nearest neighbours and they are poorly screened. Therefore both first and second nearest neighbour interactions must be used, and weighted according to how the interatomic potentials vary as a function of distance (s). In the $3d$ transition metals, the binding potential falls approximately as s^5 [15]. In the simulations of spinodal decomposition in Fe-Cr second nearest neighbours interactions were therefore weighted by a factor of 0.5.

Because the Fe-Cr phase diagram is slightly asymmetric, simulations were performed on an A-50at.%B lattice. Temperature was introduced into the dynamic Ising model simulations by setting the critical temperature to that observed in the Fe-Cr system (~900 K). Simulations were performed on a bcc grid (80x80x80 lattice units) with both first and second nearest neighbour interactions for times up to 10,000 MCS at 750 K. For the time scales examined in this paper the development of phase separation was not influenced by the simulation size.

Isosurface reconstructions are shown in Fig. 4 from the dynamic Ising model. The volume shown is a cylindrical sub-volume of the simulation which has approximately the same dimensions as analysed during a typical PoSAP experiment. As the simulation proceeds, the scale of the microstructure gradually increases.

4.4 Discussion

The development of scale with ageing from both the experimental data and simulations was determined using the zero crossing point and first minimum of the three-dimensional autocorrelation function [16]. The first minimum of the autocorrelation function is a measure of the mean domain width or half of the mean distance between domains. The results, shown in Fig. 5, indicate that in both cases no incubation period exists before the onset of phase separation. Moreover the autocorrelation analyses suggest that a time scaling regime for the development of domain size exists. A best fit to the experimental results yielded a time exponent of 0.22±0.05. The dynamic Ising model results fitted a power law relationship with a time exponent of 0.25±0.02 which is within the error bounds of the experimental results and demonstrates the linear relationship between real time and MCS during the time regimes studied. In contrast, simulations using Cahn-Hilliard-Cook model [16] for spinodal decomposition predict that a dominant wavelength will form during the early stages of decomposition, followed by a coarsening regime in which power law behaviour is observed with a time exponent close to the value of 1/3 predicted by the LSW theory for the coarsening of isolated precipitates [17, 18].

The concentration amplitude of the phase separation (concentration difference between the Fe-rich regions and Cr-enriched regions) was analysed as a function of time for both the experimental results and the dynamic Ising model. In both, the concentration amplitude was observed to increase with ageing. In order to make a comprehensive comparison between experimental data and computer simulated results, the limitations of the experimental method must be taken into account. The concentration amplitude as a function of mean domain width (first minimum of the autocorrelation function) for the dynamic Ising model correction for the finite detector efficiency and spatial resolution is shown in Fig. 6 [14]. A fairly good correlation with the experimental data is achieved, demonstrating that the dynamic Ising model not only models the kinetics of domain growth but also the development of domain concentration in the Fe-Cr system.

5. Nucleation and Growth in dilute Cu-Co alloys

5.1 Introduction

A random solid solution outside the spinodal limits but within the miscibility gap is metastable and may phase separate by nucleation and growth [9]. Small composition fluctuations are unstable and will decay, however it is energetically favourable for a sufficiently large second phase nucleus to grow. Conventionally, the free energy of the nucleus is considered to consist of a favourable volume term, due to the bonding between like atoms and an unfavourable surface term due to bonding across the interface. When the precipitates are small, the surface term dominates and the precipitates are unstable. However above a critical size, the volume term will dominate and the precipitates will tend to grow. The nucleation barrier is the free energy required to form a nucleus of critical size at which stage the precipitate is equally likely to grow or shrink. In practice the transition between spinodal decomposition and nucleation and growth is blurred because near the spinodal limits both the critical size and nucleation barrier are small.

The copper-cobalt system has been widely studied as a model for nucleation and growth [19, 20]. It exhibits a very wide miscibility gap with no intermediate phases [21]. The thermodynamics are well established and because the solid solubility limits at low temperatures are so small, dilute copper cobalt alloys nucleate almost pure cobalt particles. The precipitates are characterised by coherency, approximately spherical shape, and a lattice mismatch of less than 2%. In continuum models for phase separation the free energy is treated as a Lyapunov function. Continuum models can therefore never predict nucleation, because the free energy cannot increase. In contrast the diffusional changes in the dynamic Ising model occur through a stochastic algorithm involving only the probability of atomic exchanges. Therefore, in the dynamic Ising model, nucleation barriers can be traversed and the probability of overcoming a nucleation barrier decreases with increasing barrier size.

5.2 Alloys and Heat Treatments

The Cu-1.0%Co alloys had been prepared from oxygen-free high purity copper (5N) and electrolytic grade cobalt (4N) and homogenised at 1273 K for 48 h. The alloys were then directly quenched in a salt bath heated to the ageing temperature of 723 K and held for times up to 24 h before final water quenching. The alloys were then analysed in a position-sensitive atom probe. A series of atom maps showing the distribution of cobalt atoms observed from atom probe analyses of each specimen is shown in Fig. 7. In the earliest stages (10 min), the distribution of cobalt atoms appears to be random. After 0.5-1 h, regions clearly enriched in cobalt can be seen. After 24 h, approximately spherical precipitates 1.5nm in diameter containing over 90%Co have formed.

5.3 Dynamic Ising Model

In the Cu-Co system, both the Cu and Co-rich phases adopt a face centred cubic (fcc) lattice. In this system each atom has 12 nearest neighbours which effectively screen the second nearest neighbours. Simulations have therefore been performed using only nearest neighbour interactions. The phase diagram for the Cu-Co system is not as simple as that of Fe-Cr, due to an intervening eutectoid reaction, so matching with the Ising model is not as straightforward as for Fe-Cr. Instead, a temperature calibration can be obtained by comparing the solid solubility of cobalt from the model with experimentally determined data for the alloy. Solubility data from the copper rich side of the Cu-Co phase diagram are shown in Fig. 8a. The data have been fitted to a sub-regular solution model and extrapolated down to 723 K to estimate the solid solubility at the thermal annealing temperature of the experimental alloys. The equivalent section of the dynamic Ising model phase diagram is shown in Fig. 8b together with the phase boundary predicted by the mean field model. At low solid solubilities, the mean field is a good approximation to the Ising model because the effect of clustering outside the miscibility gap is small. A direct comparison between Figs. 8a and 8b reveals that the appropriate interaction parameter divided by kT ($\frac{\varepsilon}{kT}$) is 1.178 for the dynamic Ising model simulations of Cu-Co at 723 K. Simulations have been performed on a fcc lattice containing 100x100x100 unit cells for a Cu-1.0%Co alloy aged at 723 K.

5.4 Discussion

The results of the simulation are shown in Fig. 9. In each part, the distribution of cobalt atoms from a subsection of the simulation corresponding to volume of 15x15x20 nm is displayed. A comparison with the experimental results in Fig. 7 shows that in both small concentration fluctuations form from an initially random solid solution and that after 30000 MCS, the cobalt clusters in the model are approximately the same size as observed in the experimental alloy after ageing for 24 h.

The matrix cobalt concentration and particle density from the experimental results and model are shown in Figs. 10a and 10b, respectively. Any cluster of cobalt atoms larger than 10 was taken to be a particle and used to calculate the cluster density. All remaining cobalt atoms were assumed to be in the matrix and used to calculate the matrix concentrations. During the nucleation regime, the matrix concentration remains approximately constant and the cluster density increases. After approximately 2000 MCS, coarsening begins to dominate and so both the cluster density and matrix concentration decrease. The experimental results show similar behaviour. The observed peak cluster density of 8×10^{24} agrees well with the value of 2×10^{24} m^{-3} given by the simulation. Two further simulations were performed at temperatures of 673 K and 773 K on a Cu-1.0%Co alloy. The distribution of cobalt clusters at peak cluster density is shown in Figs. 11a and 11b, respectively. Reducing the temperature increases the supersaturation and reduces the critical nucleation size. As a result, many critical nuclei form and grow. At higher temperatures, the reduced supersaturation reduces the probability of forming critical nuclei. Furthermore, once a critical nucleus has formed and begins to grow, the matrix becomes further depleted in solute. A comparison between Fig. 11 and experimental results at peak cluster density (Fig. 7a) indicates that the energy parameter determined for use in the dynamic Ising model was approximately correct.

The results show that this simple model can reproduce an incubation period, nucleation barrier and Ostwald ripening which we expect to see in homogenous nucleation and growth. Moreover, a good quantitative agreement for cluster densities, matrix concentration and mean cluster sizes was observed. The match would appear to support the view that short-range interactions are sufficient in modelling this alloy system.

6. Conditional Spinodal in Ti-15%Al

6.1 Introduction

The pressure to improve the efficiency of aerospace engines has led to a considerable research effort in the understanding of low density Ti-Al based alloys. For instance, the composition of near-α Ti alloys has been optimised for creep and fatigue strength by the addition of elements such as Sn, Zr, Mo, Ni, Si and C. One method of maximising the strength of these alloys is to allow limited formation of the ordered Ti_3Al α_2 phase to occur [4, 22]. Diffraction studies show superlattice reflections before clear microscopic evidence of phase separation which suggest that a conditional spinodal (i.e. concurrent ordering and decomposition) [23] mechanism may be involved. Since the phase transformation is driven by atomistic diffusion and ordering occurs rapidly, the conditional spinodal in near-α Ti-Al alloys is a further type of phase transformation suitable for study using the dynamic Ising model.

6.2 Implementation of the Dynamic Ising Model

n the simulations of spinodal decomposition in Fe-Cr and nucleation and growth in dilute Cu-Co alloys, only a single energy parameter was required to define the miscibility gap. In both, phase separation occurred because the bonds between like atoms were more favourable than those between unlike atoms. However, a single interaction parameter is not sufficient to predict a conditional spinodal on a hexagonal close packed (hcp) lattice, therefore, determining the range and strength of interactions is an important issue. First principles calculations using the full-potential linear muffin tin orbital method [24] have been used to calculate the heat of formation of Ti-Al hcp super-structures at zero temperature [25, 26]. From these, volume-dependent effective cluster interactions (ECIs) can be defined which parameterise the total energy of any Ti-Al structure [25, 26]. Only nine clusters (including the empty cluster) were need to accurately characterise the formation energies of the hcp structures predicted by ab initio calculations [26]. Third and fourth neighbour pair interactions are not important, but linear triplets and clusters containing first and second-nearest neighbour pair interactions are important. Since the range of volumes observed in Ti-Al alloys is not too great the volume dependence of the ECIs can simply be expanded to a quadratic form on the average atomic volume [27]. In theory, after each swap an energy minimisation should be performed to calculate a more accurate value of the average atomic volume. However, it is found in practice that performing the minimisation after each MCS is sufficient during the early stages and less frequent energy minimisations are required during the later stages of phase separation as the mean atomic volume of the alloy tends towards its equilibrium value.

6.3 Simulations

Simulations have been performed by quenching a random solid solution containing 15 at. %Al in to a supersaturated state at 700 K that can decompose to give a two phase microstructure consisting of a coherent ordered Ti_3Al α_2 phase and a disordered α phase [28]. The DO_{19} structure of the α_2 phase is shown in Fig. 12. Each unit cell contains 2 Al atoms and 6 Ti atoms. There are four possible positions for the pair of aluminium atoms within the unit cell as shown in

Fig. 11. The simulations were performed on a 200x200x100 regular hcp lattice with periodic boundary conditions.

6.4 Data Analysis

Several parameters need to be defined to enable the development of microstructure to be characterised. An ordered site is defined as either an aluminium atom surrounded by 12 titanium atoms or a titanium atom surrounded by 8 other titanium atoms and 4 aluminium atoms as shown in Fig. 12. A random solid solution contains ordered sites, however these do not constitute second phase ordered domains. To make the distinction, a working definition was used in which ordered domains (Ti_3Al) were defined as 10 or more adjacent ordered sites. Since any sites not belonging to an ordered region are considered to be matrix, it is possible to calculate the short range order in the matrix as

$$SRO = \frac{\text{Observed Ordered Sites in Matrix} - \text{Ordered Sites Expected in Random Solid Solution}}{\text{Maximum Ordered Sites in Matrix} - \text{Ordered Sites Expected in Random Solid Solution}}$$

In the analysis of spinodal decomposition in Fe-Cr and nucleation and growth in Cu-Co alloys, parameters such as domain concentration, scale and cluster density could be defined. The corresponding parameters for studying a conditional spinodal reaction are matrix concentration, and size and number density of ordered domains.

6.5 Results

Some ordered sites exist in the random solid solution immediately following the quench (Fig. 13a). Many of these act as nucleation sites for ordered domains. During the first MCS, the rapid increase in SRO is shown by the increased number of ordered sites (Fig. 13b) and the increase in the matrix SRO parameter from 0 to ~0.07 (Fig. 15). During the next four MCS, the matrix SRO parameter continues to increase as local rearrangements form many more ordered sites. At this stage there are only a few ordered domains and so the matrix concentration remains near the alloy composition (Fig. 15) and there is only a small increase in the mean cluster size (Fig. 14). The matrix SRO reaches a maximum after ~7 MCS. On further ageing up to 20 MCS, the mean size of the ordered domains begins to increase and there is a dramatic increase in the number density of ordered domains (Fig. 14). By this stage, the reduction in matrix solute concentration is apparent, reducing the supersaturation and therefore there is a rapid decrease in the rate of formation of new ordered regions. On further ageing, ordered domains grow in size and may begin to impinge upon each other. If the neighbours have the same orientation then the free energy will be reduced if they coalesce to form a larger ordered domain. If they have a different orientation then an APB will form. The result is a rapid increase in mean domain size and the formation of a ramified interconnected structure. Ostwald ripening behaviour is observed on further ageing as larger ordered domains grow at the expense of smaller ordered domains and the interface area between matrix and α_2 is reduced.

6.6 Conclusions

The simulations on Ti-15at.% Al quenched into the α-α_2 region predict the rapid development of SRO with no large scale composition modulations. Rapid growth in the ordered domains occurs when neighbouring domains begin to coalesce to form a complex interconnected microstructure. The simulations correspond only to the earliest stages of phase separation. After 400 MCS, the mean domain width is only 5 nm.

7. Summary and Conclusions

The position-sensitive atom probe technique is unique in giving three-dimensional real space compositional information on a sub-nanometre scale. The experimental results clearly demonstrate that important concentration variations can exist on a scale of only a few lattice spacings where the definition of a coarse-grained composition parameter becomes difficult. However, the small scale of the concentration variations makes these systems ideal for study using atomistic models.

The dynamic Ising model has been implemented to simulate three phase transformations: spinodal decomposition, nucleation and growth, and a conditional spinodal reaction. In each case the kinetics predicted by the dynamic Ising model have been analysed. Modelling spinodal decomposition in Fe-Cr was achieved by selecting the critical energy parameter of the model so that the miscibility gap from the model matched the Fe-Cr phase diagram. Modelling nucleation and growth in Cu-1.0%Co alloys was achieved by ensuring that the supersaturation in the model matched the supersaturation in the experimental alloy. More complex interaction potentials for the conditional spinodal reaction in Ti-15%Al were obtained from published first principles calculations.

Despite the simplicity of the algorithm, the dynamic Ising model predicts the correct kinetics of phase separation for spinodal decomposition in Fe-Cr and shows that there is a linear relationship between real time and MCS. The dynamic Ising model also predicts the various phenomena observed during homogeneous nucleation and growth and there was good quantitative agreement with the experimental results from Cu-1.0%Co alloys. The model predicts a conditional spinodal reaction in Ti-15%Al alloys at low temperatures. An analysis of the results shows the rapid development of SRO with no large-scale composition modulations followed by growth of ordered regions to form a complex interconnected microstructure. On further ageing, large ordered domains grow at the expense of smaller ordered domains and form equiaxed ordered regions.

With suitable choices of interaction potentials it should be possible to model other, more complicated, alloy systems and, with improvements in computational algorithms it will be possible to study later stages of phase separation.

Acknowledgements

The Cu-1.0%Co alloys were supplied by Dr M. Chisholm formerly of Carnegie Mellon University. The authors would like to thank Professor D. Pettifor for the provision of laboratory facilities. J.M.H would like to acknowledge the Engineering and Physical Sciences Research Council (EPSRC) for financial support and Wolfson College for the provision of a Fellowship. A.C. thanks The Royal Society for financial support and Wolfson College for the provision of a Fellowship. R.P.S. thanks EPSRC for a studentship. This research was funded by the EPSRC under grant number GR/H/38485 and by the Division of Materials Sciences, U.S. Department of Energy, under contract DE-AC05-84OR21400 with Lockheed Martin Energy Systems. The simulations were performed in the Materials Modelling Laboratory, Oxford University which is funded by the EPSRC under grant number GR/H/58278.

References

1. "International. Workshop on Intermediate Temperature Embrittlement Processes in Duplex Stainless Steels", ed. P.H. Pumphrey, G. D.W. Smith and M. Prager, Mater. Sci. Technol. 6 (1990), pp. 209-324.
2. M.J. Blackburn, Trans. AIME 239 (1967) 1200.
3. J.C. Williams, "Precipitation processes in solids" ed. K. C. Russell and H. I. Aaronson (Warrendale, Pennsylvania: Metallurgical Soc. of AIME) (1976), pp. 191-221.
4. D. Eylon, S. Fujishiro, P.J. Postans and F.H. Froes, J. Metals 36 (1984) 55.
5. M.K. Miller and G.D.W. Smith, "Atom-Probe Microanalysis: Principles and Applications in Materials Science" (Materials Research Society, Pittsburgh, Pa, 1989).
6. A. Cerezo, T.J. Godfrey and G.D.W. Smith, Rev. Sci. Instrum. 59 (1988) 862.
7. N. Metropolis, A.W. Rosenbluth, M.N. Rosenbluth and A.H. Teller, J. Chem. Phys. 21 (1953) 1087.
8. K. Yaldram and K. Binder, J. Stat. Phys. 62 (1991) 161.
9. K. Binder, "Alloy phase stability" ed. G.M. Stocks and A. Gonis (Kluwer Academic Publishers, 1989) pp. 233-262.
10. R.O. Williams and H.W. Paxton, J. Iron Steel Inst. 185 (1957) 358.
11. R.O. Williams, Trans. AIME 212 (1958) 497.
12. J. Andersson and B. Sundman, CALPHAD 11, (1987) 83.
13. H. Kuwano, Trans. of the Japan Institute of Metals 26, (1985) 473.
14. M.K. Miller, J.M. Hyde, M.G. Hetherington, A. Cerezo, G.D.W. Smith and C.M. Elliott, Acta Metall. Mater. 43, (1995) 3385.
15. V. Heine, Phys. Rev. 153, (1967) 673.
16. J.M. Hyde, M.K. Miller, M.G. Hetherington, A. Cerezo, G.D.W. Smith and C.M. Elliott, Acta Metall. Mater. 43 (1995) 3403.
17. C. Wagner, Z. Electrochem. 65 (1961) 581.
18. I.M. Lifshitz and V.V. Slyozov, J. Phys. Chem. Solids 19 (1961) 35.
19. H.I. Aaronson and F.K. LeGoues, Met. Trans. A 23A (1992) 1915.
20. P. Haasen and R. Wagner, Met. Trans. A 23A (1992) 1901.

21. T. Nishizawa and K. Ishida, Bulletin of Alloy Phase Diagrams 5, (1984) 161.
22. G. Lütjeromg and S. Weissmann, Acta. Metall. 18 (1970) 785.
23. W.A. Soffa and D.E. Laughlin, in "Solid to Solid Phase Transformations", ed. by H.I. Aaronson, D.E. Laughlin, R.F. Sekerka and C.M. Wayman (Warrendale, Pennsylvania : Metallurgical Society of AIME, 1982) pp. 159-183.
24. M. Methfessel, Phys. Rev. B 38 (1988) 1537.
25. M. Asta, D. de Fontaine, M. van Schilfgaarde, M. Sluiter and M. Methfessel, Phys. Rev. B 46 (1992) 5055.
26. M. Asta, D. de Fontaine and M. van Schilfgaarde, J. Mater. Res. 8 (1993) 2254.
27. M. Sluiter, D. de Fontaine, X. Q. Guo, R. Podloucky and A.J. Freeman, Phys. Rev. B 42 (1990) 10460.
28. J.L. Murray, "The Al-Ti (aluminium-titanium) system" in "Phase Diagrams of Binary Titanium Alloys" (ASM, 1987) pp. 12-24.

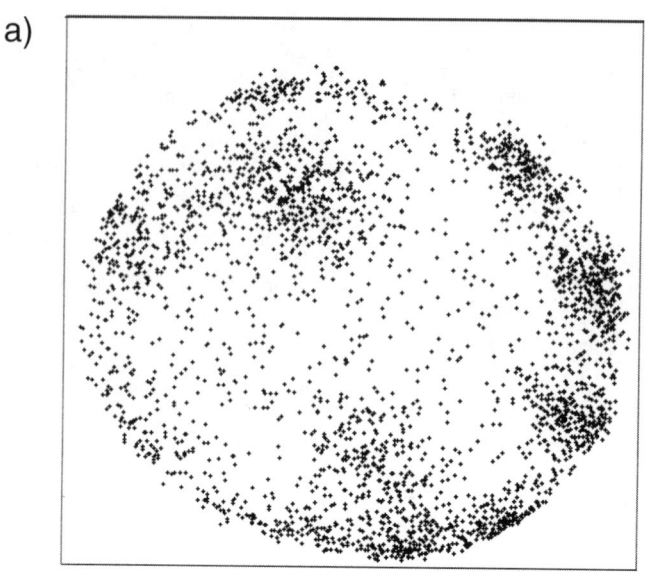

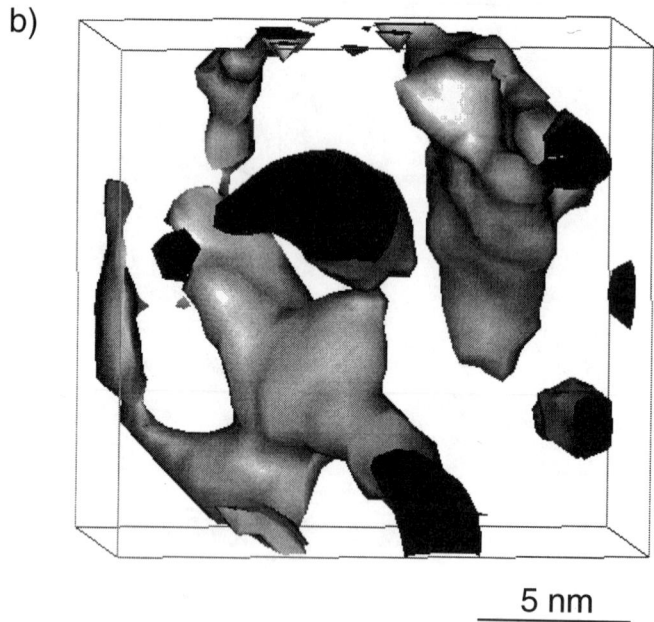

Fig. 1 Position-sensitive atom probe microanalysis of the ferrite phase in a CF3 duplex stainless steel aged for 10^4 hours at 400°C. In (a), each sphere represents the position of a Cr atom located within a slice of material 2nm in depth. In (b) the shaded regions represent those areas enriched in Cr. The volume of material analysed is 15 nm in diameter and 10 nm deep.

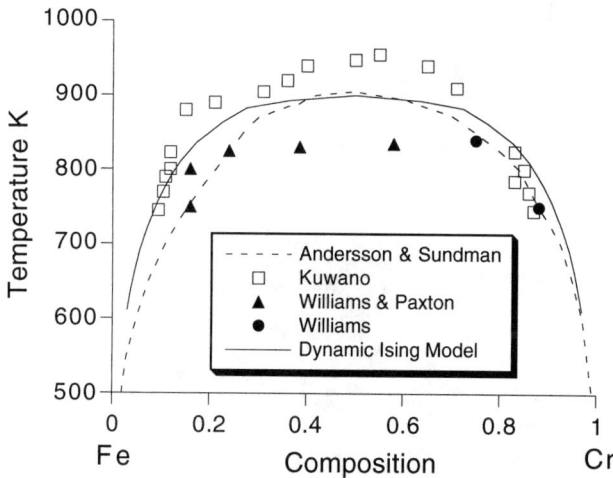

Fig. 2 The solid state miscibility gap in the Fe-Cr phase diagram. The dotted lines represent thermodynamic calculations and the symbols are from experimental investigations. The solid line is from the dynamic Ising model with a critical temperature of 900 K.

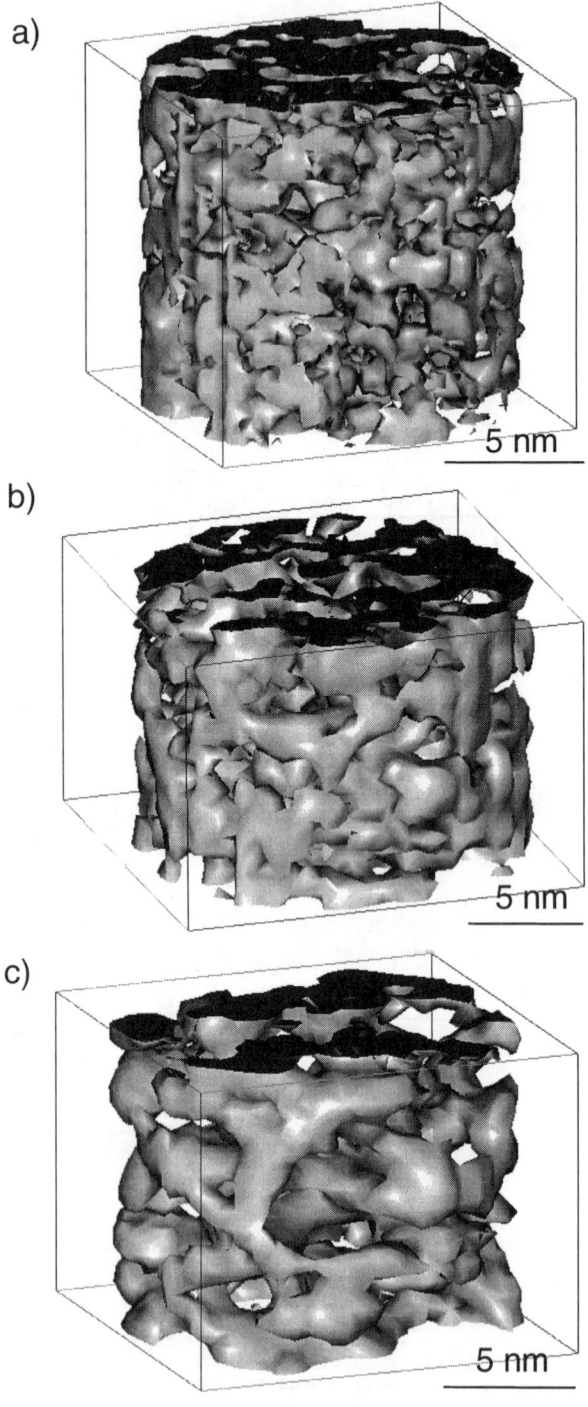

Fig. 3 Isosurface reconstructions from PoSAP analyses showing the morphology of the Cr-enriched regions from an Fe-45%Cr alloy after thermal ageing at 773 K for (a) 24 h, (b) 100 h and (c) 500 h.

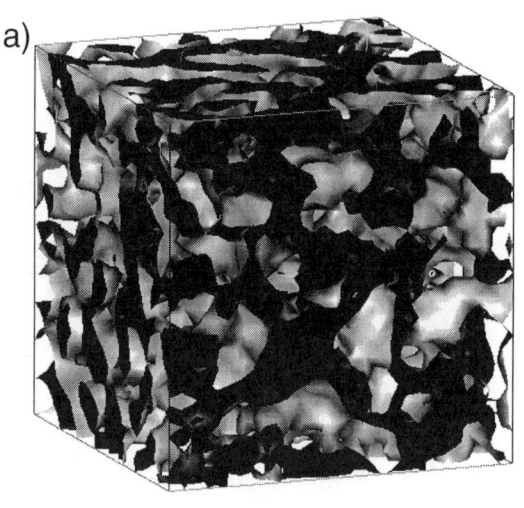

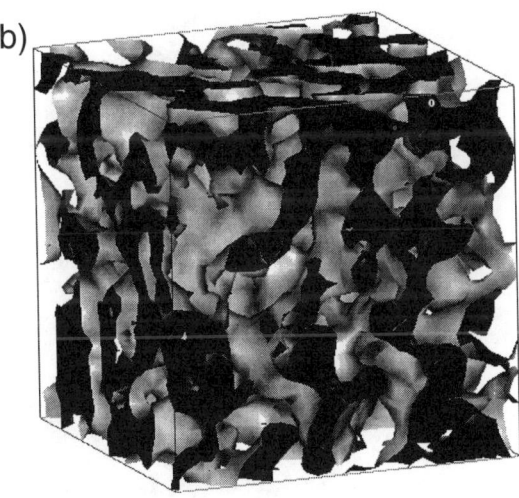

Fig. 4. Isosurface reconstructions showing the developing Cr-enriched regions from the dynamic Ising model of spinodal decomposition in an A-50%B alloy aged at 750 K for (a) 10 MCS, (b) 100 MCS, (c) 1000 MCS and (d) 10000 MCS.

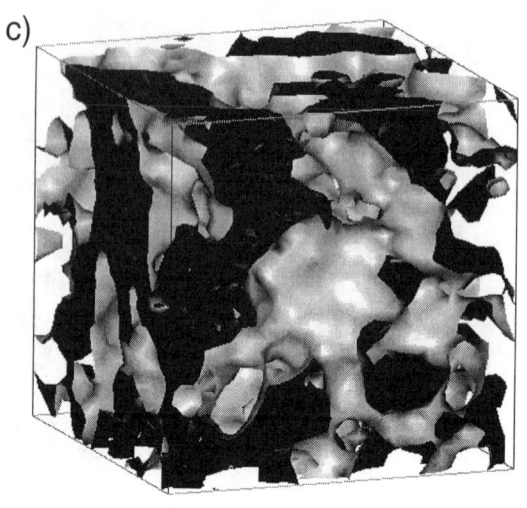

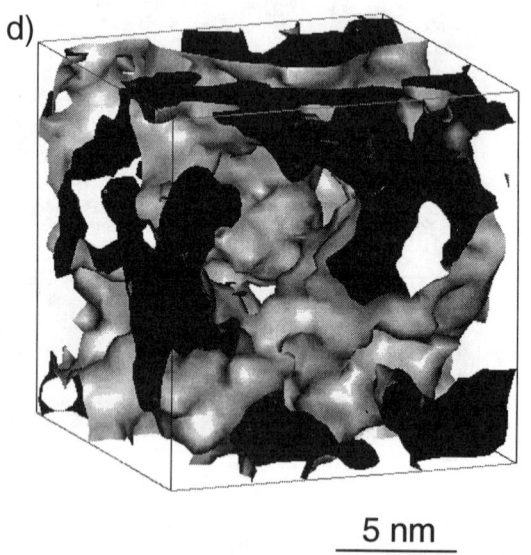

5 nm

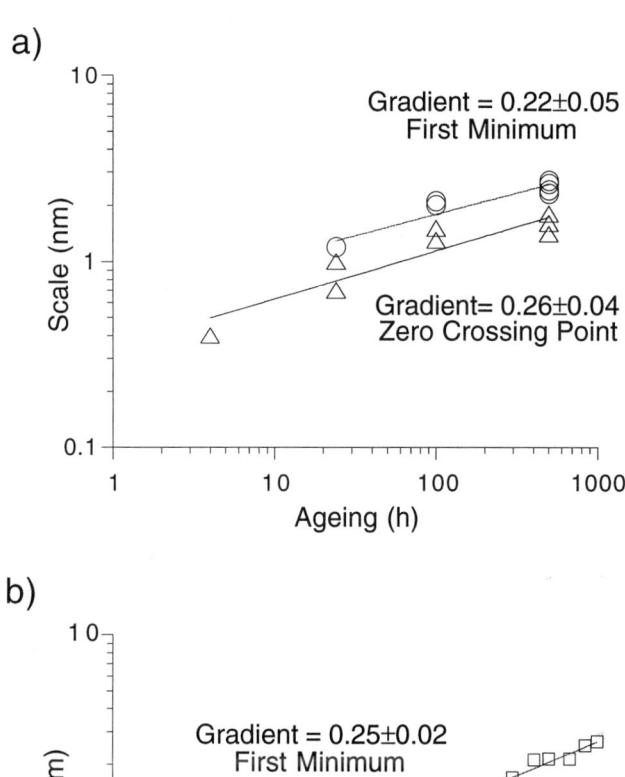
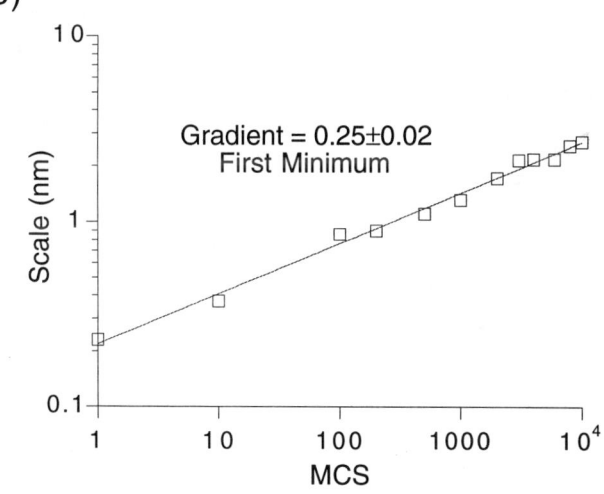

Fig. 5. Development of microstructural scale during spinodal decomposition determined from (a) PoSAP analysis of an Fe-45%Cr alloy and (b) dynamic Ising model of an A-50%B alloy.

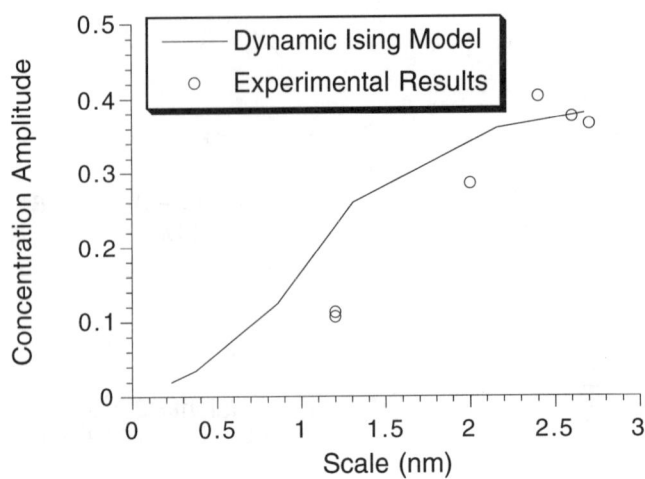

Fig. 6. Development of composition amplitude as a function of microstructural scale for both the PoSAP results and dynamic Ising model.

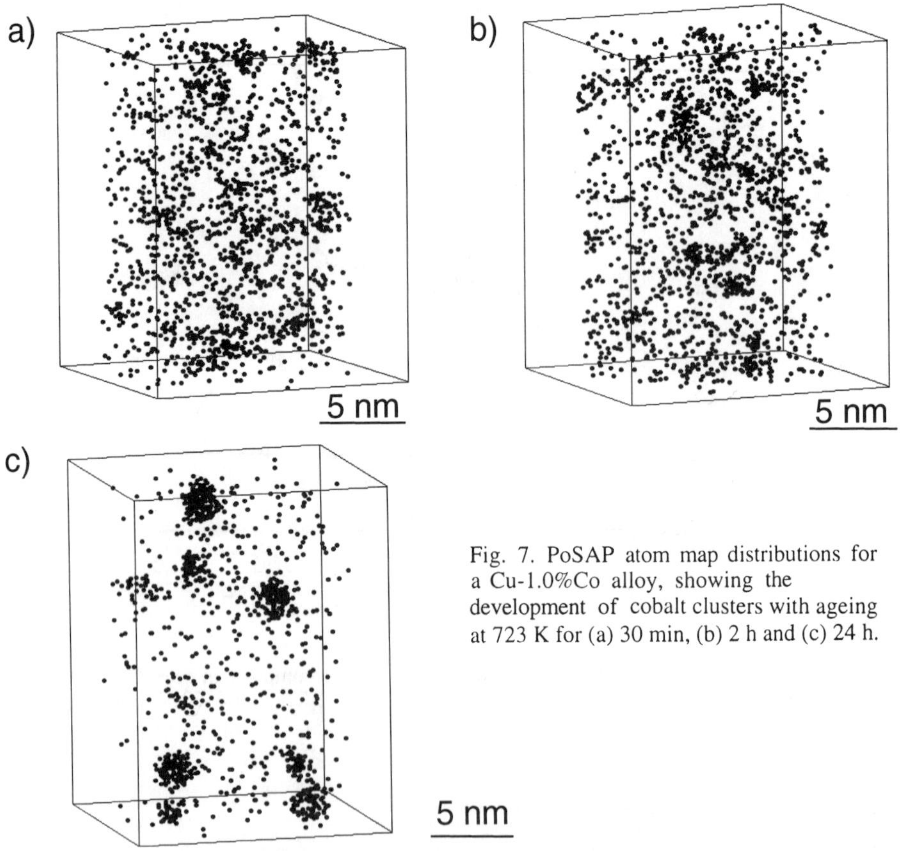

Fig. 7. PoSAP atom map distributions for a Cu-1.0%Co alloy, showing the development of cobalt clusters with ageing at 723 K for (a) 30 min, (b) 2 h and (c) 24 h.

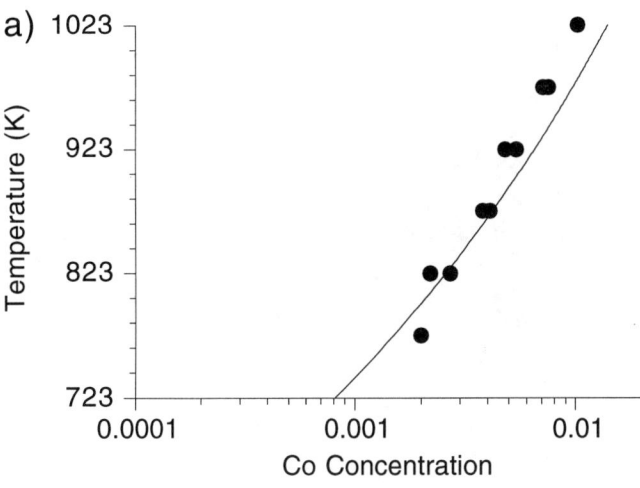

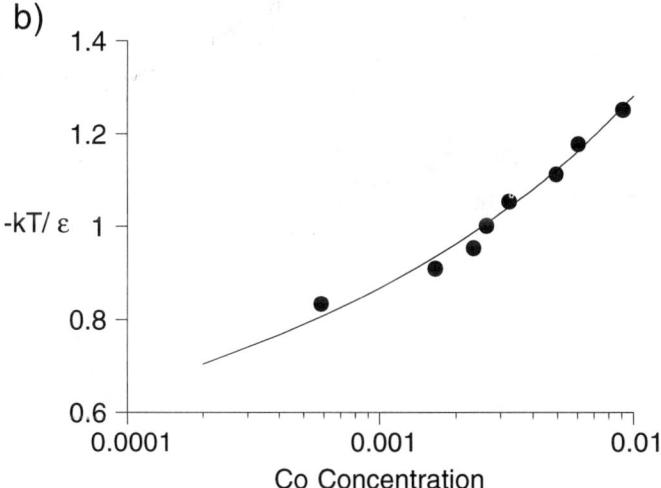

Fig. 8. Fits using the sub-regular solution model to (a) the experimentally determined solid solubility of Co in Cu at low temperatures and (b) the dynamic Ising model.

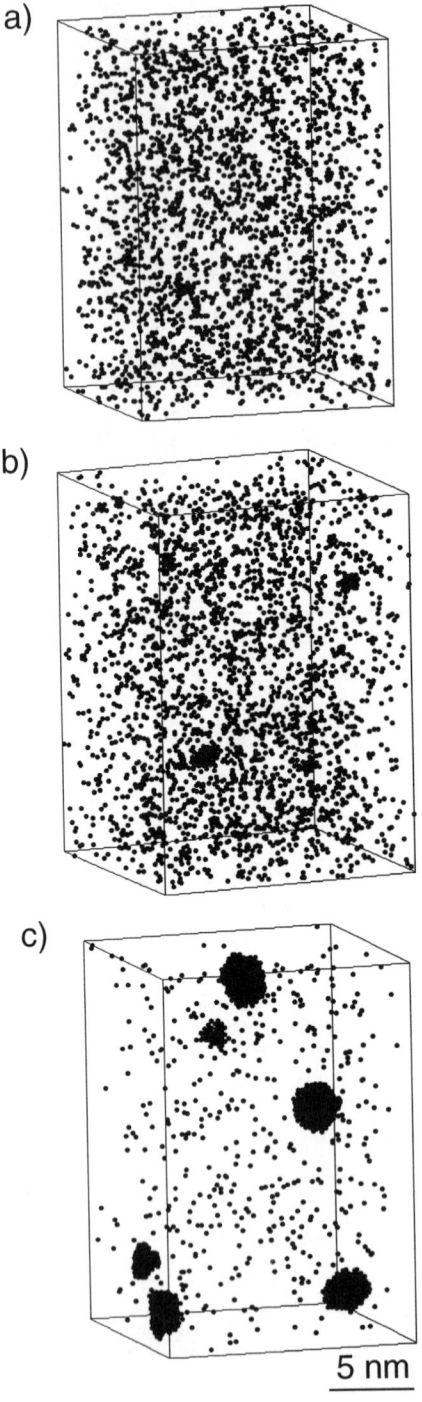

Fig. 9. Atom maps showing the positions of cobalt atoms from a 15x15x20nm section of a dynamic Ising model simulation of a Cu-1.0%Co alloy thermally aged at 723 K for (a) 600 MCS (b) 1200 MCS (c) 29000 MCS.

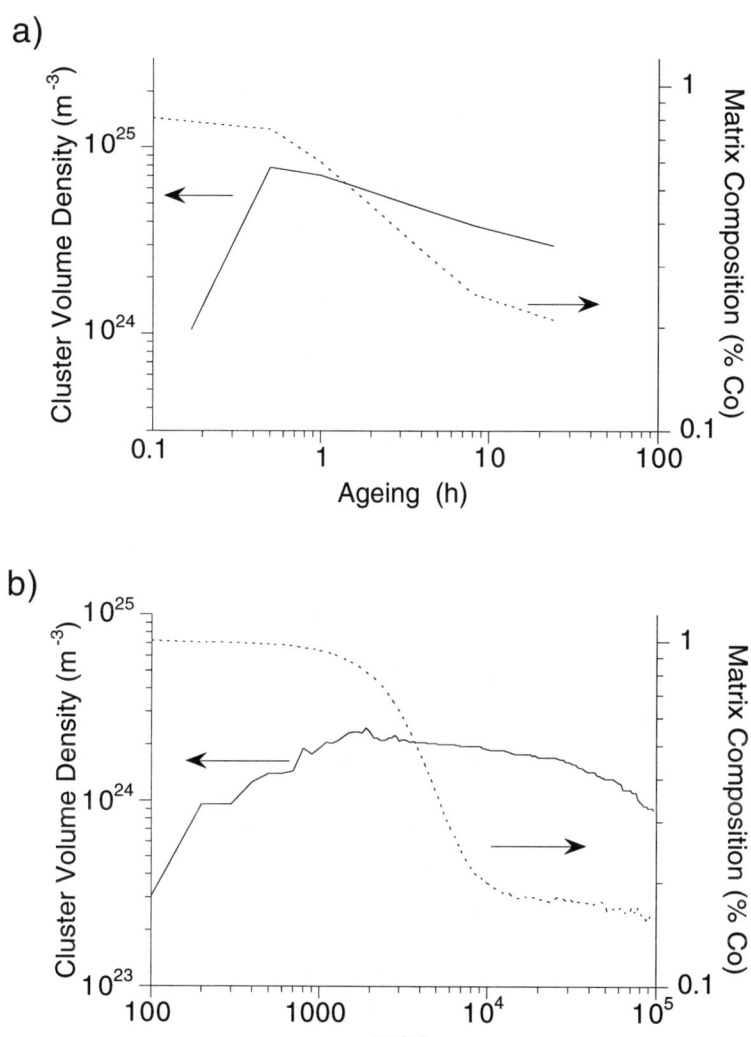

Fig. 10. Development of matrix composition and cluster density from (a) PoSAP analyses of thermally aged Cu-1%Co specimens and (b) dynamic Ising model simulation of nucleation and growth in Cu-1%Co alloys at 723 K.

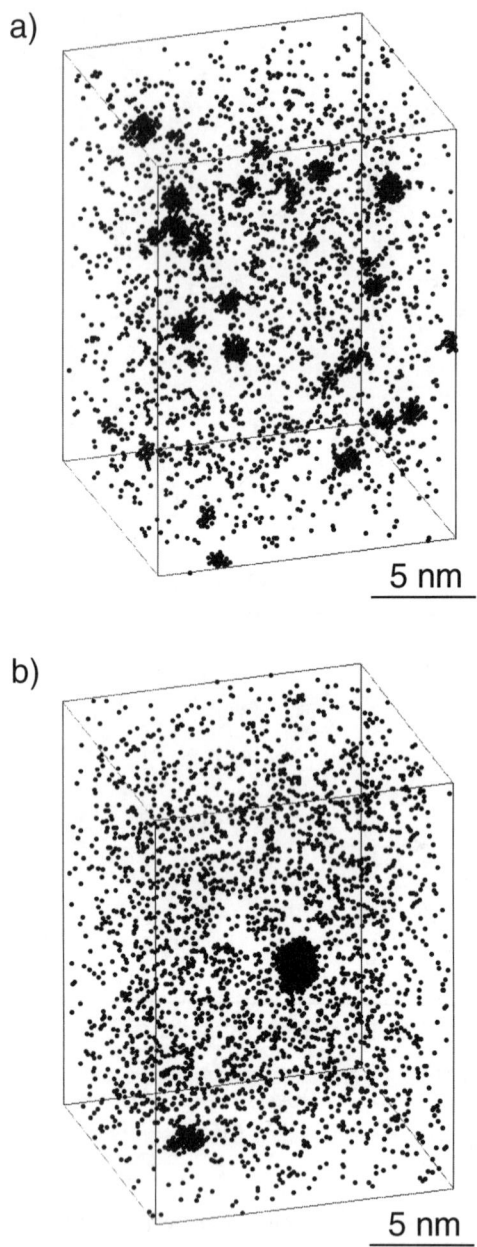

Fig. 11. Atom maps showing the distribution of Co atoms at peak cluster density from dynamic Ising model simulations at (a) 673 K (1000 MCS) and (b) 773 K (5000 MCS).

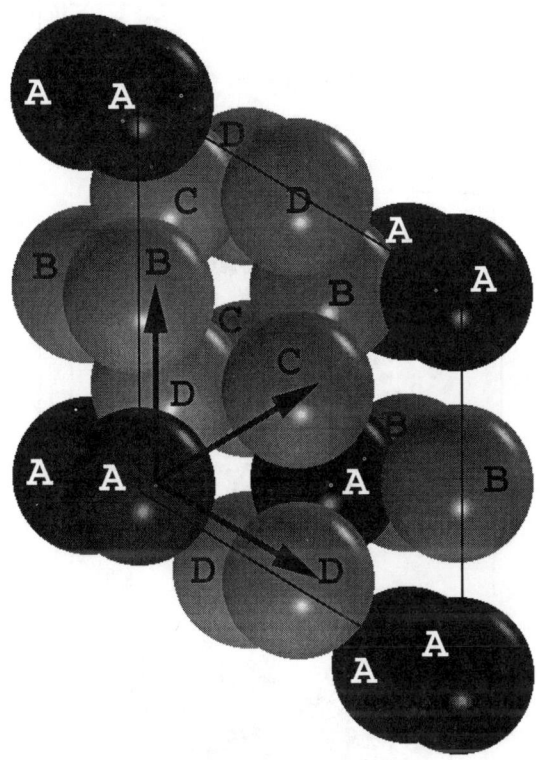

Fig. 12. Structure of the α$_2$ phase (DO19). The unit cell consists of 2 Al atoms (dark spheres) and 6 Ti atoms (light spheres). Equivalent configurations could be achieved with the Al atoms positioned on the sites marked B, C or D.

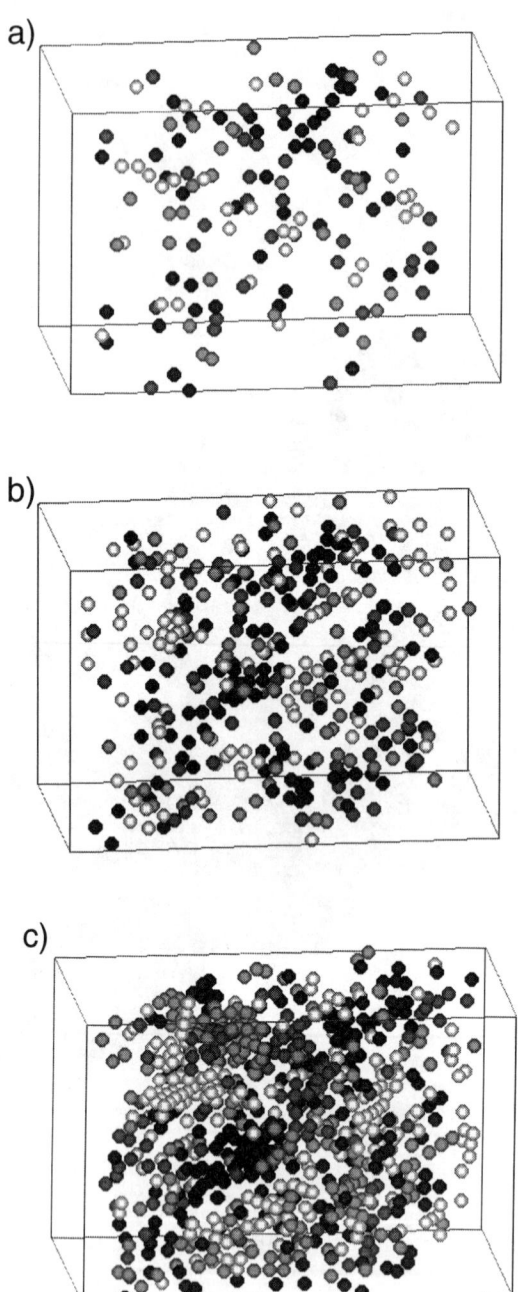

Fig. 13. 20x20x20 section of a simulation showing the development of ordered regions at 700 K after ageing (a) as quenched, (b) 1 MCS, (c) 4 MCS, (d) 15 MCS, (e) 60 MCS and (f) 240 MCS.

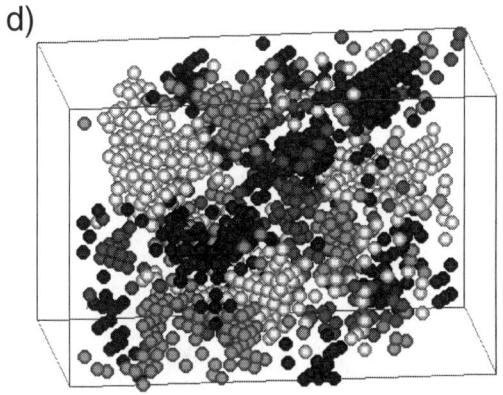

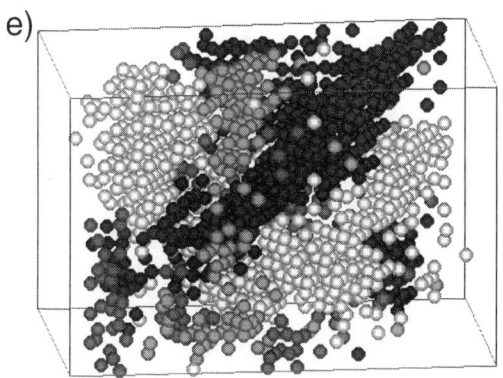

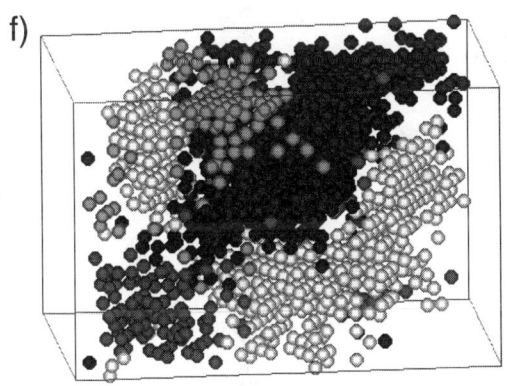

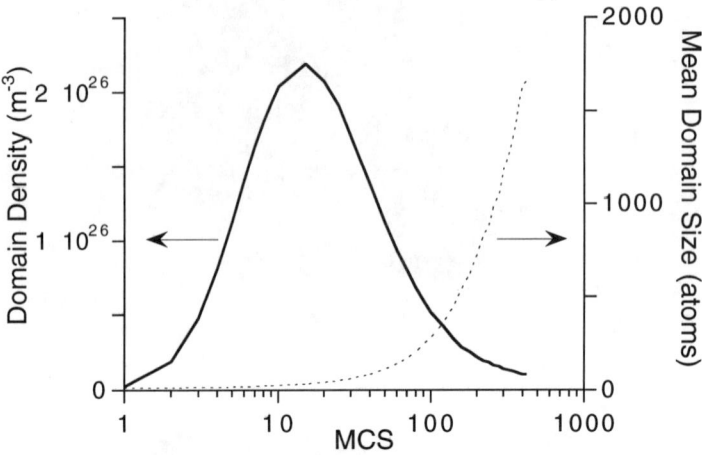

Fig. 14. Increase in mean ordered domain size and development of ordered domain density as a function of ageing.

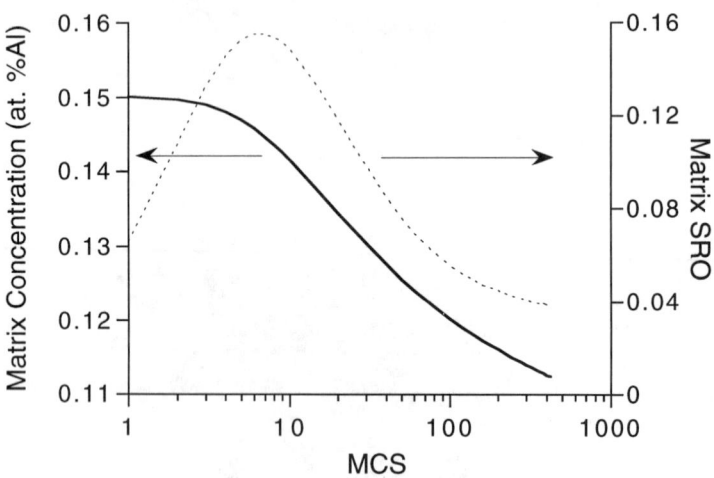

Fig. 15. Decrease in matrix concentration and development of matrix SRO as a function of ageing.

NON-CLASSICAL NUCLEATION AND GROWTH OF ORDERED DOMAINS IN A DISORDERED MATRIX

R. Poduri and L. Q. Chen

Department of Materials Science and Engineering
The Pennsylvania State University
University Park, PA 16802

ABSTRACT

Nucleation and growth of ordered domains in a homogeneous disordered matrix were studied using computer simulations based on the microscopic Langevin diffusion equations and the continuum diffuse-interface theory of Cahn and Hilliard. A particular example, precipitation of δ' (Al_3Li) ordered domains from disordered Al-Li alloys (α), was considered. It was found that a critical nucleus consists of fluctuations of both composition and long-range order parameter profiles. Only when the composition of the initial disordered matrix is near the phase boundary of the disordered phase, the composition and order parameter values inside the critical nucleus are close to those of the equilibrium ordered phase, and the critical profiles become increasingly diffuse as the composition of the disordered matrix approaches the ordering instability line.

Introduction

Precipitation of an ordered intermetallic phase from a disordered matrix is a basic process that underlies the processing of many advanced alloys such as high-temperature superalloys and ultralight aluminum alloys. Important examples include Al-Li alloys hardened by δ' ordered precipitates and Ni-Al alloys with γ' ordered precipitates. Since an ordered precipitate and a disordered matrix have not only different compositions but also structures of different symmetries, precipitation of an ordered intermetallic in a disordered matrix requires both ordering and phase separation. Therefore, the initial stage of precipitation during aging of a disordered phase within a two-phase field of ordered and disordered phases involves simultaneously many different processes including the relaxation of short-range order, development of long-range order, nucleation and growth of ordered domains, compositional decomposition, and formation of transient and metastable phases. However, in this paper, we will only consider the regime in which an ordered domain precipitates by a nucleation and growth mechanism.

Nucleation and growth takes place when the initial uniform disordered phase is metastable, i.e., it is stable with respect to small fluctuations in the independent variables describing its free energy and unstable with respect to large fluctuations. At a given temperature, the free energy F of a system exhibiting order-disorder transitions depends both on the composition c and the long-range order parameter η, such that $F(c, \eta)$ describes a surface on which the global minimum corresponds to the equilibrium state while the local minima describe metastable states. The kinetics of the system are then described in terms of the evolution of its free energy from any point on this surface towards the global minimum. However, if a system is initially at a local free energy minimum, then it has to be "jerked" out of this local minimum to reach a saddle point before it can evolve towards the equilibrium state. The focus of this paper will be on the properties of a system at a saddle point on the free energy hypersurface in the coordinates of composition and order parameter profiles, i.e., the properties of a critical nucleus.

We will discuss a particular case, precipitation of the $L1_2$ ordered (Al_3Li) particles (δ') in a FCC disordered matrix (α) in the Al-Li alloy system, by using computer simulations based on microscopic Langevin equations [1], and a non-classical nucleation theory [2], which is an extension of the original non-classical nucleation theory of Cahn and Hilliard for isostructural decomposition [3] to the case in which the precipitate and matrix phases differ in both structure and compositions.

Microscopic Langevin Diffusion Equations

In the microscopic diffusion theory, the atomic configurations and the morphologies of an alloy are described by single-site occupation probabilities, $P(A\mathbf{r},t)$, which give the probability that a given lattice site, $\mathbf{r}$, is occupied by a given atom of type A (e.g. Li in Al-Li alloys), at a given time t. The rates of change of these probabilities are described by Önsager-type diffusion equations [4], with a Langévin thermal noise term, $\xi(\mathbf{r},t)$,

$$dP(A\mathbf{r},t)/dt = \left(c_A(1-c_A)/k_BT\right)\sum_{\mathbf{r}'} L(\mathbf{r}-\mathbf{r}')\left(\delta F/\delta P(A\mathbf{r}',t)\right) + \xi(\mathbf{r},t) \qquad (1)$$

where the summation is carried out over all N crystal lattice sites of a system, L(**r-r'**) is related to the probability of an elementary diffusion jump from site **r** to **r'** per unit of time, T is temperature, k_B is the Boltzmann constant, c_A is the atomic fraction of the A component, and F is the total free energy of the system, which is a functional of the single-site occupation probability function. $\xi(\mathbf{r},t)$ is assumed to be Gaussian-distributed with average zero, and uncorrelated with respect to both space and time, i.e., it obeys the so-called fluctuation dissipation theory [5, 6],

$$\langle \xi(\mathbf{r},t) \rangle = 0$$
$$\langle \xi(\mathbf{r},t)\xi(\mathbf{r'},t') \rangle = -2k_B TL(\mathbf{r}-\mathbf{r'})\delta(t-t')\delta(\mathbf{r}-\mathbf{r'}) \qquad (2)$$

where <...> denotes an averaging, $\langle \xi(\mathbf{r},t) \rangle$ is the average value of the noise over space and time, $\langle \xi(\mathbf{r},t)\xi(\mathbf{r'},t') \rangle$ is the correlation and δ is the Kronecker delta function. The noise term is similar to that introduced by Cook to the Cahn-Hilliard equation [7]. With the noise term, Equation (1) becomes stochastic and is, in fact, the microscopic version of the continuum Langévin equation [8].

In the mean-field approximation, the total free energy of a system is given by,

$$F = (1/2)\sum_{\mathbf{r}}\sum_{\mathbf{r'}} W(\mathbf{r}-\mathbf{r'})P(\mathbf{r})P(\mathbf{r'}) + k_B T \sum_{\mathbf{r}}[P(\mathbf{r})\ln P(\mathbf{r}) + (1-P(\mathbf{r}))\ln(1-P(\mathbf{r}))] \qquad (3)$$

where W(**r-r'**) is the effective interchange interaction energy given as the sum of the A-A and B-B pairwise interaction energies, minus twice the A-B pairwise interaction energy.

To generate random numbers which satisfy the fluctuation-dissipation theorem, first, a random number, μ, is generated at any given lattice point at a given time step from a normal distribution, a Gaussian with average 0.0, and standard deviation 1.0. The random numbers are then multiplied by a factor to obtain the desired variance,

$$\xi(\mathbf{r},t) = p_f \sqrt{2k_B T L_1 \Delta t / a_s^2} \mu(\mathbf{r},t) \qquad (4)$$

where Δt is the timestep increment, a_s is the lattice parameter of the square lattice and L_1 is the exchange probability defined earlier. The coefficient, p_f, is a constant which is introduced as a correction factor that takes into account the fact that the correlation equations have been derived from linearized kinetic equations which are valid only at infinitely high temperatures, whereas equation (1) is non-linear [9]. In our simulations, the value of p_f used depended on the composition, and varied from 0.012 for $c_{Li} = 0.10$ to 0.015 for $c_{Li} = 0.078$. For more details, the reader is referred to refs. [1, 10].

Numerically, equation (1) is solved in reciprocal space, and a simple Euler technique is used to advance the probabilities in reciprocal space by one timestep. The real space probabilities are then recovered by a back Fourier transform. Following the spatial distribution of these probabilities with time yields the requisite microstructural evolution [1,10].

The Non-Classical Nucleation Theory

In the original diffuse-interface theory of Cahn and Hilliard [3], the total free energy, F, of a system is expressed as a function of composition c only, which is, in principle, valid only for systems in which the precipitate and the matrix have the same crystal structure. For precipitation of an ordered phase from a disordered matrix, we have to consider both the composition and structural differences between the nucleus and the matrix. Therefore, we have to consider the free energy functional as a function of both composition and long-range order parameter profiles. In this case, the increase in free energy, upon the formation of a fluctuation described by both composition and long-range order parameter profiles, is given by

$$\Delta F = \int \left[\Delta f + (K_\eta/2)(\nabla \eta)^2 + (K_c/2)(\nabla c)^2 \right] d^3x \tag{5}$$

where

$$\Delta f = f(\eta, c) - f(0, c_0) - (c - c_0)(\partial f/\partial c)_{0, c_0} \tag{6}$$

where c_0 is the average matrix composition, K_η and K_c are the gradient energy coefficients for order parameter and composition, respectively, and η and c are the order parameter and composition of the system, respectively. Here, f is the local free energy density, and for an $L1_2$ ordered phase in a disordered matrix in the mean-field approximation and with a second-neighbor interaction model, is given by [11]

$$f(c, \eta) = (1/2v_a)\left[(12W_1 + 6W_2)c^2 + 3(-4W_1 + 6W_2)\eta^2\right] + (k_B T/4v_a)\left[(c + 3\eta)\ln(c + 3\eta) + (1 - (c + 3\eta))\ln(1 - (c + 3\eta)) + 3(c - \eta)\ln 3(c - \eta) + 3(1 - (c - \eta))\ln(1 - (c - \eta))\right] \tag{7}$$

where W_1 and W_2 are the first- and second-nearest neighbor interactions, and v_a is the volume per atom.

If one assumes that nucleation in a metastable solution takes place by overcoming the minimum energy barrier, a critical nucleus is defined as the spatial composition fluctuation which has the minimum free energy increase among all fluctuations which lead to nucleation. If the interfacial energy is isotropic, and the critical fluctuation has a spherical symmetry, the composition profile corresponding to the critical nucleus can be obtained by solving the following Euler equations,

$$\begin{aligned} K_c\, d^2c/dr^2 + (2K_c/r)(dc/dr) &= \partial \Delta f/\partial c \\ K_\eta\, d^2\eta/dr^2 + (2K_\eta/r)(d\eta/dr) &= \partial \Delta f/\partial \eta \end{aligned} \tag{8}$$

subject to the boundary conditions,

$c(r) = c_0$ and $\eta(r) = 0$ at $r = \infty$, and dc/dr and $d\eta/dr = 0$ at $r = 0$.

These are coupled, non-linear second-order differential equations, which have to be solved numerically for a given set of f, K_c and K_η.

Application to Al-Li Alloys

As an example, we studied the nucleation of δ' (Al_3Li) ordered phase from a FCC disordered

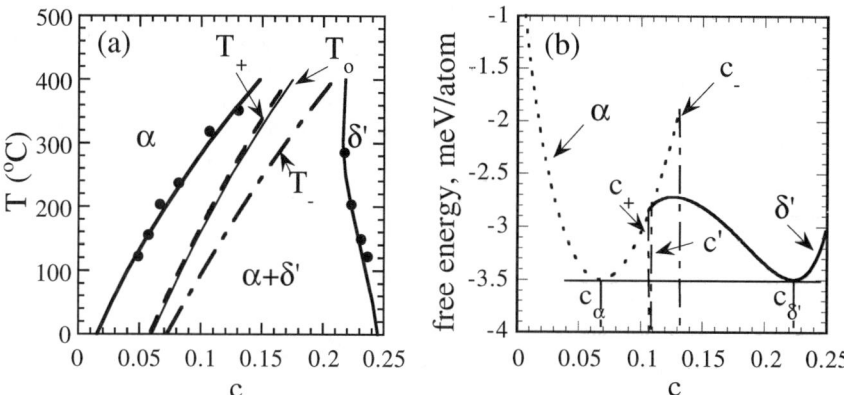

Figure 1. a) The Al-Li phase diagram and b) the free energy-composition curves at T = 465 K.

matrix in Al-Li alloys. A second-nearest neighbor interaction model is assumed for the Al-Li system and the values for the interaction parameters W_1 and W_2 were calculated from the V(0) and V(k_o) values reported by Schmitz et al. [12], as 40.435 meV/atom and -31.59 meV/atom, respectively. These give K_η and K_c as 14.422 and -2.83 meV(nm)2/atom, respectively [4]. The low-temperature part of the $\alpha + \delta'$ two-phase field using these interaction parameters is reproduced in Fig. 1a, in which the dot-dashed line (T_-) represents the ordering instability line below which a disordered phase is absolutely unstable with respect to ordering, the thin solid line (T_0) is the locus along which the ordered and disordered phases have the same free energy, the dashed line (T_+) is the disordering instability line above which an ordered phase is absolutely unstable with respect to disordering, and the thick solid lines are equilibrium phase boundaries. The free energy curves for the ordered and disordered phases as a function of composition T = 465 K are shown in Fig. 1b. According to Figs. 1a and 1b, at this temperature the equilibrium composition (of Li in atomic or mole fraction) of the disordered phase (α), c_α, is ~ 0.068; that of the metastable ordered phase δ', $c_{\delta'}$, is ~ 0.224; the ordering instability composition, c_-, is ~ 0.131; the disordering instability composition, c_+, is ~ 0.106; and the composition at which the ordered and disordered phases have the same free energy, c', is ~ 0.109.

Computer simulations were performed for several representative compositions within the $\alpha + \delta'$ two-phase field, namely 0.078, 0.10 and 0.12 at T = 465 K. To facilitate the representation of the simulated microstructures, all the results reported in this paper were obtained using 2-D projections of a 3-D system [1, 10]. All employed supercells consisting of 64x64 unit cells on the projected 2-D square lattice. In our simulations, a nucleus is considered to be critical if it grows when the Langévin noise term is switched off and if it disappears when the noise term is switched off one timestep too soon, as was done previously in refs. [5,8,13].

In order to compare with the simulation results, a cylindrical symmetry of the critical nucleus was assumed in the non-classical nucleation theory, and there is a difference of a factor of 2 for the second terms of the left-hand side of equations (8). Composition and order parameter profiles across a critical nucleus were obtained by solving equations (8) numerically, using the subroutine COLSYS developed by Ascher *et al.* [14], which is designed to solve ordinary differential equations for systems of non-linear boundary-value problems. It uses the method of spline collocation at Gaussian points, in conjunction with a damped Newton's method, to solve non-linear problems. The subroutine subdivides the boundary interval into finer and finer grids until a user-specified tolerance is satisfied. To speed up the calculation, an initial guess of the profile was supplied. In our case, we assumed the initial profiles to be Gaussian with average at r = 0.

$\underline{c}' < \underline{c} < \underline{c}$.

Fig. 2 shows a series of simulated pictures for $c_{Li} = 0.12$, a composition which lies between c' =

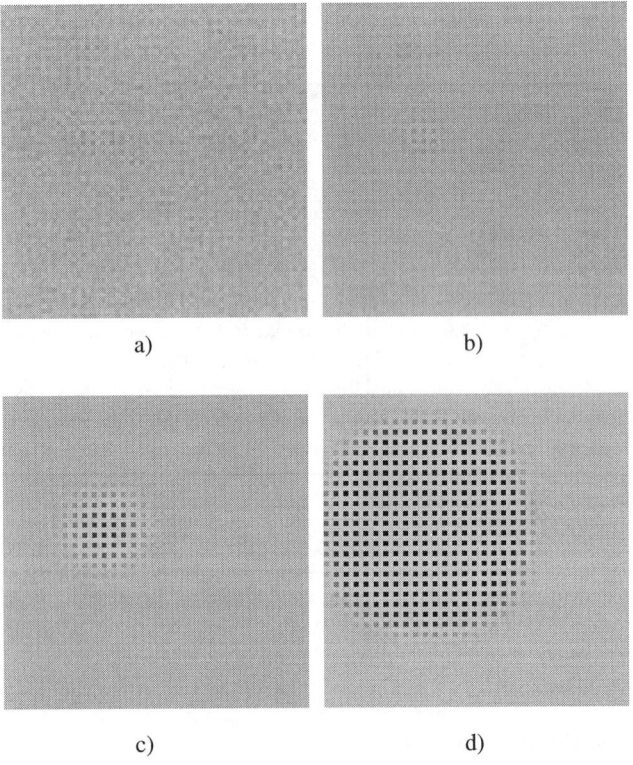

Figure 2. Computed microstructures for Li = 0.12 at different times t.
a) t = 4.40, b) t = 5.40, c) t = 14.40 and d) t = 194.40.

0.109 and $c_- = 0.131$. The occupation probability of Al is depicted on a gray scale on which black indicates occupation probability of Al is 0, and white indicates occupation probability of Al is 1.0. Previous thermodynamic analysis by Khachaturyan et al. [11] indicated that, in this region, the disordered phase is expected to undergo congruent ordering by a nucleation and growth mechanism, i.e., the critical nucleus only involves fluctuations in order parameter but not in composition.

It is apparent from fig. 2 that one does not observe a congruently ordered single-phase microstructure before compositional phase separation. This fact is also shown in Fig. 3, in which the average long-range order parameter and average compositional fluctuation are plotted versus time. The sudden drop in the values of the composition and long-range order parameter before their significant growth is due to the switch-off of noises and due to the fact that the local averaging contains artificial contributions from the random noise. The fact that both curves increase at the same time, after the noise term has been switched off, indicates that decomposition and ordering take place simultaneously.

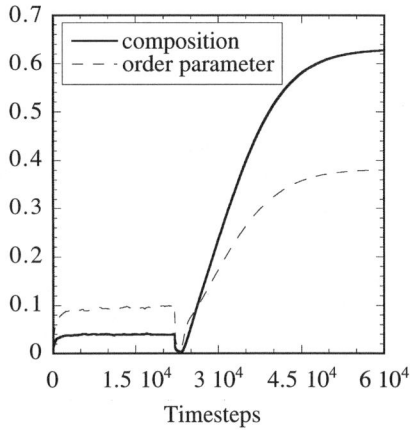

Figure 3. The variation of order parameter and 10 times composition deviation with time.

The critical composition and order parameter profiles across a critical nucleus are shown in Fig.

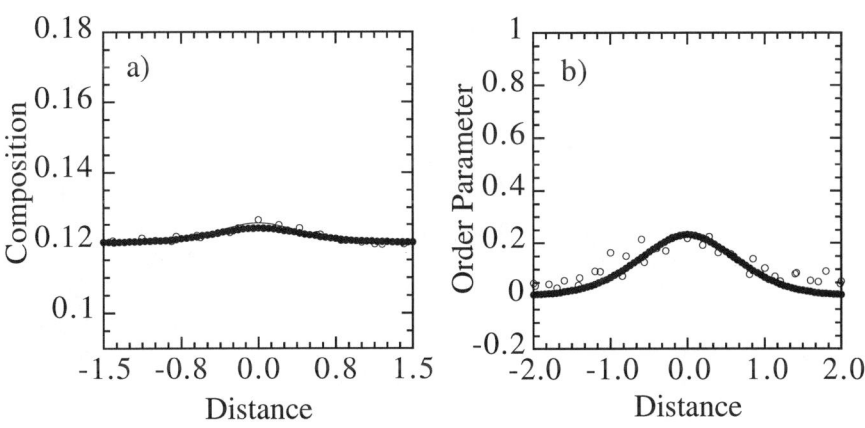

Figure 4. a) Composition and b) order parameter profiles across a critical nucleus at $c = 0.12$, using the microscopic (open circles) and continuum equations.

4a and 4b. It is quite clear that the critical nucleus contains fluctuations in both composition and long-range order. For comparison, the composition and long-range order profiles obtained from the non-classical nucleation theory are also plotted.

$c < c'$

We performed simulations for $c_{Li} = 0.10$ and $c_{Li} = 0.078$ in this region. Compared to composition $c = 0.12$, nucleating an ordered particle for lower compositions takes much longer time. In order to speed up the formation of critical nucleus, we added the same spatial distribution of noise at each time step for composition. It is argued that since our main interest is in the proper-

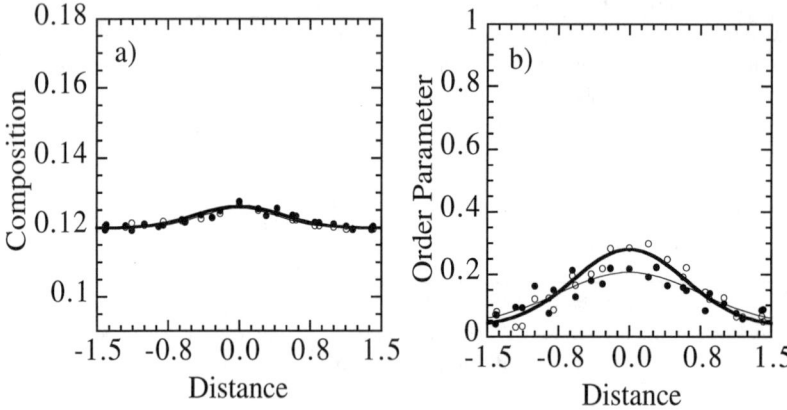

Figure 5. a) Composition and b) order parameter profiles across a critical nucleus for $c = 0.12$, using the two different methods of introducing noise.

ties of the critical nucleus instead of the nucleation rate, how a critical nucleus is formed is not very important. In fact, our test showed that for composition $c = 0.12$, the properties of the critical nucleus obtained by this more artificial method are similar to those obtained by thermal noises which are uncorrelated in both space and time, figs. 5a and 5b.

The variation of the local composition and local order parameter across the nucleus just after it is formed are plotted in figs. 6a and 6b for $c_{Li} = 0.10$, and in figs. 7a and 7b for $c_{Li} = 0.078$. The corresponding profiles obtained using the continuum equations are also included for comparison. Both the composition and the order parameter near the center of the critical nucleus are significantly higher than those for the critical nuclei at $c = 0.12$. However, they are much lower than those of the equilibrium ordered phase for $c_{Li} = 0.10$. For $c_{Li} = 0.078$, these values resemble those of the equilibrium phase, indicating that classical nucleation theory applies with greater accuracy at this composition than at $c_{Li} = 0.10$ and $c_{Li} = 0.12$. For comparison, the profiles across an equilibrium interface ($c_{Li} = 0.068$) is shown in figs. 8a and 8b. It can be seen that the profiles match very well, and the fact that the profiles using continuum equations are slightly more steep may be accounted for by the fact that higher order gradient terms have been ignored.

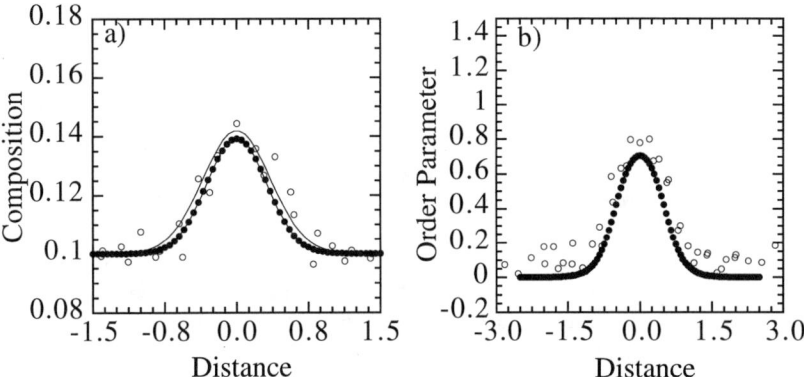

Figure 6. a) Composition and b) order parameter profiles across a critical nucleus for c = 0.10. Circles have same meaning as in fig. 4.

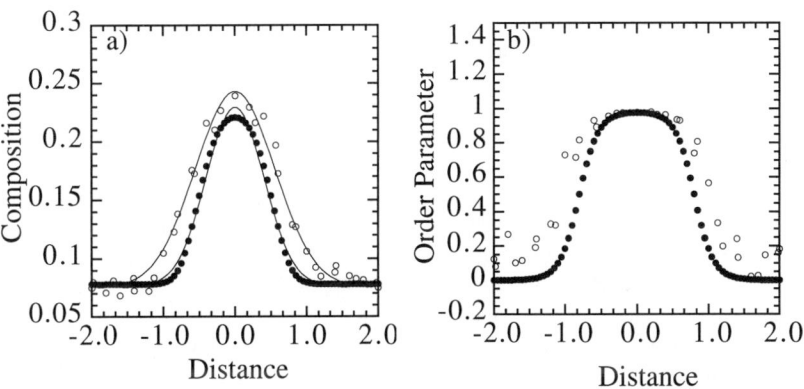

Figure 7. a) Composition and b) order parameter profiles across a critical nucleus for c = 0.078. Circles have same meaning as in fig. 4.

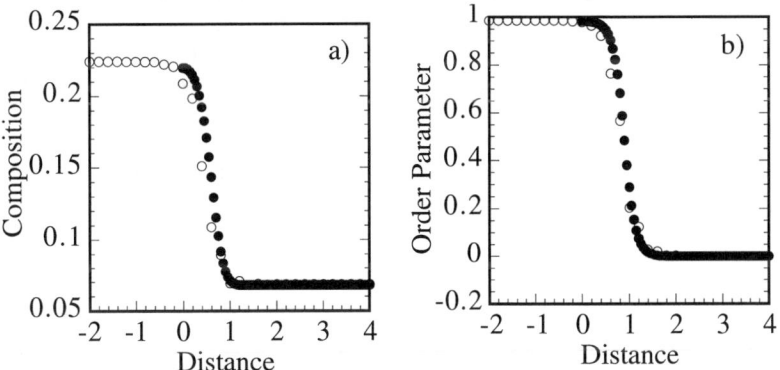

Figure 8. a) Composition and b) order parameter profiles across a critical nucleus for c = 0.068. Circles have same meaning as in fig. 4.

Discussion

Our simulation results show that congruent ordering does not take place when $c' < c < c_-$, and that ordering and decomposition take place simultaneously through a nucleation and growth process, as demonstrated by the concurrent growth of the both the order parameter as well as the compositional fluctuations. A close look at the cross-sections of the critical nuclei reveals that they contain both composition and order parameter fluctuations. These profiles are very similar to those obtained using the non-classical nucleation theory in cylindrical coordinates (for the sake of comparison with the simulation results on the projected plane) (Fig. 4). Also, note that in fig. 4a, the critical nucleus composition is such that it is very difficult to make a distinction between "bulk" and "interface". Therefore, the concepts of "bulk" free energy and "interfacial" free energy from classical nucleation theory break down at this composition.

As the average composition of Li is decreased to 0.078, the composition and order parameter values become increasingly closer to those of the equilibrium ordered phase. Therefore, it is possible to distinguish between "bulk" and "interface" for this nucleus, although even here the interface is diffuse, and not sharp. This is consistent with the non-classical nucleation theory of Cahn and Hilliard [15, 3] for isostructural decomposition.

Conclusions

A kinetic model has been developed for simulating the nucleation and growth of an ordered particle from a disordered matrix, using microscopic diffusion equations, coupled with a Langévin thermal noise term to simulate the nucleation event. Composition and order parameter profiles across critical nuclei have been obtained at different compositions using non-classical nucleation theory, and compared with those obtained from the kinetic simulations. When applied to the δ' precipitation reaction in Al-Li alloys, it shows that below c_-, there is no evidence of congruent ordering, i.e., ordering without compositional fluctuation. δ' forms by a nucleation and growth process, the nature of which changes from classical to non-classical nucleation, as the composition increases from the disordered phase boundary. The formation of a critical nucleus cannot be described using only composition fluctuations, as it requires critical fluctuations in both composition and order parameter profiles. These fluctuations within the critical nucleus decrease, as the average composition of the system approaches the ordering instability line.

Acknowledgments: The work is supported by the Office of Naval Research under the grant number N-00014-95-1-0577. All the simulations were performed on the Cray-90 supercomputer at the Pittsburgh Supercomputing Center.

References

1. R. Poduri and L.-Q. Chen, submitted to Acta. Met.
2. R. Poduri and L.-Q. Chen, submitted to Acta. Met.
3. J. W. Cahn and J. E. Hilliard, J. Chem. Phys., 31, 688 (1958)

4. A. G. Khachaturyan, Theory of Structural Transformations in Solids, Wiley, New York (1983)
5. Y. Wang, L.-Q. Chen and A. G. Khachaturyan, Solid-> Solid Phase Transformations, W. C. Johnson, J. M. Howe, D. E. Laughlin and W. A. Soffa, eds., TMS-AIME, p245 (1994)
6. Shang Keng Ma, Modern Theory of Critical Phenomena, W. A. Benjamin Inc., Reading, MA (1976)
7. H. E. Cook, Acta Met., 18, 297 (1970)
8. A. G. Khachaturyan, Y. Wang and H. Y. Wang, Materials Science Forum Vols. 155-156, 345, (1994)
9. A. G. Khachaturyan, Private Communication.
10. R. Poduri, Master's Thesis, The Pennsylvania State University, (1995)
11. A. G. Khachaturyan, T. F. Lindsey and J. W. Morris, Metall. Trans., 19A, p249 (1988)
12. G. Schmitz and P. Haasen, Acta Met., 40, 2209 (1992)
13. H. Y. Wang, Y. Wang, T. Tsakalakos and A. G. Khachaturyan, "Indirect Nucleation of a lower symmetry phase from a higher symmetry matrix, preprint (1993).
14. U. Ascher, J. Christiansen and R. D. Russell, *ACM Trans. Math. Software*, 7, 209 (1981)
15. H. I. Aaronson and K. C. Russell, Solid->Solid Phase Transformations, H. I. Aaronson, D. E. Laughlin and C. M. Wayman, eds., TMS-AIME, p371 (1982)

The Mathematics of Processing for Material Microstructure

John A. Simmons
Metallurgy Division
National Institute of Science and Technology
Gaithersburg, MD 20899

Abstract

A framework for stochastic control of material processing is presented. The formulation is based on a new Master Equation/Markov chain approach to nonequilibrium statistical mechanics where the separation of physics and statistics is maintained and time dependent control variables generalize the ideas of "bath" and Carnot cycles. A new development of cluster coordinates as marginal probabilities is given using combinatorial methods based on Mœbius function techniques and is applied to diffusive ordering on coherent lattices to generate the Fully Clustered Master Equation (FCME), A specific illustration is given for ordering on a square binary lattice.

Global ergodicity is not assumed, and it is shown how the equilibrium states for a time-independent system can split into many components when the evolution matrix is not irreducible. A general algorithm for determining equilibrium microstates is given. The idea of ergodic radius is introduced, and it is suggested that while irreducibility and homogeneity hold for small systems with periodic boundary conditions, inhomogeneous equilibrium microstates will be otherwise obtained and their relative likelihood summarized for stochastic control through the use of Master Equation methods.

Using the Jordan representation of the Markov transition matrix, marathon mappings for reducing the dimensionality and increasing the time step, and an eigenfunction expansion of the gradient free energy field are presented. The need for dimensional reduction is central, and here the concepts of marginal mappings, computable and evolution invariant subspaces are introduced. Using computable subspaces, a constructive criterion for a minimum set of physically relevant cluster variables is presented; these variables explain why expansions beyond point probabilities are needed.

It is suggested that cluster projection mappings can be used in at least a significant portion of ensemble space to achieve a global dimensional reduction and produce a purely linear stochastic formulation replacing the partially nonlinear Microscopic Master Equation method. This formulation gives rise to Clustered Master Equations(CME), which go beyond mean field methods by using probability superposition and truncated closure. The CME for ordering on a square binary lattice is calculated from compatible perturbations near the initial ensemble. Time series—and general complex linear filter—techniques are described and used to construct ergodic mappings, which are projections into invariant subspaces.

This theory is used to address stochastic control in the processing of materials undergoing diffusive ordering and phase transformations: (i) To find processing regimes which optimize the probability of producing high performance microstructures, and (ii) To predict the risk of structural breakdown of potentially important microstructures as a functional of the environmental conditions.

Mathematics of Microstructure Evolution
Edited by L. Q. Chen, B. Fultz, J. W. Cahn,
J. R. Manning, J. E. Morral, and J. A. Simmons
The Minerals, Metals & Materials Society, 1996

1. Introduction

Material scientists know that real materials have a very inhomogeneous microstructure. These "metastable" microstructures cannot be predicted on the basis of the classical statistical mechanics of solids; both the kinetics of nucleation and evolution must be an integral part of the theory. The purpose of this paper is to present the mathematical foundations for such an integrated theory determining as a function of processing environment, the statistical prediction of microstructures in multicomponent systems undergoing diffusive ordering on a coherent lattice.

For mathematicians interested in the conceptual analysis of complex physical systems, materials science offers a treasurehouse of examples. The systems to be analyzed lie in a probability space over a finite Boolean structure with an attached Markovian chain in which the primitives, called herein "pure states" are described in conjunctive normal form. It is moot whether the underlying physical system is finite or infinite, since any linguistic structure in necessarily finite; and with the expansion of computer technology, it is necessary to reexamine methods for dealing with structures which, while conceptually finite, are numerically infinite. Multicomponent atomic diffusion, particularly as applied to the evolution of ordering and diffusional phase transformations, provides a technologically important example of such a system which is at the edge of tractability.

Gibbs' original concept of stochastic ensembles without the ergodic hypothesis when combined with subsequent ideas from linear algebra and signal analysis, points towards a new paradigm for modeling nonlinear systems suitable for stochastic control. The specific system characteristics, which can be highly nonlinear, are clearly separated from the linear stochastic analysis. This mode of description is best suited to systems describable by a Markov process with a very large number of degrees of freedom, where uncertainty is a "way of life". In that case detailed causal nonlinear analysis and infinitistic formalisms not only have fundamental physical limitations, but fail to provide the capability for dealing with such complexity. Finite dimensional linear algebra offers the tools that can catalog such diversity.

Before proceeding to an example illustrating the structure of this formulation, let me summarize the underlying physical hypotheses:

i) *The set of all possible system states is finite, and can be enumerated in such a way that an ensemble probabilistic description may be utilized.*

The goal of the ensemble probabilistic description is the use of sets of independent probability distributions, usually the multinomial distribution. The idea of system states, called **pure states** in the ensemble context is a concatenated list of every contingency the system can undergo given a fixed set of **control parameters**. Although a different notation is adopted below in the more detailed examples, for this general discussion, we can call this set of states, Σ, and individual states, σ_1, σ_2, etc. At an initial time, 0, each pure state is considered to be "duplicated" some large number of times, and an ensemble constructed which is the collection of all such duplicated pure states. In the case of equal *a priori* probabilities the number of duplicates of each pure state is the same for all states, but this is not required.

During the time step, Δt, there is a finite **transition probability**, $p_{ij}(\sigma_i, \sigma_j)$ governing the transition from the pure state, σ_i, to the pure state, σ_j. The ensemble concept assumes that the control parameters remain fixed during the time step from t to $t+\Delta t$ and that the probability distributions for these transitions can be considered statistically independent during the same time step. The joint probability for many such transitions during Δt is, then, the product of the individual probabilities. Since the total number of systems in the evolving ensemble is considered fixed, conservation of probability holds and superposition holds for results at $t+\Delta t$ arising from different initial ensembles at t.

Because of the form of this assumption, the evolving ensemble is only defined at integer multiples of Δt. Each of its pure components, that is the number of each type of pure state in the evolving ensemble, is then also a function of integer multiples of Δt. That is, it is a time series with integer coefficients; or, if we divide by the total number of states in the ensemble, a time series with rational coefficients. The ensemble, itself, can then be considered as a vector of time series. In order for this vector time series to accurately reflect a continuous physical process, we invoke the sampling theorem of Shannon [1] and further assume that:

ii) *Any continuous time representation of the ensemble vector is band limited in frequency with an upper bound of nonzero frequency components given by the Nyquist condition:*

$$f_{\max} \leq \frac{1}{2\Delta t}. \tag{1.1}$$

In addition we assume that each of the control parameters considered as a continuous function of time is band limited with a cutoff frequency much less than that given by Eqn (1.1). Finally, we assume that any ensemble vector can also be considered time limited; that is, infinite times are ruled out.

In quantum statistical mechanics, the uncertainty principle in phase space demands that $\Delta E . \Delta t_\infty \geq h$. However, beyond this, the physical characteristics of real environments preclude the possibility of holding a system in a fixed environment for an infinite time:

The use of ensemble probability concepts in a kinetic context allows one to connect the ideas of numerical analysis with those of the predictive capability of a model (*i.e.* the maximum time for which the model is usable). For instance, suppose one chooses Δt as 10^{-8} secs, and one chooses N copies of a pure state, σ_1, with a transition probability of $p_\delta(\sigma_1, \sigma_2)$ for going to the state σ_2. Then, for example, 3 years is approximately 10^8 seconds, or 10^{16} time steps. Suppose that one calculates a transition probability of 10^{-26} in a theory which is "accurate" for a three year period. Select N to be 10^{20}, which certainly allows a Gaussian approximation. The expected number of transitions from that ensemble in a 3 year period is then 10^{10}. If one wishes to keep track of even the expected number of ensemble members in any state to, say, five significant figures, then one must be able to track transition probabilities—with roundoff buildup—to values of at least 10^{-30}. If, as typically happens in these types of calculations, computed probabilities occur of the order of, say 10^{-50}, such numbers must be considered physically meaningless unless a theory of far greater predictive capability is used. Thus, in analyzing any kinetic model, no matter how conceptually precise the formulation is, one must take into account the degree of approximation of the model as well as roundoff buildup during system evolution. Increasing the size of the ensemble to "narrow" the variance produces no effect beyond a certain point, just as concepts involving "epsilon" probabilities between two states must be considered physically meaningless without connecting epsilon to the predictive capability of the model.

In addition to these consideration, the variance of each of the transition probability distributions must be taken into account. In some cases these may arise from the underlying physics of the microscopic processes giving rise to the transition probabilities; at other times they are due to fluctuations in the control parameters or modeling errors in assuming a control parameter to be unchanged by a physical transition. From the macroscopic view, these microscopic fluctuations are often unimportant, but at other times they will affect the long term stability, and thus predictability, of any macroscopic system.

iii) *If there is a positive* **Hamiltonian** *function over* Σ *which is a Lyapunov function in the sense that:*

$$\sum_j p_\emptyset(\sigma_i, \sigma_j) \mathcal{H}(\sigma_j) \leq \mathcal{H}(\sigma_i), \qquad (1.2)$$

we say the system, Σ, *is* **separable**.

Although Eqn. (1.2) will not be explicitly used in this paper, it underlies the diffusion example discussed later, where a localized Hamilition is envisioned. When (1.2) holds, it provides a conceptual tool for relating the Markov approach used here with equilibrium statistical mechanics, although while equilibrium statistical mechanics permits interchangés between states having the same Hamiltonian, the Markov formalism may provide no pathway connecting those states. The existence of Hamiltonian functions raises issues directly connected to the open-ended nature of the time scale, and lies beyond the scope of this paper.

In emphasizing the kinetics of a finite system in a time dependent control environment, this theory departs from the conventional viewpoint of classical statistical mechanics. In both cases superposition inherent to the ensemble approach leads to a linear statistical theory based on the statistical expectation of an ensemble system. This formulation can be given in terms of Markov chains or the Master Equation. The ergodic hypothesis, that there is a system "perturbation" connecting any two states, takes the form that the Markov transition matrix is irreducible. The length of time necessary to achieve such perturbations is not an issue in equilibrium statistical mechanics; but to use equilibrium statistical concepts for real system kinetics requires additional assumptions about the length of time necessary to achieve such perturbations.

While it may be mathematically possible to construct fluctuations connecting any two system states, I suggest that when the probability of such fluctuations are taken into account, the numbers involved in many cases are completely overshadowed by approximations in the formulation of the system model and/or statistical noise, so that while equilibrium statistical mechanics gives lower bounds on the free energy and describes microstructures as metastable, the kinetic approach used here, which limits the "allowable system fluctuations" to those that have a realistic probability of occurrence, permits reducible Markov chains and gives a different meaning to the limiting "metastable" microstructures. In static control environments multiple stable inhomogeneous microstructures can exist which possess a localized free energy and between which there is no possibility of interchange within the time frame of the formalism, while in time dependent control environments, more complex outcomes can occur; the stability of a microstructure can be expressed in terms of its environmental controls after processing, and the description of macroscopic nucleation will take a different form from that usually taken in terms of classical fluctuation theory.

The "pragmatic" difficulty in this approach lies in the enormous dimension of the linear system required to describe the physical system. This high dimensionality introduces uncertainty and necessitates techniques of dimensional reduction to ameliorate these intractable dimensions. In this paper a variety of ways to address this issue are presented.

We turn, now, to the specific case of thermally activated diffusion in multicomponent systems, which will be illustrated throughout by means of a simple—but clearly generalizable—two dimensional binary system with Kawazaki type interchange.

2. Terminology: Joint Probability Space and Cluster Expansions

The material system is assumed to occupy a set of w atomic sites, Ω, in some lattice configuration. On this set there can be s species drawn from a species set, $\mathcal{S} = \{A, B, \cdots\}$. There are s^w possible atomic assignments, called $\Gamma_i, i = 1, \cdots, s^w$, and an ensemble of N such assignments is described as a vector, $\vec{P}$, whose components, $p^i = \dfrac{n^i}{N}$, represent the probability of each assignment type in $\vec{P}$. The index is written above to indicate that ensembles are to be represented as column vectors. The set of all possible ensembles forms a set $\mathfrak{P}$ which is a convex hypertetrahedron or **simplex**,

$$p^i \geq 0 \qquad (2.1\text{a})$$

$$\langle \Sigma | \vec{P} \equiv \Sigma_i p^i = 1, \qquad (2.1\text{b})$$

lying on the unconstrained hyperplane,

$$\Xi\mathfrak{P} \equiv \left\{ \vec{P} \in \Re\mathfrak{P} ; \langle \Sigma | \vec{P} = 1 \right\} \qquad (2.1\text{c})$$

which is parallel to the s^w-1 dimensional hyperspace,

$$\Xi\mathfrak{P}^\Delta = \left\{ \vec{P}' \in \Re\mathfrak{P}; \langle \Sigma | \vec{P}' = 0 \right\}, \qquad (2.2)$$

in the embedding Euclidean space $\Re\mathfrak{P}$ of dimension s^w. $\mathfrak{P}$ offers a way of visualizing the combinatorics of this model.

The linear form $\langle \Sigma |$ is called the **marginal form** for $\Re\mathfrak{P}$. The additional condition that no p_i can exceed one is a consequence of (2.1).

One basis for $\Re\mathfrak{P}$ is the set of ensembles,

$$^0\vec{P}_i, i = 1, \cdots, s^w, \qquad (2.3)$$

called the **pure state** basis, where $^0\vec{P}_i$ is the ensemble where all the assignments are the same, $p^j = \delta^{ij}$, (all the systems are in the same pure state).

The pure state basis offers a concatenated description of the system which can be paraphrased as, "species s_{i_1} lies at site one, and species s_{i_2} lies at site two, and", etc. It has a unique role in this formalism. In addition to the basic *a priori* assumptions given in (2.1), it is assumed that in the absence of any additional *a priori* information, the pure states conform to the Bayesian hypothesis of being equally likely. This hypothesis is used for connecting the Master Equation approach with the entropy maximization approach of information theory, which is one mathematical way of expressing conventional statistical mechanics computations; it is not intrinsically necessary for stochastic control.

The linear convex constraints expressed in Eqn (2.1a) are important; heuristically valuable linear algebra results must often be drastically modified because of their presence. As indicated above, the geometry of $\mathfrak{P}$ is that of an s^w-1 dimensioned simplex with a different subface bounded by every combination of vertices. Any ensemble which has zero probability component for a pure state, or set of pure states, lies in the boundary of this convex simplex "on the opposite side". This idea automatically suggests that as the probabilities of certain configurations become "virtually" impossible, the position of the ensemble vector can be projected onto a subface of lower dimension. However, linear relations developing among pure state probability components will not have this simple interpretation and must be treated differently.

Another basis for $\Re\mathfrak{P}$ is the dual to the space of marginal probabilities. In this diffusion model, it will be referred to as the **pigeonhole** cluster basis. This basis can be formulated using combinatorial Mœbius function techniques; it is closely related to the correlation basis of Sanchez *et al.*[1]

---

[1] In terms of probability measures on finite Boolean algebras the pure states correspond to the conjunctive normal form, while the marginal probability basis corresponds to the panoply of all mixed disjunctive/conjunctive forms with the point probabilities connecting to the disjunctive normal form.

The components of an ensemble given in terms of the pigeonhole basis,

$$^0\vec{C}_i, i = 1, \cdots, s^w, \qquad (2.4)$$

are the marginal probabilities for the first s–1 species over the first s–1 components for all disjoint partitions of the w indices in Ω into s components. The term marginal probability, apparently drawn from gambling, refers to the total probability that the given set of species will lie on the given sites independent of whatever happens on other sites. The term "pigeonhole" as used here states that once s–1 species have been assigned, the pigeonhole principle tells you where the last species will go. In the pigeonhole basis the coordinate value associated to a subset assignment is the marginal probability on that assignment; in the pure state basis the subset assignment is extended by adding the last species, and the coordinate value is the joint probability for the ensemble.

To quantify for s=2, choose all subsets of w, starting with the empty set, and proceeding with subsets having 1 element, then 2 elements, etc., which can be written

$$^0\vec{C}_j = {^0\vec{C}_{kl}}, j = \sum_{k'<k}\begin{bmatrix}w\\k'\end{bmatrix} + l ; k = 0, \cdots, w, l = 1, \cdots, \begin{bmatrix}w\\k\end{bmatrix}, \qquad (2.5)$$

where $\begin{bmatrix}w\\k\end{bmatrix} = \dfrac{w!}{k!(w-k)!}$ is the binomial coefficient. Here, k=1 corresponds to point probabilities, k=2 to pair probabilities, etc. To find the transformation matrix, η, between the pigeonhole and pure state bases, use a numbering of all the subsets of Ω from zeroth order (empty set) to Ω for both types of bases and set $\eta_j^i = 1$ if every atomic position index of the ith cluster is occupied by an A atom in the jth pure state; otherwise, $\eta_j^i = 0$. In the notation of [3], this can be written:

$$\eta_i^j = \begin{cases} 0 & \Gamma_{A|S_j} \nleq \Gamma_i \\ 1 & \Gamma_{A|S_j} \leq \Gamma_i \end{cases}. \qquad (2.6)$$

The pigeonhole cluster coordinates, or any "related" coordinates which are simply connected to marginal probabilities, will be called a **cluster coordinates**. Such coordinates are connected to the joint probabilities by

$$\begin{aligned} c^j &= \eta_i^j\, p^i \\ p^i &= \mu_j^i\, c^j, \end{aligned} \qquad (2.7)$$

where $\mu_j^i = \left(\eta_j^i\right)^{-1}$ is the inverse of η. If the joint probability basis is arranged in alphabetical order (using letters for species names), with site number used as the second sort index, the relation between the Mœbius function and pigeonhole coordinates is evident for binary systems. The arrangement for systems with more species is only a little more complicated, since the first ordering has to be on the rank, or total number of s–1 species, then subranks, then alphabetical, then sitewise. In the binary case η and μ are obviously superdiagonal matrices, and the components of μ are well-known to be $\mu_j^i = (-1)^{\#j-\#i}$, for $S_i \subseteq S_j$ and zero otherwise. Here $\#j$ are the number of elements in S_j and $\#i$ counts the number of A's in the set S_i associated to the pure state basis element,^0P_i. Symmetry is taken into account by group theoretic methods not discussed in this paper.

The energy of an ensemble can be written:

$$\langle E|\vec{P}\rangle = \Sigma p^i E_i = \Sigma c^j \mathcal{E}_j, \qquad (2.8)$$

where $E_i = \text{Energy}(\Gamma_i)$ is, presumably, calculated from first principles, and where

$$\mathcal{E}_j = \mu_j^i E_i, \qquad (2.9)$$

represents the marginal or cluster expansion of the energy of the ensemble. *The fundamental working hypothesis of cluster theory is that you only need a few of the lower order marginal energy terms to get an accurate estimate of the ensemble energy.* This is not the Hamiltonian in Connolly-Williams form [4], but linearly related to it.

It is important to note that, without taking account of symmetries, the entries in μ are all zero, +1 or –1, so that in large systems, especially, it is practically impossible to get from an E representation to an $\mathcal{E}$ representation because of cancellation of terms. However, the reverse is OK. In some sense, this makes the $\mathcal{E}$ representation, which embodies "screened" energy terms, more fundamental. The half ill-conditioned relationship between joint and marginal probabilities is universal in large systems.

By the same token it is clear that although all cluster probabilities must be nonnegative, choosing an arbitrary set of nonnegative cluster probabilities need not give rise to a corresponding joint probability distribution with all positive components. This issue, which was studied in detail in [3], will be discussed later.

3. The Fully Clustered Master Equation and Thermokinetic Statistics

Given a pure state $^0\vec{P}_i$ at time t, assume a Master Equation matrix $\boldsymbol{A}(\vec{\phi}(t))$, such that the Master Equation:

$$\Delta {^0\vec{P}_i} = \boldsymbol{A}(\vec{\phi}(t)){^0\vec{P}_i}\Delta t \qquad (3.1)$$

holds. The vector, $\vec{\phi}(t)$, is called the control parameter vector whose first "nonzero frequency components"[2] must occur at a frequency, f, where

$$\frac{1}{f\Delta t} \gg 1, \qquad (3.2)$$

and

$$\boldsymbol{A}^i_j = \begin{cases} p^i_j & i \neq j \\ -\sum_{k \neq i} p^i_k, \; |\boldsymbol{A}^i_i| < \frac{1}{2} & i = j \end{cases}, \qquad (3.3)$$

where p^i_j is the transition probability from the jth pure state to the ith pure state which scales with the time step: $p^i_j \equiv \dot{p}^i_j \Delta t$, so long as Δt is chosen small enough statistical independence of the transitions into different states, i, so that the sum of the off-diagonal transition probabilities in any column is less than one-half.[3] The value of one-half for each diagonal term is a mathematical requirement to insure nonsingularity of the associated Markov process, even though this value must be very close to one in order to insure the statistical independence of the diffusion processes involved in the Master Equation.[4]

The transfer matrix,

$$\mathbf{T} = \mathbf{I} + \mathbf{A}, \qquad (3.4)$$

the sum of whose row vectors is the row vector with all one's, is the generator of a Markov chain. Because of the restrictive inequality in Eqn (3.3), $\mathbf{T}$ is nonsingular. The evolution of the system is determined by the powers of $\mathbf{T}$, $\mathbf{T}^k$ giving the system state at time $k\Delta t$.

---

[2] Since the functional connection between $\vec{\phi}(t)$ and $\mathbf{A}$ is unspecified, and since this is likely to be nonlinear, one really can only look at the frequency components of $\boldsymbol{A}(\vec{\phi}(t))$. This semantic identification will be assumed throughout.

[3] The index convention used throughout is common in tensor analysis and has the upper index referring to columns and the lower index referring to rows. Thus, a vector given in terms of its coordinates is a column vector and a linear form is a row vector.

[4] The value of one-half does become of some interest if one formally tries to increase the time step by using multiples of a power of the Markov matrix, $\mathbf{T}$.

Mathematically, the transition probabilities p_j^i are assumed to derive from a statistical analysis of microscopic states; and, as is customary in information theory type analyses, it will be considered that they represent expected values for the transition probabilities. Fluctuations in these transition probabilities—which must be interpreted as noise in this theory—are connected to the variance of the distribution, which aside from ensemble size must be estimated from microscopic considerations. Our results will always be in terms of expected probabilities, and since the noise is never zero, statistical judgement can be exercised to consider that probabilities of sufficiently small value are effectively zero.

In this paper the physical transition probabilities are obtained from thermal activation for interchange between nearest neighbor pairs using a cluster expansion for the activation energy. To illustrate this concept segregation of vacancies will be ignored and a binary system employed on a two-dimensional square lattice with periodic boundary conditions having atomic interchange controlled only by nearest-neighbor (nn) and second nearest-neighbor (2nn) bonds. Figure (3.5) illustrates a possible snapshot of a part of such a system. A possible AB exchange pair is shown connected by the large arrows and all possible sites influencing the interchange are shown inside the heavy box with nn bonds and 2nn bonds displayed. The twelve site set consisting of the interchange pair and its nn and 2nn is called the **influence set** for the interchange pair. There are two influence set motifs depending on whether the interchange is horizontal or vertical. In this example one need not be concerned with the functional form of the interchange probability—that has been discussed elsewhere. [5]—only the assumption that the probability, p_j^i, is determined from knowledge of the species occupying sites in the appropriate influence set through information on the nn and 2nn bond configurational energy. This assumption aids in connecting the Master Equation approach with the usual classical statistical mechanics calculations, such as CVM. The value of Δt must be chosen sufficiently small that individual interchanges can be considered statistically independent. If this is not the case, then pair interchange configurations are too simple, and one must look at larger complexes in determining atomic interchanges. Further restrictions on Δt are discussed in conjunction with Eqn (3.6)

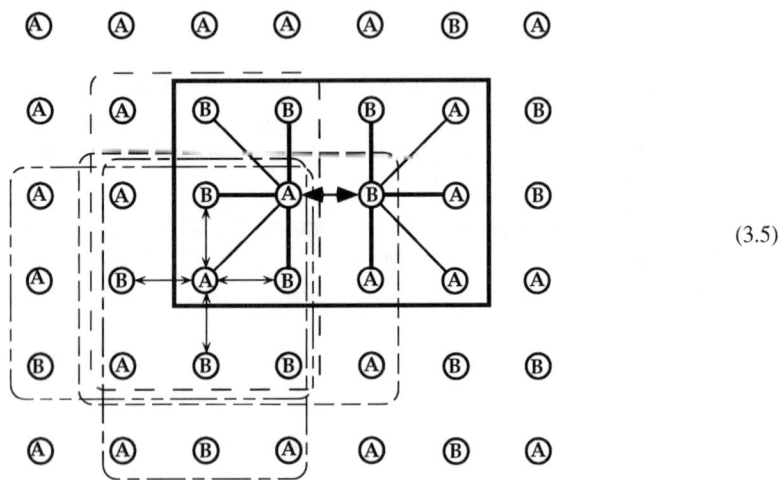

(3.5)

Figure (3.5) illustrates how the change in the influence set joint probability distribution can be obtained from the joint probabilities of all influence sets. Focussing on the A atom at the lower left corner of the influence set, one sees that this atom can only move to one of four places, and the probability of its doing so can be calculated using other influence sets centered about the possible interchange sites. Then, all joint probability distributions can be updated by a spreadsheet process, a form of the Master Equation called the Fully Clustered Master Equation (FCME):

$$p(\Gamma_S; t + \Delta t) = p(\Gamma_S; t)$$
$$- \sum_{\chi \stackrel{\cap}{=} \Gamma_S} \sum_{\Gamma_{\mathcal{I}_\chi} \stackrel{\cap}{=} \chi \uplus \Gamma_{(S-\chi)}} p(\Gamma_{\mathcal{I}_\chi}; t) \cdot p_{\emptyset}(\chi|\Gamma_{\mathcal{I}_\chi}) \Delta t$$
$$+ \sum_{\chi_\emptyset \stackrel{\cap}{=} \Gamma_S} \sum_{\Gamma_{\mathcal{I}_\chi} \stackrel{\cap}{=} \chi_\emptyset \uplus \Gamma_{(S-\chi)}} p(\Gamma_{\mathcal{I}_\chi}; t) \cdot p_{\emptyset}(\chi|\Gamma_{\mathcal{I}_\chi}) \Delta t \qquad (3.6)$$
$$+ \sum_{\substack{\chi^1{}_\emptyset \stackrel{\cap}{=} \Gamma_S \\ \chi^2{}_\emptyset \stackrel{\cap}{=} \Gamma_S}} \sum_{\Gamma_{\mathcal{I}_{\chi^1}} \stackrel{\cap}{=} \Gamma_{\mathcal{I}_{\chi^2}} \stackrel{\cap}{=} \chi^1_\emptyset \uplus \chi^2_\emptyset \uplus \Gamma_{(S-\chi^1-\chi^2)}} p\left(\Gamma_{\mathcal{I}_{\chi^1}} \uplus \Gamma_{\mathcal{I}_{\chi^2}}; t\right) \cdot p_\emptyset(\chi^1|\Gamma_{\mathcal{I}_\chi}) \cdot p_\emptyset(\chi^2|\Gamma_{\mathcal{I}_\chi}) \Delta t$$
$$+ \ldots$$

Here $\mathcal{X}$ ranges over all interchange site pair, a pair of adjoining sites with either AB or BA configuration given (since AA and BB interchanges accomplish nothing). The particular assignment of AB or BA to $\mathcal{X}$ is denoted by the Greek variable χ. $\mathcal{I}_x$ is the influence set for an interchange site pair; Γ indicates a particular assignment of species to a set of sites; $\stackrel{\cap}{=}$ means that where two sets of sites intersect, the species assignments are the same; $p_\emptyset$ is the probability of an interchange at χ; and $\chi_\emptyset$ is the interchange site pair with the species interchanged (BA or AB). Finally, $\uplus$ represents the combination or union of assignments over disjoint domains (or equivalently over two domains where the assignments are consistent ($\stackrel{\cap}{=}$) on their intersection). Thus, Eqn (3.6) expresses the fact that the number of occurrences of Γ_S is reduced by all AB interchanges which replace an atom at some site in S with one of opposite type, and is increased by all interchanges coming from assignments differing at one site plus double interchanges from assignments differing at two sites, etc. In all cases the value of Δt is chosen small enough that not only are the interchanges statistically independent, as discussed above, but double and higher order interchange effects can be neglected. The higher order interchange probabilities are of the order of powers of first order interchange probabilities. Since the choice of probability values that can be considered negligible is a critical aspect in kinetic theories, corrections for double and even triple interchanges may be of physical significance. Their inclusion would make the calculation of the Master Equation matrix, **A**, more complex, but would not otherwise alter the structure of the mathematical theory.

The formulation given in (3.6) avoids the use of conditional probabilities wherever possible to give a marginal probability form of the Master Equation. It is important to note that the FCME conserves probability in such a fashion that changes in probability are proportional to the existing probability, and (3.6) can be applied to the entire set, Ω; therefore, Eqn (2.1a) demanding nonnegative probabilities is satisfied.

In Eqn (3.6) one sees that sets of the type $\mathcal{I}_\chi \cap S$, $\mathcal{X} \cap S \neq \emptyset$ occur in the first two terms. From this observation one can conclude that the maximum set, S, whose probability distribution can be updated without need of probability distributions on larger sets than influence sets is the intersection of all influence sets associated to interchanges from a given site in any direction. Such a set, which in Fig (3.5) consists of all first and second neighbors of any site, is seen surrounding the lower left A atom in the influence set. Put in another way, this shows that one must use sets larger than influence sets in order to update the joint distribution on influence sets, and one cannot, in general, obtain closure in the FCME without knowing the joint probability distributions over all of Ω.

When Ω is chosen as the influence set, the FCME is the same as the ordinary Master Equation. The difficulty is the dimensionality of the probability space. While we can examine the global properties of the FCME, fully general computations are only possible when Ω has very few sites, and because of the exponential dimensions involved, this limitation appears likely to prevail for some time. Anyway, a space of such enormous dimensions contains too much information. Instead we shall seek ways of dimensional reduction in which the linear probabilistic concepts of ensemble theory prevail, so that we can still deal with very complex systems using linear analysis while at the same time dealing with, possibly, highly nonlinear constitutive laws, which, in this context translate into highly nonlinear transition probabilities.

These nonlinearities appear in the relationship of the control parameters with the matrix coefficients. To remove these nonlinearities, or when the control parameters interact with the system, the definition of state has to be expanded. Time dependent control parameters extend the concepts of "bath" and Carnot cycle in thermodynamics. This viewpoint is comparable to the grand canonical ensemble of Gibbs, where the control parameters represent enormous reservoirs whose state is considered unaffected by any changes in the system, Ω. During processing, $\vec{\phi}(t)$ can be expected to change with time, with the restriction that the frequency components occurring in the Fourier transform of $\vec{\phi}$ obey Eqn (3.2). If (3.2) is not obeyed, the linear FCME is without justification. On the other hand if all of the frequency components of $\vec{\phi}$ are much lower than any associated to the physical system, the control system be said to be constant with respect to the evolving system, and the system can approach equilibrium; however, low frequency components in the control parameter regime can be expected to produce nonlinear effects. Complete separation of linearity and nonlinearity requires that the eigenvalues of the matrix $\mathbf{A}$ must be separated from zero by a gap sufficient to allow the exponentially decaying modes associated to $\mathbf{A}$ to decay to zero.

4. Marginal Projections and Clustered Master Equations

In this section projection methods are developed which allow one to reduce the dimension of the FCME by restricting the description of the system to a class of physically important parameters for which an analysis can be performed.

Let $\mathbf{Y}$ be an $n \times n$ matrix of the form:

$$\mathbf{Y} = \begin{pmatrix} \mathbf{Q}_{i_1} & 0 & \cdot & \cdot & 0 \\ 0 & \mathbf{Q}_{i_2} & & & \cdot \\ \cdot & & & & \cdot \\ \cdot & & & \mathbf{Q}_{i_{n-1}} & 0 \\ 0 & , & \cdot & 0 & \mathbf{Q}_{i_n} \end{pmatrix}, \quad (4.1\mathrm{A})$$

where each matrix $\mathbf{Q}_{i_j}$ is of the form:

$$\mathbf{Q}_{i_j} = \begin{pmatrix} 0 & 0 & \cdot & \cdot & 0 & 0 \\ 0 & 0 & & & & 0 \\ \cdot & & & & & \cdot \\ \cdot & & & & & \cdot \\ 0 & & & & 0 & 0 \\ 1 & 1 & \cdot & \cdot & 1 & 1 \end{pmatrix} \quad (4.1\mathrm{B})$$

with the row of ones occurring at the $i_j{}^{\text{th}}$ row of $\mathbf{Y}$. Such a matrix is a special form of a projection matrix (i.e. $\mathbf{Y}^2 = \mathbf{Y}$). $\mathbf{Y}$ provides an example of a marginal projection matrix, since marginal probabilities are conserved under $\mathbf{Y}$ with a variety of "similar" states at one time instant being aggregated into the projected image state.

A **retractive marginal projection**, or just **marginal projection** is defined by:

$$\begin{aligned} \mathbf{Y}^2 &= \mathbf{Y} \\ \mathbf{Y}\mathfrak{P} &\subseteq \mathfrak{P}, \end{aligned} \quad (4.2)$$

and a **marginal projective mapping**, or just **marginal mapping** is defined by

$$\mathbf{Y}^2 = \mathbf{Y}$$
$$\mathbf{Y} \, \Xi\mathfrak{P} \subseteq \Xi\mathfrak{P}, \tag{4.3}$$

where $\Xi\mathfrak{P}$ is the hyperplane introduced in (2.1c). Without the convexity constraints, one need make no distinction between these two ideas; but the distinction is required here.

The general idea of a **probability mapping**, $\mathbf{Y}$, can be defined as a mapping of the n–dimensional embedding Euclidean space, $\Re^n$, with basis vectors, say, $\vec{B}_1, \cdots, \vec{B}_n$ into the k dimensional embedding Euclidean space, $\Re^k$, with basis vectors, $\vec{B}^*_{i_1}, \cdots, \vec{B}^*_{i_k}$ with the property that if $\vec{P} \in \Xi\mathfrak{P}$ in $\Re^n$, then $\mathbf{Y}\vec{P} \in \Xi\mathfrak{P}$ in $\Re^k$. This can be simply expressed as:

$$\mathbf{Y}^\dagger \langle \Sigma |_{\Re^k} = \langle \Sigma |_{\Re^n}, \tag{4.4}$$

where $\mathbf{Y}^\dagger$ is the adjoint of $\mathbf{Y}$. The distinction between probability mappings and marginal mappings is principally semantic, since each range element should be thought of as the superposition of all its preimages, and in finite dimensional spaces isomorphisms between projections and probability mappings can always be developed.

The case of marginal projections—(5.2)—occurs, for instance, when a fixed set of sites are distinguished, marginal probabilities are computed for these sites, and a new joint probability distribution defined which is zero unless the occupants of the complementary sites (over which the sum was carried out) are all B atoms in which case it equals the marginal probability. In this case, $\mathfrak{P}$ is mapped onto one of its subfaces.

An example of marginal mappings—(5.3)—can be given in terms of cluster coordinates. Noting that the projective property is coordinate independent, if the marginal probabilities for k atom clusters are given to complete assignments with at least k A atoms and zero probability given to all ensemble components with more than k A atoms, the resulting **cluster projection** mapping is clearly a projection when expressed in cluster coordinates, since the low order basis elements map to themselves, and the high order ones map to zero. Since $\langle \Sigma |$ is dual to the zeroth order basis element, the mapping conserves $\Xi\mathfrak{P}$. In this case, however, the image of a point, $\vec{P}$, in $\mathfrak{P}$, while still lying on $\Xi\mathfrak{P}$, may no longer lie inside $\mathfrak{P}$. However, one knows that

$$\vec{P} \in \left(\mathbf{Y}\vec{P} + \mathfrak{N}\mathbf{Y} \right) \cap \mathfrak{P} \equiv \yen\vec{P}, \tag{4.5}$$

where $\mathfrak{N}\mathbf{Y}$ is the null space of $\mathbf{Y}$. Here, $\yen\vec{P}$, the **preimage** of $\vec{P}$ inside $\mathfrak{P}$, is a polytope of dimension no greater than n, where $n=s^w-r$ is the dimension of the null space of $\mathbf{Y}$ and r is the rank of $\mathbf{Y}$. Clearly both marginal projections and marginal mappings are closed under matrix multiplication, and only the points of $\mathbf{Y}\mathfrak{P}$ can correspond to an ensemble probability.

The set of ensembles, $\yen\vec{P}$, can be physically more important than its representative projected image. Indeed, we show in the next section how the concept of ergodic set can be phrased in these terms. Also, one interpretation of the entropy maximization algorithm of [3] is to find a "best" representative in $\yen\vec{P}$, thereby defining a nonlinear inverse mapping to $\mathbf{Y}$.

When $\mathbf{T}$ can be considered constant in time, a more detailed understanding of the class of marginal mappings arises from studying the full Jordan decomposition of $\mathbf{T}$, which, although not generally computable, can still be studied in principle. A summary of the Jordan decomposition and its relationship to equilibrium statistical mechanics is given in the Appendix.

We know that $\mathbf{A}$ has at least one zero eigenvector on $\mathfrak{P}$ which is a fixed point of $\mathbf{T}$. For our example we choose the barycenter of the equilibrium polytope. Let $_\bullet\vec{P}$ be such a fixed point of $\mathbf{T}$. $\mathbf{A}$ and $\mathbf{T}$ map the s^w-1 dimensional hyperspace $\Xi\mathfrak{P}^\Delta$ into itself. Choose a basis of $\Xi\mathfrak{P}^\Delta$ yielding the Jordan decomposition of $\mathbf{T}$. This basis, together with $_\bullet\vec{P}$ forms a basis for $\Re\mathfrak{P}$, and in terms of this basis with $_\bullet\vec{P}$ chosen first, T^1_1 is one and the rest of the first row and column are zeroes. The remainder of the s^w-1 by s^w-1 matrix is the Jordan decomposition of **T** on $\Xi\mathfrak{P}^\Delta$. The translation of $\mathfrak{P}$ to $\Xi\mathfrak{P}^\Delta$ by using the vector $_\bullet\vec{P}$ gives rise to a convex set, $\mathfrak{P}^\Delta$ containing the origin with which the entire system evolution can be described.

To develop marginal mappings consistent with **T**, it's easy to see that for a simple real eigenspace of **T**, the idempotent property of **Y** suggests that **Y** either map the eigenvector onto itself or onto the zero vector. On simple complex eigenspaces, almost the same holds except that both the real and complex parts of the eigenvector have to be mapped together. On Jordan spaces, **Y** must either leave all base vectors fixed or map all base vectors onto the same vector as is the eigenvector for that block. If such a **Y** leaves $_\bullet\vec{P}$ fixed, then it clearly satisfies (5.3) and (5.6).

Think of selecting some initial state, $_I\vec{P}$, "near disorder" and watch the evolution of $_I\vec{P} - {_\bullet\vec{P}}$ in $\mathfrak{P}^\Delta$ as a function of $k\,\Delta t$, i.e. monitor $\mathbf{T}^k\left(_I\vec{P} - {_\bullet\vec{P}}\right)$ as a function of k.

Other than the real value, one, the Jordan decomposition of **T** on $\Xi\mathfrak{P}^\Delta$, has eigenvalues with absolute value less than one, as seen from the Gershgorin theorem (A.3). Therefore, the $\lambda_j{}^{\text{th}}$ component associated to the eigenvalue λ_j, has been reduced by an amount $|\lambda_j|^k$. For λ_j of small modulus, this component rapidly becomes sufficiently small that it represents a probability which can effectively be set equal to zero. Thus, approximately exponential modal decay occurs after a delay period, k_J associated to the disappearance of Jordan blocks.[5] Choose a value $k^\blacktriangle$, so that for $k \geq k_J + k^\blacktriangle$ and $|\lambda_j| \leq \Lambda^\blacktriangle$, the probability can be effectively set to zero when the $\lambda_j{}^{\text{th}}$ component of $\left(\vec{P} - {_\bullet\vec{P}}\right)$ is no more than $M^\blacktriangle$, which represents a bound to all the components of $\left(_I\vec{P} - {_\bullet\vec{P}}\right)$.

If we define a marginally projective mapping, **Y**, in which the modes associated to modulus less than $\Lambda^\blacktriangle$ are projected to zero, referred to as **marathon mappings**, the longer term behavior of $\mathbf{T}^k\left(_I\vec{P} - {_\bullet\vec{P}}\right)$ can be found by following the mapping on the range space of **Y**. Image points of this mapping may not lie in $\mathfrak{P}^\Delta$ (although they ultimately will, since $\mathfrak{P}^\Delta$ surrounds the equilibrium polytope image minus $_\bullet\vec{P}$ to which the image point converges), but the image point itself is a solution to:

$$p^i + {_\bullet p^i} \geq 0$$
$$\langle E^{\lambda_j}|\vec{P} = e_i^{\lambda_j} p^i = e_i^{\lambda_j}{}_k p^i = \langle E^{\lambda_j}|\mathbf{T}^k\left(_I\vec{P} - {_\bullet\vec{P}}\right), \quad |\lambda_j| \leq \Lambda^\blacktriangle. \tag{4.6}$$

Here $\vec{E}_{\lambda_j}$, $\lambda_j=1,...,J$ is the set of such eigenvectors with eigenvalues, λ_j and $\langle E^{\lambda_j}|$ are their associated dual linear forms, $_\bullet p^i$ are the joint probability components of $_\bullet\vec{P}$, and $_k p^i$ are the joint probability components of $\mathbf{T}^k\left(_I\vec{P} - {_\bullet\vec{P}}\right)$, i.e., $(\mathbf{T}^k{}_I P)^i - {_\bullet p^i}$. The set described by Eqn (4.5) (with an obvious shift by $_\bullet\vec{P}$) is, then, non-empty. The members of this set can be thought of as all states with the same long term behavior as $\mathbf{T}^k\left(_I\vec{P} - {_\bullet\vec{P}}\right)$, differing from each other only in terms of transient behavior with a decay time less than the order of $k^\blacktriangle \Delta t$. If one now works in multiples of $k^\blacktriangle \Delta t$ on the image space of **Y**, one is working in a space of reduced dimension with larger time steps.

---

[5] One can argue that by using only a tiny perturbation of the coefficients of **A** (or **T**), one could construct a physically equivalent evolution matrix with all distinct eigenvalues, so that all systems should be considered to be without Jordan blocks. As can be seen by looking at the structure of a Jordan block, there is associated to each Jordan block of size k in **T** a set of basis vectors through which the system cycles with time, only "dropping down" with λ after k time steps. Part of the reason for this arises from the relationship between the vector basis associated to a Jordan block and an eigenvalue basis when the eigenvalues are distinct but only slightly perturbed from a central eigenvalue. The coordinate transformation involves products of differences of eigenvalues, rendering this transformation very ill-conditioned and sensitive to noise. The situation is particularly acute when the eigenvectors are all almost parallel, since in this case the "metric" of the space associated to the eigenbasis is greatly distorted *vis-a-vis* that of the Jordan basis. The physical argument cited above implies that the vectors in a Jordan block are associated to such degeneracies in resolving clumped eigensystems.

A particularly interesting example of the above is the marginal projection which can be constructed on a subset of $\mathfrak{P}^\Delta$. This is done by taking the set

$$\mathfrak{E} \cap \frac{1}{\sqrt{2}} \mathfrak{P}^\Delta , \qquad (4.7)$$

where $\mathfrak{E}$ is the subspace of $\Xi \mathfrak{P}^\Delta$ generated by the set of all eigenvectors in $\Xi \mathfrak{P}^\Delta$, including those in Jordan blocks, of $\mathbf{T}$ (but not other block basis vectors). The projection of any vector $\vec{P}$ into $\mathfrak{E}$ is given by:

$$\mathbf{Y}\vec{P} = \sum \vec{E}_{\lambda_j} \langle E^{\lambda_j} | \vec{P} . \qquad (4.8)$$

If $\mathbf{Y}\vec{P}$ lies inside $\frac{1}{\sqrt{2}}\mathfrak{P}^\Delta$, then, since all the eigenvalues of $\mathbf{T}$ have real part $\leq$ one, $\mathbf{T}\,\mathbf{Y}\vec{P}$ also lies in $\mathfrak{E} \cap \frac{1}{\sqrt{2}}\mathfrak{P}^\Delta$. The square root factor has been used to ensure that the set is also retractive for complex eigenvectors, since the curvature of the harmonic orbit has to taken into account *vis-a-vis* the linear boundary of $\mathfrak{P}^\Delta$. The Markov process on the range space of such a projection has all simple eigenvalues, so that on the range space, the Markov mapping and its Master Equation matrix are said to be semisimple:

$$\mathbf{A}_{|\mathfrak{R}\mathbf{Y}} \equiv {}^\nabla\mathbf{A} = {}^\nabla\mathbf{U}\,{}^\nabla\mathbf{D}\,{}^\nabla\mathbf{U}^{-1} . \qquad (4.9)$$

Here ${}^\nabla\mathbf{D}$ is a diagonal matrix with possibly complex eigenvalues and eigenvectors, which occur in complex conjugate pairs.[6] The eigenvectors of ${}^\nabla\mathbf{A}$ on $\mathfrak{E} \cap \frac{1}{\sqrt{2}}\mathfrak{P}^\Delta$, have the form:

$${}^\nabla\vec{E}_j = {}^\nabla\mathbf{U}_j^i \,{}^\nabla\vec{P}_i , \qquad (4.10)$$

with associated eigenvalues, λ_j. The system evolution on such an image space is a "gradient" system controlled by the linear function,

$$\dot{F} = -\Sigma_j \lambda_j \langle {}^\nabla E^j | , \qquad (4.11)$$

where $\langle {}^\nabla E^j |$ are the linear forms dual to the eigenvectors, ${}^\nabla\vec{E}_j$. Setting $\vec{P}(t) = \sum_j a^j(t)\,{}^\nabla\vec{E}_j$ gives $\frac{d\dot{F}}{dt} = -\sum_j \lambda_j \frac{da^j}{dt}$, which has its maximum absolute value for the directions, $\frac{da^j}{dt}$, determined by $\frac{1}{\lambda_j}\frac{da^j}{dt}$ =constant, *i.e.*, using an eigenvector coordinate system, the system path is parallel to the normal of the surfaces $\dot{F}$ equals a constant. $\dot{F}$ represents the **probability flux** and decreases with time for systems approaching an equilibrium fixed point.

To deal with realistic models, where the exponential dependence of w on the dimension of Ω quickly produces matrices intractable size, requires methods to limit the amount of information to be processed. This will be attempted by abandoning the full Jordan representation, and seeking, instead, partial representations for which modeling computations can be performed. The technique of dimensional reduction introduced here forms the bridge between the general ensemble description of the Master Equation and computable algorithms. The particular construction used herein employs marginal probabilities in a fully linear context, although a brief discussion will be given of a nonlinear alternative. For simplicity of presentation, no explicit time dependence is used in this development; such dependence can readily be introduced using the notation of the next section where time dependence is explicitly discussed.

Since the FCME is a linear system, we can consider representations of $\mathbf{T}$ in other coordinate systems, such as , for example, a cluster basis where $\mathbf{T}_C = \eta \mathbf{T} \mu$. Suppose we can distinguish a new coordinate system where $\mathbf{T}$ takes one of the forms:

$$\begin{bmatrix} M_1 & 0 & M_3 & 0 \\ N_1 & N_2 & N_3 & 0 \\ 0 & 0 & P_3 & 0 \\ 0 & 0 & P_4 & Q_4 \end{bmatrix} \begin{bmatrix} c_1 \\ c_2 \\ c_3 \\ c_4 \end{bmatrix} \qquad (4.12A)$$

or
$$\begin{bmatrix} M_1 & M_2 & M_3 & 0 \\ 0 & N_2 & N_3 & 0 \\ 0 & 0 & P_3 & 0 \\ 0 & 0 & P_4 & Q_4 \end{bmatrix} \begin{bmatrix} c_1 \\ c_2 \\ c_3 \\ c_4 \end{bmatrix}. \quad (4.12B)$$

In this case we refer to the components within c_1 are **physically relevant** parameters, those within c_2 are **ignorable** parameters, those within c_3 are **transient** parameters, and those within c_4 as **inaccessible** parameters. Systems where **T** has either of these types of structure can be dealt with by projection methods, where we need only know the form of M_1 and some information about M_2 and M_3 in order to track the behavior of the physically relevant parameters.

The marginal projective mapping, **Y** is said to be **evolution consistent** when:

$$YTY = YT$$

or (4.13A)

$$YAY = YA \, ;$$

and **evolution invariant** when

$$YTY = TY$$

or (4.13B)

$$YAY = AY \, .$$

Note that, since **Y** is a projection, commutativity of **Y** and **T** implies evolution consistency and invariance. Select out a basis, of dimension K, for the range of **Y**, list them at the beginning of the basis list and designate them physically relevant. In the case of (4.13A) we shall describe the associated linear space of ensembles as a **computable subspace**, while for (4.13B) it will be referred to as an **invariant subspace**. The dimension of computable subspaces, which is the number of linearly independent forms directly determined by (4.13A), is called its **complexity**. Call the other basis elements "secondary".

When transient ensembles are present, the nonzero matrix M_3 means that probability is flowing into the physically relevant space. Since the defining characteristic of transient components is that no probability flows into these components, then there are a set of characteristic times, and, in particular a maximum characteristic time past which negligible probability flows into the physically relevant ensemble space from the transient space. Those ensembles which lie in the range of **Y** can be considered "predictable" ensembles. Those which have transient components in them can be considered to have "noise sensitive" nucleation processes. The subspace of $\mathfrak{RP}$ with transient and inaccessible components set to zero is an invariant subspace on which (4.13A) or (4.13B) can be employed, depending on the form of (4.12). Algorithms for finding invariant or approximately invariant subspaces will not be considered here. Instead, we turn to computable subspaces.

Given the complete distribution, $c(t)$ in term of a new basis set at time t, We can represent **T** in the block form:

$$c(t + \Delta t) = \begin{bmatrix} M_1 & M_2 \\ N_1 & N_2 \end{bmatrix} c(t), \quad (4.14)$$

where blocks have K or $s^w - K$ rows or columns as appropriate. Then, (4.13A) is equivalent to $M_2 = 0$ while (4.13B) N_1 is equivalent to $N_1 = 0$ on the remainder of the basis elements outside the range of **Y**

The prototype of invariant subspaces arises from taking the linear space generated by all powers of **T** acting on an ensemble or set of ensembles. The simplest invariant spaces are those in which total concentrations of individual species are conserved, but these spaces can, also, be treated using computable subspaces. Invariant subspaces are closely related to ergodicity, which is discussed in the next section.

The prototype of computable subspaces consists of any ensemble space parametrized by coordinates generated by all linear combinations of powers of the adjoint of **T** applied to a linear form (or set of linear forms).[6] The relationship between possible computable spaces and probability superposition for marginal probabilities is apparent. We now focus on this type of construction.

From an output oriented point of view, one is seldom interested in the complete description of a system, but only seeks certain characteristics of the system. In stochastic systems, these are often determined by linear properties, often referred to as the expectation of a random variable. The free energy is one such linear form which is generated from marginal probabilities when the cluster expansion of the Hamiltonian (2.8) is used.[7] We can merely refer to such functions as members of the dual space to $\mathfrak{RP}$. Eqn (4.13A) describes the condition of closure starting from any set of linear forms needed for the modeling process. **Closure** means that all ensembles, including pure states, sharing the same coordinates, retain those coordinates throughout the evolution process. For many systems, and certainly for diffusional processes, marginal probabilities are much closer to output considerations than pure state joint probabilities. For instance, again using the example of the overall concentration of each species, $\sigma(i) = \Sigma_{\omega \in \Omega} c_{\Sigma_i}(\omega)$, obtained from the point probabilities of each species throughout Ω, one sees that each $\sigma(i)$ is invariant under the adjoint of **T**, so that the s dimensional space formed by all $\sigma(i)$ is also invariant. In this case the identification of dual computable spaces which are also invariant spaces is evident and affords a genuine mass-conserved reduction of **T**. With control variables describing a reservoir with no mass transfer, these are natural dimensional reductions with which to begin.

To study inhomogeneity, more information is needed. If one includes each of the $c_{\Sigma_i}(\omega)$ throughout Ω, closure does not generally occur. Adding higher order cluster components improves matters, but practical considerations usually force truncation of the expansion before closure is obtained. However, if the starting set could be modified and closure obtained, Eqn (4.13A) asserts that the secondary components of an ensemble have no effect upon the evolution of the physically relevant components of that ensemble. In such cases the full FCME represented by **T** can be replaced by a reduced linear system represented by

$$\mathbf{M}_1 = \mathbf{YTY} . \tag{4.15}$$

With a cluster expanded activation energy, this observation provides one criterion for cluster selection; the exact form of the cluster basis used will depend on **T**, but can be expected to be analogous to correlation type bases used in CVM.[2]

The remarkable feature of such dimensional reductions is that they retain the linear stochastic properties of the full FCME while simultaneously permitting descriptions of transition probabilities with both highly nonlinear constitutive laws and nonlocal dependence. For instance, if the physically relevant components of two ensembles are equal at time t, they will remain the same for all subsequent times when a representation of the type (4.13A) is valid. Indeed, the physically relevant components of any eigenvector of **T** (including those of equilibrium microstructures) are conserved in the eigenvectors of the Markov matrix, $\mathbf{M}_1$. On the other hand, **T** is faithfully reproduced on the range of **Y** when (4.13B) holds.

Techniques such as the Microscopic Master Equation (MME) or the Path Probability Method typically consist of a nonlinear and a linear part. In such systems the physically relevant components are chosen as correlations (coordinates) associated to certain localized clusters.

---

[6] Linear forms can be thought of as row vectors, in which case one merely computes powers of T and multiplies in from the left. Otherwise, for real T the adjoint is the transpose.

[7] The connection of the free energy to the probability flow functional, $\dot{F}$, should occur through the form of the activation energy.

Given the physically relevant components of the system ensemble at time t, the nonlinear part consists in estimating the remaining components to achieve a physically realizable ensemble in $\mathfrak{P}$; a process called **probability superposition**.[3] Such methods as the Kirkwood-Kikuchi-Barker (KKB) or maximum entropy approximations can be used in this case. Either of these methods yield an ensemble in $\mathfrak{P}$ whose physically relevant components have been fixed—this is typically a problem in linear convex analysis—however, any more information about the "exact" distribution at time t is, at best, statistical. While there is some justification for using the maximum entropy estimate (a multinormal distribution assuming a uniform distribution over the preimage), the KKB approximation (which coincides with maximum entropy for point probabilities) can only be justified on the grounds that the higher order cluster components don't have a significant effect on the evolution of the physically relevant components;*i.e.*, (4.13A) is approximately valid. Having estimated the ensemble in $\mathfrak{P}$, the FCME is applied to it, but only the change in the physically relevant components is calculated.

If these different superposition methods were to yield the same evolution path, Eqn (4.13A) would be implied, while if the cluster projected space were invariant under **T**, Eqn (4.13B) would hold. Unfortunately, neither of these contingencies is, generally, valid, although, as indicated approximations using larger clusters can be expected to better conform to (4.13A). The statistical errors introduced by truncated closure must be investigated case by cases. Despite this deficiency, we have seen that there is, ultimately, no need to use nonlinear superposition methods; cluster projection, which is a form of linear superposition, can be used instead. We can still write down an approximation to the full FCME by using the matrix **YAY**. Such a matrix is called a Clustered Master Equation (CME). The associated Markov matrix is

$$\mathbf{YTY} = \mathbf{Y} + \mathbf{YAY} = \mathbf{Y}(\mathbf{I} + \mathbf{AY}) = (\mathbf{I} + \mathbf{YA})\mathbf{Y}. \quad (4.16)$$

We illustrate these ideas in terms of the specific example shown in Fig. (3.5). Suppose we choose an $N{\times}N$ square array of atomic sites and specify cyclic boundary conditions for simplicity. The dimension of $\mathfrak{RP}$, is given by $w = 2^{N^2}$. This figure typifies w, which is finite in principle, but usually numerically "infinite". Let us denote the barycenter of $\mathfrak{P}$ by

$$_{\odot}\vec{P} = \frac{1}{2^{N^2}} \sum_{k=0}^{N^2} \binom{N^2}{k} {}^0\vec{P}_{kl} = \sum_{k=1}^{N^2} \frac{1}{2^k} \binom{N^2}{k} {}^0\vec{C}_{kl}, \quad (4.17)$$

where the pure state basis has been counted in a manner similar to (2.5). Symmetry implies that the coefficients of all translationally equivalent (with respect to the "activation energy" on Ω) cluster components of $\mathbf{T}\left({}_{\odot}\vec{P}\right)$ must be the same, so ${}_{\odot}\vec{P}$ is an (unstable) fixed point in the mass-conserved point probability approximation, but, usually not in higher order approximations. Here, we note that ${}_{\odot}\vec{P}$ is interior to $\mathfrak{P}$ as is every perturbation of ${}_{\odot}\vec{P}$. Thus, $\mathbf{T}\left({}_{\odot}\vec{P}\right)$ can be calculated using Eqn. (3.6) for any desired partial cluster basis, and then **YTY** calculated for any localized cluster projection operator. For instance, suppose we attempt a dimensional reduction based on all clusters contained in any localized nn/2nn square array of sites. Although the dimension of the space of all four point clusters is $\Sigma_{k=0}^{4}\binom{N^2}{k}$, the dimension of the space associated to square clusters is $10N^2+1$; still a large value, but tractable in some instances. If only pair correlations are used, the dimension is reduced to $5N^2+1$ (however, a valid comparison KKB approximation cannot be directly used on this space). To provide a detailed comparison to the MME, a space of larger dimension which exploits the full form of the local activation energy may be useful. (*cf* [5]) In that case—depending on simplifying approximations—the dimension is no more than $785N^2+1$.[8] However, the general observation that higher dimensioned approximations should be "better" requires that they are chosen wisely. It's perfectly possible in this example, for instance, that increasing the dimension produces extensive closure error among the "finer" states. Indeed, while we have postulated here that local cluster coordinates can be used, it may be that some or many global parameters are also required. The drop in dimension produced by such global constraints as fixed concentrations is usually nominal; and other modest dimensional reductions can be obtained from global symmetries involving both the local form of the transition probability and the geometry of the particular pure state. However, it is to be expected that selecting an "optimal" physical coordinate system for most applications will be a major technological hurdle. Typically, there is a variety of time and correlation scales associated to any complex system, and the selection of physical coordinates can be expected to be a bootstrap process.

Let the cluster projected basis of this space be $\vec{C}i$, $i = 1, \cdots, 10N^2 + 1$, and calculate:

$$(\mathbf{YT})_j^i = \langle C^i | \left[\mathbf{T}\left(_\odot \vec{P} + \varepsilon \delta_{j0}^k \vec{P}k\right) - \mathbf{T}\left(_\odot \vec{P}\right) \right], \quad \begin{array}{l} i = 1, \cdots, 10N^2 + 1 \\ j = 1, \cdots, 2^{N^2} \end{array} \quad (4.18)$$

$$(\mathbf{M_1})_j^i = (\mathbf{YTY})_j^i = \langle C^i | \left[\mathbf{T}\left(_\odot \vec{P} + \varepsilon \delta_j^k \vec{C}k\right) - \mathbf{T}\left(_\odot \vec{P}\right) \right], \quad i,j = 1, \cdots, 10N^2 + 1 .$$

Except for $i=1$, where the value of $\mathbf{T}\left(_\odot \vec{P}\right)$ is needed, either matrix is nonzero only for those i clusters one of whose sites is attached to an interchange pair which has an influence set completely containing the perturbed cluster or pure state (*cf* Eqn. (3.6)). Clearly, **YT** still represents an enormous computational challenge, but several means are available for calculating **YTY**.

In principle, the difficulty involves having to calculate conditional joint probability distributions (large cluster distributions) over the union of all influence sets affecting the perturbed cluster. However, in the linear superposition case, for instance, all subcluster probabilities of any influence set are fixed except the particular perturbed one, so we need only carry out linear superposition with respect to the individual influence set knowing that construction was given globally and must have a unique contraction to any subset; and this result can be found by the representation shown in (2.7) restricted to each influence set.[3] In the case of $_\odot \vec{P}$ it's particularly easy to see this because of randomness outside the influence set. If we use a nonlinear construction such as the KKB algorithm, for instance, which will work for the full square, [3]—which is why this example was chosen—then, again, because of the structure of the KKB approximation, the individual influence set approximations will be compatible with that over the whole of Ω. In the KKB case, this result can be used to give joint probabilities in $\mathfrak{P}$ even at starting points which lie on the boundary of $\mathfrak{P}$. The first order Clustered Master Equation giving **YTY** is, then, obtained from the first three terms in Eqn (3.6) by setting $S = \vec{C}_j$.

Both MME's and CME's provide global dimensional reduction. The concept of truncated closure means that reapplying superposition, as in the Microscopic Master Equation, is unnecessary. When the equilibrium, or "long time" states have been identified, they may have negative probabilities, but application of (4.5) using some form of superposition will identify a valid state which is determined only up to its reduced dimension cluster expansion. This is required because of the lack of naturally invariant subspaces.

---

[8] The exchange probabilities are determined by the horizontal and vertically oriented influence sets. If one seeks to determine the physical coordinates required to compute the exchange probabilities with no approximations, it is not necessary to use the full $7680N^2+1$ dimensional space which describes both horizontally and vertically oriented influence sets less overlap. Using the activation energy form given in [5], and looking at the horizontal influence set, there can be 0, 1, 2, 3 or 4 A atoms from the 2nn diagonal square and 0, 1, 2 or 3 A atoms from the nn positions for each of the two atoms in the interchange positions. Taking account of the fact that two 2nn positions are also nn positions, rules out 6 combinations from the 20 possible for each of two sets of left-right nn/2nn configurations. This gives 196 coordinates obtained by summing over marginal probabilities in an influence set with an AB configuration at the center which are required to determine the jump probability; and 196 coordinates for those with a BA configuration at the center. The vertically oriented influence sets require no more than $392N^2$ additional coordinates (since there may be some dimensional reduction among the vertical and horizontal coordinates). Including the single point probabilities, to be able to find the pair probabilities, gives the dimension sited. It may be required to add other coordinates: For example, particular Fourier components of global sums over pair probabilities to provide details of evolution of short-range order spectra in kinematic approximation.

In carrying out the perturbation analysis or choosing the initial point for a MME, care must be taken to start at the *a priori* weighted barycenter of all pure states (including zero probability for excluded pure states) and to include all, but no more than, the perturbations compatible with any constraints imposed by the controls. For example, assuming species conservation and starting from the pure state with all sites occupied by B atoms, say, begins from a simple fixed point of **T**, as there are no perturbations available which preserve the global concentrations. However, even without species conservation, adding all perturbations of the type indicated in (4.18) would give preimages involving only a "small" subspace of $\Re\mathfrak{P}$.[9]

To get some physical feeling, consider a CME operating on an initial state near that of the random mixture. At time zero we have the most information about the system state, and we have, also, the complete CME eigenfunction expansion of the ensemble. If the CME is the FCME, the information is "perfect", and we can tell exactly how many of each kind of pure state exists in the ensemble. Such information is never available in large systems. In fact, if we had it and could predict behavior with absolute certainty, we would not need ensemble theory. We can only expect the initial state to be given by a set of physical parameters, each of which describes a hyperplane in $\Re\mathfrak{P}$. The intersection of these hyperplanes with $\mathfrak{P}$ determines the polytope characterizing all of our information about the ensemble; in fact the ensemble may as well be taken as the uniform superposition of all pure states inside the polytope. If (4.13A) is valid, these ensembles will remain coherent. Since there only a finite set of pure states, only a finite set of possible values can be taken by the physical parameters without using the linear algebra of $\Re\mathfrak{P}$. We can relate to this through the dual device of *projecting in $\Re\mathfrak{P}$ through what we don't know onto what we do know*—a form of Occam's razor—to give the intersection of a dual linear space in $\Re\mathfrak{P}$ with $\mathfrak{P}$. With increasing time the number of eigenfunctions representing the ensemble decreases as does our information about the system state. The dimension of the polytope describing what we don't know increases accordingly. The CME eigenfunction expansion is an alternate characterization of this polytope; however the system occupies only one point in the reduced dimension space associated to the CME (called the **CME space**).

Consider first the eigenfunction components associated to relatively rapid decay. We can imagine in our example that even though a global free energy doesn't exist, a spatially localized free energy as a function of the "frozen" state of the surrounding medium can be assumed, *i.e.* there is a minimum neighborhood size—the **ergodic radius**—for which a localized ergodic hypothesis is valid. This radius is not constant with temperature, but, rather, decreases with decreasing temperature, and if Ω is small enough *vis-a-vis* the ergodic radius, **T** will be irreducible. Lack of space precludes a figure, but one can crudely envision nuclei of statistically independent local ordering occurring at spatially separated locations together with complex intermediate regions. In terms of the Master Equation, these short term fluctuations can be thought of as connected to the rapidly decaying eigenfunctions of **A**. The global basis representation given by the cluster expansion keeps track of fine graining issues. These fluctuations lower localized free energies. This is not an "elliptic" system nor is it a global process. Stated another way: Associated to any perturbation of the system there is a change in the local free energy function, but in general, the system is not a gradient system; that is, the free energy differential is anholonomic. However, after the decay of the "Jordan block terms", the "free energy gradient" will exist and have an eigenfunction expansion given by Eqn (4.11).

Returning to the probability picture, note the CME is not a mean field approximation, but rather one of truncated closure, where the linearity arises from the stochastics, not the physics. In the full pure state space $\mathfrak{P}$, an initial ensemble, $\vec{P}_0$ is set out—in our example, $_\odot \vec{P}$, which contains all pure states with equal probability and reflects the assumption of equal *a priori*

---

[9] It's important to recognize, here, that statistical superposition is not the same as physical superposition. So, an ensemble with a pure state containing A atoms at both sites one and two is not the same as the superposition of ensembles which have A atoms only at site 1 or A atoms only at site 2.

probability. With other *a priori* knowledge, such as concentrations, an alternate starting point should be chosen. Perturbing a coordinate set basis element by a perturbation compatible with any system constraints and employing linear superposition determines a nearby element of $\mathfrak{P}$, *i.e.* a different distribution of pure states. The perturbation lies in $\Xi\mathfrak{P}$ and has the interpretation modifying the ensemble by trading the number of representatives of some of the pure states in $\vec{P}_0$ with other pure states. Each of the set of pure states in the ensemble has its own differential ensemble evolution vector (which is a column of $\mathbf{T}$ in the pure state representation). The CME determines the effect of the change in pure state occupancy on the evolution vector for each pure state component *individually*, and *then* superimposes them. The calculation does not employ the mean field approximation which uses only the reduced coordinates independent of the inverse images in $\mathfrak{P}$. The trick utilized in the example lies in the use of localized cluster coordinates combined with a localized influence set to make the CME matrix more easily computable. In so far as two different superpositions into $\mathfrak{RP}$ give rise to different evolution expressions among the physically relevant coordinates, closure does not hold and statistical errors occur. In this case the CME is some linear "approximation" to a nonlinear system; but, unless the correct superposition formula is known (which is tantamount to "perfect knowledge" in the spirit of Laplace), it is impossible to determine the correct linear approximation; Clustered Master Equations as linear approximations to Microscopic Master Equations comprise another alternative.

In the physical model reference was made to the frozen state of the surrounding medium. That effect is ignored in the localized influence example used in this paper. Long range effects, particularly stresses, need to be included for most technologically important systems. Mean field approximations can be attempted, but lead to nonlinear evolutionary equations which fail at bifurcations, while attempting to expand the state space by including atomic displacements is a major undertaking. The most promising approach is to calculate the stress field at each interchange site location throughout Ω for each pure state—anything from MD to mean field approximations based on a state-dependent tensor Green's function can be employed here—and to express the transition probability as a function of the influence set and the interchange site stress. As above, a perturbation in the cluster coordinates modifies the probability distribution, and at those few interchange sites which are directly affected by the perturbation, the modified interchange probabilities, which involve both influence set and stress effects, can be calculated. In addition to these changes, however, the cluster perturbation produces a different stress throughout Ω which can be approximated using a multipole expansion with the Green's tensor for the unperturbed state. Then, the effect on the interchange probabilities at *all* other interchange sites in Ω have to be taken into account and the evolution vector change calculated. Despite the local perturbation, this is a global problem; efficient algorithms for such calculations are beyond the scope of this paper.

After the decay of the short-time eigenfunctions: if a Monte Carlo simulation were performed, we would observe one example of localized ordering lying on a "narrow" path in $\mathfrak{P}$ as determined by the characteristics of the Monte Carlo algorithm. If a Microscopic Master Equation simulation is performed, we would be following one specific path through physical parameter space, and would reach after the decay time a new state containing the same amount of information as the original state, since at each time step we chose a particular superposition with which to choose the next step. In the CME space we still have all the long term information about the physical parameters coded in the remaining eigenfunction expansion, but we have *no* information about the short term characteristics, since this information has decayed away—the MME, on the other hand, chose one arbitrary set of coefficients for the short term eigenfunctions. In the former case the dimension of the subspace about which we have information decreases, while that about which we have no information increases in a complementary fashion. In the latter case this distinction is lost.

Physically, after the initial transient period, eigenfunctions of **A** characteristic of longer range diffusion become more dominant. Localized order will be present to a greater or lesser degree. Ordinary bulk diffusion is less likely to occur in locally ordered systems. Instead, we can imagine bulk diffusion occurring through "surface diffusion" around the boundaries of ordered microregions characterizing the different states making up the ensemble. The ensemble will continue to evolve with any given boundary localized subdistribution further subdividing (depending on the reducibility of **T**) as longer range correlations develop and more pure states are excluded. The system continues to become more elliptic near each equilibrium microstructure. Ultimately, a free energy function is definable for states localized to an individual equilibrium microstructure; with several such microstructures the free energy can be extended linearly between them, but if the system is not ergodic, there is no physical way to compare the free energies of different microstructures, and without further information all such free energies can be taken equal. The number of equilibrium microstates associated with any ensemble is, of course, obtained from its projection onto the null space of **A**.

In the dual probability picture the amount of information continues to erode until only the long term information associated to the null space of **A** remains. If the system is irreducible, this is a one dimensional space in $\Re\mathfrak{P}$ giving rise to a equilibrium ensemble in $\mathfrak{P}$. All other information is gone. If the system is reducible, the appropriate intersection with $\mathfrak{P}$ is the equilibrium polytope whose vertices are the possible equilibrium microstates lying in different facets of $\mathfrak{P}$. This picture can be almost indistinguishable from the irreducible one when the time has not been extended long enough to reach equilibrium.

This Master Equation formalism rests on the invariance of the linear form, $\langle\Sigma|$, during both change of coordinate system and projections. A framework is, thereby, provided for describing "transient" modes. When the coefficients in **T** are sufficiently constant that an equilibrium polytope can be identified, an alternate formalism can be set up using the vertices of the equilibrium polytope as "pure states". In this case one can demand the invariance of the natural quadratic form associated to the equilibrium microstates, and one can incorporate global symmetries among the classes of equilibrium microstates. Unitary changes of coordinate system are natural in this context because of the "enduring character" of the equilibrium microstates, but the interpretation of transient states will be artificial.

All previous inhomogeneous modeling methods of which the author is aware use only point probability type cluster coordinate systems. Although a global free energy function is often postulated in these models, multiple microstructures are obtained which are not ergodically connected. In these methods the path to the final state is typically ill-conditioned, so that although "deterministic" field equations describe the evolution of the point probability density, the final outcome is not unique; numerical and other uncertainties lead to unpredictable bifurcations. This suggests that the basins of attraction are extraordinarily complicated, being in some sense an approximate and degenerate mapping of the CME on the computable cluster space. Although the situation may be simpler on ensembles which are characteristic of the later stages of evolution, it has been shown in homogeneous Path Probability computations that, even in irreducible situations, point probability formulations are inadequate to lead to the correct equilibrium structure.

5. Ergodic Mappings and Time Series

One form of ergodicity asserts that if **T** is irreducible,

$$E_N = \frac{1}{N}\sum_{k=1}^{N} T^k \Rightarrow \lim_{N\to\infty} E_N \vec{P} = {}_\infty\vec{P}, \tag{5.1}$$

${}_\infty\vec{P}$ is the equilibrium ensemble, which is a fixed point of **T**.[7] Equation (5.1) is interpreted as any ensemble vector passing arbitrarily near any other system state during system evolution with multiple passbys for the more probable states. But, there is another way of viewing this result which is consistent with a kinetic viewpoint and doesn't require the ergodic hypothesis: Using the more modern theory of time series, Eqn (5.1) is equivalent to the statement that the FFT of an N-term time series with all ones is $N\delta(\omega)$, where $\delta(\omega)$ is the zero frequency component of the time series.

One can think of the sequence:

$$\mathbf{I}, \mathbf{T_1}, \mathbf{T_2T_1}, \mathbf{T_3T_2T_1}, \cdots$$
$$\mathbf{T_i} \equiv \mathbf{T}(\phi(k\Delta t)) \tag{5.2}$$

as an s^w by s^w matrix set of time series windows, one for each component of the matrix product matrix shown in (5.2); so that each component represents an individual time series. Additionally, the algebra of polynomials in $\mathbf{T}$ is formally identical with that of z transforms—actually $y \equiv z^{-1}$ transforms.

Define:

$$\mathbf{T_0} \equiv \mathbf{I}$$
$$\mathbf{F_i} = \sum_{k=0}^{K} a_k \mathbf{T_i}^k , \tag{5.3}$$

where $\{a_k\}$ is a sequence of real or complex numbers to be discussed shortly. The sequence:

$$\mathbf{F_0I}, \mathbf{F_1T_1}, \mathbf{F_2T_2T_1}, \cdots, \mathbf{F_iT_iT_{i-1}}, \cdots, \mathbf{T_1} \tag{5.4}$$

forms a new time series which can alternately be thought of as a modified time series of locally ergodic states, or as a digitally filtered description of the system evolution.

In order to get a physical interpretation of the time series in (5.4), assume that the coefficients of $\mathbf{T_i}$ are constant over the span of $\pm K\Delta t$, i.e. the coefficients of $\mathbf{T}$ are changing slowly vis-a-vis Δt. Then $\mathbf{F_i}$ commutes with any power of $\mathbf{T_{i\pm k}}$, $k \leq K$. Depending on the nature of $\{a_k\}$, the time series in (5.4), then, has different interpretations. However, in all cases one can consider that there is a local superposition of states associated to each of the matrix windows. It is important to note that even if the coefficients of $\mathbf{T_i}$ do not change with time, the "moving picture" in each matrix window may have a very intricate time behavior. Suppose we choose the sequence $\{a_k\}$ to be real with $K=2L$, where

$$a_{L-l} = a_{L+l}, \ 0 \leq l \leq L$$
$$\sum_{k=0}^{2L} a_k = 1,$$
$$\tilde{a}_\kappa = \begin{cases} 1 & \kappa \leq K_0 \\ 0 & \kappa > K_0 \end{cases}, \tag{5.5}$$

where $\{\tilde{a}_\kappa\}$ refers to the discrete Fourier transform with base $K^* \gg K$ of $\{a_k\}$ with extensive zero padding (K^*-K zeroes added on the end), and K_0 lies between zero and K^*. Then, since the frequency spectrum of $\mathbf{F}^2$ derives from the convolution of the time series, and is the square of the (real) frequency spectrum of $\mathbf{F}$, each of the $\mathbf{F}$'s are projections mapping time sequences from $\Re\mathfrak{P}$ with arbitrary coefficients to those with "low frequency" coefficients, and we are dealing with consistent marginal mapping over a time extended state space. If all $a_k \geq 0$, we have consistent marginal projections.

The technology for highly efficient low-pass filters is very advanced.[8] The coefficients of such a filter satisfy (5.5); indeed almost all are positive with only very small negative terms in the tails. The frequency spectrum of a well designed filter is essentially one up to a cutoff value; then has a very steep descent; and is essentially zero over the remaining set of discrete frequencies. The lower the cutoff frequency, the longer the filter must be to achieve any fixed degree of precision. Consequently, except for the narrow transition frequency band, **F** behaves like a projection which commutes with **T**, and thus approximately satisfies both conditions in (4.13). If one uses functions $\mathbf{F_i}$ derived from such low pass filters, the associated marginal mappings will be referred to as **ergodic mappings**. The preimages of these mappings approximately consist of all ensemble time series with high frequency fluctuations of the given low frequency projected image involving states within a given time window of the image state. As is known in the time series field, such series can be **decimated** by increasing the time step by a factor associated to the cutoff frequency of $\{a_k\}$.[9] If $\mathbf{T_i}$ is independent of time, it is clear that not only should there be fewer time steps, but the dimension of the effective state space should be reduced.

6. Process Control

From the Markov point of view presented in this paper, the goals of process control can be summarized as:

> **I.** To find processing regimes, $\vec{\phi}_P(t)$, which optimize the probability (maximize production yield) of producing high performance microstructures.
> **II.** To predict the risk of structural breakdown (stability) of potentially important microstructures as a functional of the environmental conditions, $\vec{\phi}_E(t)$, in which the material will be used.

The product

$$\mathbf{T_k T_{k-1}}, \cdots, \mathbf{T_1} \,_0\vec{P} \qquad (6.1)$$

represents the expected probability of each of the states p_i (where the p_i's here refer to any basis in which the matrices are expressed) at time $k\Delta t$ given the initial probabilities at time 0 expressed by $_0\vec{P}$. Eqn (6.1) contains the ensemble probability information on both macroscopic nucleation and growth— microscopic aspects are coupled to the effects of noise occurring in each Δt interval. When no new information is added during the time interval $(0, k\Delta t)$, this must serve as the basis for nonlinear stochastic control, the nonlinearity arising from the dependence of **T** on $\vec{\phi}(t)$.

If new information is added, or if one merely wants to attempt to circumvent the issue of large variance, the probability distribution of the current state can be revised. Realistic algorithms require mathematical statistics; however, in the ideal case, the information theory method resting on the Bayesian hypothesis in pure state coordinates provides one formula for updating:

$$_\star \vec{P} = \max_\Theta \left(\sum_{i=1}^{s^w} -p^i \log \left(_0 p^i(k)\, p^i\right) \right), \qquad (6.2\text{A})$$

$$p^i \geq 0$$

where

$$_0 p^i(k) = (\mathbf{T_k T_{k-1}}, \cdots, \mathbf{T_1} \,_0\vec{P})^i \qquad (6.2\text{B})$$

is the input from the Markov process and Θ consists of a set of—hopefully, linear—constraints[10] of the form

$$\alpha^i_j p^j = \beta^i, \qquad (6.2\text{C})$$

---

[10] One clear case where linearity does not obtain occurs when incorporating results from diffraction intensity; in that case, even in kinematic diffraction, the constraint is quadratic.

incorporating the new information. Unfortunately, one cannot simply change the entropy to other than the joint probability coordinate system because of the nonlinear logarithm term.

It's informative to illustrate the use of Eqn (6.2) in the context of constant **T**. If one applies (6.2) at the outset ($t=0$) and uses the cluster form of the configurational energy together with the overall concentrations as the only input information, one obtains the classical statistical mechanics expression for equilibrium, which in a binary material takes the form:

$$\begin{aligned}
_\infty \vec{P} &= \min_{\vec{P}} \left[\left(\Sigma_i p^i \ln \left(_l p^i p^i \right) \right) - \frac{1}{kT} \Sigma_i p^i E_i \right] \\
&= \min_{\{\Sigma | \vec{P} = 1, \, p^i \geq 0\}} \left[\frac{1}{kT} \Sigma_i p^i \left(\ln \left(_l p^i p^i \right) - E_i \right) \right] \\
&\quad \sum_{l=1}^{w} c^{1l} = \alpha \\
&= \min_{\{\Sigma | \vec{P} = 1, \, p^i \geq 0\}} \left[\frac{1}{kT} \Sigma_i p^i \left(\ln \left(_l p^i p^i \right) - \eta_i^j \mathcal{E}_j \right) \right] \\
&\quad \sum_{l=1}^{w} c^{1l} = \alpha
\end{aligned} \tag{6.3}$$

$_l p^i$ are the initial state probabilities (only needed in the reducible case), α is the A concentration obtained from point probabilities, and the double index on the c's puts the cluster size first. There are no approximations in this formula, only the use of the cluster expansion for the configurational energy. As previously discussed (6.3) is, generally, only valid when the Markov matrix **T** is irreducible.

7. Summary

- Because the ergodic hypothesis doesn't hold in most solid systems, the global free energy must be considered a mathematical abstraction which can be used as a minimizing reference for the actual phase-space-localized free energies characterizing equilibrium microstructures. However, a spatially localized free energy as a function of the frozen state of the surrounding medium can often be assumed, so that local ergodicity holds giving rise to a minimum neighborhood size—the **ergodic radius**—for which the ergodicity holds.

- The kinetics of diffusional transformations on discrete lattices can be modeled using Master Equation (or Markov) techniques accessible to first principles calculations. The Master Equation describes system evolution in an ensemble space of conceptually finite, but numerically infinite dimension; it yields the expected probabilities for any input ensemble as a function of time, and includes macroscopic nucleation and metastability. The number of nonzero independent terms needed to describe any ensemble (system complexity) generally drops dramatically as the system evolves.

- System evolution is subject to (possibly time varying) controls which are insensitive to the internal system state. When such parameters have a frequency content involving frequencies whose inverse gives times far greater than the discretization time, Δt, they yield a generalization of the ideas of "bath" and Carnot cycles used in thermodynamics.

- The powerful technique of marginal probability coordinates gives rise to cluster expansions as found in such theories as the Cluster Variation Method and Connolly-Williams expansion method.

- The Jordan decomposition of the time-invariant Master Equation matrix, **A**, yields a basis of modes all of which decay with time except for the null eigenspace, which characterizes the equilibrium microstructures. The dimensional reductions required for applications are expressed in terms of marginal projections and marginal projective mappings. In the Jordan case, marathon mappings are those which discard the "transient" modes to follow the low frequency response of the system. For ensembles containing no Jordan block terms, a free energy gradient function can be defined. Its eigenfunction expansion is presented.

- The concepts of computable and evolution invariant subspaces are introduced in terms of partial Jordan decomposition. Computable subspaces can be interpreted in terms of linear functions on probability space, and the process of dimensional reduction as projecting through what we don't know onto a set of physically relevant coordinates.
- It is suggested that cluster projection mappings, using clusters beyond point probabilities, can be used to achieve a global dimensional reduction and produce a purely linear stochastic formulation called a Clustered Master Equation (CME) forming an alternative—known to be correct at sufficiently high dimension—to the Microscopic Master Equation, which is the only other valid statistical alternative.
  - CME's are linear Markov equations with possibly negative terms operating in a cluster projected space where the high order cluster components are zero.
  - By making use of advanced computational capability, a CME maps out every contingency to the degree of detail afforded by the physically relevant coordinates.
  - CME's approximate the full FCME in the sense of Eqn (4.5) which says that the CME approximation has the same low order cluster coordinates as the exact FCME solution.
  - Because of probability superposition, CME's are not mean field. In so far as two different superpositions into $\mathfrak{RP}$ give rise to different evolution expressions among the physically relevant coordinates, closure does not hold and statistical errors occur.
  - The physically relevant components of any eigenvector of **T** (including those of equilibrium microstructures) are conserved in the eigenvectors of a CME with closure.
- An example is developed in conjunction with binary systems on a square lattice with nn/2nn interactions. The CME for this system is calculated from perturbations near the uniform random mixture ensemble—which is the barycenter of the probability hypersimplex, $\mathfrak{P}$.
- Signal processing techniques can be used on the polynomial algebra generated by the powers of the Markov transition matrix. Approximate marginal projective mappings can be constructed which isolate various frequency bands. These low frequency filters permit dimensional reduction and time decimation and comprise a generalization of the concept of ergodicity. They are used to construct ergodic mappings into invariant subspaces.
- Using the Perron-Frobenius theory discussed in the Appendix, it has been shown how the reducibility of a Markov transition matrix constant in time leads to a split in the set of equilibrium states. For diffusive ordered systems, irreducibility and homogeneity undoubtedly hold for small systems with periodic boundary conditions; this has been numerically observed by L.Q. Chen.[10] In larger systems without the artificial restrictions of homogeneity, and even more when periodic boundary conditions are relaxed, as, for instance, in the modeling of microstructure evolution in thin films, it is expected that inhomogeneous equilibrium microstates will be obtained, and their relative likelihood summarized for stochastic control by the Markov chain method.
- This theory addresses stochastic control for processing materials undergoing diffusive ordering and phase transformations. Here the goals are: (i) To find processing regimes which optimize the probability of producing high performance microstructures, and (ii) To predict the risk of structural breakdown of potentially important microstructures as a functional of the environmental conditions.

8. Acknowledgments

The author wishes the thank the ARPA Applied and Computational Mathematics Program for continuity of support. This project would never have been attempted without the insights and inspiration of John Cahn and Ryoichi Kikuchi, and would have proceeded at a far slower pace without the prodding and numerical grounding provided by L.Q. Chen. Thanks, also, to Oliver Penrose and Pierre Cenedese for valuable discussions, and to S.R. Coriell, G.B. McFadden and Y. Oono for their comments on the manuscript.

9. References

1. *cf* S.A. Tretter, Introduction to Discrete-Time Signal Processing, Wiley (1976)
2. J.M. Sanchez, F. Ducastelle and D. Gratias, "Generalized Cluster Description of Multi-component Systems", Physica, **128A**, pp. 334–350, (1984)
3. J.A. Simmons, "On the Superposition of Probabilities", in Proceedings of the International Workshop on the Theory and Applications of the Cluster Variation and Path Probability Methods, (6/18–22/95) To be published.
4. J.W.D. Connolly and A.R. Williams, "Density-functional theory applied to phase transformations in transition-metal alloys", Phys. Rev. B, **27** (1983), pp. 5169–5172
5. L.Q. Chen and J.A. Simmons, "Microscopic Master Equation Approach to Diffusional Transformations in Inhomogeneous Systems—Single-Site Approximation and Direct Exchange Mechanism", Acta Met, **42**, #9, (1994), pp. 2943–2954
6. M.W. Hirsch & S. Smale, Differential Equations, Dynamical Systems, and Linear Algebra, Academic Press (1974)
7. S.L. Campbell and C.D. Meyer, Jr., Generalized Inverses of Linear Transformations, Chap. 8, Pitman (1979)
8. J.H. McClellan, T.W. Parks, and L.R. Rabiner, "A computer program for designing optimum FIR linear phase digital filters", IEEE Trans. Audio Electroacoust., **AU-21** (1973), pp. 506–526
9. R.E. Crochiere and L.R. Rabiner, "Optimum FIR Digital Filter Implementations for Decimation, Interpolation, and Narrow-Band Filtering", IEEE Trans ASSP, **ASSP-23** (1975), pp. 444–464
10. L.Q. Chen, Private Communication
11. G. Birkoff and S. MacLane, A Survey of Modern Algebra, Chapt. 10, MacMillan (1953)
12. A. Graham, Nonnegative Matrices and Applicable Topics in Linear Algebra, Halsted Press (1987), Gershgorin Theorem is in transposed form.
13. A. Berman and R.J. Plemmons, Nonnegative Matrices in the Mathematical Sciences, Academic Press (1979)
14. H. Flanders, Differential Forms, Academic Press (1963)
15. O. Penrose, Foundations of Statistical Mechanics, Pergamon Press (1970)
16. L. Stratonovich, Nonlinear Nonequilibrium Thermodynamics, Vols. I and II, Springer-Verlag (1994)
17. R.D. Levine, "The Theory and Practice of the Maximum Entropy Formalism", in Maximum Entropy and Bayesian Methods in Applied Statistics, J.H. Justice, ed., Cambridge Univ. Press (1986), pp. 85–94
18. T. Beelen & P. VanDooren, "Computational aspects of the Jordan canonical form", in Reliable Numerical Computation, M.G. Cox & S. Hammerling, eds. Oxford Science Publishers (1990)

Appendix: Equilibrium and the Perron-Frobenius Theorem

In this appendix it is assumed that $\vec{\phi}(t)$ can be considered so that T defines a constant Markov chain. The goal is to analyze the fixed points of T and to connect them with the customary ideas of equilibrium states in classical statistical mechanics.

Some important theorems from matrix theory essential for understanding the probability flow process described by the Master Equation follow:

I. **Jordan Canonical Form**[10]: Let H_j^i be any $n \times n$ matrix with complex coefficients expressed in the pure state coordinate system. Then there exists a new "generalized normal mode" coordinate system,

$$^0\vec{Q}_j = U_j^i \, ^0\vec{P}_i$$
$$^0\vec{P}_j = (U^{-1})_j^i \, ^0\vec{Q}_i , \qquad (A.1a)$$

so that, in the new coordinate system, where H has the form:

$$\mathbf{H_U} = \mathbf{U^{-1} H U} . \qquad (A.1b)$$

$\mathbf{H_U}$ consists of a set of Jordan blocks;

$$\begin{pmatrix} \mathbf{J}_1 & 0 & \cdot & & \cdot & 0 \\ 0 & \mathbf{J}_2 & & & & \cdot \\ \cdot & & \cdot & & & \cdot \\ \vdots & & & \cdot & & \\ & & & \mathbf{J}_{b-1} & 0 \\ 0 & \cdot & \cdot & & 0 & \mathbf{J}_b \end{pmatrix}, \qquad (A.1c)$$

where each of the Jordan blocks has the form:

$$\mathbf{J}_s = \begin{pmatrix} \lambda_s & 0 & \cdot & & \cdot & & 0 \\ 1 & \lambda_s & 0 & & & & \cdot \\ \cdot & 1 & \lambda_s & \cdot & & & \cdot \\ \cdot & 0 & & \cdot & & & \cdot \\ \cdot & & & \cdot & \lambda_s & 0 & 0 \\ \cdot & & & 0 & 1 & \lambda_s & 0 \\ 0 & \cdot & \cdot & & 0 & 1 & \lambda_s \end{pmatrix}, \qquad (A.1d)$$

with the **eigenvalue**, λ_s,[11] a root of the characteristic polynomial for H obtained from the determinant equation:

$$|\lambda \mathbf{I} - \mathbf{H}| = 0 . \qquad (A.1e)$$

The total number of times λ_s appears in the Jordan representation of H is equal to its root multiplicity in the characteristic Eqn (A.1e), and if all the coefficients of H are real, then the eigenvalues occur in complex conjugate sets with associated complex conjugate eigenvectors.

---

[11] In conventional matrix notation the ones appear above the diagonal.

II. **Cayley-Hamilton Theorem**[6,11]: If the characteristic polynomial of Eqn (4,1E) is written in polynomial form:

$$\chi(\lambda) \equiv |\lambda \mathbf{I} - \mathbf{H}| = \sum_{k=0}^{n} a_k \lambda^k, \text{ where } a_n = 1, \text{ then:}$$

$$\sum_{k=0}^{n} a_k \mathbf{H}^k = \mathbf{0}.$$

(A.2)

This result is a direct consequence of the Jordan representation given above.

III. **Gershgorin Theorem**[12]: Let **H** be any $n \times n$ matrix with complex coefficients, and let

$$\Lambda_j \equiv \sum_{\substack{i=1 \\ i \neq j}}^{n} |h_{ij}|, \quad j = 1, \cdots, n.$$

(A.3a)

Then, all the eigenvalues, λ, of **H** lie in the union of the disks:

$$|z - h_{j,j}| \leq \Lambda_j, \quad j = 1, \cdots, n.$$

(A.3b)

IV. **M matrices**[12 (Lemma 5.4),13]: These are nonsingular matrices with all negative off-diagonal elements and positive diagonal elements. This is very similar in form to the negative of the Master Equation matrix, **A**, except that **–A** has some zero eigenvectors. The result we need is that all principal minors of any order of a nonsingular **M** matrix are positive.

V. **Multivector probability flow and the generalized Liouville's Theorem**: The minors of a matrix are connected to its Grassmann algebra, which in modern algebraic terminology is called the exterior algebra. The matrix **A** on $\Re\mathfrak{P} \otimes \Re\mathfrak{P}^\dagger$, for example, lifts to the **compound matrix** $\bigwedge^p \mathbf{A} : \bigwedge^p \Re\mathfrak{P} \otimes \bigwedge^p \Re\mathfrak{P}^\dagger$, where $\bigwedge^p \Re\mathfrak{P}$ ($\bigwedge^p \Re\mathfrak{P}^\dagger$) is the $\binom{\aleph}{p}$ dimensioned space of p–vectors (p–forms) over $\Re\mathfrak{P}$, with $\bigwedge^1 \mathbf{A} = \mathbf{A}$. If $(\bigwedge^p)_\mathbf{i} = {}^0\vec{P}_{i_1} \wedge \cdots \wedge {}^0\vec{P}_{i_p}$ and $(\bigwedge^p)_\mathbf{j} = {}^0\vec{P}_{j_1} \wedge \cdots \wedge {}^0\vec{P}_{j_p}$, then if $\mathbf{i} \neq \mathbf{j}$, $(\bigwedge^p \mathbf{A})_\mathbf{j}^\mathbf{i}$ has the physical interpretation of $(-1)^p$ times the probability flow from the simplex generated by ${}^0\vec{P}_{j_1}, \cdots, {}^0\vec{P}_{j_p}$ into the smallest simplex generated by ${}^0\vec{P}_{j_1}, \cdots, {}^0\vec{P}_{j_p}, {}^0\vec{P}_{i_1}, \cdots, {}^0\vec{P}_{i_p}$, while if $\mathbf{i} = \mathbf{j}$, $(-1)^{p-1}(\bigwedge^p \mathbf{A})_\mathbf{j}^\mathbf{j}$ gives the total probability flow out of the simplex generated by ${}^0\vec{P}_{j_1}, \cdots, {}^0\vec{P}_{j_p}$. This interpretation, which is a finite dimensional generalization of Liouville's Theorem, follows directly from the interpretation of **A** and the definition of the compound matrix given in Flanders monograph.[14]

Applying the Gershgorin Theorem to **A**, gives the remarkable result that if λ is an eigenvalue of **A**, $\lambda = \mathrm{Rl}(\lambda) + i\,\mathrm{Im}(\lambda)$, then

$$\exists \varepsilon > 0 \ni |\lambda + \tfrac{1}{2} - \varepsilon| \leq \tfrac{1}{2} - \varepsilon \text{ so that } \begin{array}{l} -1 < \mathrm{Rl}(\lambda) \leq 0, \\ |1 + \lambda| \leq 1. \end{array}$$

(A.4)

This, is turn, implies that the eigenvalues of **T**, which (as is easily seen in any triangular representation) are 1 plus those of **A**, all lie in the unit circle and have positive real part, so that **T** is, also, nonsingular.

Ergodic properties: Define an ordering relation among the s^w basis vectors by:

$${}^0\vec{P}_i \leq {}^0\vec{P}_j \iff V\left({}^0\vec{P}_i\right) \supseteq V\left({}^0\vec{P}_j\right).$$

(A.5)

This transitive ordering relation contains the graph theory arguments usually used to discuss irreducibility. Here $V\left({}^0\vec{P}_i\right)$ is the set of pure state basis vectors with nonzero components in $^0\vec{P}_i, \mathbf{T}\left(^0\vec{P}_i\right), \mathbf{T}^2\left(^0\vec{P}_i\right), \cdots, \mathbf{T}^{s^w}\left(^0\vec{P}_i\right), \cdots$. The Cayley-Hamilton theorem implies that the linear space generated by the powers of $\mathbf{T}$ has some dimension, $n_i \leq s^w$, with a basis: $^0\vec{P}_i, \mathbf{T}\left(^0\vec{P}_i\right), \mathbf{T}^2\left(^0\vec{P}_i\right), \cdots, \mathbf{T}^{n_i}\left(^0\vec{P}_i\right)$. Because of (3.4), n_i has the same value for both $\mathbf{T}$ and $\mathbf{A}$.

The elements of $\mathbf{V}\left(^0\vec{P}_i\right)$ are found from the monotone increasing set of elements with nonzero values lying in the ith column for $^0\vec{P}_i$ in the matrix $\mathbf{T}^k$ as k goes from zero to n_i. This sequence is monotonic as seen by looking at matrix multiplication for a matrix with nonzero diagonals, and all of whose entries are non-negative. As soon as state j appears (nonzero component in the jth row of the ith column), it will appear in the next iteration of $\mathbf{T}$ and remain thereafter. Since all the powers of $\mathbf{T}\left(^0\vec{P}_i\right)$ are linearly dependent on the first $n_i + 1$ powers (including zero), the "development" of $\mathbf{V}\left(^0\vec{P}_i\right)$ stops after $\mathbf{T}^{n_i}\left(^0\vec{P}_i\right)$, that is, $\mathbf{V}\left(^0\vec{P}_i\right)$ consists of all states which have nonzero entries in the ith column of $\mathbf{T}^{n_i}\left(^0\vec{P}_i\right)$. Also, if $^0\vec{P}_j$ appears in $\mathbf{V}\left(^0\vec{P}_i\right)$, then $\mathbf{V}\left(^0\vec{P}_i\right) \supseteq \mathbf{V}\left(^0\vec{P}_j\right)$.

Maximal elements in this ordering (i.e. minimal $\mathbf{V}\left(^0\vec{P}_i\right)$) must exist. Such states are called **ergodic**. If $^0\vec{P}_i$ and $^0\vec{P}_j$ are ergodic states, then either

$$\mathbf{V}\left(^0\vec{P}_i\right) = \mathbf{V}\left(^0\vec{P}_j\right) \text{ or}$$
$$\mathbf{V}\left(^0\vec{P}_i\right) \cap \mathbf{V}\left(^0\vec{P}_j\right) = \emptyset. \quad (A.6)$$

Those states for which $\mathbf{V}\left(^0\vec{P}_i\right) = \mathbf{V}\left(^0\vec{P}_j\right)$ are called **locally ergodic pure states**, since the pure states any of the $\mathbf{V}$'s are ergodically connected with each other. Denote the different linear spaces generated by the locally ergodic pure states, called **ergodic subspaces**, by Φ_ρ. The number of pure states, and, therefore, the dimension of Φ_ρ will be designated as n_ρ, and if $\rho = 1$, $\mathbf{T}$ is an **irreducible** matrix. By permuting the ordering of pure states so that equivalent inertial states are contiguous, one sees:

$$\mathbf{T} = \begin{pmatrix} \mathbf{T}_1 & 0 & \cdot & \cdot & 0 & * & \cdot & * \\ 0 & \cdot & \cdot & 0 & \cdot & & & \\ & \cdot & \cdot & \cdot & \cdot & & & \\ & \cdot & \cdot & \cdot & 0 & \cdot & & \\ & & & \cdot & \mathbf{T}_{n_\rho} & & & \\ & & & \cdot & 0 & \cdot & & \\ & \cdot & & \cdot & \cdot & & & \\ & & \cdot & \cdot & \cdot & & & \\ 0 & \cdot & \cdot & \cdot & 0 & * & \cdot & * \end{pmatrix} \text{ so that} \quad (A.7)$$

i) All the columns of $\mathbf{T}$ restricted to Φ_ρ are zero outside of Φ_ρ.

ii) If n_i is the order of $^0\vec{P}_i$, then there are no zeroes in the Φ_ρ block of $\mathbf{T}^{n_i}$. $\quad (A.8)$

The asterisks in (A.7) reflect unknown values; however, this expression fails to reflect the fractal-like structure that $\mathbf{T}$ may exhibit in large dimensioned systems. This structure is entirely determined by the positions of the zeroes in $\mathbf{T}$, and not by the specific values of the nonzero entries in $\mathbf{T}$.

Φ_ρ is a space which is invariant under **T** and has a dimension equal to the number of pure states in Φ_ρ. We also know that for any locally ergodic pure state $^0\vec{P_i}$, the set of vectors, $^0\vec{P_i}, \mathbf{T}\left(^0\vec{P_i}\right), \mathbf{T}^2\left(^0\vec{P_i}\right), \cdots, \mathbf{T}^{n_i}\left(^0\vec{P_i}\right)$, are linearly independent and generate a subspace of Φ_ρ, but there is no reason to expect **T** to be a single Jordan block, so there can be many such subspaces of Φ_ρ.

One can now construct the **equilibrium fixed point basis** for each ergodic set, Φ_ρ and show that there is only one such element associated to each ergodic set: As observed, the matrix $\mathbf{T}^{n_i}$ has no nonzero entries over the Φ_ρ block, and obviously has all diagonal elements less than one (unless Φ_ρ consists of a pure isolated state) so the matrix $\tilde{\mathbf{A}} = \mathbf{I} - \mathbf{T}^{n_i}$, whose eigenvectors are of the form $1 - \lambda^{n_i}$, also has no nonzero terms. But, because $\mathbf{T}^{n_i}$ is still a Markov matrix, the sum of the rows in $\tilde{\mathbf{A}}$ are zero, so that $\tilde{\mathbf{A}}$ has a nontrivial null space.

If one strikes any row and corresponding column from $\tilde{\mathbf{A}}$ to make a principal minor matrix, one sees from the Gershgorin Theorem that all the spectral radii, Λ_i are now strictly less than the diagonal element, so that all the eigenvalues of the principal minor matrix are all positive, whence the principal minor itself is positive.

Thus $\tilde{\mathbf{A}}$ is an **M** matrix in the sense of IV and has all positive eigenvalues, so that **A** must have all strictly negative eigenvalues, and the principal minors of –**A** must all be positive. There is *exactly* one zero eigenvector of **A** in Φ_ρ, corresponding to a fixed point of **T**; that is, there are no nilpotent Jordan blocks in **A**, and that fixed point satisfies (2.1a).

One can find the zero eigenvector for the irreducible matrix **A** restricted to Φ_ρ, for example, by striking out the n_ρ^{th} column in **A** and looking at the vector, $\vec{A}_\rho = \mathfrak{a}^i\,^0\vec{P_i}$, $i = 1, \cdots, n_\rho$, generated by the $n_\rho - 1 \times n_\rho - 1$ minors obtained by striking out the ith row and multiplying by $(-1)^{i-1}$ (the overall sign of $\vec{A}_\rho$ is of no importance, since it generates a linear subspace and the only physically important point lies on the state-space, $\mathfrak{P}$). These minors are easily related to the principal minors of **A** by making use of the fact that the sum of the rows is zero. Thus for the ith row, replace the n_ρ^{th} row with the negative sum of all the other rows and make use of the fact that the only nonzero contribution to the determinant will be that furnished by the ith row, since all other terms will have two rows the same.

To find the sign, one must move the row from the n_ρ^{th} position to the ith position which cancels out the alternation in sign used to define $\vec{A}_\rho$. Thus, the component $\mathfrak{a}^i$ can be taken as the ith principal minor of **A** which is always nonzero The vector, $\vec{A}_\rho$, serves to define the **partition function** for Φ_ρ, and its ith component represents the probability flow from the simplex of all other pure states in Φ_ρ towards the ith state, and is, therefore, positive. Once the vector $\vec{A}_\rho$ is known, all other partition function related variables for the ergodic state Φ_ρ can be defined. The set of vectors $\vec{A}_\rho$ are the **equilibrium microstates** determined by **T** (or generating the null space of **A**). When there is only one equilibrium microstate, the matrix **T** is irreducible, and the Markov theory coincides with that of classical statistical mechanics.[15]

This type of viewpoint has been discussed, for instance, by Stratonovich.[16] These types of equilibrium states are often considered "metastable", but from the control point of view, they represent the expected outcomes of a system subjected to controls whose values become constant beyond a certain time.[17] Since the set of equilibrium microstates are, themselves, ensemble probabilities, they can be thought of as the vertices of a convex polytope, the **equilibrium polytope**

The relationship between the coefficients and roots of the characteristic polynomial of **T** play a fundamental role in the understanding of the system evolution associated to **T**. In order to apply Perron-Frobenius techniques, the dimension of the system must be sufficiently small *viv-a-vis* the noise in the matrix elements of **T** that the root structure of the characteristic polynomial for **T** can be clearly resolved. A quantitative result for this resolution gives rise to a very difficult algebraic problem.[18]

Mathematically, this relates to high powers of the Markov matrix, $\mathbf{T}^k$ when k is chosen as the maximum number of iterations for which the model is valid. Assume that k is sufficiently large that nontrivial Jordan blocks can be ignored, that "infinite precision" arithmetic is used, and that **T** is "in principle" irreducible. Then, when examining $\mathbf{T}^k$ we may find, depending on the system size, large numbers of probabilities which are "physically" zero. If we set these values to zero, the irreducible structure of **T** can change dramatically. Aside from now having many zero eigenvalues, the equilibrium space for **T** associated with eigenvalue one can also be changed destroying the hypothesis of physical irreducibility. Because the modification to $\mathbf{T}^k$ has arisen through a "noise reduction" projection on a very high power of **T**, determining the changes to **T** which yield the projected $\mathbf{T}^k$ is an impossibly ill-posed problem. Instead, this paper suggests dimensional reduction of the model using projection methods.

APPEAL TO MATHEMATICIANS
ON THE HIERARCHY IN CVM AND PPM

Ryoichi Kikuchi

Department of Materials Science and Engineering, UCLA

Los Angeles, CA 90095-1595

Abstract

Many phenomena can be understood based on the density function $c(x)$. However, in a large number of cooperative phenomena, correlations among particles are essential and the theory needs probability functions for particles in various clusters. Theories based on correlations are always approximate, except Onsager's treatment. In several cases discussed in the paper, approximations lead to qualitatively misleading information. Because of this nature of the approximation, mathematical treatments leading to accurate results are greatly desired.

Introduction

In materials science, the local density $c(x)$ plays the central role. In many cases, the theory based on $c(x)$ could explain observed results successfully. Examples are phase diagrams of various alloys, Fick's theory of diffusion, Cahn's theory of spinodal decomposition [1] and Kobayashi's calculation of dendritic growth [2]. The success seems to indicate that the models, although simple, contain the essence of the phenomena.

On the other hand, there are examples in which the treatments based on only $c(x)$ are inadequate. There are two groups of such cases. One group is when the correlation among particles are essential, and the other is when the rigorous results, which take into account correlations fully, lead to properties qualitatively different from approximations based on limited range of correlations. These cases present challenging problems to materials scientists, but also are considered intriguing from mathematical point of view. This paper presents problems in which correlations play the essential role, with the emphasis on mathematical aspects, to contribute to this symposium which is co-sponsored by the mathematics community.

Phase Diagrams

When the theory of order-disorder was pioneered by Bragg and Williams in 1934 [3], they used the point approximation without correlations among particles. The theory was successful for bcc binary alloys. However, Shockley immediately noticed [4] that for fcc alloys the technique then available led to the phase diagram in Fig. 1, which was qualitatively different from experiments. Even Bethe's method [5], which took into account the correlation among nearest-neighbor pairs, did not lead to a stable ordered phase for fcc alloys. Li [6] tried to improve the theory for fcc by introducing correlations among the four particles in a tetrahedron, but his theory was insufficient and led to Fig. 2, which was still qualitatively different from experiments.

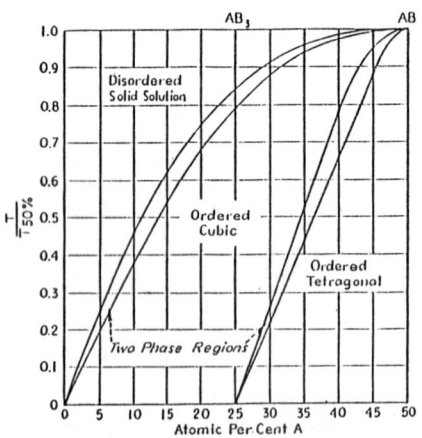
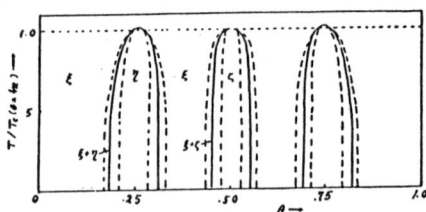

Fig. 1 Binary fcc phase diagram based on the point approximation [4]

Fig. 2 Binary fcc phase diagram calculated by Li [6]

It was only when Golosov et al. in Soviet [7] and van Baal [8], independently, took into account correlation fully among local tetrahedron clusters that the qualitatively satisfactory phase diagrams of fcc-based order-disorder systems were derived. This is shown in Fig. 3 [9], and indicates the need for correlations in the fcc ordered alloys which have frustrated structures.

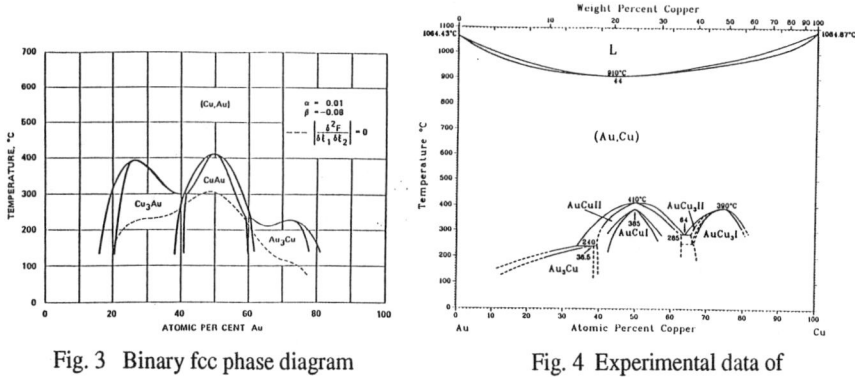

Fig. 3 Binary fcc phase diagram
due to de Fontaine and Kikuchi [9]

Fig. 4 Experimental data of
Cu-Au [10]

These successful calculations opened the field of theoretical phase diagram computations using tetrahedron and larger clusters. In minimizing the free energy, the entropy formula is usually based on the Cluster Variation Method (CVM) [11]. The CVM is a hierarchical structure and each treatment is built on a local cluster of a certain size and shape. The free energy is written in terms of the variables for probability of appearance of the basic cluster, and then is minimized to obtain the basic equations (non-linear, algebraic) to determine the variables. These basic equations can always be written without difficulty however large the cluster may be, but naturally the computation time for solving the equations becomes longer for larger clusters.

From the mathematical point of view, the entire field of statistical mechanics can be regarded as a branch of probability theory. When the theory of cooperative phenomena, for which the correlation is essential, is investigated by mathematicians, there is no reason why a revolutionary new approach could not be discovered.

Spinodal Decomposition

The theory of spinodal decomposition leads us to understanding of importance of correlations. In the original theory of Cahn [1], his arguments were based on the free energy curve calculated by the point approximation, which has the double-well shape similar to the "pair" curve in Fig. 5. Following his success, many attempts were made to locate the spinodal points using calculations of better accuracy. However, these efforts resulted in a contradiction. It was found that spinodal points do not exist in the rigorous theory. It is demonstrated in Fig. 5 [12], which plots the free energy $F \equiv E-TS$ for series of improved approximations and their extrapolation to the rigorous limit, the latter being of the flat bottom with no inflexion points. The improved approximations take into account correlations among atoms in larger local clusters.

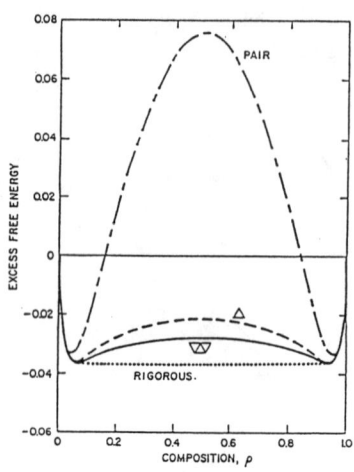
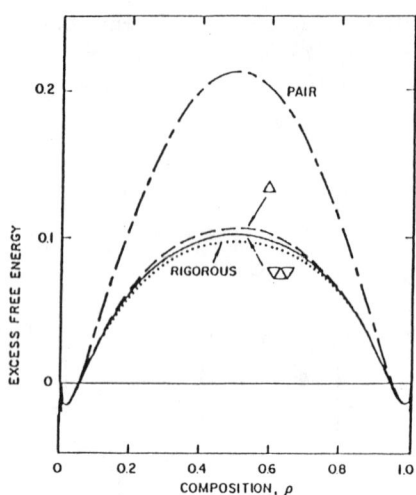

Fig. 5 Free energy against composition

Fig. 6 Free energy right after quenching

The solution of the puzzle lies in understanding the nature of the spinodal decomposition experiments, which are not done in equilibrium but after quenching. Figure 6 shows F right after quenching, and indicates that even the rigorous limit shows the central hump of the shape of the original Cahn curve. This is the F which is relevant to the spinodal decomposition analysis.

As the system is quenched, T in F follows the quenching, while E and S depend on atomic configurations which change much more slowly. Therefore, T_qS right after the quench is smaller than T_hS at $T_h(>T_q)$ before the quench. This is the reason for the hump in Fig. 6. On the other hand, as the approximation in CVM improves, S becomes larger, and hence TS in an approximation is less than TS in the rigorous treatment. This is the reason for the hump in approximations. The still unanswered question is whether we can interpret a certain cluster approximation as corresponding to a rigorous treatment for a certain quenching condition.

The free energy with the central hump is another expression of the van der Waals loop in the p-v diagram of the gas-liquid transition. The loop is a result of approximation, and does not exist in the rigorous limit. However, here again we meet a puzzling situation, namely the loop portion looks natural, to be used in explaining the hysteresis of the transition, i. e. the overheating and undercooling. The question is whether such an explanation has validity or not.

Puzzle of the Magnetization Curve

The free energy curves in Fig. 5 lead to one more puzzle which has received little attention. In the Ising model, the magnetization is derived from a minimum of the free energy. When we are interested in the rigorous limit, we may extrapolate approximations, and can reach a reasonable limiting curve. This leads to a dilemma, because Onsager's rigorous treatment [13] contains equal numbers of + and − spins, and hence cannot give any information about the magnetization. Therefore, strangely enough, the approach from approximations can give more information than the rigorous theory, when both are for an infinitely large system. The question we ask is whether the approximation result leading to magnetization is spurious and should not be accepted, or the approximate treatment has its own justification.

Diffusion Theory

We can prove that as long as the local free energy depends on only the composition $c(x)$ of A in an AB binary alloy and not on other variables (for example the variables for correlation for neighboring atoms, schematically to be written as $\xi(x)$), the change of the free energy F leads, after a partial integration, to

$$\frac{dF}{dt} = A \int dx \, \frac{\partial F}{\partial c(x)} \frac{dc(x)}{dt} = A \int \frac{d\mu(x)}{dx} J(x) dx \leq 0 \quad (1)$$

where A is the cross-sectional area, $\mu(x) \equiv \partial F/\partial c(x)$ is the chemical potential of A, $J(x)$ the flux of A such that $dc(x)/dt = -dJ(x)/dx$. The inequality in (1) represents the second law. Since (1) holds for any function of $c(x)$, it leads to

$$\frac{d\mu(x)}{dx} J(x) \leq 0 \quad (2)$$

When the flux is negatively proportional to the chemical potential gradient, the relation (2) is satisfied.

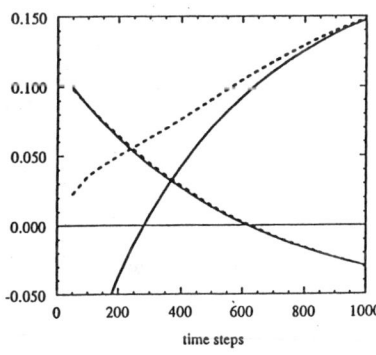

Fig. 7 Atom flux and $\Delta\mu/\Delta x$ near a steep junction

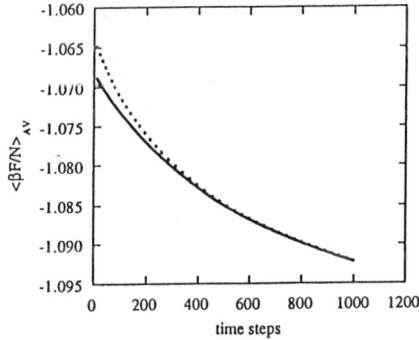

Fig. 8 Change of free energy with time, corresponding to Fig. 6

On the other hand, the Path Probability Method (PPM) [14] treatment of interdiffusion leads to a different results around a steep composition gradient [15]. Results of numerical computation for a region near a steep junction are plotted against time in Fig. 7, in which the decreasing curves are for J(x) and the increasing curves are for the local value of $\Delta\mu/\Delta x$. The solid curves are from the initial distribution satisfying local equilibrium, while in the initial state of the dotted curves the probability of neighboring points is given as a product of the two point probabilities.

In Fig. 7, we used the μ expression which had been derived beforehand in the equilibrium CVM for a one-dimensionally inhomogeneous system by differentiating the free energy by the space variable, and the expression is

$$2\beta\mu_n = -6\beta(\varepsilon_{11} - \varepsilon_{22}) - 11\ln\frac{x_n(1)}{x_n(2)} + 2\ln\left(\frac{y_{v-1}(21)y_{v-1}(11)y_n(11)y_v(11)y_v(12)}{y_{v-1}(12)y_{v-1}(22)y_n(22)y_v(22)y_v(21)}\right) \quad (3)$$

Here, n is the coordinate of the lattice plane in units of the inter-plane distance, v the coordinate of a center of inter-plane bond, 1 and 2 are the A and B species, and ε_{ij} the interaction energy between i and j. Our μ is actually the difference of chemical potentials of A and B atoms. It may be noted is that μ in (3) is not at a lattice point only but it takes into account the local variation of composition around a lattice position. When we use μ_n in (7) for the present relaxation studies, the system is not locally in equilibrium so that the pair values y's are independent of the point variables x_n. This is the fundamental difference between the present method and the kinetic formulation based on x's only. In other words, our calculation is more general than the assumption leading to (1), so that (1) is to be modified by schematically adding the term dependent on correlations $\xi(x)$ among atoms around x as

$$\frac{dF}{dt} = A\int dx \frac{\partial F}{\partial c(x)}\frac{dc(x)}{dt} + A\int dx \frac{\partial F}{\partial \xi(x)}\frac{d\xi(x)}{dt} \quad (4)$$

Figure 7 shows that J and $\Delta\mu/\Delta x$ become zero at different times, leading to the conclusion that they are not proportional to each other, and hence do not satisfy (2). Since we noted above that (2) holds as long as the density function f(x) depends on only c(x), the behavior in Fig. 7 indicates that the correlations among particles, $\xi(x)$, is functioning. The difference between the solid and the dotted curves in Fig. 7 originates in whether the nearest neighbor atoms are correlated or not in the initial distribution, and supports the significance of correlations in the region near a sharp junction. We verify in Fig. 8 that the free energy always decreases with time even during the period shown in Fig. 7.

The behavior of J and $\Delta\mu/\Delta x$ in Fig. 7 does not contradict Onsager's theory [16] that the driving force of J is $-\Delta\mu/\Delta x$, because Onsager's proof is under the condition that the state is near equilibrium and hence $\xi(x)$ is a function of c(x) even in kinetic processes.

Scalar Product Treatment of Boundary Tension σ

A recent theory of boundary tension [17], i.e. the excess free energy σ due to a boundary, clearly indicates the significance of correlation, how to work it out, and also suggests the nature of the future challenge. It was proved [17,18] that σ between two coexisting phases I and II can be calculated from the equation:

$$\exp(-a\beta\sigma) = \sum_v \sqrt{p^I(v)} \sqrt{p^{II}(v)} \qquad (4)$$

where $p^I(v)$ is the probability that a plane in the bulk phase I takes a configuration v, and $p^{II}(v)$ in II, and a is a unit area along the boundary. Since (4) is a scalar product (S.P.) of two vectors, we see that σ is a measure of how much the two phases deviate from each other.

The sum in (4) has the form of a partition function with the Boltzmann factors replaced by *a priori* probabilities. Therefore, we can calculated σ using the standard technique of CVM, by writing $a\beta\sigma$ in terms of probability variables for a chosen basic cluster, and then minimizing σ with respect to these variables. Figure 9 shows results of σ for the interphase boundary of the 2-D square lattice Ising model, placed in the <10> orientation [17]. To see how well the method works, we compared our results with σ* of Onsager's [13]. The four curves from top are the S.P. results using the basic clusters of n-squares (n=1, 2, 3, 4) which are placed linearly. The points near the 0 line and with error bars are the extrapolation of the four n's to n→∞. The extrapolated values are very close to Onsager's rigorous results, and verify the reliability of the technique. The three curves from the bottom are by the Sum method which sums the

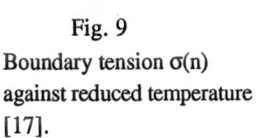
Fig. 9
Boundary tension σ(n) against reduced temperature [17].

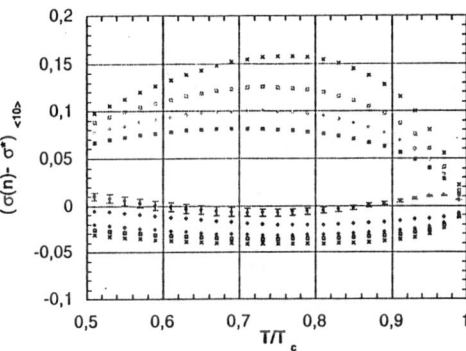

Although our extrapolation results look good, the extrapolation method is not unique. The hierarchy of approximations is like a series in the calculus, and it is highly desirable to explore a method of treating such hierarchy of approximations. We want to present this as a challenge to the mathematical community.

We may comment on an additional puzzle in the σ calculation based on the Sum method, although it is not described explicitly above. The width of a boundary is infinite in the rigorous theory, whereas it is finite in the approximation used in the Sum method. Nevertheless, the σ values calculated using the Sum method are reasonable. The reason why it is reasonable has not been clarified. The S. P. approach and Onsager's rigorous theory do not contain the width information, and are free from the puzzle.

Proof of Convergence of 3-D Hierarchy

The argument of the previous section is based on the property that the n→∞ limit leads to the rigorous Onsager's result. This convergence in 2-D square lattice was proved by Schlijper [19]. In the 3-D cases, the dimension reduction property [20] implies that approximation can be improved by choosing 2-dimensionally extended clusters. However, no proof exists that the rigorous limit can be achieved by extending the size of 2-dimensional clusters, although attempts have been made by Finel [21] and myself [22]. The claim of the latter is that the desired proof depends on the validity of a conjecture that the CVM entropy varies monotonically as the size of the cluster increases. Contribution from mathematicians would be highly appreciated.

Concluding Remarks

Statistical mechanics connects microscopic models with macroscopic observations. The physics in its foundation lies in the epitaph equation written on Boltzmann's tomb stone in Vienna:

$$S = k.\log W \quad (\text{"." after k is on the tomb stone.}) \quad (5)$$

This equation connects a physics concept, entropy S, with a mathematical concept W, the number of ways which is proportional to the probability. The mathematical aspect of statistical mechanics is to work on W and hence this discipline is a branch of the probability theory.

There are many experiments which can be understood based on the density function $c(x)$. On the other hand, there also are a vast quantity of cooperative phenomena for which $c(x)$ is not adequate and correlations among particles are essential. Significance of correlation is demonstrated in this paper by the hierarchy of the Cluster Variation Method. Such cooperative phenomena are believed challenging from pure mathematics point of view, and more contributions from mathematicians are highly desirable.

References

1. J. W. Cahn, *Acta Met.*, **9**, 795 (1961).
2. R. Kobayashi, private communication (1994).
3. W. L. Bragg and E. J. Williams, *Proc. Roy. Soc. (London)*, **A145**, 699 (1934).
4. W. Shockley, *J. Chem: Phys.*, **6**, 130 (1938).
5. H. Bethe, *Proc. Roy. Soc. (London)*, **A150**, 552 (1935).
6. Yin-Yuan Li, *J. Chem. Phys.*, **17**, 447 (1949).
7. N. S. Golosov, L. Ya. Pudan, G. S. Golosova and L. E. Popov, *Soviet Phys. - Solid State* **14**, 1280 (1972).

8. C. M. van Baal, *Physica (Utrecht)*, **64**, 571 (1973).
9. D. de Fontaine and R. Kikuchi, *National Bureau of Standards SP-496*, p.999 (1977).
10. H. Okamoto, D.J. Chakarabarti, D.E. Laughlin and T.B. Massalski, "Binary Alloy Phase Diagrams," (ASM International, 1990) p.360
11. R. Kikuchi, *Phys. Rev.*, **81**, 988 (1951).
12. R. Kikuchi, *J. Chem. Phys.* **47**, 1664 (1967).
13. L. Onsager, *Phys. Rev.*, **65**, 117 (1944).
14. R. Kikuchi, *Prog. Th. Phys., Suppl.*, **35**, 1 (1966).
15. R. Kikuchi, L.-Q. Chen and A. Beldjenna, *Nanostructured Materials*, **5**, 269 (1995).
16. L. Onsager, *Phys. Rev.*, **37**, 405 (1931); **38**, 2265 (1931).
17. P. Cenedese and R. Kikuchi, Proceedings of *International Workshop on Theory and Applications of the Cluster Variation and Path Probability Methods*, Teotihuacan, Mexico (June 18-22,1995) to be published (Prenum Press).
18. R. Kikuchi, *J. Chem. Phys.* **57**, 777 (1972).
19. A. G. Schlijper, *J. Stat. Phys.*, **35**, 285 (1984).
20. R. Kikuchi and S. G. Brush, *J. Chem. Phys.*, **47**, 195 (1967).
21. A. Finel, Thèse de Doctor d'etat, Université Paris VI, 1987.
22. R. Kikuchi, *Prog. Th. Phys., Suppl.*, **115**, 1 (1994).

Mean-Field Equations for Configurational Kinetics of Alloys at Arbitrary Degree of Nonequilibrium

V. G. Vaks, S. V. Beiden, and V. Yu. Dobretsov

Russian Research Centre "Kurchatov Institute", Moscow 123182, Russia

Abstract. We apply the mean-field approximation to the master equation describing the time evolution of various alloy states to obtain the kinetic equation for local concentrations at arbitrary degree of nonequilibrium. We discuss applications of this equation to several problems of phase transformations.

Problems of structural evolution of nonequilibrium alloy states attract much attention. Theoretically, they are studied using either approximate kinetic equations (AKE) [1-9] for local concentration or mean occupation number $c_i = c(\mathbf{r}_i, t)$ at lattice site i (say, for A-atoms in the binary alloy A-B), or direct simulation, e.g. Monte-Carlo methods [10-14]. AKE seem to be more transparent and universal than simulation methods, and many new qualitative results in this field have been obtained using AKE [1-4]. However, the currently used phenomenological forms of AKE [1-5] are based on the Onsager-type linear relations [15] between time derivatives dc_i/dt and thermodynamic driving forces $\partial F/\partial c_j$ (where F is the free energy of an inhomogeneous alloy), which hold only for the near-equilibrium states [16]. Thus application of these AKE to states far from equilibrium, such as those obtained by deep quenching, transient states at spinodal decomposition (SD), etc, can hardly be justified, while kinetic phenomena just in these states seem to attract most interest [3,4,13,17].

To more consistently investigate these phenomena, in this work (being an extended version of earlier communication [18]) we use the approach based on the fundamental master equation determining evolution of probabilities of various alloy states. A similar approach has been discussed by Gouyet [19]).

The master equation for probability $P\{\alpha,t\}$ to find the occupation number set $\alpha = \{n_i\}$ (where $n_i = 1$ if atom A is at site i and $n_i = 0$ otherwise) has the form

$$dP(\alpha)/dt = \sum_{\beta}[W(\alpha,\beta)P(\beta) - W(\beta,\alpha)P(\alpha)] \equiv \hat{S}P \qquad (1)$$

where $W(\alpha,\beta)$ is the $\beta \to \alpha$ transition probability. For probabilities $W(\alpha,\beta)$ we accept the "thermally activated direct exchange model" [6,9]. Then we can express the transfer matrix $\hat{S}$ in (1) in terms of the microscopic inter-site jump probabilities $w_{kl} = \omega_{kl}\exp[-\beta(E_{kl}^s - \hat{E}_{kl}^{in})]$ where ω_{kl} is the attempt frequency, $\beta = 1/T$ is the reciprocal temperature, E_{kl}^s is the saddle point energy, while the initial (before jump) energy $\hat{E}_{kl}^{in}$ for the alloy A-B with the hamiltonian $H\{n_k^A, n_l^B\}$ can be written as the variational derivative: $\hat{E}_{kl}^{in} = n_k^B \delta H/\delta n_k^B + n_l^A \delta H/\delta n_l^A - n_k^B n_l^A \delta^2 H/\delta n_k^B \delta n_l^A$ [9].

To treat equation (1), in this work we apply the mean field approximation (MFA) which was earlier developed for description of steady states in irradiated alloys [8,9]. We suppose that interatomic potentials v_{ij} obey conditions for the applicability of MFA, i.e. their interaction range includes a sufficiently large number of sites $N_0 \gg 1$, which for actual alloys is sometimes true even quantitatively [4].

Multiplying eq. (1) by operators n_i, $n_i n_j$, etc, and summing over all configurations $\{n_k\}$, i.e. averaging these relations over distribution $P\{n_i,t\}$, we obtain an exact system of equations for mean occupations $c_i(t) = <n_i>$ and correlators $g_{i...j}(t) = <n_i \ldots n_j>$ [9]. For inhomogeneous alloys under consideration such averaging, e.g. that for $<n_i> = c_i = c(\mathbf{r}_i,t)$, has a clear physical meaning only if characteristic lengths l for space variations of c_i exceed the intersite distance a. Since l is usually not lower than the interaction radius r_0 being supposed to be large, the above condition in our case is satisfied.

In the case of applicability of MFA $N_0 \gg 1$, the above-mentioned equations for the mean occupations c_i and correlators $g_{i...j}$ in the main approximation in $1/N_0$ decouple [8,9] and can be written explicitly. Thus we obtain the mean-field kinetic equation (MFKE) for the A-B alloy

$$dc_i/dt = \sum_s \gamma_{is}\left(c_i' c_s e^{\varphi_s^A + \varphi_i^B} - c_s' c_i e^{\varphi_i^A + \varphi_s^B}\right) \qquad (2)$$

Mathematics of Microstructure Evolution
Edited by L. Q. Chen, B. Fultz, J. W. Cahn,
J. R. Manning, J. E. Morral, and J. A. Simmons
The Minerals, Metals & Materials Society, 1996

Here $\gamma_{is} = \gamma_{si} = \omega_{ij}\exp(-\beta E_{ij}^s)$ is the c_i-independent part of the jump probability, $c'_i = 1-c_i$, $\varphi_i^p = \beta(v^p c)_i$, $(v^p c)_i = \sum_j v_{ij}^p c_j$ is the MFA potential acting on a p-species atom at site i, and v_{ij}^A and v_{ij}^B are related to the configurational potential $v = V^{AA} + V^{BB} - 2V^{AB}$ and the "asymmetric" one $u = V^{AA} - V^{BB}$ as: $v^A = \frac{1}{2}(u+v)$, $v^B = \frac{1}{2}(u-v)$.

Writing the reduced MFA free energy $f = \beta F$ for an inhomogeneous alloy as

$$f = \sum_i (c_i \ln c_i + c'_i \ln c'_i) + \frac{1}{2}\sum_{ij}\beta v_{ij} c_i c_j \qquad (3)$$

we can rewrite eq. (2) in the form

$$dc_i/dt = 2\sum_s M_{is} \sinh\left[(\partial f/\partial c_s - \partial f/\partial c_i)/2\right] \qquad (4)$$

where $M_{is} = \gamma_{is}\{c_i c'_i c_s c'_s \exp[\beta(uc)_i + \beta(uc)_s]\}^{1/2}$.

For fully equilibrium states all derivatives $f_i = \partial f/\partial c_i$ are equal to the chemical potential μ, so rhs of (4) vanishes. For near-equilibrium states we can expand this rhs in powers of $f_s - f_i$, which converts it into the linear Onsager form $\sum_s L_{is} f_s$ [16]. Comparing eq. (4) with the arbitrary extrapolation of the Onsager equation to the nonlinear region used by Khachaturyan et al. [3,4,15], we see that their "partly linearized" kinetic equation (PLKE) corresponds to (i) linearizing MFKE in $f_s - f_i$ (while this difference is, generally, not small in the nonlinear cases), and (ii) neglecting the c_l-dependence of the generalized kinetic coefficient M_{is} in eq. (4). The latter neglection can lead to the nonuniform in time and space significant distortion of temporal evolution of nonequilibrium alloys, particularly at early stages of phase transformations when inhomogeneities in c_i are large.

Multiplying eq. (4) by f_i, summing over i and denoting $\exp(f_i)$ as ξ_i, we obtain an analog of H-theorem for alloys (earlier discussed by Penrose [7]):

$$df/dt = \frac{1}{2}\sum_{is} M_{is}(\xi_i\xi_s)^{-1/2}(\xi_s - \xi_i)\ln(\xi_i/\xi_s) \leq 0 \qquad (5)$$

Relations (2)-(5) are immediately generalized to many-component alloys. In particular, eqs. (2) and (3) for them take the form

$$dc_{pi}/dt = \sum_{qs}\gamma_{is}^{pq}\left[\exp(f_{ps} + f_{qi}) - \exp(f_{pi} + f_{qs})\right] \qquad (6)$$

$$f = \sum_{pi} c_{pi}\ln c_{pi} + \frac{1}{2}\sum_{pi,qj}\beta v_{ij}^{pq} c_{pi} c_{qj} \qquad (7)$$

Here $c_{pi} = <n_{pi}>$ is the mean occupation number of site i by a p-species atom, and f_{pi} is the formal partial derivative $\partial f/\partial c_{\mu i}$ of function (7), disregarding the normalizing condition $\sum_p c_{pi} = 1$. Eq. (6) can be applied, in particular, to the alloy A-B with vacancies (ABv-alloy) being treated as a 3-component alloy. As actually intersite jumps in alloys are mainly realized via vacancies, eqs. (6) enable us to find expressions for the effective "direct jump probabilities" γ in (2) in terms of the microscopic vacancy-atom jump probabilities.

Intersite jumps in eqs. (2), (6) occur mainly between nearest or next-nearest sites i and s. For disordered alloys the relevant values of c_i and c_s in our case, as mentioned, are supposed to be close to each other. Expanding differences $c_s - c_i$ in powers of $\mathbf{r}_{si}\nabla c$ where $\mathbf{r}_{si} = \mathbf{r}_s - \mathbf{r}_i$, we obtain a continuous version of eq. (2):

$$\partial c_i/\partial t = \text{div}\left\{M(c)[\nabla c/cc' + \nabla(\beta vc)_i]\right\} \qquad (8)$$

Here $c = c_i$, $M(c) = \gamma cc' \exp[\beta(uc)_i]$ is the c_i-dependent mobility, and the tensor γ is $\frac{1}{2}\sum_s \gamma_{is}\mathbf{r}_{is}\mathbf{r}_{is}$ (becoming a scalar for cubic lattices). If we suppose the characteristic length l for space variations of c_i to exceed not only a but also the interaction radius r_0, we can approximate the last term of (8) as $(vc)_i \simeq v_0 c + \epsilon\Delta c$ where $v_0 = \sum_j v_{ij} = -4T_c$ determines the critical temperature T_c for SD, and ϵ is $r_0^2 v_0/6$ (for cubic lattices).

Then linearizing (8) in deviations $c_i - c_0$ from the initial constant value c_0, we obtain the Cahn equation [1] describing early stages of SD, with the Martin's expression [6] for the mobility $M_C = M(c_0)$. At later stages of SD, the local concentrations c_i approach their "liquid" or "gas" values at the binodal curve, while the interface widths l decrease and become of the order of r_0. Therefore, to describe these stages of SD (as well as effects of long-ranged elastic interactions for which inequality $l \gg r_0$ is violated [4]) we should employ the full equation (8).

For the ordered alloy with ν different sublattices α, it is convenient to take averages $<n_{\mathrm{p}i}^\alpha>= c_{\mathrm{p}i}^\alpha$ for each sublattice separately. Then the site numbers i, j in eqs. (2) - (7) are replaced by pairs of indices: $i \to i\alpha$, $v_{ij}^{\mathrm{pq}} \to v_{ij}^{\mathrm{pq},\alpha\beta}$, etc, where i or j now number different ν-atomic cells. Instead of ν quantities $c_{\mathrm{p}i}^\alpha$ we can consider the mean concentration in the cell, $c_{\mathrm{p}i} = \nu^{-1} \sum_\alpha c_{\mathrm{p}i}^\alpha$, and $\nu - 1$ local order parameters $\eta_{\mathrm{p}i}^\lambda$. Expanding differences $c_{\mathrm{p}s}^\alpha - c_{\mathrm{p}i}^\alpha$ in eqs. (6) in powers of $\mathbf{r}_{si}\nabla c_{\mathrm{p}i}^\alpha$, we obtain instead of eq. (8) the set of equations

$$\partial c_{\mathrm{p}i}^\alpha / \partial t = \sum_{\mathrm{q}\beta} \left\{ D_{\mathrm{p}i,\mathrm{q}i}^{\beta\alpha} + \left[\xi_{\mathrm{q}i}^\alpha \left(\gamma_{\alpha\beta}^{\mathrm{pq}} \nabla + \gamma_{\alpha\beta}^{\mathrm{pq}} \nabla^2 \right) \xi_{\mathrm{p}i}^\beta - \{\mathrm{p} \to \mathrm{q}\} \right] \right\} \qquad (9)$$

Here $\xi_{\mathrm{p}i}^\alpha = \exp(f_{\mathrm{p}i}^\alpha)$, $f_{\mathrm{p}i}^\alpha = \partial f/\partial c_{\mathrm{p}i}^\alpha$ is defined analogously to $f_{\mathrm{p}i}$ in eq. (6), symbol $\{\mathrm{p} \to \mathrm{q}\}$ means the interchange of indices p and q,

$$D_{\mathrm{p}i,\mathrm{q}i}^{\beta\alpha} = \bar{\gamma}_{\alpha\beta}^{\mathrm{pq}}(\xi_{\mathrm{p}i}^\beta \xi_{\mathrm{q}i}^\alpha - \xi_{\mathrm{p}i}^\alpha \xi_{\mathrm{q}i}^\beta), \qquad \gamma_{\alpha\beta}^{\mathrm{pq}} \nabla = \sum_s \gamma_{i\alpha,s\beta}^{\mathrm{pq}} \mathbf{r}_{si} \nabla,$$

$$\gamma_{\alpha\beta}^{\mathrm{pq}} \nabla^2 = \frac{1}{2} \sum_s \gamma_{i\alpha,s\beta}^{\mathrm{pq}} (\mathbf{r}_{si} \nabla)^2, \qquad \bar{\gamma}_{\alpha\beta}^{\mathrm{pq}} = \sum_s \gamma_{i\alpha,s\beta}^{\mathrm{pq}} \qquad (10)$$

The second, diffusive term in the rhs of eq. (9) is, generally, by a factor a/l or a^2/l^2 lower than the first one, $D^{\beta\alpha}$. However, in equations for the concentration derivatives $\partial c_{\mathrm{p}i}/\partial t$, after summing over α, terms $D^{\beta\alpha}$ vanish being antisymmetric in indices α and β. Thus the time evolution of the concentration $c_{\mathrm{p}i}$ is determined just by diffusive terms and proceeds much slower than that for the order parameters $\eta_{\mathrm{p}i}^\lambda$ [20], while in equations for $\partial \eta_{\mathrm{p}i}^\lambda/\partial t$ the diffusive terms within accuracy $\sim a/l$ can be neglected. For example, for the binary alloy A-B with two equivalent sublattices 1 and 2 (e.g. for the B2 or L1$_0$ phase), defining c_i and η_i as: $c_{1i} = c_i + \eta_i$ and $c_{2i} = c_i - \eta_i$ and neglecting all terms of the order of a/l or a^2/l^2, we obtain

$$\frac{\partial \eta_i}{\partial t} = \bar{\gamma}_{12} e^{\beta(uc)_i} \left[c'_{1i} c_{2i} e^{-\beta(v\eta)_i} - c'_{2i} c_{1i} e^{\beta(v\eta)_i} \right] = -2 M_i \sinh\left(\frac{1}{2} \frac{\partial f}{\partial \eta_i} \right) \qquad (11)$$

where $(uc)_i = \sum_j (u_{ij}^{11} + u_{ij}^{12}) c_j$, $(v\eta)_i = \sum_j (v_{ij}^{11} - v_{ij}^{12}) \eta_j$, $M_i = \bar{\gamma}_{12}(c_{1i} c'_{1i} c_{2i} c'_{2i})^{1/2} \exp[\beta(uc)_i]$, and f is the reduced free energy (3) expressed via c_i and η_i.

Comparing eq. (11) with the Allen-Cahn equation proposed to describe the antiphase boundary (APB) motion [2], we see that their equation corresponds to linearization of eq. (11) in $\partial f/\partial \eta_i$, neglecting $c-$ and η-dependence of the generalized mobility M_i, and consideration of only smooth functions $\eta_i = \eta(\mathbf{r}_i)$ with $l \gg r_0$ for which $(v\eta)_i$ can be approximated as $v_0\eta_i + \epsilon\Delta\eta_i$. Equations (9) also show that the space variations of η_i and c_i are directly related to each other. It results, in particular, in depletion of the alloy minority component near APB and in a number of peculiar effects in the kinetics of APB motion in non-stoichiometric alloys [20,21].

Equations (2)-(11) can be applied to most different problems of the nonlinear kinetics of alloys. Below we discuss several examples.

1. *Nonlinear concentration waves at first stage of SD*. According to the linearized Cahn equation [1], after a quench of an alloy below the spinodal curve (SC), an approximately periodic distribution of concentrations $c(\mathbf{r},t)$ develops. It has a characteristic wavelength $2\pi/k_C$, while the concentration wave amplitudes increase with a characteristic time $t_C \sim k_C^{-2}$, where the wavenumber $k_C = k_C(c_0, T)$ corresponds to most fast growing waves and decreases when the initial concentration c_0 approaches SC [1].

However, the experimentally observed microstructures in most cases correspond not to linear but to non-linear stages of SD, and have usually not periodic, but a "tongue-like" character, see e.g. Figs. 5(a)-5(c) in [20], or Fig. 1 (middle column) in [22]. It was also argued that non-linear effects neglected in the theory [1] may suppress development of pronounced Cahn's waves [15]. ¿From the other side, MC simulations using short-range potentials (as well as the AKE study [5]), obtained for first stages of SD rather the random cluster than wave-like patterns [10,11]. With rising the interaction range r_0, some MC results for the kinetics near SC converge to those of MFA [12]. However, even for long-range potentials, importance of the above-mentioned non-linear effects for transient morphologies under SD seems to remain non-clear [14].

To investigate this and other problems of SD, we performed two-dimensional (2D) simulations based on eq. (8). We approximated sums $(vc)_i$ by integrals and took $v(r)$ to be gaussian: $v(r) = -A\exp(-r^2/2\sigma^2)$ supposing $\sigma \gg a$, while the constant A is proportional to the MFA critical temperature T_c. The asymmetric potential u was supposed to be less long-ranged than v, thus we putted $(uc)_i \simeq u_0 c_i$. Simulations were made at a square lattice $40\sigma \times 40\sigma$ with periodic boundary conditions, using the dimensionless time variable $t' = t\gamma\sigma^{-2}$. The as-quenched distribution $c(\mathbf{r}, 0)$ was characterized by its mean value c_0 and small random fluctuations. To treat partial differential equation (8) we employed method of lines, reducing (8) to a system of ordinary differential equations solved by the Runge-Kutta method.

All characteristics of temporal evolution depend strongly on parameters c, $T' = T/T_c$ and u_{ij}/T_c where T_c is the critical temperature. For example, for the parameter values $c = 0.35$, $T' = 0.4$ and $u = 0$ (used in simulations illustrated by Figs. 1, 2, 3b and 3e), the linear stage discussed by Cahn corresponds to $t' \lesssim 5-7$. At larger t' nonlinear interactions between Cahn's concentration waves become significant. However, for all states c_0, T within SC after times $t_C \sim k_C^{-2}$ we observed a pronounced wave-like structure with a characteristic wavelength $\lambda \sim 2\pi/k_C$. The structure retains its features well within the non-linear region, which is illustrated by Fig. 1 for $t' = 10$. Thus non-linear effects don't destroy the wave-like distribution of concentration at initial stages of SD. Fig. 3b for $t' = 20$ corresponds to beginning of the next, precipitation stage of SD. Comparison of Figs. 1 and 3b shows that formation of "tongue-like" precipitates at these times is the result of coalescence of concentration waves. At larger t' effects of strong interaction between the forming precipitates ("droplets") discussed below become important.

2. *"Bridge" mechanism for droplet coalescence.* Until recently only two coarsening mechanisms were discussed for droplet phase separation (i) Lifshits-Slyozov evaporation-condensation (EC) mechanism for separated droplets [23] in which bigger droplets grow at the expence of evaporation and disappearance of smaller ones, and (ii) Binder and Stauffer mechanism [11,14] of collision and coalescence between droplets caused by their thermal diffusion without any interaction. Recently Tanaka [17] observed peculiar effects of inter-droplet interaction under their coalescence during SD in a 2D fluid mixture which he attributed to features of droplet diffusion in liquids.

In our simulations we observe similar effects for the solid alloy model, where we see no diffusion of droplets as whole, but rather a strong dynamic coupling of diffusion fluxes around droplets. This new mechanism of coalescence (to be called the "bridge" mechanism) is illustrated in Fig. 2. It dominates the first stage of coarsening of droplet pattern at intermediate c_0 values (for example, for $t' = 20 - 500$ at $c_0 = 0.35$, $u = 0$), before the EC mechanism becomes effective.

Figs. 2a-2f illustrate two main versions of the bridge mechanism: (i) that for droplets 1 and 2 when a smaller droplet is consumed by its bigger neighbor(s) that pulls it over the bridge formed by diffusion fluxes, and (ii) that for droplets 3 and 4 of similar size when the "bridge" can occur for considerable time (e.g. $\Delta t' \sim 100$ for droplets 3 and 4), before its density begins to sharply rise and droplets coalesce. These processes seem to be rather similar to those shown by Tanaka [17] in his Figs. 1 and 4.

3.*Effect of the* A-B *asymmetry of mobilities in the* A-B *alloy on SD.* Such asymmetry is always present in real alloys but it is usually disregarded in theoretical treatments. In our model it is described by the MFA value u_0 of the asymmetric potential u. Presence of factor $\exp(\beta u_0 c)$ in mobility results, firstly, in a c-dependent change of the time scale, i.e. slowing down kinetics at $u < 0$ and speeding it up at $u > 0$. Secondly, this factor changes a relative importance of the EC and droplet fusion mechanisms of coarsening. Since fusion proceeds via bridge mechanism corresponding to noticeable values $c \sim 0.1 - 0.2$ in the bridge region, negative u lead to suppression of this mechanism with respect to the EC one, as the latter is determined by diffusion over "gas" region $c \ll 1$ where the suppression is small. At positive u the situation reverses. For example, at $c_0 = 0.35$, $t = 0.4T_0$, $u_0 = -2T_0$, the coarsening proceeds via evaporation of 5 (of total 11) and coalescence of 2 droplets during $t' = 250 - 2500$; at $u_0 = 0$, seven droplets (of 13) coalesce and 2 evaporate during $t' = 50 - 5000$; and at $u_0 = 2T_0$, 13 (of 18) droplets coagulate and none evaporates during $t' = 3 - 35$.

Figs. 3(a)-3(c) illustrate effect of u on the formation of droplet pattern, Figs. 3(d) and 3(f) show typical coarsening mechanisms for $u_0 = -2T_0$ and $u_0 = 2T_0$, respectively, and Fig. 3(e) shows an advanced stage of coarsening when the EC mechanism dominates.

4.*Description of transient ordered states in phase-separating alloys.* This interesting problem was discussed by Chen and Khachaturyan (CK) [3] using PLKE, and by Reinhard and Turchi (RT) [13] using Monte Carlo methods. To compare accuracy of various approaches, we employed MFKE (2) to same model and states as those treated by CK and RT. We used the reduced time $t^* = \gamma_{is} t$ related to that of CK as: $t^*_{CK} = c_0 c'_0 t^*$.

Our treatment of states with $c_0 = 0.175$ and $c_0 = 0.25$, $T = 0.44 T_c$ has shown that main qualitative features of their evolution found by CK with PLKE are confirmed with MFKE, though at later stages the quantitative differences increase. (and t^*_{CK} values in captions to their Figs. 4 and 5 seem to be given with misprints). For $c_0 = 0.5$, $T = T_c/3$, the Monte Carlo study of RT revealed new features of evolution not seen by CK, creation of islands of disordered phase along APB and coexistence for intermediate times of three phases: gas-like, liquid-like and ordered. RT supposed that absence of such structures in the CK study was due to errors inherent to MFA. Our results illustrated in Fig. 4 show that it is not so, and MFKE reproduces all features of evolution found by RT. In particular, our Figs 4a and 4b seem to be similar to Figs. 4(b) and 4(c) of RT and may be considered as their "averaged" versions, in accordance with the averaged character of MFKE (2) and a sufficiently large interaction range in the CK model used.

This work had been initiated by stimulating discussions with G. Martin. It was supported by the International Science Foundation under Grant N MQA000, and by the Russian Fund of Fundamental Research under Grant N 93-02-14776.

References

[1] J.W. Cahn, Acta Metall. **9**, 795 (1961).
[2] S.M. Allen and J.W. Cahn, *ibid.*, **27**, 1085 (1979).
[3] L.-Q. Chen and A.G. Khachaturyan, Phys. Rev. B **44**, 4681 (1991).
[4] S. Semenovskaya and A.G. Khachaturyan, Phys. Rev. Lett. **67**, 2223 (1991); L.-Q. Chen and A.G. Khachaturyan, *ibid.* **70**, 1477 (1993).
[5] R. Petschek and H. Metiu, J. Chem. Phys. **79**, 3443 (1983).
[6] G. Martin, Phys. Rev. B **41**, 2279 (1990).
[7] O. Penrose, J. Stat. Phys. **63**, 975 (1991).
[8] V G. Vaks and V.V. Kamyshenko, Phys. Lett. A **177**, 269 (1993).
[9] V.G. Vaks and S.V. Beiden, Phys. Lett. A **182**, 140 (1994); Zh. Exp. Teor. Fiz. **105**, 1017 (1994).
[10] M. Rao *et al.*, Phys. Rev. B **13**, 4328 (1976).
[11] K. Binder and D. Stauffer, Adv. Phys. **25**, 343 (1976).
[12] D.W. Heermann, W. Klein, and D. Stauffer, Phys. Rev. Lett. **49**, 1262 (1982).
[13] L. Reinhard and P.E.A. Turchi, *ibid.* **72**, 120 (1994).
[14] J.D. Gunton, M. San Miguel and P. Sahni, in *Phase Transitions and Critical Phenomena*, ed. by C. Domb and J.H. Lebowitz (Academic, London, 1983), Vol. 8, p. 269.
[15] A.G. Khachaturyan, *Theory of Phase Transformations and Structure of Solid Solutions* (Nauka, Moscow, 1974).
[16] L.D. Landau and E.M. Lifshits, *Statistical Physics* (Nauka, Moscow,1976), Ch. 12.
[17] H. Tanaka, Phys. Rev. Lett. **72**, 1702 (1994).
[18] V.G. Vaks, S.V. Beiden and V.Yu. Dobretsov, Pis'ma v ZhETF **61**, 65 (1995).
[19] J.-F. Gouyet, Europhys. Lett. **21**, 335 (1993); Phys. Rev. E, **51**, 1711 (1995).
[20] S.M. Allen and J.W. Cahn, Acta Metall. **24**, 425 (1976).
[21] V.Yu. Dobretsov, G. Martin, F. Soisson and V.G. Vaks, Europhys. Lett. **31** 417 (1995).
[22] K. Oki, H. Sagane and T. Eguchi, J. de Phys. **38**, C7-414 (1977).
[23] I.M. Lifshits and V.V. Slyozov, J. Phys. Chem Solids **19**, 35 (1961).

FIGURE CAPTIONS

Figure 1. Profile of concentration $c(\mathbf{r})$ at spinodal decomposition in the 2D model described in text for following parameter values: $c_0 = 0.35$, $T = 0.4T_0$, $u = 0$ and $t' = 10$.

Figure 2. Temporal evolution of concentration $c(\mathbf{r})$ for the 2D model described in text at $c_0 = 0.35$, $T = 0.4T_0$, $u = 0$ and following t': (a) 120, (b) 130, (c) 140, (d) 160, (e) 180, and (f) 200. Insert in Fig. 2(f) shows relation between darkness and c values varying between 0 and 1 from bottom to top.

Figure 3. Same as in Fig. 2 at $c_0 = 0.35$, $T = 0.4T_0$, and the following values (u_0, t'): (a)(-2T_0, 200); (b) (0, 20); (c) (2T_0, 2.4) (d) (-2T_0, 1230); (e) (0, 5000); and (f) (2T_0, 16.5).

Figure 4. Temporal evolution of a 2D model from Ref. [3] for a square lattice of 128 × 128 sites at $c_0 = 0.5$, $T = \frac{1}{3}T_0$: (a) $t^* = 400$, and (b) $t^* = 1900$.

MODELING AND CONTROL OF ADVANCED CHEMICAL VAPOR DEPOSITION PROCESSES[1][2]

H.T. Banks, K. Ito, J.S. Scroggs, H.T. Tran
Center for Research in Scientific Computation
Department of Mathematics
North Carolina State University
Raleigh, NC, USA

N. Dietz
Department of Physics
Department of Materials Science and Engineering
North Carolina State University
Raleigh, NC, USA

and

K.J. Bachmann
Department of Materials Science and Engineering
Department of Chemical Engineering
North Carolina State University
Raleigh, NC, USA

Abstract

We give an overview of a research program in scientific computation concerning the control of the flow dynamics of a key semiconductor manufacturing process, namely, rapid thermal organometallic chemical vapor deposition (RTOMCVD), that operates at higher mass flow rates/densities than accessible with conventional processing tools. We outline our development to date of realistic mathematical models that accurately describe gas flow and deposition process for RTOMCVD including the utilization of a novel non-invasive method of real-time process monitoring and control, p-polarized reflectance spectroscopy. These modeling efforts will be the basis for the formulation of control and design problems for the regulation of epitaxial deposition processes operating at high flow rates and/or vapor densities. Therefore, we also include a brief discussion of the optimal reactor design problem and feedback control methodology that is applicable to on-line process monitoring and control of the growth process in the RTOMCVD reactor.

---

[1] This research was supported in part by the Advanced Research Projects Agency under grants AFOSR F49620-95-1-0437 and AFOSR F49620-94-1-0447 and in part by the Air Force Office of Scientific Research under the DoD MURI program through grant AFOSR F49620-95-1-0447.

[2] This paper was presented as an invited lecture at the ASM/TMS Materials Week '95 as part of the Symposium on "Mathematics of Thermodynamically Driven Microstructural Evolution," Oct. 29 - Nov. 2, 1995, Cleveland, OH.

1. Introduction

Recent developments in microelectronics, requiring large investments for processing tools in an environment of rapid change, have led to the development of cluster tools that retain flexibility in tool modernization at minimum compromise of the overall process flow and cost. Two criteria that dominate the development of such cluster tools are low thermal budget and high throughput. The former is driven by the increasing tightening of tolerances in microelectronics processing associated with decreasing feature size. The latter is primarily motivated by economic concerns. Here we focus on a specific process, organometallic chemical vapor deposition (OMCVD), which is widely used in optoelectronic device processing and in the fabrication of advanced transistors that are needed, for example, in fast logic applications. Future developments based on this process require real-time process monitoring and closed-loop feedback control to assure the reproducible fabrication of device features of molecular dimensions. Also, the increasing demands on the control of the compound stoichiometry, in conjunction with the requirement of high throughput, forces operation at higher vapor densities and flow rates than accessible with existing processing tools. Thus a judicious assessment of the flow dynamics and their optimization play a crucial role in tool design and process development. The purpose of this paper is to provide an outline of the mathematical part of this task. Details of the hardware design and examples of specific processes will be published elsewhere. However, to lay the foundation for the mathematical modeling, as a specific example, a brief review of a few salient features of a particular hardware design and of the kinetics of OMCVD processes that enter into the modeling are included in this paper.

2. Reactor Design

The design of a modern OMCVD reactor that fits the cluster tool concept requires attention to efficient wafer exchange, rapid thermal processing to minimize the thermal budget and the capability of handling operating pressures ranging from subatmospheric to superatmospheric pressure which mandate unconventional modifications in the gas distribution/flow control and waste management subsystems. A conceptual design of a vertical rapid thermal organometallic chemical vapor deposition (RTOMCVD) reactor is shown in Fig. 1. In this configuration, the reactants and carrier gas are introduced at the top of the reactor and flow down toward the heated substrate in three concentric input nozzles. The substrate wafer is mounted on a graphite susceptor which is heated using radio frequency (rf) heating. Radio frequency, which is a proven method, is employed to realize rapid substrate heating and cooling cycles and to assure uniform temperature distribution across the wafer surface. The susceptor is rotated to improve the film uniformity. A small flux of carrier gas is injected into the bottom port through which the rotary motions are affected in order to keep the susceptor area clear of deposits. The fused silica reactor vessel is enclosed in a sealed water-cooled stainless steel enclosure that is pressurized by flowing carrier gas to the total pressure of carrier gas and other process gases in the reactor chamber. Thus, the safety of the process is assured at process pressures below the rupture limits of the steel enclosure and within the safe operating range of flow controllers and valves. Waste gases are exhausted from the reactor vessel and the steel enclosure through the exit port at the top of the reactor.

Critical aspects in the design of the RTOMCVD reactor are the input nozzles including the shape of the nozzles and the distance between the inlet and the susceptor, the rotational rate of the

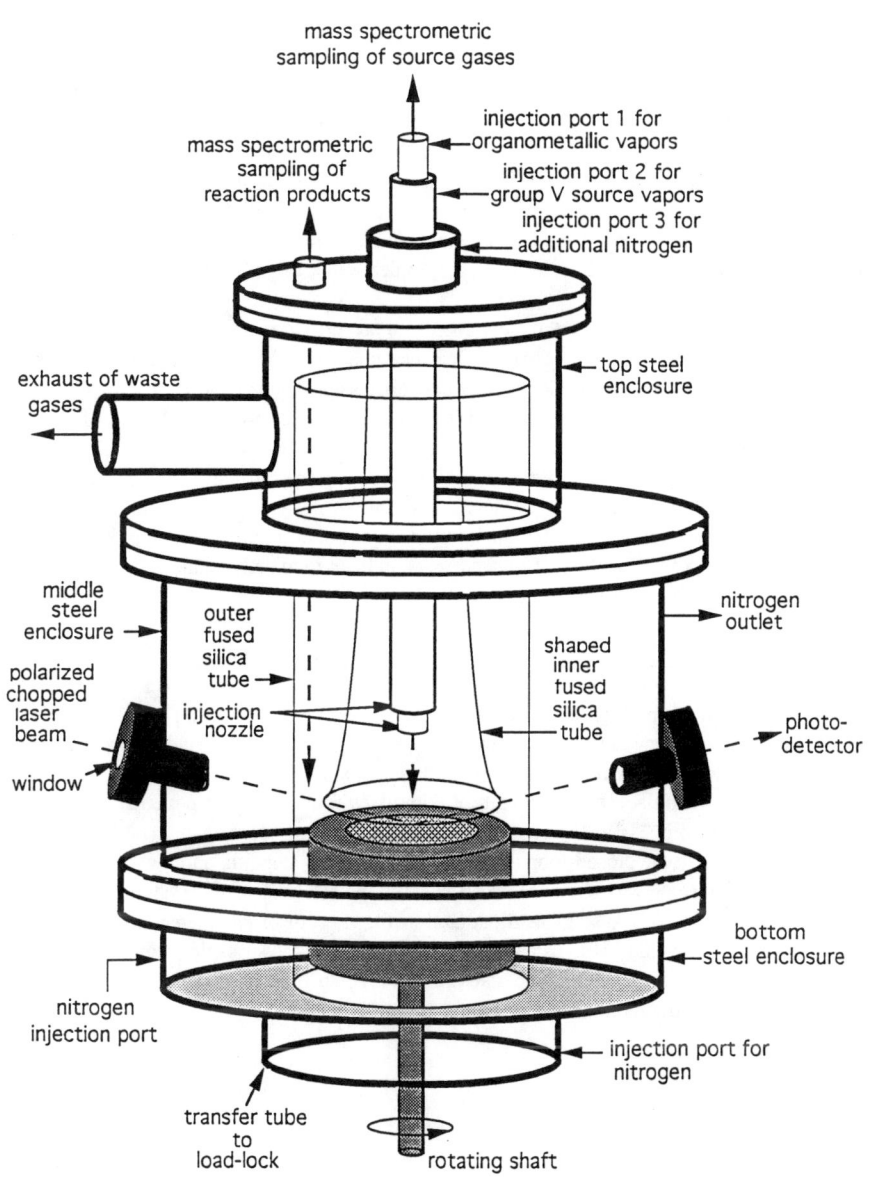

Figure 1. A schematic diagram of the RTOMCVD reactor.

susceptor, total pressure and the choice of carrier gas. In addition, the location of the susceptor is also important. A conventional vertical reactor will have the gas entering at the top, with flows directed downward impinging on the susceptor (see Fig. 1). In this configuration, the temperature increases toward the susceptor and the gas expands accordingly. However, this creates a destabilizing density gradient with the heavier gas above the lighter gas near the susceptor. An alternative design involves inverting the reactor so that the gas enters from the bottom and flows toward the susceptor at the top. In this case the density gradient would be more stable but positioning the susceptor upside-down would present a challenging design problem. Extensive support by numerical simulations of the gas flow and temperature profile under operating conditions will be required to guide the above design and optimization of the RTOMCVD reactor.

Finally, two ports on the side of the steel enclosure are provided for optical real-time process monitoring and control using p-polarized reflectance spectroscopy at the Brewster angle. This technique is based on the changes in the real part (Brewster angle position) and imaginary part (offset from zero reflectance at the minimum in the reflectance for p-polarized light) of the dielectric function upon heteroepitaxial overgrowth. Starting at the Brewster angle of the substrate, where the reflectance of highly p-polarized light is close to zero, provides for high sensitivity in the initial stage of the overgrowth. The detector output of the p-polarized reflectance spectroscopy is available as an input signal for the computer controlled regulation of the deposition rate via feedback to the flow controllers. This feedback can be used to change the pulse rate, duration, and amplitude of the input fluxes of the process gases.

3. Description of the Closed-Loop Control System

Because the control of the fluid dynamics is essential for processes operating at high flows or vapor densities, advanced methods of mathematical modeling, optimization, and control theory will be applied to guide the experimental implementation of the RTOMCVD reactor. The overall real time process monitoring and feedback control comprises four major components: chamber flow process, deposition process at the substrate wafer, real time process monitoring, and feedback control (see Fig. 2). In the closed-loop (or feedback) control system, the input or control parameters are modified in some way using information about the behavior of the system output. In the context of the RTOMCVD system as illustrated in Fig. 1, control parameters such as the pulse rate, duration, and amplitude of the source fluxes, rf heating of the susceptor region, and possibly total pressure as well as susceptor spin rate are modified so that the difference between the desired and measured growth rate and stoichiometry are minimized.

3.1 Chamber Flow Process

In the RTOMCVD reactor, the process gases flow through the input nozzle, and mixing and reaction chamber. On their way to the heated substrate wafer, the vapors undergo homogeneous and/or heterogeneous reactions due to the temperature changes and mixing in the chamber. In addition, the imposed temperature gradients cause significant buoyancy driven convection to superimpose on the forced flow. These complex fluid dynamics are described by a system of non-linear partial differential equations representing the continuity, momentum, energy, and species equations. Even though the RTOMCVD process is operating at high mass flow rates, since the reactor length is small, the flow velocity is still much less than the speed of sound. The flow, therefore, can be assumed to be laminar. Because of large temperature gradients (in our case, 500-1000 K) coupled with possibly large differences in molecular weights, the Boussinesq approxi-

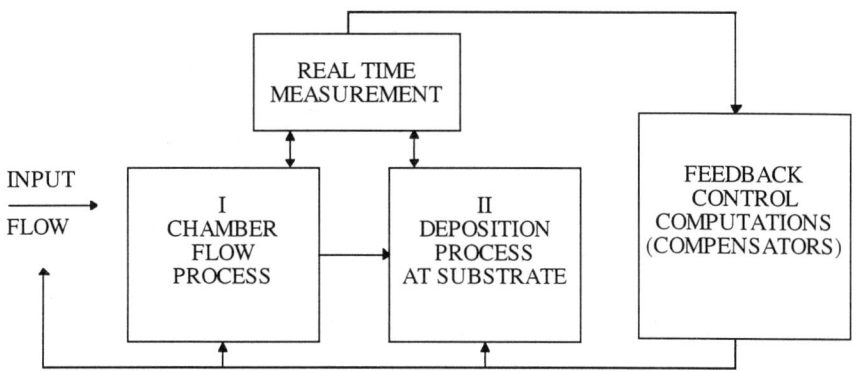

Figure 2. Schematic diagram of the overall closed-loop control system.

mation [3] is not appropriate for flow modeling. Furthermore, transport coefficients and thermophysical properties, density, viscosity, conductivity, species diffusivities, and heat capacity are highly temperature dependent. The nonlinear interactions among the transport processes, buoyancy driven convection flows and viscous terms make the flow dynamics a fully three-dimensional phenomenon. Finally, the expansion of the gas as it approaches the hot susceptor wafer plays an important role in the flow dynamics and is modeled by the ideal gas law:

$$p_i = \frac{\rho c_i RT}{m_i}$$

where p_i, c_i, m_i are the partial pressure, mass fraction, and molecular weight of the i-th species, ρ is the density, R is the universal gas constant, and T is the temperature. Furthermore, the mass concentrations c_i, $i=1,...,N$, satisfy the constraint

$$\sum_{i=1}^{N} c_i = 1$$

Based on the above observations and assumptions, the transport equations have the following forms [4]:

continuity equation:

$$\frac{\partial}{\partial t}\rho + \nabla \cdot (\rho u) = 0,$$

momentum equation:

$$\rho\left(\frac{\partial}{\partial t}u + u \cdot \nabla u\right) + \nabla p = \mu\left(\Delta u + \frac{1}{3}\nabla(\nabla \cdot u)\right) - \rho g e_3,$$

energy equation:

$$\rho C_v\left(\frac{\partial}{\partial t}T + u \cdot \nabla T\right) + p\nabla \cdot u = \nabla \cdot (\kappa \nabla T) + H,$$

species equation:

$$\rho\left(\frac{\partial}{\partial t}c_i + u \cdot \nabla c_i\right) = \nabla \cdot (\rho D_i(\nabla c_i + \alpha c_i \nabla \log T)) + r_i.$$

In these equations, $u \in R^3$ is the velocity vector, p is the total pressure, μ is the (dynamic) viscosity, g is the acceleration due to gravity, e_3 is the unit vector in the z-direction, C_v is the specific heat, κ is the thermal diffusivity, D_i is the solutal diffusivity, α is the thermal diffusion coefficient (Soret coefficient), and r_i is the chemical reaction rate of the i-th species. Finally, the term H in the energy equation denotes the heat generation term which may include the heat generation due to viscous dissipation (viscous shearing) and chemical reactions. Note also that both thermal (Soret) diffusion and buoyancy forces are taken into account in the model.

One of the most important aspects in realistic modeling is the choice of physical boundary conditions. However, it is difficult to formulate representative boundary conditions for some variables and instead, idealized boundary conditions must be used. Typically, one assumes no-slip and no-penetration at solid walls. The velocity at the inlet and at the substrate are specified. The outlet conditions can be specified by assuming that the flow at the exit port is fully developed. The thermal boundary conditions significantly influence the flow phenomena [12]. Assumptions such as constant wall and susceptor temperature, insulated sidewalls, constant wall heat-transfer flux and energy flux balance at the susceptor are often used in modeling studies. For the chemical species, molar fraction is usually specified at the inlet, zero flux on nonreacting surfaces, and species flux is specified as a first order reaction rate on reacting surfaces [12], [14]. From these observations, the formulation of the right boundary conditions is a nontrivial task which will require the interaction of all members of our interdisciplinary team.

The input to the transport model includes flow parameters such as pulse rate, duration, and amplitude of the input fluxes of the source vapors as well as the gas composition, the gas velocity, and the total pressure. These parameters enter the model through the boundary conditions and in the calculations of transport coefficients such as species diffusivities. Solutions of the transport equations will provide the temperature, gas composition and velocity in the vicinity of the substrate wafer which are used as input for the model of the deposition process (see Fig. 3 and Fig. 4).

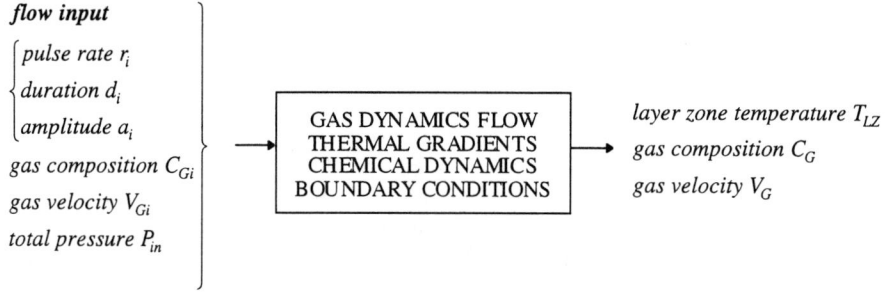

Figure 3. Schematic diagram of the chamber flow process and its input and output.

3.2 Surface Deposition Process

The overall deposition process involves complex interactions amongst the flow, thermal, and species fields in a chemically reacting system which can be described by the following steps:

a) homogeneous vapor phase reactions generating the reactants that participate in the surface reaction;
b) transport of these species in the vapor phase to the location of the substrate and across the diffusion layer to the substrate;
c) adsorption of some, or all of the reactants at the growth surface;
d) surface diffusion and reactions, and incorporation of atomic species into the growing film;
e) nucleation and growth;
f) desorption of the reaction products that are not incorporated into the film;
g) transport across the diffusion layer in the vapor phase and possible homogeneous follow-up reactions of these waste by-products.

Figure 4. Schematic representation of the deposition process with its input and output.

The deposition process in the CVD reactor is thus very complex, and the overall rate law, relating the growth rate to the concentrations of the input reactants and exhausted waste products, does not reveal the detailed growth mechanism. For example, steps (b) and (g) are directly affected by the flow dynamics. In addition, the rotational rate of the susceptor strongly influences the uniformity of the film thickness. Detailed modeling of the deposition process depends heavily on future theoretical and experimental work. Figure 4 lists the parameters which influence the model and the information it should provide. We note that the desired growth rate, R, and stoichiometry, S, will be compared against those measured by the p-polarized reflectance spectroscopy (PRS) (see section 3.3) for on-line feedback control.

3.3 On-Line Measurement by PRS

The acquisition of experimental information regarding the details of the epitaxial growth process under realistic conditions, that is, under conditions of steady state epitaxial thin film growth, requires non-intrusive probes that permit the real-time monitoring of the deposition kinetics. Optical techniques (for example, p-polarized reflectance spectroscopy (PRS) [5], [17], which is illustrated in Fig. 5), are particularly suited for the purpose. PRS is based on the changes in the reflectivity during the formation of a heteroepitaxial stack with regard to a beam of p-polarized light that

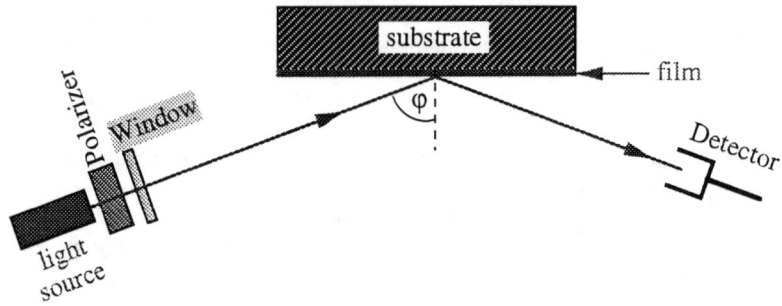

Figure 5. Schematic representation of the PRS.

impinges onto the surface at the Brewster angle of the substrate. For example, a chopped beam of HeNe laser light of 632.8nm wavelength is polarized by means of a Glan-Thompson polarizer to a ratio of the s- and p-polarized components of the incident intensity $I_s/I_p \approx 10^{-6}$ corresponding to a residual reflectivity of $\sim 10^{-4}$ for a silicon substrate. The reflected beam is detected by a photodiode, the output of which is processed through a phase sensitive amplifier and read into a computer.

In a simplified consideration of the heteroepitaxial growth of a film on a substrate, the complex reflectivity coefficient r_p of a three layer stack composed of the ambient flow region, the film and the substrate - labeled a, b, and c with interfaces labeled a/b and b/c, respectively - is given by

$$(3.1) \qquad r_p = \frac{r_{a/b} + r_{b/c}\exp(-2i\psi)}{1 + r_{a/b}r_{b/c}\exp(-2i\psi)}.$$

Here $r_{a/b}$ and $r_{b/c}$ are the Fresnel reflection coefficients for the interfaces a/b and b/c for p-polarized light,

$$\psi = \frac{2\pi t_b}{\lambda}\sqrt{\varepsilon_b - \varepsilon_a \sin^2\varphi_0}$$

is the film phase factor, and t_b, φ_0, ε_a, and ε_b designate the film thickness, the angle of incidence and the dielectric functions of the ambient flow and the film, respectively, at the frequency of the incident electromagnetic wave of wavelength λ. Due to constructive and destructive interference on the two interfaces a/b and b/c, the reflected intensity exhibits periodic variations that are represented by the exponentials in equation (3.1) and are shown in Fig. 6 for the example of GaP heteroepitaxy on an Si substrate wafer. For given parameters φ_0, λ, ε_a, and ε_b, the measurement of these interference oscillations provides real-time information on the film thickness and growth

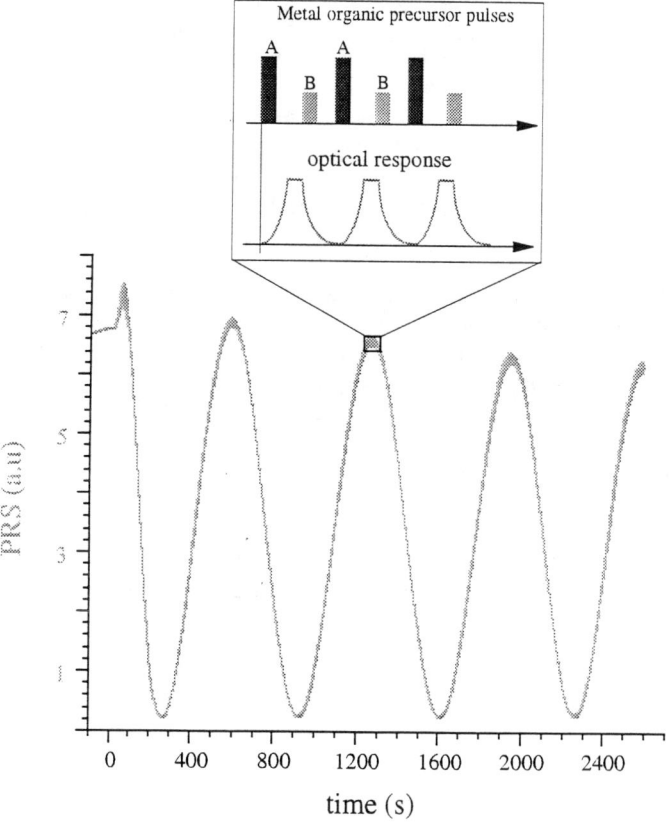

Figure 6. Correlation of fine structure in the PRS signal and timing sequence of the precursor pulses.

rate. Note that, for pulsed fluxes of the source vapors, that is, for GaP growth on Si, t-butyl phosphine (TBP) and triethyl gallium (TEG), a fine structure is superimposed on the interference oscillations. The correlation of this fine structure to the source vapor pulses is shown in the inset in Fig. 6 and contains information on the kinetics of the film growth, that is, the time constants associated with the surface reactions of the source vapor molecules. In order to describe these surface reactions the model must be expanded to include a surface reaction layer on the growing film (four layer stack). Initial modeling of the PRS response to the surface reactions/film growth has been carried out with simplifying assumptions and has provided for the first time specific information concerning the dielectric properties of the reaction layer in relation to the source vapor decomposition and film growth [18]. Presently this modeling effort is being expanded to encompass more complex reaction mechanisms and thus to provide for a closer approximation of the experimental conditions.

3.4 Design Parameters

In the last decade, there have been extensive modeling efforts to complement the on-going physical experiments (see e.g. [1]-[2], [6]-[14]). Effects of operating conditions, reactor geometry, and heat transfer characteristics on flow patterns and growth rate uniformity have been well documented among these studies. For example, in the study by Fotiadis et al. [12] for a vertical, axisymmetric reactor for metalorganic vapor phase epitaxy (MOVPE), it was shown for the first time the importance of modifying the reactor geometry and controlling heat transfer factors. They demonstrated that having an optimal reactor shape is critical to eliminating both natural convection and inertia driven cells. In addition, it was seen that the uniformity in film thickness could be improved by rotating the susceptor.

In our research efforts, numerical simulations of the physical models described above are being performed to identify desirable ranges of inlet flow rates, susceptor rotations, pressures, orientation of the susceptor and shape of the input nozzle. In addition, simulations are being used to determine input nozzle distance from the susceptor so that flow recirculations due to natural convection are minimized and good deposition rate uniformity is achieved.

4. Feedback Control Methodology

In this section we turn to a brief description of control ideas that could be applicable to the control problems described above. We first give a brief review of the linear theory for feedback control in the presence of "full state" information which is referred to as linear quadratic regulator (LQR) design. The theory rests on having a linear dynamic system

(4.1)
$$\frac{dy}{dt}(t) = Ay(t) + Bu(t)$$
$$y(0) = y_0$$

which describes the state $y(t)$ of the process one wishes to control. The desired design objectives are embodied in a cost functional or performance criterion

(4.2)
$$J(y_0, u) = \int_0^\infty \{<Qy(t), y(t)> + <Ru(t), u(t)>\}dt$$

which entails a weighting of the desired objectives (in this case keeping y small and in fact driving it to zero) and the input costs by choice of the design operators Q and R. "Hard" constraints on the inputs or controls can often be implemented effectively *via* choice of R as a "soft" constraint. Under appropriate assumptions (stabilizability, detectability, *etc.*[24]), the solution of minimizing (4.2) subject to (4.1) is given in feedback form by

(4.3) $$u(t) = -Ky(t) = -R^{-1}B^*\Pi y(t)$$

where Π is the symmetric, positive definite solution to the algebraic Riccati equation

(4.4) $$A^*\Pi + \Pi A - \Pi B R^{-1} B^* \Pi + Q = 0.$$

Use of the above control results is based on the assumption of *full* state feedback, no modeling errors, no observation errors, no disturbances to the system during the control period, and other "ideal" situation assumptions. This collection of assumptions is satisfied by almost no problems of practical interest. For example, full state knowledge is almost never available (even for finite dimensional systems) and one must design the control based on output feedback, or incomplete information (observations) on the state. In this case one must design a dynamic compensator or state estimator where one uses output or observations

(4.5) $$y_{OB}(t) = Cy(t)$$

to construct an estimate $y_c(t)$ of the state given by the compensator system

(4.6) $$\dot{y}_c(t) = A_c y_c(t) + F y_{OB}(t).$$

Once this state estimator is defined, it is used in a feedback of the form (4.3) in place of the full state, *i.e.* the control is taken as

(4.7) $$u(t) = -Ky_c(t)$$

for some gain K. Of course, the important question in this compensator design is how to choose A_c, F and K in (4.6), (4.7). One wants to choose A_c, F so that $|y_c(t)-y(t)| \to 0$ as $t \to \infty$ and K so that the closed loop system (*i.e.*, (4.1) coupled with (4.6), (4.7)) is asymptotically stable. If possible, one desires a design that also would possess some "robustness" with respect to modeling errors, noisy observations, exogenous unmodeled disturbances, *etc.* Among possibilities are LQR/LTR, H^∞, minmax, and other designs (see e.g. [21]).

Briefly, let us outline the so called H^∞/minmax control design. We assume that the basic system (4.1) is replaced by a system with disturbances $\eta(t)$ given by

(4.8) $$\dot{y}(t) = Ay(t) + Bu(t) + D\eta(t)$$
$$y(0) = y_0.$$

We also assume that disturbances affect *the measurement output* or observations

(4.9) $$y_{OB}(t) = Cy(t) + E\eta(t).$$

There is an associated *performance output* given by

(4.10) $$z(t) = Hy(t) + Gu(t).$$

The performance criterion is given by

(4.11)
$$J(y_0, u, \eta) = \int_0^\infty \{|z(t)|^2 - \gamma^2 |\eta(t)|^2\} dt$$
$$= \int_0^\infty \{\langle Qy(t), y(t)\rangle + \langle Ru(t), u(t)\rangle - \gamma^2 |\eta(t)|^2\} dt$$

where

(4.12)
$$Q = H^*H \geq 0, \quad R = G^*G > 0, \quad H^*G = 0$$
$$\tilde{Q} = DD^* \geq 0, \quad \tilde{R} = EE^* > 0, \quad DE^* = 0$$

are again design parameters to be chosen by the user.

If Π, Σ are minimal positive definite solutions of the algebraic Riccati equations

(4.13)
$$\Pi A + A^*\Pi - \Pi(BR^{-1}B^* - \gamma^{-2}\tilde{Q})\Pi + Q = 0$$
$$\Sigma A^* + A\Sigma - \Sigma(C^*\tilde{R}_{-1}C - \gamma^{-2}Q)\Sigma + \tilde{Q} = 0$$

where $\gamma > 0$ is a ("robustness") design parameter chosen so that

(4.14) $$r_{sp}(\Sigma\Pi) < \gamma^2, \quad \text{or} \quad \Pi - \gamma^2\Sigma^{-1} < 0$$

(here r_{sp} denotes the spectral radius of the operator), then the unique optimal control is given by

(4.15) $$u(t) = -R^{-1}B^*\Pi y_c(t) = -Ky_c(t).$$

In this formulation the state estimator or compensator $y_c(t)$ is defined by

(4.16) $$\dot{y}_c(t) = A_c y_c(t) + F y_{OB}(t)$$

where

(4.17)
$$A_c = A - BK - FC + \gamma^{-2}\tilde{Q}\Pi$$
$$F = (I - \gamma^{-2}\Sigma\Pi)^{-1}\Sigma C^* \tilde{R}^{-1}$$

The resulting closed loop system

(4.18)
$$\begin{bmatrix} \dot{y} \\ \dot{y}_c(t) \end{bmatrix} = \begin{bmatrix} A & -BK \\ FC & A_c \end{bmatrix} \begin{bmatrix} y(t) \\ y_c(t) \end{bmatrix}$$

with *controlled output*

(4.19)
$$z(t) = \begin{bmatrix} H & 0 \\ 0 & -GK \end{bmatrix} \begin{bmatrix} y(t) \\ y_c(t) \end{bmatrix}$$

is asymptotically stable. Moreover, we have the "robustness bound"

(4.20)
$$\|T(\cdot)\|_\infty \leq \gamma$$

where $\|\cdot\|_\infty$ is the H^∞ norm and $T(s)$ is the transfer function of disturbance to output map with

(4.21)
$$T(s) = \frac{\hat{z}(s)}{\hat{\eta}(s)}$$

for the single input and single output case.

For finite dimensional linear systems, the above summary results are valid under reasonable assumptions (see [20]). Even for infinite dimensional systems there is an analogous theory [22]. However, for the vapor deposition processes that are the focus of this note, there are substantial difficulties in applying or implementing the control methodology outlined here. Even if an infinite dimensional theory is applicable, there may be computational difficulties (*e.g.*, the approximating finite dimensional systems may be very large, very ill-conditioned, *etc.*). While there are examples of successful implementation of these ideas for infinite dimensional systems (*e.g.* see [19]), this task is by no means trivial. However, a more important difficulty lies in the *nonlinearities* that will necessarily be present in the vapor deposition model discussed in general terms in the earlier sections of this note. One can expect the nonlinearities to be an essential feature of reasonable models in this case and the usual linearization techniques in the engineering literature do not offer great promise in alleviating these difficulties. Unfortunately, an analogue or comparable framework to that outlined above for linear systems is not available to treat nonlinear system problems. One must thus turn to other possible approaches. These must involve both a way to reduce the inherently infinite dimensional nonlinear system problem to a finite dimensional nonlinear system problem and a way to implement some type of nonlinear robust feedback control design.

We first discuss briefly several ideas for the reduction of a nonlinear infinite dimensional system problem to a finite dimensional problem. We assume that our nonlinear system can be written in the form

(4.22) $$\frac{dy}{dt}(t) = Ay(t) + f(y(t)) + Bu(t)$$

where A is a linear, self-adjoint, nonnegative operator and f is nonlinear.

4.1 A Nonlinear Galerkin, Inertial Manifold Approach [15]

Let $\phi_1, \phi_2, \ldots \phi_m$ be the first m eigenfunctions of the operator A. We use these to write the identity as a sum of two orthogonal projections

(4.23) $$T + S = I$$

where range T=span$\{\phi_1, \phi_2, \ldots \phi_m\}$ is of finite dimension and S is infinite dimensional. The nonlinear system (4.22) can then be written as

(4.24) $$\frac{d}{dt}Ty = TAy + Tf(Ty + Sy) + TBu$$
$$\frac{d}{dt}Sy = SAy + Sf(Ty + Sy) + SBu.$$

For a nonlinear Galerkin approach we take $Sy = \Phi(Ty, u)$ as a graph of the approximating dynamics, with Sy representing the "residual mode" components. There are many variants involved with the choice of Φ. The standard Galerkin is given by $\Phi \equiv 0$, i.e., the residual mode components are just truncated. For an inertial manifold approach one can choose $\frac{d}{dt}Sy \approx 0$ so that the residual mode components are stationary (i.e., the dynamical system is stationary on the residual manifold). This leads to

(4.25) $$SAy + Sf(Ty + Sy) + SBu = 0.$$

Whatever the choice, one then uses finite dimensional nonlinear feedback control and compensator techniques on the first equation of (4.24).

4.2 A Reduced Basis Elements Approach [16]

To briefly illustrate these ideas, we assume for ease in exposition that we are dealing with nonlinear steady dynamics which can be parameterized

(4.26) $$E(y, u, \mu) = 0.$$

Here μ represents some physical parameter (e.g., input flow rate in the flow models discussed earlier) about which we choose to interpolate to obtain a reduced finite dimensional set of basis elements. One can use Lagrange interpolation or Taylor expansion to generate a set of basis elements in desired ranges of values of the parameter μ. In the Lagrange approach one obtains basis elements $\phi_i = y(\mu_i)$, $i = 1, 2, \ldots m_1$, by solving (4.26) at selected values μ_i of μ and at reference controls μ_i. In the Taylor approach one assumes that the solution $y(\mu)$ near some reference parameter $\bar{\mu}$ and nominal operating value of the control u can be written

$$y(\mu) = y(\bar{\mu}) + y'(\bar{\mu})(\mu - \bar{\mu}) + \frac{1}{2}y''(\bar{\mu})(\mu - \bar{\mu})^2 + \ldots \quad .$$

The approximating elements are obtained by choosing $\phi_i = y^{(i)}(\bar{\mu})$, $i = 1, 2, \ldots, m_2$. Once the approximating elements are chosen, then we take trial solutions $y^m(\mu) = \sum_{i=1}^{m} y_i \phi_i$ where the y_i are determined by the Galerkin relationship

$$\langle E(y^m, u, \mu), \phi_i \rangle = 0 \qquad i = 1, 2, \ldots m.$$

One must then apply the control design method to this nonlinear finite dimensional system.

Finally, we turn to possible robust feedback control designs for nonlinear systems [23]. Following the ideas from linear finite dimensional theory, suppose that after one of the above finite dimensional reductions (or another for that matter), one has a nonlinear system in R^m given by

(4.27) $$\frac{dy}{dt}(t) = \mathcal{F}(y(t)) + Bu(t) + D\eta(t)$$

where η is again some "disturbance" (unmodeled dynamics, residual error, *etc*.) and we wish to find solutions (u^*, η^*) to the minmax problem

$$\underset{u}{\text{Min}} \underset{\eta}{\text{Max}} \quad J(u, \eta) = J(u^*, \eta^*)$$

where

(4.28) $$J(u, \eta) = \int_0^{\tau} \{ | Cy(t) - y_d(t) |^2 + \beta | u(t) |^2 - \gamma^2 | \eta(t) |^2 \} dt.$$

Here y_d denotes some desired state of the system.

One possibility is a nonlinear programming (Hamilton-Jacobi) approach that solves the H-J equation

(4.29) $$\frac{\partial V}{\partial t} + \mathcal{F}(y) \cdot \frac{\partial V}{\partial y} - \frac{1}{2\beta} \left| B^* \frac{\partial V}{\partial y} \right|^2 + \frac{1}{2\gamma^2} \left| D^* \frac{\partial V}{\partial y} \right|^2 + \frac{1}{2} |Cy - y_d|^2 = 0$$

for the "value" $V(\tilde{t}, \tilde{y}) = J(u^*, \eta^*; \tilde{t}, \tilde{y})$ where we start the process (4.27) at time $\tilde{t}$ and state $\tilde{y}$. The parameter γ is again an attenuation bound in that we have for all $\eta \in L^2(0, \tau)$

(4.30) $$\underset{u}{\text{Min}} \int_0^{\tau} J(y(t), u(t), \eta(t)) dt \leq \gamma^2 \int_0^{\tau} |\eta(t)|^2 dt.$$

The solution of this problem is then given by the nonlinear feedback control law

(4.31) $$u^*(t) = -\frac{1}{\beta} B^* \frac{\partial V}{\partial y}(t, y(t)).$$

To implement the control methodology ideas discussed here (or others) for the CVD reactor problems introduced in this note offers significant conceptual and computational challenges. Moreover, the potential for advanced hardware development and challenging work regarding the kinetics of thin film deposition and etching processes mandate a coordinated multidisciplinary approach in our research efforts.

References

1. H. Moffat and K.F. Jensen, "Complex Flow Phenomena in MOCVD Reactors: I. Horizontal reactors", J. Crystal Growth, 77 (1986), 108-119.

2. D.W. Hess, K.F. Jensen, and T. J. Anderson, "Chemical Vapor Deposition: A Chemical Engineering Perspective", Rev. Chem. Eng., 3 (1985), 97-186.

3. D. D. Gray and A. Giorgini, "The Validity of the Boussinesq Approximation for Liquids and Gases", Int. J. Heat Mass Transfer, 19 (1976), 545-551.

4. H. Schlichting, Boundary Layer Theory, (New York, NY: McGraw-Hill, 1955).

5. N. Dietz and K.J. Bachmann, "Real Time-Monitoring of Epitaxial Processes by Parallel-polarized Reflectance Spectroscopy", Mat. Res. Soc. Bull. 20, No. 5, (1995), 49-55.

6. J. Ouazzani and F. Rosenberger, "Three-Dimensional Modeling of Horizontal Chemical Vapor Deposition I. MOCVD at Atmospheric Pressure, J. Crystal Growth, 100 (1990), 545-576.

7 J Van de Ven, G.M.J. Rutten, M.J. Raaijmakers, and L.J. Giling, "Gas Phase Depletion and Flow Dynamics in Horizontal MOCVD Reactors", J. Crystal Growth, 76 (1986), 352-372.

8. M.E. Coltrin, R.J. Kee, and J.A. Miller, "A Mathematical Model of Silicon Chemical Vapor Deposition, J. Electrochem. Soc., 133 (1986), 1206-1213.

9. M.E. Coltrin, R.J. Kee, and G.H. Evans, "A Mathematical Model of Fluid Mechanics and Gas-Phase Chemistry in a Rotating Disk Chemical Vapor Deposition Reactor", J. Electrochem. Soc., 136 (1989), 819-829.

10. J. Ouazzani, K.C. Chiu, and F. Rosenberger, "On the 2D Modeling of Horizontal CVD Reactors and its Limitations, J. Crystal Growth, 91 (1988), 497-508.

11. D.I. Fotiadis, Two- and Three-Dimensional Finite Element Simulations of Reacting Flows in Chemical Vapor Deposition of Compound Semiconductors, (PhD thesis, University of Minnesota, Minneapolis, 1990).

12. D.I. Fotiadis, S. Kieda, and K.F. Jensen, "Transport Phenomena in Vertical Reactors for Metalorganic VaporPhase Epitaxy: I Effects of Heat Transfer Characteristics, Reactor Geometry, and Operating Conditions", J. Crystal Growth, 102 (1990), 441-470.

13. L.R. Black, I.O. Clark, B.A. Fox, and W.A. Jesser, "MOCVD of GaAs in a Horizontal Reactor: Modeling and Growth, J. Crystal Growth, 109 (1991, 241-245.

14. K.F. Jensen, E.O. Einset, and D.I. Fotiadis, "Flow Phenomena in Chemical Vapor Deposition of Thin Films", Ann. Rev. Fluid Mech., 13 (1991), 197-232.

15. R. Temam, "Inertial Manifolds and Multigrid Methods", SIAM J. Math. Anal., 21 (1990), 154-178.

16. J.S. Peterson, "The Reduced Basis Method for Incompressible Viscous Flow Calculations", SIAM J. Sci. Stat. Comput., 10 (1989), 777-786.

17. K.J. Bachmann, U. Rossow and N. Dietz, "Real-time Monitoring of Heteroepitaxial Growth Processes on the Silicon (001) Surface by P-Polarized Reflectance Spectroscopy", Materials Science and Engineering, B, 37 (1995), 472-478.

18. N. Dietz and K.J. Bachmann, "P-Polarized Reflectance Spectroscopy: A Highly Sensitive Real-time Monitoring Technique to Study Surface Kinetics Under Steady State Epitaxial Deposition Conditions", Vacuum, Dec. 1995, to appear.

19. H.T. Banks, R.C. Smith, D.E. Brown, R.J. Silcox, and V.L. Metcalf, "Experimental Confirmation of a PDE-based Approach to Design of Feedback Controls", ICASE Rep. No.95-42, NASA Langley Res. Ctr., Hampton, VA, May, 1995; SIAM J. Control and Opt., submitted 1995.

20. T. Basar and P. Bernhard, H^∞ -Optimal Control and Related Minimax Design problems, A Dynamic Game Approach, (Boston: Birkhauser, 1991).

21. J. Doyle, K. Doyle, K. Glover, P. Khargonekar, and B. Francis, "State-Space Solutions to Standard H_2 and H^∞ Control Problems", IEEE Transactions on Automatic Control, AC-34 (1989), 831-847.

22. B.Van Keulen, H_∞-Control for Distributed Parameter Systems: A State-Space Approach, (Boston: Birkhauser, 1993).

23. A.J. van der Schaft, "Nonlinear State Space H_∞ Control Theory", in Perspective Control, (H.L. Trentelman and J. C. Willems, eds., Progress in Systems Control, 2nd ECC, Groningen: Birkhauser, 1993).

24. Brian D.O. Anderson and J.B. Moore, <u>Optimal Control: Linear Quadratic Methods</u>, (New Jersey: Prentice-Hall, 1990).

PATTERN FORMATION IN NONUNIFORM MEDIA:

CO OXIDATION ON MICROSTRUCTURED AND COMPOSITE Pt SURFACES

M. Bär[1,2], A. K. Bangia[1], I. G. Kevrekidis[1], G. Haas[2] and H.-H. Rotermund[2]
[1] Dept. of Chemical Engineering, Princeton University
Princeton, NJ 08544-5263, USA
[2] Fritz-Haber-Institut der Max-Planck-Gesellschaft
Faradayweg 4-6, 14195 Berlin, Germany

Abstract

This work addresses spontaneous pattern formation in nonuniform media. We study the propagation of chemical waves during CO oxidation on microstructured and composite Pt catalysts. Numerical simulations using a mechanistic reaction-diffusion model are performed to reproduce experimental observations and to obtain further insight into the dynamics of pattern interaction with the heterogeneous nature of the surface. Two examples are considered: (a) instabilities of a periodic wave train near corners of a no-flux boundary and (b) transmission through an interface between media with different wave properties (shape and speed).

1. Introduction

A variety of catalytic reactions on single crystal surfaces have been known to exhibit spontaneous pattern formation for certain operating conditions [1,2,3]. For instance, the spatiotemporal structures found during the CO oxidation on $Pt(110)$ surfaces range from fronts, pulses, spirals, and standing waves to chemical turbulence [4,5] in accordance with standard reaction diffusion phenomenology [6,7]. Recent studies explored pattern formation on microstructured surfaces (of length scale comparable to the typical pattern wavelength) fabricated by lithographic techniques [8,9]. Such a pretreatment enables detailed observation of the interaction between pattern forming processes and the size and geometry of the surface. The study of pattern formation can be extended to heterogeneous (composite) media formed by combining different catalytic surfaces on a scale where surface diffusion provides the coupling of their individual catalytic activities. Here we present selected examples of pattern formation phenomena on microstructured and composite surfaces.

2. The Experiment

The experiments were performed on $Pt(110)$ and $Pt(100)$ single crystal surfaces. The microstructured reacting domains were constructed in the following way. First, a 80 - 100 nm thick Titanium layer was deposited on the Pt surface. Then a negative photo-resist process was used to etch out microscopic regions of bare Pt. The sample was transferred into an UHV chamber and the surface was cleaned by standard procedures. The reaction temperature and the partial pressures of carbon monoxide and oxygen were kept constant throughout the experiment. The composite surfaces were laid out by depositing a submonolayer coverage of Au on parts of the clean Pt crystal, or by sputtering a thick Ti layer partially covering the surface down to submonolayer coverages. The spatiotemporal dynamics was recorded with a photoemission electron microscope (PEEM) [26]. This technique combines high temporal (25 $msec$, limited by the video frequency) and spatial (0.2 μm) resolution with a field of view up to 800 × 800 square microns. The contrast in the PEEM image stems from local differences in the work function caused by the different adsorbates and adsorbate coverages. In the PEEM images oxygen rich regions appear dark, while CO-rich areas appear lighter gray.

3. The Model

The mechanistic model of the catalytic CO oxidation on Pt surfaces incorporates Langmuir-Hinshelwood elementary adsorption, desorption and reaction steps, the interplay between adsorbates and surface structure [11] and the relevant transport mechanism (CO diffusion) [12]. A simplified model that describes the dynamics of the adsorbate coverage on the surface, u, and the surface structure (fraction w of 1 × 1 surface, lifting the reconstruction from 1 × 2 clean $Pt(110)$) reads

$$\frac{\partial u}{\partial t} = \frac{1}{\epsilon} u(1-u)(u - \frac{w + b(\mathbf{x})}{a}) + \nabla^2 u \quad (1)$$

$$\frac{\partial w}{\partial t} = f(u) - w, \quad (2)$$

with

$$f(u) = \begin{cases} 0 & \text{if } u < \frac{1}{3} \\ 1 - 6.75u(u-1)^2 & \text{if } \frac{1}{3} \leq u \leq 1 \\ 1 & \text{if } u > 1. \end{cases} \quad (3)$$

All the variables appearing in eqns. 1-3 are dimensionless quantities obtained by scaling of physical variables. For typical conditions, $u = 0$ corresponds to a CO (O) surface coverage of

0.65 (0.02) monolayers (ML), whereas $u = 1$ represents 0.2 ML (0.25 ML) of CO (O) coverage. The dimensionless space and time units correspond roughly to 1 μm and 2 sec respectively. The external control parameters – CO partial pressure p_{CO}, oxygen partial pressure p_{O_2} and surface temperature T – have been mapped into a, b and ϵ in the above equations. a indicates whether the system can be bistable ($a > 1$) or oscillatory/excitable ($a < 1$) and b describes the balance of the CO and oxygen on the surface. $b = 0$ corresponds to equal amounts of both adsorbates, while $b > 0$ ($b < 0$) indicates dominance of CO (O). The inert domain boundaries are modelled by zero-flux boundary conditions. The grain boundaries in composite surfaces are modelled by a step change in the kinetic parameters while the diffusion constants are taken to be the same for the two components of a composite medium. The color code used for simulation results displays CO (O) rich as white (black), in correspondence with the experimental images.

The results are organized as follows: In section 4 we discuss the influence of sharp corners in the shape of the no-flux boundary on the propagation of planar waves. Results on the behavior of chemical waves in composite media are then presented in section 5.

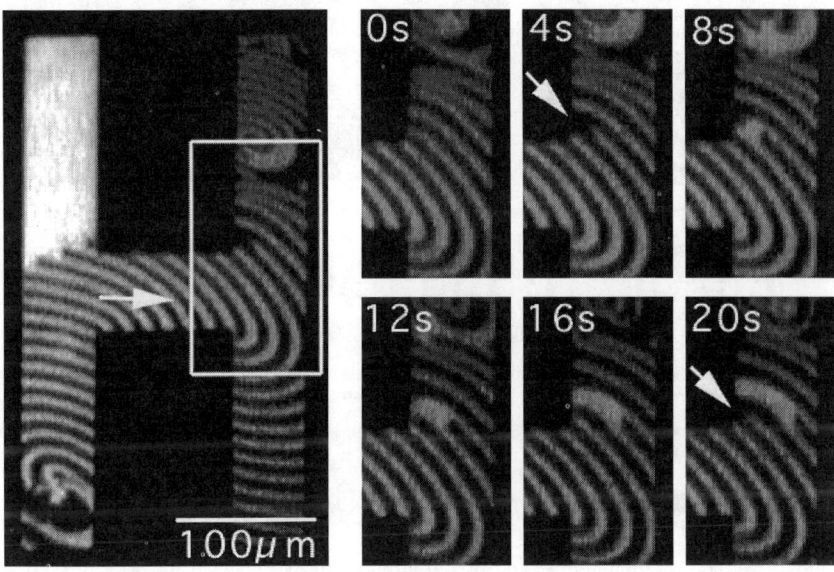

Figure 1: Concentration wave trains "rounding corners" in an H-shaped Pt domain. The sequence of images on the right shows an instability where some of the waves detach and retract from the boundary. The experimental conditions are: $p_{O_2} = 4 \times 10^{-4} mbar$, $p_{CO} = 5 \times 10^{-5} mbar$, $T = 440K$.

4. Microstructures: Wave-trains at sharp corners

The behavior of concentration fronts at the exit of narrow channels into larger domains has recently been studied [13]; these studies emphasized the importance of front curvature at the channel exits, causing slow-down and even reflection of the fronts. Planar fronts in general exist over a range of velocities $c_{min} \leq c \leq c_{max}$. Perturbations that affect the local front curvature K change the local wave velocity according to the approximate relationship $c(K) = c - DK$ [25]. Such a change may cause a drop in c below the minimum value c_{min} causing the waves to break locally or even extinguish.

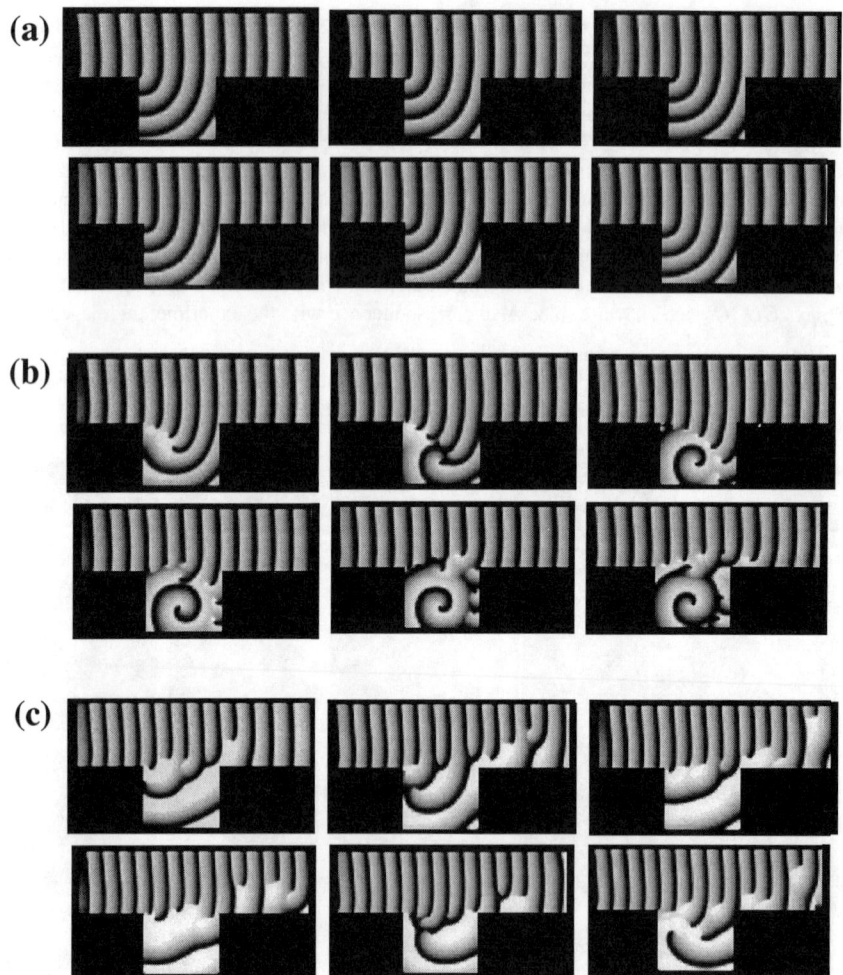

Figure 2: Numerical simulations of a wave train in a grooved channel for (a) $b = 0.16$, (b) $b = 0.18$, (c) $b = 0.20$. Other parameters for all simulations are fixed at $\epsilon = 0.025$, $a = 0.84$.

To study the effects of curvature on wave propagation computationally, we construct 90 degree corners as a model geometry that induces front curvature. Fig. 1 displays an experimental image showing a spiral in the bottom left leg of an H-shaped Pt domain surrounded by inert Ti. The spiral acts as a source of a periodic wave-train that travels around several corners in the domain. While the waves at the bottom left and right corners of the middle bar of 'H' stay attached to the boundary as they progress, some of the waves going around the upper right corner show an instability and retract from the boundary. A series of snapshots in Fig. 1 shows the time evolution of waves at this corner. The anisotropy of the diffusion constant is observed to play an important role in the instability. The waves that go around the lower right corner of the middle bar of the 'H' attain much larger curvature values compared to the ones going around the upper right corner. A second important factor is the angle of incidence the waves form with the boundary. At the upper right corner of the middle bar, the direction of the incident waves is no longer parallel to the boundary; eventually the tangential growth of the wave becomes too small to sustain attachment of the waves to the boundary.

Numerical simulations of waves in a related geometry are displayed in Fig. 2. Here, a wave source has been imposed on the system by a Dirichlet-boundary condition on the left end of the channel. The corner is realized by introducing a deep "groove" in the bottom channel wall. At higher excitability (lower values of b, Fig. 2a), the waves curve strongly at the corner while still remaining attached to the boundary. At a slightly lower excitability (Fig. 2b), the abrupt change at the corner results in the detachment of the waves from the boundary. The open end of the wave curls up to form a spiral wave in the groove that coexists with the planar wavetrain persisting in the upper channel. This phenomenon has also been observed in experiments with the Belousov-Zhabotinsky reaction [14]. At even weaker excitability, the new "mode" that forms in the groove finally invades the flow of the periodic wave-train leading to irregular behavior (Fig. 2c).

Figure 2 displays the leading modes and their corresponding "energies" extracted using principal component analysis (Karhunen-Loève expansion [15]) of simulation data. The modes come in pairs of comparable energies where members of each pair are phase shifted versions of each other (this is typical of travelling wave type solutions). The analysis confirms that the behavior seen in Fig. 2a is periodic (Fig. 3a), while the flow in Fig. 2b consists of a quasiperiodic pattern (Fig. 3b) with competing localized coherent modes of different frequencies. Such a sustained coexistence of wave patterns with different temporal frequencies is rare in reaction-diffusion systems; usually, the source with higher frequency suppresses the one with lower frequency (see *e. g.* [5,10]). But here, the spiral mode in the groove survives, because it interacts only laterally with the open ends of the wavetrain initiated at the left groove boundary. The "corner instability", as should be expected, depends strongly on the frequency of the incoming waves: the higher the frequency the more unstable the wavetrain. A more systematic theoretical investigation should also take into account diffusion anisotropy as well as variations in the geometry. The analysis for the case shown in Fig. 2c does not produce a stationary set of modes (Fig. 3c) indicating presumably "turbulent" (spatiotemporally chaotic) behavior. In addition, the number of modes needed to capture the space-time dynamics increases considerably. The observed phenomena are important in the propagation of reaction-diffusion waves in porous media or random networks, because the discussed instabilities will completely change the "flow" in such non-trivial geometries.

5. Microcomposites: Wave transmission across grain boundaries

The basic idea here is to couple, on a micrometer scale, two different media which individually support wave-type solutions. The presence of "grain boundaries" has a significant effect on wave propagation as they have to adapt in speed and shape to the new conditions of the different medium. An experimental realization of a composite catalyst is given by the coupling of a pure $Pt(110)$ surface with another medium consisting of Pt surface covered with submonolayer levels (up to 5%) of Gold [16,17] or Titanium [18]. We look at cases where a spiral in one of the media "sends" waves towards the interface. Fig. 4 shows the case of a Ti-based composite surface where complete and partial transmission of waves is observed under identical conditions at different locations on the boundary. Similar observations of complete transmission and a preferred formation of spirals near a boundary of pure Pt and Au-covered catalyst were reported by Asakura *et. al.* [17].

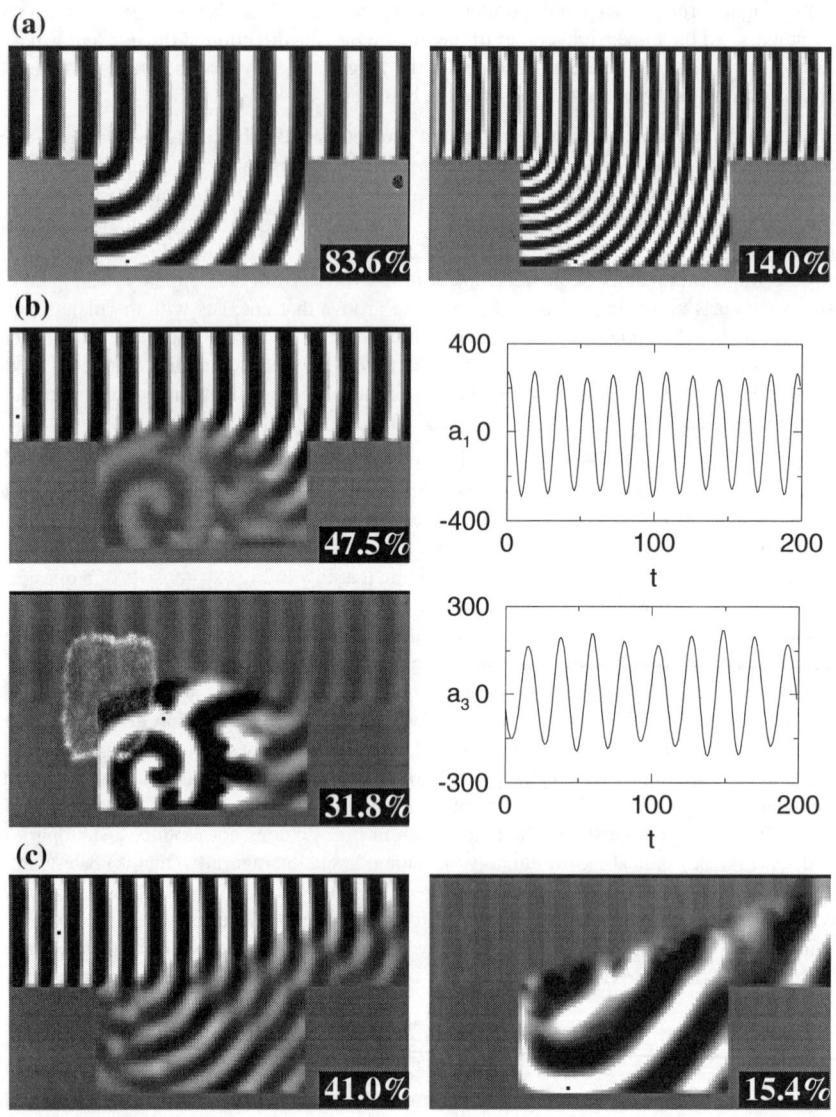

Figure 3: Karhunen-Loève modes obtained from grooved channel simulations at (a) $b = 0.16$ (periodic), (b) $b = 0.18$ (quasiperiodic), (c) $b = 0.2$. The time history of the modal coefficients is shown in (b) for $b = 0.18$.

Model calculations with such media have been conducted in one- and two- dimensions [19]. Upon increase in the difference in the kinetic parameter b (a measure of excitability) between the two halves of the composite medium, complete as well as partial (every second wave) transmission of wave trains has been observed for planar, one-dimensional waves. In two dimensions, the scenario is even richer: partial transmission may lead to formation of new spirals at the boundary, that either suppress the original source or they are constantly forced by the incident waves forming a state resembling "spiral turbulence" [20,21]. Only for very weak excitability of the second surface component, the case of 1:2 transmission (where every second wave manages to pass across the interface), easily found in one spatial dimension, is recovered. The experimental picture in Fig. 4 shows such a situation. However, if the boundary is not normal to the propagation direction of the incoming waves as in the middle part of the F in Fig. 4, complete transmission is still possible. This geometry related effect is reproduced in a simulation displayed in Fig. 5.

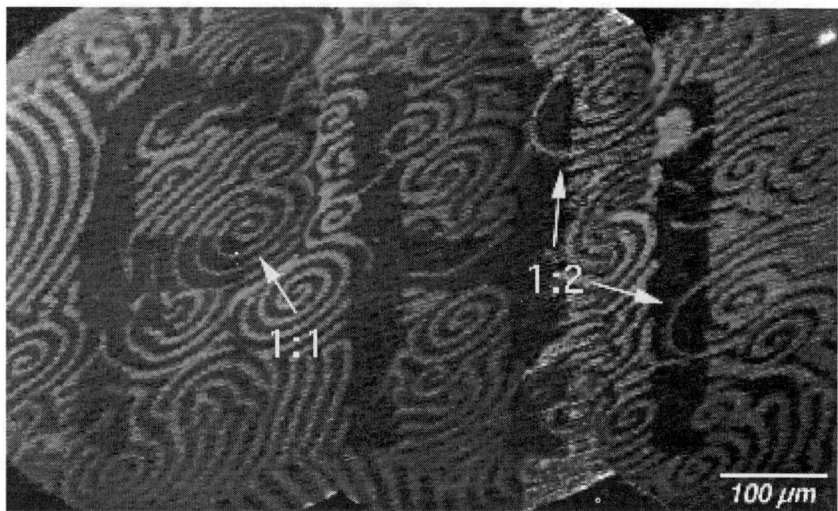

Figure 4: PEEM image showing the interaction of spiral waves with grain boundaries on a composite surface. The dark region in the shape of FHI is clean platinum surface. The rest of the surface is covered by submonolayer levels of Ti. The experimental conditions are: $p_{O_2} = 4 \times 10^{-4} mbar, p_{CO} = 5.5 \times 10^{-5} mbar, T = 440K$.

An important further goal is the study of composite media with spatially periodic property variations. In such situations, the length scale of the variation λ_v interacts with the length of diffusion λ_d and the pattern wavelength λ_p. In the case where λ_v is smaller than λ_d and λ_p, indications for a nonlinear effective medium behavior have been observed in initial computational studies [22] and recent experiments [23].

6. Conclusions

We have investigated pattern formation during the catalytic oxidation of CO on microstructured and composite surfaces. The influence of the domain boundary shape leads to novel effects when one of the length scales of the microstructures becomes comparable to or smaller than the intrinsic length scales (diffusional length, wavelength) of the reaction-diffusion system [9,13,22-24] or when the perturbation of the waves due to the boundary leads to the excitation of new dynamical modes (e. g. spiral waves) as demonstrated in the examples given here. A systematic variation of this length scale by appropriate microdesign is one of our future research goals. Obviously, the type of pattern can be controlled by the choice of material, geometry and scale of the microstructures and composite. Other forms of spatial control that will be explored

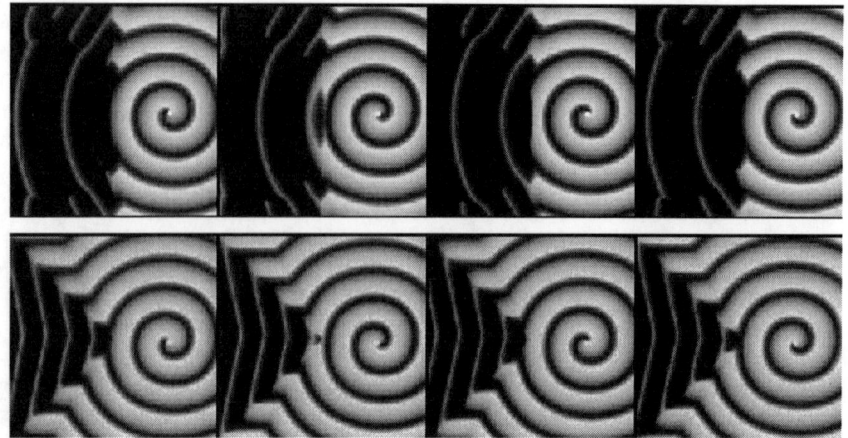

Figure 5: Transmission of waves across boundaries at different angles of incidence: 1:2 transmission (top) across an interface normal to the propagation direction; 1:1 transmission (bottom) for a different incidence angle. The model parameters are $\epsilon = 0.025$, $a = 0.84$, $b_1 = 0.07$, $b_2 = -0.27$.

in the future are spatially inhomogeneous and temporally varying illumination of the surface by laser light, resulting in localized heating. This method also allows for real-time feedback between the spontaneous patterns and the imposed structure. Methods used to study catalytic pattern formation at high (atmospheric) pressure [27] should also combine fruitfully with techniques for constructing microdesigned surfaces. The study of composite catalysts has already been extended to other surface reactions e.g. to the NO reduction with CO [24]. A long term goal is the finding of optimal scales and geometries which, through pattern selection, may influence and optimize the overall selectivity and reactivity of the catalysts.

Acknowledgements. We would like to acknowledge the support by the National Science Foundation-NSF and ARPA/ONR (IGK, AKB) and the Deutsche Forschungsgemeinschaft-DFG (MB).

References

1. G. Ertl, "Oscillatory Kinetics and Spatiotemporal Self-organization in Reactions at Solid Surfaces", Science, 254 (1991), 1750.

2. R. Imbihl and G. Ertl, "Oscillatory Kinetics in Heterogeneous Catalysis", Chemical Reviews, 95 (1995), 697.

3. M. Eiswirth and G. Ertl, in Chemical Waves and Patterns, ed. R.Kapral and K. Showalter (Kluwer, Dordrecht, Netherlands, 1995).

4. S. Jakubith, H. H. Rotermund, W. Engel, A. von Oertzen and G. Ertl, "Spatiotemporal Concentration Patterns in a Surface Reaction: Propagating and Standing Waves, Rotating Spirals and Turbulence", Phys. Rev. Lett., 65 (1990), 3013.

5. S. Nettesheim, A. von Oertzen, H.-H. Rotermund and G. Ertl, "Reaction Diffusion Patterns in the Catalytic CO-oxidation on Pt(110): Front Propagation and Spiral Waves", J. Chem. Phys., 98 (1993), 9977.

6. A. S. Mikhailov, Foundations of Synergetics I (Springer-Verlag, Berlin, 1990).

7. M. C. Cross and P. C. Hohenberg, "Pattern Formation outside of Equilibrium", Rev. Mod. Phys., 65(1993), 851.

8. M. Graham, I. G. Kevrekidis, K. Asakura, J. Lauterbach, K. Krischer, H. H. Rotermund and G. Ertl, "Effects of Boundaries on Pattern Formation: Catalytic Oxidation of CO on Platinum", Science, 264 (1994), 80.

9. Graham, M., Bär, M., Kevrekidis, I.G., Asakura, K., Lauterbach, J., Rotermund, H.-H., and Ertl, G., "Catalysis on Microstructured Surfaces: Pattern Formation during CO Oxidation in Complex Pt Domains", Phys. Rev. E, 52 (1995), 76.

10. T. Engel and G. Ertl, "Elementary Steps in the Catalytic Oxidation of Carbon Monoxide on Platinum Metals", Adv. Catalysis, 28 (1979) 1.

11. K. Krischer, M. Eiswirth and G. Ertl, "Periodic Perturbations of the Oscillatory CO Oxidation on Pt(110): Model Calculations", J. Chem. Phys., 97 (1992), 303-319.

12. M. Bär, N. Gottschalk, M. Eiswirth and G. Ertl, "Spiral Waves in a Surface Reaction: Model Calculations", J. Chem. Phys., 100 (1994), 1202-14.

13. G. Haas, M. Bär, I. G. Kevrekidis, P. B. Rasmussen, H. H. Rotermund and G. Ertl, "Observation of Front Bifurcations in Controlled Geometries: From One to Two dimensions", Phy. Rev. Lett., 75 (1995), 3560.

14. K. Agladze, J. P. Keener, S. C. Müller and A. Panfilov, "Rotating Spiral Waves Created by Geometry", Science, 264 (1994), 1746.

15. K. Fukunaga, Introduction to Statistical Pattern Recognition (Academic Press, New York, 1990).

16. K. Asakura, J. Lauterbach, H. H. Rotermund and G. Ertl, "Modification of Spatiotemporal Pattern Formation in an Excitable Medium by Continuous Variation of its Intrinsic Properties: CO oxidation on Pt(110)", Phys. Rev. B, 50 (1994), 8043.

17. K. Asakura, J. Lauterbach, H. H. Rotermund and G. Ertl, "Spatiotemporal Concentration Patterns Associated with the Catalytic Oxidation of CO on Au Covered Pt(110)", J. Chem. Phys., 102 (1995), 8175.

18. J. Lauterbach, K. Asakura, H.-H. Rotermund and G. Ertl, unpublished results.

19. M. Bär, I. G. Kevrekidis, H.-H. Rotermund, and G. Ertl, "On Pattern Formation in Composite Excitable Media", Phys. Rev. E. (in press, 1995).

20. P. Coullet, L. Gil and J. Lega, "Defect Mediated Turbulence", Phys. Rev. Lett., 62 (1989), 1619.

21. M. Hildebrand, M. Bär and M. Eiswirth, "Statistics of Topological Defects and Spatiotemporal Chaos in a Reaction-Diffusion System", Phys. Rev. Lett., 75 (1995), 1503.

22. A. K. Bangia, M. Bär, I. G. Kevrekidis, M. D. Graham, H. H. Rotermund and G. Ertl, "Catalysis on Microcomposite Surfaces", submitted to 14th International Symposium on Chemical Reaction Engineering (Brugge, Belgium, May 1996).

23. O. Steinbock, P. Kettunen and K. Showalter, "Anisotropy and Spiral Organizing Centers in Patterned Excitable Media", Science, 269 (1995), 1857.

24. N. Hartmann, M. Bär, I. G. Kevrekidis, K. Krischer and R. Imbihl, "Rotating Chemical

Waves in Small Circular Domains", Phys. Rev. Lett. (submitted, 1995).

25. J. J. Tyson and J. P. Keener, "Singular Perturbation Theory of Travelling Waves in Excitable Media", Physica D, 32 (1988), 327.

26. H. H. Rotermund, W. Engel, S. Jakubith, A. von Oertzen and G. Ertl, "Methods and Application of UV Photoelectron Microscopy in Heterogeneous Catalysis", Ultramicroscopy, 36 (1991), 164.

27. H. H. Rotermund, G. Haas, R. U. Franz, R. M. Tromp and G. Ertl, "Imaging Pattern Formation in Surface Reactions from Ultra High Vacuum to Atmospheric Pressure", Science, 270 (1995), 608.

STRUCTURAL STABILITY

OF SPINODAL DECOMPOSITION

Yoshitsugu Oono

Department of Physics and Beckman Institute
University of Illinois at Urbana-Champaign
405 N Mathews, Urbana, IL 61801

Abstract

The Cahn-Hilliard equation is now the standard equation to describe the spinodal decomposition process in binary alloys. The equation turns out to be quantitatively correct asymptotically in time. I will explain how this conclusion can be reached with the aid of a discrete space time modeling method called the cell dynamical system (CDS) approach. The reason for the quantitative success of the Cahn-Hilliard equation is the universality or the structural stability, i.e., its insensitivity to details. However, this universality has been established only empirically. Theoretically relevant facts are the universality of the interface dynamics equation, and the dynamical universality of the interface (universality in the dispersion relation) that exists even when the static universality does not. A method to derive the interface equation of motion (e.g., the Allen-Cahn equation) as a renormalization-group equation is described. These theoretical universalities seem to be not enough to support the universality observed empirically. Other systems, such as binary fluid systems, block copolymers, etc., are also briefly discussed, because they shed considerable light on the nature of the Cahn-Hilliard equation. CDS modeling of spinodal decomposition has taught us how to think flexibly of the numerical schemes to solve partial differential equations. Consequences of the flexible approach are discussed towards the end. A hyperbolic equation solver and a Navier-Stokes equation solver based on the physics-motivated algorithms are mentioned as byproducts of our study.

Introduction
The Cahn-Hilliard equation [1] is the standard equation to describe spinodal decomposition of symmetric binary alloys:
$$\frac{\partial \psi}{\partial t} = \Delta(-\psi + \psi^3 - \Delta\psi). \tag{1}$$
Here all the numerical parameters are absorbed into the dependent and independent variables through scaling. The equation was initially introduced to understand the early stages of spinodal decomposition.[1] Certainly, it is a semiquantitatively good model for early stages.

In this paper I wish to stress that the true excellence of the equation lies in its asymptotic long time behavior. As everybody knows the most often studied quantity is the form factor defined by
$$S(k) = \langle \psi_k \psi_{-k} \rangle, \tag{2}$$
where ψ_k is the Fourier transform of the field ψ and $\langle \ \rangle$ is the ensemble average, which can often be replaced by the spatial average over the sample. It is widely established that the form factor at different times can be superposed onto a single master curve [2] through scaling:
$$F(x) = \langle k \rangle^d S(x\langle k \rangle), \tag{3}$$
where $\langle k \rangle$ is, e.g., the peak position of the form factor (the reciprocal of the pattern size). Let us call this the *master form factor*.

I assert that the Cahn-Hilliard equation is THE equation for the symmetric binary alloy in the sense that it *quantitatively* describes the master form factor of 'any' equation of the form
$$\partial_t \psi = \Delta(\mu(\psi) - \Delta\psi), \tag{4}$$
where $\mu(\psi)$ has a symmetric double well potential. To understand this claim, probably the easiest way is to trace how I reached this conclusion.

Cell Dynamical System Modeling
Initially, I wished to study the long time asymptotic behavior of critically deeply quenched binary alloys. No one can solve the Cahn-Hilliard equation analytically. Hence, we must use computers. Consequently, discretization is inevitable. The time increment and spatial increment cannot be taken large, if I wish to claim that the partial differential equation is actually solved. It should clearly be recognized that conscientious numerical simulation in the standard sense of this word of the Cahn-Hilliard equation for a long time has never been performed. We must also clearly recognize that no one can derive the equation from the underlying microscopic dynamical picture. Hence, why do I have to respect the Cahn-Hilliard equation? After all, Nature gives us phenomena, not equations. I should try to capture the essence of physics directly on the computer, so I should invent a space-time discrete model *de novo*. The model consists of a lattice (in the physicists' sense) of cells in each of which lives a continuous scalar variable $\psi_t(n)$, where n denotes the spatial address of the cell and t the time. I call this type of models cell dynamical system (CDS) models.

The CDS modeling consists of three steps. Let us consider a symmetric binary alloy consisting of atoms A and atoms B.
(i) When one looks at a cell containing more A (resp., B), at later times it would contain more A (resp., B) further. If $\psi = 1$ (resp., -1) corresponds to the pure A (resp., B) phase, then the local dynamics governing the relation between time t and $t + 1$ is governed by a symmetric hyperbolic map[2] with two attractors at $\psi = \pm 1$ and a repeller at 0. For

---
[1]During my talk, Dr. Cahn confirmed this. In short, he did not fully recognize the excellence as well as the difficulty of the equation initially.

[2]Hyperbolicity means that around any fixed point the slope of the function defining the map does not vanish.

example,
$$\psi_{t+1}(n) = 1.3\tanh\psi_t(n). \tag{5}$$

(ii) Once segregated, a bulk phase never spontaneously produces large concentration fluctuations. Hence, there must be a stabilizing mechanism of the uniform pure phases. The easiest way is to add a diffusion coupling:

$$\psi_{t+1}(n) = 1.3\tanh\psi_t(n) + 0.5[\langle\langle\psi\rangle\rangle - \psi_t(n)]. \tag{6}$$

Here $\langle\langle\ \rangle\rangle$ is the average of the concentration field surrounding the cell n. The average must be carefully chosen to maintain the isotropy during long computations.
(iii) Alchemy is impossible. Hence, if a cell wishes to increase its A, it must get A from its surrounding cells. Let us rewrite (6) in the following form:

$$\psi_{t+1}(n) = \psi_t(n) + \mathcal{I}_t(n). \tag{7}$$

Here $\mathcal{I}$ may be interpreted as a fictitious increment allowing alchemy. This increment is the 'budget demand' of the cell n. However, its surrounding cells also have their own demands. Therefore, the cell n can get, on the average, the difference between its own demand and the average demand of its neighbors $\langle\langle\mathcal{I}\rangle\rangle$. In this way the CDS model of spinodal decomposition is reached [3]:

$$\psi_{t+1}(n) = \psi_t(n) + \mathcal{I}_t(n) - \langle\langle\mathcal{I}_t\rangle\rangle, \tag{8}$$

where, in our case

$$\mathcal{I}_t(n) = 1.3\tanh\psi_t(n) + 0.5[\langle\langle\psi\rangle\rangle - \psi_t(n)] - \psi_t(n). \tag{9}$$

Universality in Spinodal Decomposition
The model was, naturally, not well received initially; they claimed that the model was very arbitrary. However, we have empirically found that the master form factor obtained from the CDS spinodal model is highly insensitive to the actual choice of the map so long as it is symmetric with two hyperbolic sinks and one hyperbolic source [4]. In any case with the numerical precision we can afford, the difference between the results for different maps is not distinguishable from that due to the difference in initial uniform random patterns. Adding noise to the system does not change its asymptotic behavior, although dynamics is considerably slowed down [5].

Now, let us introduce a special discretization scheme for the Cahn-Hilliard equation.[3] The overall Laplacian in (1) is to impose the conservation law. Hence, I wish to discretize the semilinear parabolic operator inside the overall Laplacian. We solve the ordinary differential equation

$$\frac{d\psi}{dt} = \psi - \psi^3 \tag{10}$$

in the following form with the aid of the semi-group operator $\mathcal{F}_s$ (this is an algorithm to construct a map from a continuum model; I need not be able to do it analytically)

$$\psi_{t+s} = \mathcal{F}_s(\psi_t). \tag{11}$$

The discretization scheme, which I call a semi-group scheme, is

$$\psi_{t+\delta t}(x) = \mathcal{F}_{\delta t}(\psi_t(x)) + \delta t[\Delta]\psi_t(x), \tag{12}$$

---
[3] My colleague Nigel Goldenfeld insisted that I should find a relation between the standard approach and my heterodox approach. This scheme owes its existence to his insistence.

where $[\Delta]$ is a discretized Laplacian. Notice that $[\Delta]f \propto \langle\langle f \rangle\rangle - f$. The semigroup can be explicitly computed in our case as

$$\mathcal{F}_{\delta t}(f) = f/[e^{-\delta t} + f^2(1 - e^{-\delta t})]^{1/2}, \qquad (13)$$

which has exactly the form of the map required in (i) above. We impose (iii) above to (12) as before. If δt and the spatial increment are small enough (with some stability conditions satisfied), the resultant scheme converges to the simple Euler scheme. Thus, the empirical (computational) universal result mentioned above combined with this observation implies that the Cahn-Hilliard equation and the CDS models must have indistinguishable master form factors.

This of course implies that the Cahn-Hilliard equation gives the same asymptotic form factor for any choice of symmetric double well potentials. Miraculously, the Cahn-Hilliard equation is asymptotically quantitatively correct. We can even design a free energy function which can accelerate CDS simulations without altering results.

In Asymptopia
However, it is impossible to check the above conclusion directly, because an honest numerical solution of the fourth order nonlinear partial differential equation for very long time for very large system is still prohibitively difficult. All the published numerical long time results on the Cahn-Hilliard equation should be interpreted as the numerical results due to inefficient or not optimized CDS models.

In order to observe a truly asymptotic form factor, the interface thickness must be negligible relative to the average pattern size (domain width). I know at least this ratio must be 30 [6]. I must have at least 7 layers in the system to have a reliable near peak structure of the form factor. The interface thickness cannot be less than two lattices even with the optimized CDS model. This implies that the system size (measured in the cell size) must be at least 500^3 and at least a 80,000 time step computation is needed. This is still prohibitively difficult. My system was 196^3 with up to 40,000 time steps.[4] Hence, I have not reached the asymptopia.

However, the hardening trick $\psi \to sgn(\psi)$ (that is, $\psi > 0$ (resp., < 0) is mapped to $+1$ (resp., -1)) before constructing the form factor gives much faster convergence of the form factor. In this sense, our result is enough to fix the asymptotic form factor. Therefore, in this sense, we have reached the asymptopia [7].

How can I be sure? Unhardened form factors give lower bounds, and hardened ones give upper bounds of the form factor, so that I can check the convergence numerically. Also the form factor obtained from the hardened result satisfies the following asymptotic laws: Porod's law ($S(k)/k^{d+1} \to C_1$ (a positive constant) in the $k \to \infty$ limit) [8], Yeung's law ($S(k)/k^4 \to$ const. for $k \to 0$) [9], and Tomita's sum rule ($\int dk[k^{d+1}S(k) - C_1)] = 0$) [10]. Porod's law implies that the interface curvature is much larger than the interface thickness, Yeung's law implies the mesoscopic uniformity of the pattern, and Tomita's sum rule implies the smoothness of the interface.

Why Is There Universality?
The Cahn-Hilliard equation is quantitatively good, because there is a sort of universality, at least empirically. Thus, the question we must answer is why there is this universality. The naivest guess physicists make in these days is that this is due to some underlying renormalization group structure. Indeed this argument was used to explain why there is a master form factor for each system; this is due to the statistical self-similarity in the

---

[4] With the honest Euler scheme, δt and δx must be much much smaller than in the CDS model, so the system 1000^3 with a million time steps with the Euler scheme should still be too small to do the computation matching to ours.

segregation pattern [11]. The universality discussed here is, however, a much stronger version independent of each system. It is usually claimed that the asymptotic behavior is determined by the interface dynamics; its universality is the reason for the universality we observe. How much is this claim reliable?

Interface Dynamics Derived by RG

There are many ways to derive interface dynamics equations, but the RG approach explained here suggests the time window where the equation is reliable. For simplicity, I will explain the main idea using the following simple semilinear parabolic equation:

$$\frac{\partial \psi}{\partial t} = \psi - \psi^3 + \Delta \psi. \tag{14}$$

It has a plane kink solution ψ_K. I split (14) into two parts as

$$\frac{\partial \psi}{\partial t} = \left(\psi - \psi^3 + \frac{\partial^2 \psi}{\partial z^2}\right) + \Delta_2 \psi, \tag{15}$$

where Δ_2 is the Laplacian acting only on the variables x and y. The last term is regarded as a perturbation. Consider the following general codimension one kink solution as the unperturbed solution:

$$\psi_0(\boldsymbol{r}) = \psi_K(z - f(x, y)). \tag{16}$$

Here $z = f(x, y)$ describes the interface shape.

To treat the Δ_2 term as a perturbation, f must be a gentle function of x and y. Assuming $\psi = \psi_K + \varphi$, we get to the lowest nontrivial order

$$\frac{\partial \varphi}{\partial t} = L\varphi - H\psi'_K(z - f) + (\nabla_2 f)^2 \psi''_K(z - f), \tag{17}$$

where L is the linearized operator:

$$L = 1 - 3\psi_K^2(z - f) + \partial_z^2. \tag{18}$$

H is twice the mean curvature of the interface, and ∇_2 is the gradient acting on the functions of x and y only. Notice that $\psi'_K(z - f)$ is the zero eigenfunction of the operator L (corresponding to the Nambu-Goldstone mode). Notice further that ψ''_K is orthogonal to this function. Hence, the bare perturbation series reads

$$\psi(\boldsymbol{r}) = \psi_K(z - f(x, y)) - tH\psi'_K(z - f(x, y)) + \cdots. \tag{19}$$

This result is meaningful, as it is, only for the time range such that $tH \ll 1$. A general method to make this uniform over the time range up to $tH \sim 1$ is the renormalization group method [12]. I introduce renormalized interface $f_R(x, y, \tau)$, and absorb the secular term into this as

$$\psi(\boldsymbol{r}) = \psi_K(z - f_R(x, y, \tau)) - (t - \tau)H\psi'_K(z - f_R(x, y, \tau)) + \cdots, \tag{20}$$

which is the renormalized perturbation result. Since τ is arbitrary, and is not in the problem, the solution ψ should not depend on this parameter. Hence, partially differentiating the above equation w.r.t. τ, I obtain

$$-\frac{\partial f_R}{\partial \tau}\psi'_K(z - f_R) + H\psi'_K(z - f_R) = 0 \tag{21}$$

to the lowest order. Identifying τ and t, I arrive at

$$\psi(\boldsymbol{r}) = \psi_K(z - f_R(x, y, t)) \tag{22}$$

and f_R is governed by
$$\frac{\partial f_R}{\partial \tau} = H, \qquad (23)$$
which is the Allen-Cahn interface equation of motion [13]. This is reliable up to the time scale of $tH \sim 1$. Since $H \sim t^{-1/2}$, the equation cannot be reliable forever.

The case of the Cahn-Hilliard equation is not this simple, because the Nambu-Goldstone mode is not isolated from the essential spectrum. The equation corresponding to (17) is
$$\frac{\partial \varphi}{\partial t} = -\Delta[L\varphi - H\psi'_K(z-f) + (\nabla_2 f)^2 \psi''_K(z-f)]. \qquad (24)$$

I have not been able to construct the most singular solution of this equation, but if $(-\Delta)^{-1}$ is applied to the equation, then
$$\frac{\partial (-\Delta)^{-1}\varphi}{\partial t} = L\varphi - H\psi'_K(z-f) + (\nabla_2 f)^2 \psi''_K(z-f)] + h, \qquad (25)$$
where h is a harmonic function needed to satisfy the boundary conditions. This can be solved perturbatively as before:
$$(-\Delta)^{-1}\varphi = -tH\psi'_K(z-f) + \cdots. \qquad (26)$$

Again, I have ignored harmless nonsecular terms. From this, we can recover the standard result as an RG equation, which was first derived by Kawasaki and Ohta [14]. The result is reliable up to the time scale of $tH \sim 1$. In this case it is believed that $H \propto t^{-1/3}$, so the reliable time window of the equation is narrower than the Allen-Cahn case discussed above.

In the above RG derivation, the RG idea is used to extract the invariance of the interface dynamics on the initial interface shape (provided it is not excessively wrinkled). However, the outcome is model-free, i.e., the initial-shape independence implies model independence.

Unexpected Universality of Interface Dynamics

Shinozaki and I have studies the interface dynamics, or more precisely, the interface dispersion relation (the relaxation spectrum) of the kink solution [15]. The interface is perturbed to have a transversal wave of wave number k, which decays exponentially with the decay rate $\omega(k)$. For various free energy functions, we have empirically obtained the following universal relation
$$\omega(k) = Mk^3(1 + 2.9k\xi)/\xi \qquad (27)$$
correct up to $k\xi \sim 3$. Here M is the mobility of the system and ξ is the interface thickness defined by
$$\xi \equiv \left(\int_{-\infty}^{\infty} dz \psi'_K(z)\right)^2 / 2 \int_{\infty}^{\infty} dz (\psi'_K(z))^2, \qquad (28)$$
where ψ_K is the interface profile. A remarkable feature of this universality is that it holds for any potential even though we cannot superpose the interface profiles. That is, although there is no static or structural universality, there is a dynamical universality. This supports the universality observed when the pattern width is about one order or more larger than the interface thickness.

In the above universal dispersion relation, the $k \to 0$ limit behavior, that is, $\lim_{k\to\infty}\omega(k)/k^3$ = const.(> 0) can be proved.[5] Since $k = 0$ is the bottom of the essential spectrum, the

---

[5] With one assumption that the second lowest eigenvalue which lies deep inside the essential spectrum of the linearized Cahn-Hilliard operator around the kink solution does not change the feature of the essential spectrum near $k = 0$ when the kink solution is perturbed to have a transversal wave.

Nambu-Goldstone mode is touching it in this limit. Hence, I could not think of any elegant proof. A short and elegant proof of this limiting behavior without any assumption is strongly desired.

The local structure, or the scaled order parameter correlation function up to $r\langle k\rangle \simeq 10$, can be explained in terms of the interface equation [16], but we have not yet succeeded in obtaining the longer length scale features of the correlation function or the main features of the scaled form factor around its main peak. This may suggest the following difficulty.

Weak Point of the Interface Argument
As we have seen the interface is dynamically very universal. Can we conclude from these results that the scaled form factor is universal? There are still two difficulties in this argument. One is of course that we do not know the real long time equation of interface dynamics. Hence, the interface equation may not be the right equation to describe the asymptotic behavior of the system. The other difficulty is the nonuniversal early stages of evolution. Even if the interface equation is truly asymptotic, it is meaningful only after the pattern size is sufficiently large. I believe that the equation has many different solutions. Hence, we need a 'correct' initial condition to solve the interface equation, which is derived from the initial condition at $t = 0$ for the Cahn-Hilliard equation. Thus it may well be the case that the right solution (or set of solutions) is already selected by the early stage before the interface dynamics dominates the system dynamics.

Although empirically (computationally) it is very likely that the Cahn-Hilliard equation enjoys a vast universality, theoretically, our understanding of the universality is still rather poor.

Nonuniversality in Reality
What do the real experiments tell us about the universality? For solid phase spinodal decomposition, no two master form factors agree with each other. That is, there is no empirical universality. The form factor obtained from the Cahn-Hilliard equation agrees with none.

This should be due to the differences in phase ordering dynamics,[6] anisotropy, defects, elasticity effects, etc. For 'ideal' symmetric binary alloys, there must be the universality discussed above. How can I be so sure about the universality against all these empirical counter examples?

Still the Cahn-Hilliard Equation Must Be Correct.
To support that the nonexistence of the universality is due to many complications in solid state, we must study a system free of such complications. If the system is fluid, and the two segregated phases have similar densities and viscosities, then many of the difficulties plaguing the solid phase should not exist. We must, however, pay the price of long-range hydrodynamic interactions, which is computationally costly (the CPU time needed becomes about one order longer than the binary alloys). We may assume that the fluid velocity field relaxes very quickly, so that we may adiabatically eliminate the velocity field. This was done for the first time by Kawasaki [17], and the resultant model is described by the following modification of the Cahn-Hilliard equation:

$$\frac{\partial \psi}{\partial t} = \left[\Delta - \nabla\psi \cdot \int dr' \mathcal{T}(r - r') \cdot \nabla\psi(r')\right] (-\psi + \psi^3 - \Delta\psi), \qquad (29)$$

where $\mathcal{T}$ is the Oseen tensor. I do not know any well-controlled argument to justify this. Probably, the leading order singular behavior is correctly captured by this equation, so to understand the global behavior, the equation may be used. This statement looks empirically correct.

---

[6]For ternary alloys such should be the case. This was pointed out by Dr. Cahn during my talk.

For binary symmetric fluid cases, there is a fairly good universality in the experimentally obtained form factors. Polymers [18] and small molecular fluids [19] both give the same universal form factors, which agree well with the asymptotic result obtained from (29) [7]. The asymptotic form factor does not depend on the free energy functions (or maps) nor viscosity of the fluid.

The interface equation of motion for binary fluids can be derived in a similar fashion as above (formally, because the inversion of the operator in the square bracket is needed). This time it is believed that $H \sim t^{-1}$ [20], so that there is a chance for the interface equation to be correct for all large t. However, the early time problems discussed before remain, and is in a certain sense more serious than the binary alloy case. In some systems a clear crossover of the growth exponent from 1/3 (solid like) to 1 (fluid like) is observed, but in other systems not. That is, the preasymptotic behavior is markedly non-universal. This nonuniversality can be understood as the viscosity effect. If the viscosity is large, we see a clear crossover. This behavior is reflected on the dispersion relation of the interface [21]. The Nambu-Goldstone eigenvalue is away from the essential spectrum for large k but deep inside it for small k. If the viscosity of the system is large, then this eigenvalue plunges into the essential spectrum at relatively large k, but if the viscosity is small, the eigenvalue stays away from the essential spectrum for wider range of k. This implies that the dispersion relation is solid-like for wider range of k, if the viscosity is larger, so that for early stages of the segregation, the system can exhibit solid like dispersion relation for a relatively long time.

In this case the limiting behavior $\lim_{k\to 0} \omega(k)/k = \text{const.}(> 0)$ is believed to be true, but I have no idea to prove it, because the relevant eigenvalue is deep inside the essential spectrum which scales like k^2.

Other Complications
If we pay respect to the Kawasaki exchange dynamics, then the mobility (or the Onsager coefficient) becomes order-parameter dependent. The simplest equation in this case was first discussed by Langer et al. [22] and by Kitahara and Imada [23]:

$$\frac{\partial \psi}{\partial t} = \nabla \cdot (1 - c\psi^2)\nabla(-\psi + \psi^3 - \Delta\psi), \tag{30}$$

where c is a positive constant. A corresponding CDS model was studied in conjunction to the paradox of uniform field applied to the Cahn-Hilliard equation. Note that the uniform bodily force field like gravity appears as an addition of a linear potential (say, $-gz$) in the Cahn-Hilliard equation. However, the overall Laplacian annihilates this term, so that there cannot be any bulk gravity effect however large the density difference between different phases may be. This is, of course, physically absurd. This absurdity comes from the composition independence of the mobility. The above Langer-Kitahara modification of the Cahn-Hilliard equation is the minimal sensible modification to correct this defect [24].

How serious is this modification when there is no external field? An extensive numerical study was performed long ago by Yeung (with the aid of a CDS version) [25]. If the free energy has a very steep penalty for the order parameter to go beyond its maximum or minimum values, and the mobility vanishes near the extremal values of the order parameter (in this case $c = 1$), then the bulk diffusion is prohibited, so that the growth is solely due to surface diffusion and the growth exponent becomes 1/4. For this model noise effect is important, and there is a tendency of recovering 1/3 with larger noise as expected. If c is less than one, then, the crossover to the usual 1/3 growth would occur in very late times, recovering the ordinary Cahn-Hilliard-like behavior. In this case extensive and truly asymptotic simulation has never been done due to its slow dynamics, but the currently available results strongly suggest that the form factor is not affected by the Langer-Kitahara modification.

What Can We Conclude about the Cahn-Hilliard Equation?
Thus, although many complications occur in reality, we may conclude that the Cahn-Hilliard equation quantitatively correctly describes the asymptotic behavior of *ideal* binary alloys. The conditions for the ideality are: (a) Static and dynamical isotropy (without elasticity effects); (b) Symmetrically double-welled free energy without long-range interactions (which gives the symmetric hyperbolic map mentioned above in the CDS); (c) Non-vanishing mobility which may depend on the order parameter (composition) but must be a symmetric function of the composition; (d) Impossibility of spontaneous formation of inhomogeneous structures in the equilibrium segregated phases; (e) Conservation of number of particles of each species.

The CDS modeling with its ease of construction and simulation allows us to explore various cases in which the above conditions are violated. An interesting case is an extreme asymmetric potential (map) case, for which the CDS local map has only one sink at $\psi = -1$, and $\psi = 1$ is a repeller. In this case uniform phase A (+1-phase) is unstable, but uniform B phase is stable. Which, A or B, aggregates more tightly? The answer is A. Being excluded out makes the excluded gather more tightly than the spontaneous gatherers. Readers wonder whether such an alloy of A and B exists. There cannot be, because pure solid A is stable. However, if A is a vibrating grain (granular materials lacking spontaneous cohesive tendency) and B is space (vacuum), the 'alloy' or the mixture is a vibrating non-dense granular system. With this special map (or corresponding single minimum potential at $\psi = -1$) the Cahn-Hilliard equation describes the behavior of grains on a vibrating plate [26].

The ease of CDS modeling could motivate new PDE models as the formal continuum limit of CDS models. A typical and perhaps the most successful example is the following CDS model [27]:

$$\psi_{t+1}(n) = (1 - b)\psi_t(n) + \mathcal{I}_t(n) - \langle\langle\mathcal{I}\rangle\rangle, \qquad (31)$$

where $\mathcal{I}$ is exactly the same as (8), and b is a small non-negative constant. This equation sheds considerable light on the Cahn-Hilliard equation. If $b \neq 0$, then this is drastically different from (8). This is the simplest model of diblock copolymer melt phase segregation; b is interpreted as the inverse square of the molecular weight. Puri later noted that this equation can model a phase segregating system undergoing a first order chemical reaction $A \leftrightarrow B$. From this we have proposed a continuum version

$$\frac{\partial \psi}{\partial t} = \Delta(-\psi + \psi^3 - \Delta\psi) - b\psi. \qquad (32)$$

The salient feature of the equation is that its asymptotic solution has a layered structure whose layer thickness is proportional to $b^{-1/3}$ [28]. With modification of the double well free energy function, we can make a solvable model, which exactly gives this power law [29]. This 1/3 and the 1/3 in the growth law of spinodal decomposition are mathematically the same [30]. Thus, if we demonstrated that there is only one length scale (pattern size) other than the interface thickness in the asymptotic solution, we could rigorously prove that the growth law is 1/3. In other words, if the master form factor really exists mathematically, then the power law must be 1/3.

Generalization of the CDS approach to multicomponent systems such as Al-Li [31] has been pursued extensively. For the formation of δ'-phase of Al-Li even a quantitative reproduction of experimental features can be achieved with a minimal modification of a very abstract model close to the minimal model. A nontrivial extension of the CDS approach is given recently by Zapotocky et al. to study the ordering process of nematic liquid crystals [32].

Lesson Learned: Physics-motivated Algorithms
What did I learn from the study of phase ordering except for respecting the Cahn-Hilliard

equation? Our study has demonstrated two things: (I) If you ask a right question, you can answer it quantitatively correctly, even if you use a crude numerical scheme; (II) direct discrete space-time modeling of physics can be a useful way to invent new PDEs as seen above and numerical algorithms to solve existing PDE (purportedly) describing the same physics. (I) is not surprising after the advent of renormalization group theory: to this end CDS scheme is almost the optimal numerical approach to the problem. However, we still need some 'art' to make an efficient CDS model as is exemplified well by the liquid crystal CDS.

The basic idea of devising a new algorithm (physics motivated algorithm) for a PDE is to capture the salient physics presumably described by the PDE directly on discrete space-time. For example, the difficulty of the Navier-Stokes equation or the advection-reaction equations lies in the advective term, which is a manifestation of the translational symmetry of the continuum space. Thus from my point of view the essence of numerical scheme for these equations is how to describe this symmetry in terms of discrete space. With the aid of interpolation formulas, we can overcome this difficulty. In this way we can describe flow dynamics directly on the computer, and the discrete scheme has turned out to be a good numerical scheme to solve the Navier-Stokes equation [33]. Similarly, an accurate representation of Huygens' principle gives a dispersion-free wave equation solver [34].

My idea about the physics-motivated PDE solver is actually slightly more radical: if our physics-motivated PDE solver cannot solve the PDE accurately, the PDE is wrong, i.e., it does not describe the phenomenon correctly.

Acknowledgements.
The results summarized above have been obtained by my collaboration with many coworkers, esp., Sanjay Puri, Chuck Yeung, Ari Shinozaki, and Monica Bahiana. Conversations with Hiroyuki Tomita and Shin-ichi Sasa were helpful in writing this manuscript. This work was, in part, supported by the National Science Foundation Grant NSF-93-14938.

References
1. J. W. Cahn and J. I. Hilliard, "Free Energy of a Nonuniform system I. Interface Free Energy," J. Chem. Phys., 28, 258 (1958); J. W. Cahn, " Phase Separation by spinodal Decomposition in Isotropic Systems," J. Chem. Phys., 42, 93 (1965).
2. J. L. Lebowitz, J. Marro and M. H. Kalos, "Dynamical Scaling of Structure Function in Quenched Binary Alloys" Acta. Metall., 30, 297-310 (1982).
3. Y. Oono and S. Puri, "Computationally Efficient Modeling of Ordering of Quenched Phases," Phys. Rev. Lett., 58, 836-839 (1987).
4. Y. Oono and S. Puri,"Study of Phase-separation Dynamics by Use of Cell Dynamical Systems. I. Modeling," Phys. Rev. A, 38, 434 (1988).
5. S. Puri and Y. Oono, "Effect of noise on spinodal decomposition," J. Phys. A, 21, L755-L762 (1988).
6. Y. Oono and S. Puri, "Large Wave Number Features of Form Factors for Phase Transition Kinetics," Mod. Phys. Lett. B, 861-867 (1988).
7. A. Shinozaki and Y. Oono, "Spinodal Decomposition in 3-space," Phys. Rev. E, 48, 2622-2654 (1993).
8. G. Porod, in "Small Angle X-ray Scattering," edited by O. Glatter and L. Kratky (Academic Press, New York, 1983).
9. C. Yeung, "Scaling and the Small -wave Vector Limit of the Form Factor in Phase-ordering Dynamics," Phys. Rev. Lett., 61, 1135-1138 (1988); H. Furukawa, Phys. Rev. A, 40, 2341 (1989); H. Tomita, Prog. Theor. Phys., 85, 47 (1991).
10. H. Tomita, "Sum Rule for Small Angle Scattering by Random Interface," Prog. Theor. Phys., 72, 656-658 (1984).
11. C. Roland and M. Grant, "Monte Carlo Renormalization-Group Study of the Late Stage of Phase Separation Process," Phys. Rev. Lett., 60, 2605 (1988).

12. L. Y. Chen, N. Goldenfeld and Y. Oono, "Renormalization Group Theory for Global Asymptotic Analysis" Phys. Rev. Lett., 73, 1311-1315 (1994); Phys. Rev. E to appear.
13. S. M. Allen and J. W. Cahn, Acta. Metall., 27, 1085 (1979).
14. K. Kawasaki and T. Ohta, "Kinetic Drumhead Model of Interfaces I," Prog. Theor. Phys., 67, 147-163 (1982).
15. A. Shinozaki and Y. Oono, "Dispersion Relation around the Kink Solution of the Cahn-Hilliard Equation," Phys. Rev. E, 47, 804-811 (1993).
16. C. Yeung, Y. Oono and A. Shinozaki, "Possibilities and Limitations of Gaussian-closure Approximations for Phase-ordering Dynamics," Phys. Rev. E, 49, 2693-2699 (1994); H. Tomita, Prog. Theor. Phys., 90, 521 (1993).
17. K. Kawasaki, "Theory of Early Stage Spinodal Decomposition in Fluids near the Critical Point I," Prog. Theor. Phys., 57, 826-839 (1977).
18. F. S. Bates and P. Wiltzius, "Spinodal Decomposition of a Symmetrical Critical Mixture of Deuterated and Protonated Polymers," J. Chem. Phys., 91, 3258 (1989); M. Takenaka and T. Hashimoto, "Scattering Study of Self-assembling Processes of Polymer Blends in Spinodal Decomposition II. Temperature Dependence," J. Chem. Phys., 96, 6177-6190 (1992).
19. N.-C. Wong and C. M. Knobler, J. Chem. Phys., 69, 725 (1978); Y. C. Chou and W. I. Goldburg, Phys. Rev. A, 23, 858 (1981).
20. E. D. Siggia, "Late Stages of Spinodal decomposition in Binary Mixtures," Phys. Rev. A, 20, 595-605 (1979).
21. A. Shinozaki, "Dispersion Relation around a Kink Solution in Binary Fluids Undergoing Spinodal Decomposition," Phys. Rev. E, 48, 1984-1988 (1993).
22. J. S. Langer, M. Bar-on, and H. D. Miller, "New Computational Method in the Theory of Spinodal Decomposition," Phys. Rev. A, 11, 1417-1429 (1975).
23. K. Kitahara and M. Imada, "On the Kinetic Equation for Binary Mixtures," Prog. Theor. Phys. supp., 64, 65-73 (1978).
24. K. Kitahara, Y. Oono and D. J. Jasnow, "Phase Separation Dynamics and External Force Field," Mod. Phys. Lett. B, 2, 765-771 (1988).
25. C. Yeung, "Some Problems on Spatial Patterns in Nonequilibrium Systems," (PhD Thesis, University of Illinois at Urbana-Champaign, 1989).
26. Y. Oono, "Cell-dynamics Modeling of Vibrating Powder," Int. J. Mod. Phys. B, 7, 1859-1864 (1993).
27. Y. Oono and Y. Shiwa, "Computationally Efficient Modeling of Block Copolymer and Benard Pattern Formations," Mod. Phys. Lett. B, 1, 49-55 (1989).
28. T. Ohta and K. Kawasaki, Macromolecules, 19, 2621 (1986).
29. Y. Oono and M. Bahiana, "Block Copolymer Lamellar Thickness; an Exactly Solvable Model," J. Phys. Condens. Matter, 1, 5297-5299 (1989).
30. Y. Oono and M. Bahiana, "2/3-Power Law for Copolymer Lamellar Thickness Implies a 1/3-Power Law for Spinodal Decomposition," Phys. Rev. Lett., 61, 1109-1112 (1988).
31. R. S. Goldstein, M. F. Zimmer, and Y. Oono, "Modeling Mesoscale Dynamics of Formation of δ'-phase in Al-Li Alloys," Mod. Phys. Lett., 7, 1083-1094 (1993).
32. M. Zapotocky, P. M. Goldbart, and N. Goldenfeld, "Kinetics of Phase Ordering in Uniaxial and Biaxial Nematic Films," Phys. Rev. E, 51, 1216-1235 (1995).
33. Y. Oono and A. Shinozaki, "Cell Dynamical Systems," Forma, 4, 75-102 (1989); T. Yabe and T. Aoki, Comp. Phys. Comm., 66, 219 (1991).
34. L. San Martin and Y. Oono, "Physics Motivated Numerical Solvers for Partial Differential Equations," submitted to Phys. Rev. E.

Cascades of spinodal decompositions in the ternary Cahn-Hilliard equations

David J. Eyre
Department of Mathematics
University of Utah
Salt Lake City, UT 84112
eyre@math.utah.edu

Abstract

The dynamics of the Cahn-Hilliard equations for a ternary alloy are studied during the phase separation process. The results show a primary spinodal decomposition followed by coarsening and then possibly a secondary spinodal decomposition.

1 Introduction.

In this paper, the Cahn-Hilliard (CH) equations for a ternary alloy will be used to study a sequence of spinodal decompositions. The equations model a ternary alloy quenched into a three phase region of the phase diagram.

In the numerical results presented below, it is observed that the solution of the CH equations, under mild restrictions, undergoes a primary spinodal decomposition and becomes a nearly binary material. This pseudo-binary then coarsens and may undergo an additional phase separation. This sequence of spinodal decompositions will be referred to as a cascade.

It is the goal of this paper to present results from numerical simulations that may be of interest to materials scientists, and to provide an intuitive understanding of the resulting observations rather than a rigorous analysis of them.

In the next section, the equations and the initial value problem for spinodal decomposition are given. Then the numerical simulations are presented in §3. A brief review of the primary spinodal decomposition is given in §4. Then the pseudo-binary solutions are studied in §5, and the secondary separation and coarsening processes are studied in §6 and §7.

2 The Cahn-Hilliard Equations.

Consider a ternary alloy held in a closed vessel Ω. Let $u_i(\vec{x}, t)$ for $i = 1, 2, 3$, represent the mole fraction of the components of the alloy as a function of space and time, then by mass conservation and positivity

$$\sum u_i(\vec{x}, t) = 1, \quad u_i(\vec{x}, t) \geq 0. \tag{1}$$

This restriction implies that (u_1, u_2, u_3) lies on a triangular region of the first orthant known as the Gibbs Triangle, which will be denoted by $\mathcal{G}$.

The Cahn-Hilliard energy functional [2] for this material is

$$\mathcal{F}(u_1, u_2, u_3) = \int_\Omega \left[\Phi(u_1, u_2, u_3) + \frac{\varepsilon^2}{2} \sum_i |\nabla u_i|^2 \right] d\vec{x}. \tag{2}$$

The function $\Phi(u_1, u_2, u_3)$ is the Gibbs free energy per unit volume, ε is the gradient energy coefficient, and ∇ is the spatial gradient operator. The average composition of the alloy is conserved because the alloy is held in a closed vessel, so

$$\frac{1}{|\Omega|} \int_\Omega u_i(\vec{x}) d\vec{x} = \bar{u}_i \quad \text{for } i = 1, 2, 3. \tag{3}$$

The alloy being modeled is assumed to be a regular solution [12], so the Gibbs free energy will have the form

$$\Phi(u_1, u_2, u_3) = \sigma \sum_{i \neq j} u_i u_j + \sum_{i=1}^{3} u_i \log u_i, \tag{4}$$

where σ is inversely proportional to temperature. If $\sigma \geq 3$, then $\Phi(u_1, u_2, u_3)$ has three minima and one local maximum in $\mathcal{G}$. This energy allows for one, two and three phase materials, along with spinodal and metastable regions of $\mathcal{G}$.

The dynamic CH equations are derived from (1)-(3) and can be represented as

$$\frac{\partial}{\partial t}\mathbf{u} = \Delta\left[-\varepsilon^2 \Delta \mathbf{u} + \mathbf{f}(\mathbf{u})\right] \tag{5}$$

where

$$\mathbf{u} \stackrel{\text{def}}{=} (u_1, u_2), \quad u_3 = 1 - u_1 - u_2, \quad \vec{x} \in \Omega, \quad t > 0. \tag{6}$$

The nonlinear function $\mathbf{f}(\mathbf{u})$ is

$$\mathbf{f}(\mathbf{u}) = P\nabla_u \Phi(u_1, u_2, u_3)|_{u_3 = 1 - u_1 - u_2} \tag{7}$$

where P is appropriate projector from $\mathbf{R}^3$ into the span of all vectors containing $\mathcal{G}$. The operator Δ is the Laplacian. As in the binary case, $\mathcal{F}(u_1, u_2, u_3)$ is a nonincreasing energy functional of the dynamic equations.

The natural and average conserving boundary conditions for the system are

$$\nabla \mathbf{u} \cdot \vec{n}\,|_{\vec{x} \in \partial \Omega} = \nabla(\Delta \mathbf{u}) \cdot \vec{n}\,|_{\vec{x} \in \partial \Omega} = \mathbf{0}. \tag{8}$$

respectively. The simulations presented in this paper will employ periodic boundary conditions. These boundary conditions also leave $\bar{\mathbf{u}}$ invariant in time.

The initial conditions used in this paper will model the nearly homogeneous pre-quenched material as

$$\mathbf{u}(\vec{x}, 0) = \bar{\mathbf{u}} + \tilde{\mathbf{u}}(\vec{x}), \quad |\tilde{\mathbf{u}}(\vec{x})|_\infty \ll 1, \tag{9}$$

where $\tilde{\mathbf{u}}(\vec{x})$ is a zero average, small perturbation of $\bar{\mathbf{u}}$.

Mathematical issues related to the Cahn-Hilliard equation have been reviewed recently by Elliott [4]. Generalization of the CH equations to multicomponent systems appears first with deFontaine [3], and with Morral and Cahn [11]. Subsequently the studies reappear recently in the metallurgical literature with Hoyt [10]. Mathematical results include the existence and uniqueness theory of Elliott and Luckhaus [5] and a survey by Eyre [6].

In particular, Elliott and Luckhaus [5] proved the well-posedness of the equations, and proved that the solution remains within $\mathcal{G}$ for all time, i.e. that each component remains positive.

3 Cascades of Spinodal Decomposition.

Numerical simulations are now given that show the spinodal decomposition processes of the ternary solution.

In the simulations, the average composition of the first two metals are equal and the average composition of the third metal is varied. The precise average for each simulation is given below. The parameters in the model were chosen to be

$$\sigma = 3.5, \quad \varepsilon^2 = 5 \times 10^{-4} \tag{10}$$

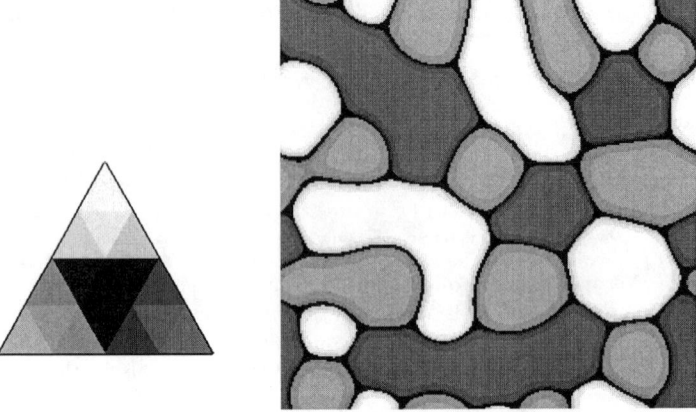

Figure 1: Gray scale representation of the Gibbs triangle and the spatial features of a solution.

and periodic boundary conditions were employed. In all the simulations presented the initial data (9) was random, generated uniformly on $(\bar{u} - 0.01, \bar{u} + 0.01)$.

The simulations are presented as time snapshots using a gray scale representation of **u** by mapping **u** to the shaded Gibbs triangle shown in Figure 1. Specifically, $\mathcal{G}$ was subdivided into 16 equal area triangles, and each subtriangle given a unique shade. Furthermore, notice that $\mathcal{G}$ in Figure 1 can be coarsely divided into four subtriangles where the top triangles are the lightest, the lower left triangles are dark, the lower right triangles are darker still, and central triangles are the darkest.

Figure 1 also presents a snapshot of the solution of the CH equations. The solution is shown in the physical domain on a 256^2 grid. The most striking feature of the solution is that it appears to be roughly piecewise-constant over large regions of space, defining three distinct phases. The three phases indicate the solution takes values near each vertex in $\mathcal{G}$ over large regions of space.

Between the distinct phases the solution has a graduated shading that represents the interfaces. Notice that the width of the interfaces is measurable and that the interfaces are diffuse. The solution in the interface can be estimated by comparing the interface shading with the gray scale map.

In Figures 2, 3, and 4 below, three different simulations are presented, and six snapshots are given for each simulation. Time increases across the rows and down the columns, so the earliest snapshot is on the top left and the latest snapshot is on the bottom right. The exact times for each simulation are given below, but in every case the snapshots are equally spaced on a logarithmic time scale. The first frame in every case shows the spinodally decomposed pseudo-binary solution. This solution has already evolved from the nearly constant initial conditions.

In all the simulations presented below, equilibrium theory given in [6] predicts that the lowest energy state of the material will have three phases.

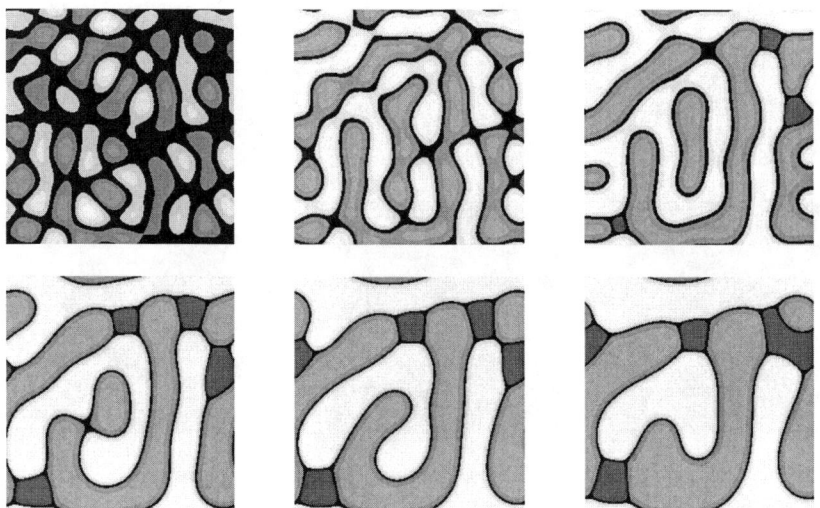

Figure 2: Three phase separation with $\bar{\mathbf{u}} = (0.43, 0.43)$.

The average composition of the simulation presented in Figure 2 is $\bar{\mathbf{u}} = (0.43, 0.43)$. The time of the earliest snapshot is $t_1 = 0.001$ and $t_{j+1} = 2.154\, t_j$. The early stages of the evolution are marked by the finest scale perturbations of $\mathbf{u}(x,0)$ smoothing rapidly followed by separation into a two phase material or a pseudo-binary. The pseudo-binary solution undergoes coarsening and then phase separates a distinct third phase. The third phase appears in the interfacial regions.

In Figure 3, the average composition of the solution is $\bar{\mathbf{u}} = (0.25, 0.25)$. The time of the earliest snapshot is $t_1 = 0.001$ and $t_{j+1} = 1.778\, t_j$. This simulation also shows a detailed pseudo-binary phase separation, followed by pattern coarsening, and finally a secondary phase separation.

The secondary phase separation processes shown in Figure 2 and 4 are quite different. Notice that in Figure 3, the primary separation leaves two distinct phases, and then in the secondary separation, one of those phases decomposes into two different phases. In the end, one phase of the pseudo-binary observed after primary separation is gone.

In Figure 2, during secondary separation, small inclusions of the third phase appear in regions where there was an interface between the phases of the pseudo-binary. In the end, both of the phases of the pseudo-binary are still present.

In Figure 4, the average composition of the solution is $\bar{\mathbf{u}} = (0.46, 0.46)$. The time of the earliest snapshot is $t_1 = 0.001$ and $t_{j+1} = 1.778\, t_j$. This simulation also shows a detailed pseudo-binary phase separation, followed by pattern coarsening, but this solution does not undergo secondary phase separation.

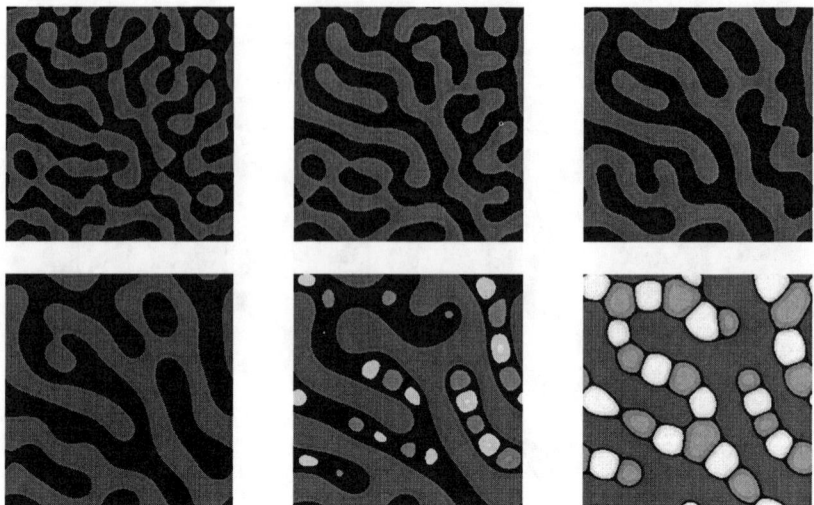

Figure 3: Three phase separation with $\bar{\mathbf{u}} = (0.25, 0.25)$.

4 Primary spinodal decomposition.

In this section, the transition from a nearly constant solution to a pseudo-binary solution is reviewed. This analysis is due to Morral and Cahn [11].

It can be easily seen that any constant value $\bar{\mathbf{u}}$ is a stationary solution of (5), and a natural question to consider is the stability of the average. For simplicity, the stability analysis will be carried out on $\Omega = [0, \ell_1] \times [0, \ell_2]$. In this case, the linearized CH equation is

$$\mathbf{u}_t = -\Delta \left[\varepsilon^2 \Delta \mathbf{u} - D(\bar{\mathbf{u}})\mathbf{u} \right] \quad (11)$$

where D is the Jacobian of $\mathbf{f}$. If n_1 and n_2 are integers, then by separation of variables, (11) admits a solution of the form

$$\mathbf{u}(\vec{x}, t) = \sum_k \tilde{\mathbf{v}}_k(t) \sum_{n_1, n_2 \in \mathcal{A}} \cos \frac{n_1 \pi x_1}{\ell_1} \cos \frac{n_2 \pi x_2}{\ell_2} \quad (12)$$

where the index set of the second sum is

$$\mathcal{A} = \left\{ n_1, n_2 \left| \pi \sqrt{\frac{n_1^2}{\ell_1^2} + \frac{n_2^2}{\ell_2^2}} = k \right. \right\}.$$

Since the terms of the summation are orthogonal with respect to the standard L^2 inner product, each term of the sum must satisfy the equation. A short calculation of the derivatives yields the following differential equation for $\tilde{\mathbf{v}}_k(t)$,

$$\frac{d\tilde{\mathbf{v}}_k(t)}{dt} = -k^2 \left(k^2 \varepsilon^2 I + D \right) \tilde{\mathbf{v}}_k(t). \quad (13)$$

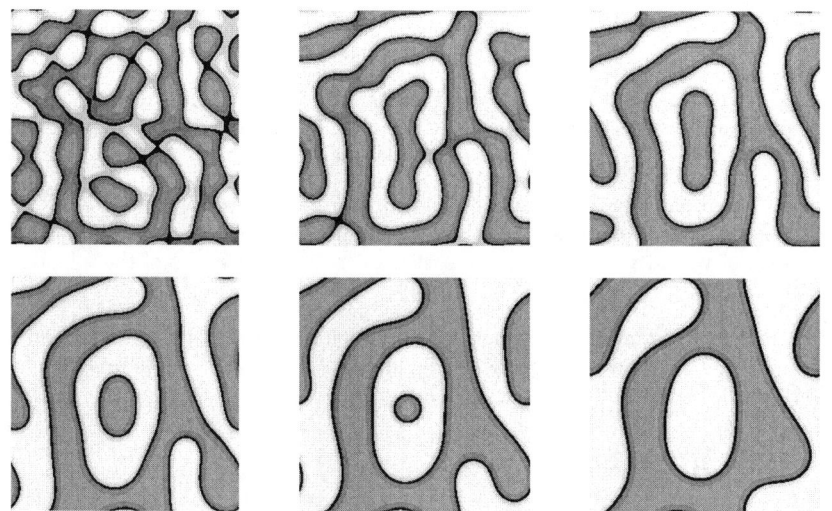

Figure 4: Metastable binary intermediate with $\bar{\mathbf{u}} = (0.46, 0.46)$.

To express the solution of (13) analytically, let ξ_1 and ξ_2 be the eigenvalues of D, and let $\mathbf{v}_1$ and $\mathbf{v}_2$ be the corresponding eigenvectors, then the eigenvalues and eigenvectors of the ode (13) satisfy

$$-k^2(k^2\varepsilon^2 I + D)\mathbf{v}_i = \left[-k^2(k^2\varepsilon^2 + \xi_i)\right]\mathbf{v}_i. \qquad (14)$$

By denoting $\lambda_i(k) = -k^2(k^2\varepsilon^2 + \xi_i)$, the solution of the linear odes for $\tilde{\mathbf{v}}_k(t)$ are

$$\tilde{\mathbf{v}}_k(t) = c_k e^{\lambda_1(k)t}\mathbf{v}_1 + d_k e^{\lambda_2(k)t}\mathbf{v}_2 \qquad (15)$$

where c_k and d_k are determined from the initial conditions. This solution can be easily studied as a function of k and D.

As a function of k, there are two interesting cases. First notice that for large k, $d\tilde{\mathbf{v}}_k/dt$ is dominated by $-\varepsilon^2 k^4$, therefore, in this case $\tilde{\mathbf{v}}_k \to \mathbf{0}$. This occurs if the wavenumbers n_i is sufficiently large or if the domain sizes ℓ_i are sufficiently small. Furthermore, in a closed domain, the eigenvalues of (11) are discrete, so for small enough ℓ_i, all the eigenvalues are negative and $\tilde{\mathbf{v}}_k \to \mathbf{0}$.

For small k, $d\tilde{\mathbf{v}}_k/dt$ is dominated by the eigenvalues of D, and there are three interesting cases:

- If D is positive definite, then $\xi_1, \xi_2 > 0$, and every $\lambda_i(k) < 0$. Consequently $\bar{\mathbf{u}}$ is locally stable with respect to small perturbations. This case occurs where Φ is convex, i.e. in the stable one-phase region of $\mathcal{G}$, in the classical metastable regions of $\mathcal{G}$ and in all of $\mathcal{G}$ at high temperatures.

- If D is indefinite, then $\xi_i < 0 < \xi_2$, and $\lambda_i(k) > 0 > \lambda_j(k)$ when $k < k_c$. Consequently, $\bar{\mathbf{u}}$ is unstable with respect to generic perturbations in the direction

$\mathbf{v}_i$ and $\bar{\mathbf{u}}$ is stable with respect to all perturbations in the $\mathbf{v}_j$ direction. For short-times, the solution can be written as a perturbation of the average in the direction of $\mathbf{v}_i$ as

$$\mathbf{u}(\vec{x}, t) = \bar{\mathbf{u}} + \Theta(\vec{x}, t)\mathbf{v}_i \qquad (16)$$

where $\Theta(\vec{x}, t)$ is the superposition of all the fourier modes corresponding to the direction $\mathbf{v}_i$. This case occurs where $\bar{\mathbf{u}}$ is in a subset of the spinodal region of $\mathcal{G}$.

- If D is negative definite, then $\xi_1, \xi_2 < 0$, and $\lambda_2(k), \lambda_2(k) > 0$ when $k < k_c$. Consequently, $\bar{\mathbf{u}}$ is unstable with respect to all generic initial conditions in the span of $\mathbf{v}_1$ and $\mathbf{v}_2$. This case also occurs where $\bar{\mathbf{u}}$ is in a subset of the spinodal region of $\mathcal{G}$.

An important result of [11] is that if the solution separates, then it will separate into a pseudo-binary because generically the linearized equations will have a unique largest growth mode, and that mode will dominate.

5 Pseudo-binary solutions on lines.

In this section, the pseudo-binary (16) is examined. It will be assumed that $\bar{\mathbf{u}}$ is in the spinodal region, and that $\mathbf{v}_1$ is the direction of fastest separation.

The ternary Cahn-Hilliard equation can be projected onto $\mathbf{v}_1$ to give a binary Cahn-Hilliard equation for $\Theta(\vec{x}, t)$

$$\Theta_t = -\Delta \left[\varepsilon^2 \Delta \Theta - f(\Theta) \right] \qquad (17)$$

where $f(\Theta) = \langle \mathbf{f}(\bar{\mathbf{u}} + \Theta \mathbf{v}_1), \mathbf{v}_1 \rangle$. This equation is found by substituting (16) into (5) and taking the dot product of the resulting equation with $\mathbf{v}_1$. The solution of (17) will not generically be a solution of (5), but it is interesting to study. Furthermore, Figure 5 shows that (5) it is a reasonable representation of the solutions presented in Figures 2 and 3. In Figure 5, the solution $\mathbf{u}(\vec{x}, t)$ is mapped onto $\mathcal{G}$. The solution for the simulation in Figure 2 lies on a curve while the solution for the simulation in Figure 3 lies on a line.

By construction, the mass of (16) is $\bar{\mathbf{u}}$ if $\bar{\Theta} = 0$. However, there exist families of $\bar{\mathbf{u}} \in \mathcal{G}$ that have identical eigenvectors and the pseudo-binaries lie on identical subsets of $\mathcal{G}$, so to study the effects of $\bar{\mathbf{u}}$ on the solutions of this form, a different average can be selected from this family. An example of this is shown in Figure 6 where the pseudo-binary for three different average compositions that each satisfy (16) for different Θ's are shown. The relative fractions of the separated phases change between simulations because the position of $\bar{\mathbf{u}}$ relative to the unstable zero of $f(\Theta)$ is changing.

The nonlinearity $f(\Theta)$ is the directional derivative $\partial \Phi / \partial P^T \mathbf{v}_1$ since $\mathbf{f}(\mathbf{u})$ is given by (7), and

$$f(\Theta) = \langle P \nabla_u \Phi(u_1, u_2, u_3), \mathbf{v}_1 \rangle = \langle \nabla_u \Phi(u_1, u_2, u_3), P^T \mathbf{v}_1 \rangle. \qquad (18)$$

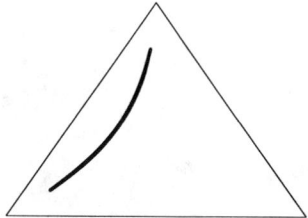 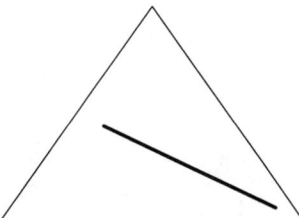

Figure 5: Pseudo-binary solutions of the CH equations with average compositions (0.43,0.43) and (0.25,0.25) mapped into $\mathcal{G}$.

Furthermore, since it has been assumed that $\bar{\mathbf{u}}$ is in the spinodal and $\mathbf{v}_1$ is the direction of fastest separation, then it must be that $f(\Theta)$ is a bistable type nonlinearity. So it is expected that Θ will vary between the stable zeros of $f(\Theta)$. If Θ_1^* and Θ_2^* are the stable zeros of $f(\Theta)$, then

$$\mathbf{u}_i = \bar{\mathbf{u}} + \Theta_i^* \mathbf{v}_1 \qquad (19)$$

are the locations of the stable zeros within $\mathcal{G}$, and the values of (16) will take values in $\mathcal{G}$ between $\mathbf{u}_1$ and $\mathbf{u}_2$.

It is also interesting to consider the properties of the pseudo-binary solution as a function of $\mathbf{u}_1$ and $\mathbf{u}_2$. There are three interesting distinct possibilities for these locations; Both of the zeros may be in the spinodal region of $\mathcal{G}$, both of the zeros may be in a stable (or metastable) region of $\mathcal{G}$, or one zero may be in the spinodal region and the other in the stable region.

The numerical simulations presented above provide examples of these possibilities. The pseudo-binary presented in Figure 2 is an example of spinodal-spinodal pair, i.e. the zeros of $f(\Theta)$ are in the spinodal region of $\mathcal{G}$.

The pseudo-binary presented in Figure 3 is an example of spinodal-stable pair, i.e. one stable zero of $f(\Theta)$ is in a stable region of $\mathcal{G}$ and the other is in the spinodal region of $\mathcal{G}$.

The pseudo-binary presented in Figure 4 is an example of stable-stable pair, i.e. the zeros of $f(\Theta)$ are in the stable region of $\mathcal{G}$.

As mentioned above, this representation of the pseudo-binary is not generically a solution of (5), but in a forthcoming paper [7], a pseudo-binary that is a solution (5) for any $\bar{\mathbf{u}}$ will be given and studied.

6 Predicting secondary separation.

If the pseudo-binary (16) is not too far from the solution of (5), then predicting and understanding the secondary separation process is equivalent to studying the stability of (16) with respect to perturbations in the direction of $\mathbf{v}_2$.

Given the observations of the previous section, it should not be surprising that secondary separation occurs in the first two simulations. To understand why, consider

 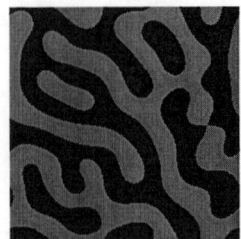 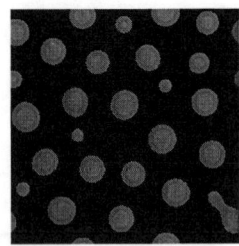

Figure 6: Different average compositions of the pseudo-binary.

a homogeneous solution $\mathbf{u} = \mathbf{u}_i$ given in (19). If $\mathbf{u}_i$ is in a spinodal region of $\mathcal{G}$, this solution is generically unstable. The direction of maximal instability is uncertain, but the direction orthogonal to $\mathbf{v}_1$ is unstable.

Now consider the implications of this on (16) and (17). The solution of (17) will be near the zeros of $f(\Theta)$ except in the interfaces between phases. Therefore, much of the pseudo-binary will be near $\mathbf{u}_i$ and consequently, within the spinodal region of $\mathcal{G}$. So it seems reasonable to believe that the pseudo-binary is unstable.

On the other hand, if $\mathbf{u} = \mathbf{u}_i$ is in a stable or metastable region of $\mathcal{G}$, then much of the pseudo-binary will be within a locally stable region of $\mathcal{G}$, and consequently the pseudo-binary should be locally stable.

These observations are seen in the numerical simulations. In Figure 4, the pseudo-binary has both zeros of $f(\Theta)$ near minima of Φ. Consequently, the solution is locally stable and does not undergo secondary separation. It is worth stressing however, that the global energy minimizer for this solution has three phases, and therefore this pseudo-binary is metastable.

In Figure 3, the pseudo-binary has one zero of $f(\Theta)$ near a saddle point of Φ while the other zero is near a minimum of Φ. The saddle is unstable with respect to separation into two additional phases, therefore the material in one of the tube-like spatial structures is near this saddle, and decomposes into two distinct phases. The phase within the other tube-like structure is near the minimum and appears unchanged after secondary separation.

In Figure 2, both stable zeros of $f(\Theta)$ are in a spinodal region of $\mathcal{G}$ and consequently a third phase appears. The obvious question that remains is why this secondary separation results in an inclusion-type geometry. This is fairly easy to understand. In this case, primary decomposition does give pseudo-binary, but not of the form (16). The true pseudo-binary lies on a curve in $\mathcal{G}$ as opposed to a straight line. The endpoints of this curve are near the minima of Φ, and consequently the interface between these minima is curved towards the centroid of $\mathcal{G}$ to balance the mass of the solution. As the interface becomes more curved it eventually enters the spinodal region of $\mathcal{G}$ that is unstable with respect to separation in the third phase. At this point there is no longer an activation like energy barrier to creating the third phase, and the solution can lower its free energy by fully separating. Consequently, the third phase appears within the interface itself. A detailed analysis of this process will be included in [7].

7 Coarsening of the pseudo-binary.

In this final section, the coarsening between primary and secondary separation is examined and a geometric argument is given to justify the observations made from the simulations given in Figures 3 and 4. A detailed mathematical description of this process has been completed and will appear in [7].

To understand the coarsening phenomena, it is useful to recall the following two concepts. First, it is known that immediately after spinodal decomposition, solutions of the binary Cahn-Hilliard equation are highly oscillatory, nearly periodic and the separation length between like phases is $O(\varepsilon)$ [1, 9]. Pseudo-binary solutions of the ternary alloy CH equations are very similar to solutions of the binary CH equation, and so it is reasonable to expect that that they too will be highly oscillatory and nearly periodic. The simulations presented in Figures 3, 4 and 5 confirm this expectation.

Second, it was shown in §5, that if the domain size in either direction was small, spinodal decomposition can be inhibited because the eigenvalues of the spatial operator are discrete.

Now the simulation given in Figure 2 will be used to give an intuitive argument for the coarsening. The pseudo-binary in this case has one stable phase and one unstable phase. Both phases are highly oscillatory and occupy narrow tube-like regions of space, and the unstable phase does not decompose until the width of the tubes has increased significantly.

If these observations are considered in light of the concepts just discussed, the immediate conclusion that can be drawn is that the initial domain size of the spinodally decomposed pseudo-binary is so small that it inhibits secondary separation. Thus the unstable phase of the pseudo-binary has an effective domain, in this case the tubes, and the effective domain must be sufficiently large for secondary separation to take place.

Acknowledgments

This work was partially supported by an ARO contract with the University of Minnesota's AHPCRC.

References

[1] J. W. CAHN, *On spinodal decomposition*, Acta Metallurgica, 9 (1961), pp. 795–801.

[2] J. W. CAHN AND J. E. HILLIARD, *Free energy of a nonuniform system - I. Interfacial free energy*, J. Chem. Phys., 28 (1958), pp. 258-267.

[3] D. DEFONTAINE, Ph.D. Thesis, Northwestern University, 1967.

[4] C. M. ELLIOTT, *The Cahn-Hilliard model for the kinetics of phase separation*, in Mathematical Models for Phase Change Problems, J. F. Rodrigues, ed., Birkhäuser Verlag, Basel, 1989, pp. 35–73.

[5] C. M. ELLIOTT AND S. LUCKHAUS, *A generalised diffusion equation for phase separation of a multi-component mixture with interfacial free energy*, IMA preprint 887.

[6] D. J. EYRE, *Systems of Cahn-Hilliard Equations*, SIAM J. Appl. Math., 53, (1993), pp. 1686-1712.

[7] D. J. EYRE, in preparation.

[8] P. C. FIFE, *Models for phase separation and their mathematics*, in Proceedings, Taniguchi International Symposium on Nonlinear Partial Differential Equations and Applications, Kinokuniya Publishing, 1990.

[9] C. P. GRANT, Ph.D. Thesis, University of Utah, 1991.

[10] J. J. HOYT, *Spinodal decomposition in ternary alloys*, Acta Metall., 37 (1989), pp. 2489-2497.

[11] J. E. MORRAL AND J. W. CAHN, *Spinodal decomposition in ternary systems*, Acta Metall., 19 (1971), pp. 1037-1045.

[12] D. A. PORTER AND K. E. EASTERLING, *Phase Transformations in Metals and Alloys*, Van Nostrand Reinhold (International), London, 1981.

Chemical Potential Approach to Diffusion-Limited Microstructure Evolution

M.A. Fradkin and J. Goldak
*Department of Mechanical & Aerospace Engineering
Carleton University, Ottawa K1S 5B6 Canada*

R.C. Reed
*Department of Materials Science & Metallurgy
Cambridge University, U.K.*

November 13, 1995

Abstract

A general approach to solving the diffusion problem is developed for chemical potential flows in multicomponent system. The 'partial equilibrium' type of interface boundary conditions of equal chemical potential is applied to the interface between contacting phases. The resulting non-linear diffusion equation is solved numerically. Thermodynamic data from a phase diagram optimization package have been used. The redistribution of carbon between austenite and ferrite (martensite) in low-carbon steel (Hillert's problem) has been studied.

MODERN METHODS FOR MODELING MICROSTRUCTURAL EVOLUTION
AMS INTERNATIONAL MATERIALS WEEK'95 CLEVELAND, OH

1 Introduction

There are some established thermodynamical criteria to determine the state of equilibrium in a heterophase system containing many different chemical components. The general principle is that there is no mass transfer across the interface in the state of thermodynamical equilibrium for a heterophase system and it means that the chemical potentials in contacting phases are equal for all the chemical constituents. That gives the conditions to calculate the interface chemical composition for all phases. As mass transfer processes for different components often have time scales differing in the orders of magnitude, the 'partial equlibrium' concept can be naturally introduced that corresponds to a zero mass flux for the particular component that has the fastest diffusion kinetics.

The diffusion-limited evolution of the heterophase microstructure is determined by the growth/dissolution process for individual phases as well as change in their chemical composition through multi-component diffusion. The proper choice of boundary conditions for a mass-transfer equation is crucial for any attempt to simulate this evolution by means of analytical or numerical solution of the kinetic equation. Usual method is to solve a linear diffusion equations for individual phases subject to boundary conditions of prescribed compositions at the interfaces and continuity of the mass fluxes across the interfaces. In the present paper we suppose a different approach based on the kinetic equation for a chemical potential distribution rather then for compositional field. As the inhomogeneity of chemical potential fields is the driving force for diffusion and, generally speaking, for microstructure evolution toward equilibrium configuration, this method corresponds to the enthalpy formulation of the heat transfer problem.

This algorithm can utilize wide variety of thermodynamical and thermochemical data currently available for computer calculation of equilibrium phase diagram in various systems. The approach is particularly convenient when dealing with boundary condition at the interface, because there is no boundary singularities in the equilibrium chemical potential fields in a contrast with the compositional fields which have step-like behavior at the interface. We present in this paper the results of analysis of the kinetics of a carbon redistribution between ferrite (martensite) and supercooled austenite in a low-carbon steel (Hillert problem) which has been carried out in the proposed chemical potential approach.

2 Kinetic Equation for Chemical Potential

The diffusion flux of component A can be written in the form [1]

$$\mathbf{j}_A = -\Gamma_A x_A \nabla \mu_A = -D_A \nabla x_A , \qquad (1)$$

where x_A is the content of component A, μ_A is its chemical potential, Γ_A is mobility of component A and D_A is its 'tracer' diffusion coefficient. Then we have relationship between diffusion coefficient and mobility

$$\Gamma_A = \frac{D_A}{x_A} \left(\frac{\partial \mu_A}{\partial x_A}\right)^{-1} = D_A \left(\frac{\partial \mu_A}{\partial \ln x_A}\right)^{-1} . \qquad (2)$$

Multiplying the continuity equation for the mass transfer of component A

$$div(\mathbf{j}_A) + \frac{\partial x_A}{\partial t} = 0 \qquad (3)$$

by the derivative of its chemical potential with respect to composition we get the kinetic equation for the distribution of μ_A

$$\frac{\partial \mu_A}{\partial x_A} div(\mathbf{j}_A) + \frac{\partial \mu_A}{\partial t} = 0 \qquad (4)$$

Supposing Γ_A being constant we can write

$$div(\mathbf{j}_A) = -\Gamma_A \left(x_A \nabla \mu_A + \nabla \mu_A \cdot \nabla x_A \right) \qquad (5)$$

and substituting this into Eq.(4) we can obtain

$$\frac{\partial \mu_A}{\partial t} = \Gamma_A \frac{\partial \mu_A}{\partial x_A} \left(x_A \nabla \mu_A + \nabla \mu_A \cdot \nabla x_A \right) \ .$$

Finally we can write the kinetic equation for the chemical potential in the form of non-linear 'diffusion' equation

$$\frac{\partial \mu_A}{\partial t} = \Gamma_A \left(\frac{\partial \mu_A}{\partial \ln x_A} \nabla \mu_A + |\nabla \mu_A|^2 \right) \qquad (6)$$

3 Boundary Conditions

Partial equilibrium condition with regards to component A

$$\mu_A^\alpha = \mu_A^\beta \qquad (7)$$

means no flux of component A across the interface between the phases α and β. This condition should be satisfied for all phases present in the system. In order to determine the (partial) equilibrium concentration of the component A at both sides of the α/β interface some thermodynamical model is needed which can provide the chemical potential of component A in all phases.

There is the only general interface boundary condition for non-linear diffusion equation (6) which is the conservation of mass for component A at the α/β interface expressed through the chemical potential of component A

$$D_A^\alpha \left(\frac{\partial \mu_A^\alpha}{\partial \ln x_A^\alpha} \right)^{-1} \nabla \mu_A^\alpha - D_A^\beta \left(\frac{\partial \mu_A^\beta}{\partial \ln x_A^\beta} \right)^{-1} \nabla \mu_A^\beta = M_A^{\alpha/\beta} \qquad (8)$$

where superscript α, β stands for a particular phase and right-hand side describes 'effective' sink of a component A due to the movement of $\alpha\beta$ interface. We do not have to set up explicitly the boundary condition for the value of chemical potential of A on both sides of the interface.

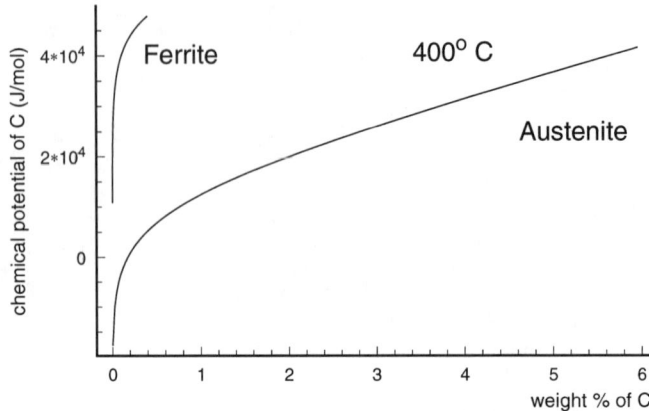

Figure 1: Carbon chemical potential in ferrite and austenite at 400°C *vs* carbon concentration

4 Example: Carbon Redistribution Between Ferrite and Austenite

As an example of this algorithm let us consider the carbon diffusion in a mixture of ferrite (martensite) and supercooled austenite. This problem, known as the Hillert' problem, has been the subject of numerous analytical as well as computational studies in recent years (see [2] and references therein). This microstructure can appear as a result of fast transformation upon rapid cooling of austenite when the slow carbon diffusion process does not have enough time to change the carbon distribution during transformation. It is supposed that during bainite formation in a low-carbon steel at low temperatures the ferrite plates growth so fast into austenite that carbon does not diffuse from ferrite into austenite. The supersaturation of austenite with respect to carbon takes place and the diffusion process, which starts after growth is practically over, lead to an 'escape' of carbon from the ferrite plates. Initially carbon has a homogeneous distribution with composition equal to average steel composition, however the carbon chemical potential has step-like distribution. The diffusion leads to homogeneous chemical potential and step-like distribution of carbon.

4.1 Interface Carbon Concentration

The carbon content on both sides of the ferrite/austenite interface can be found through the solution of Eq.(7) for carbon chemical potential in both phases. In order to calculate the dependence of carbon chemical potential on the composition some thermodynamical data are required and we have used the steel thermodynamical model developed previously for phase diagram optimization and described elsewhere [3]. A compositional dependence of the carbon chemical potential for ferrite μ_C^α and austenite μ_C^γ at 400°C is shown in the Fig.1. It

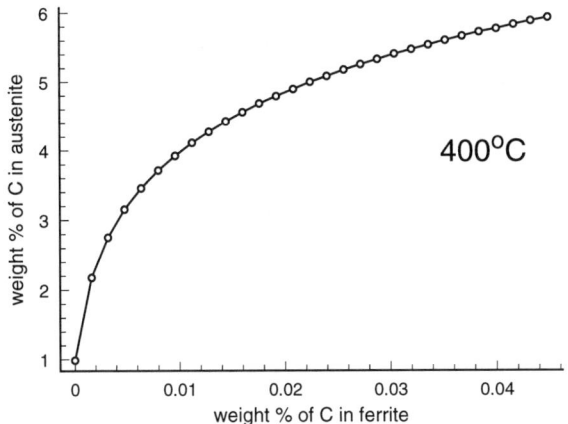

Figure 2: Interface carbon concentration in ferrite and austenite at 400°C calculated from the partial equilibrium condition of equal chemical potential

can be approximated by empirical dependencies

$$\mu_C^\alpha = 70.25 + 5.64 \log x_C \tag{9}$$
$$\mu_C^\gamma = 6.495 + 153.8 x_C \tag{10}$$

where chemical potential is measured in kJ/mol and x_C is an atomic fraction of carbon. From chemical potential calculated in both phases one can find (partial) equilibrium interface composition of carbon shown in Fig.2. We suppose the problem geometry as a periodical arrangement of parallel ferrite plates in austenite and this transforms the diffusion equation into one-dimensional form with coordinate axis going along a normal plate direction.

4.2 Carbon Redistribution Kinetics

For the carbon chemical potential in ferrite described by Eq.(9) the kinetic equation takes very simple form. If
$$\mu_C = A + B \log x_C$$
then
$$\frac{\partial \mu_C}{\partial \ln x_C} = B$$
and Eq.(6) takes a form
$$\frac{\partial \mu_C}{\partial t} = D_C \left(\Delta \mu_C + \frac{|\nabla \mu_C|^2}{B} \right) \tag{11}$$

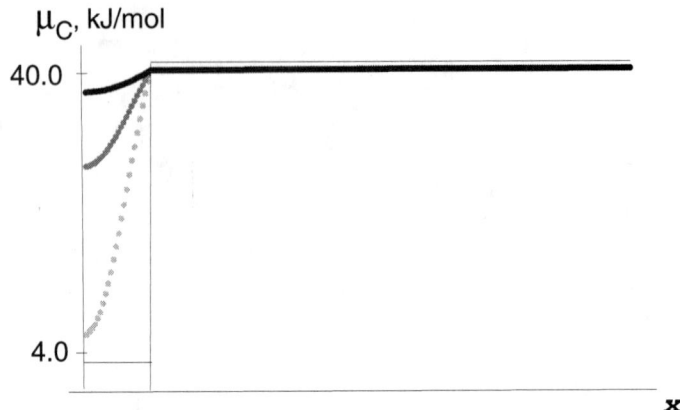

Figure 3: Evolution of the carbon chemical potential in ferrite/austenite bilayer at 400°C

In this case $\Gamma_C = D_C/B$ and assumption of constant Γ leads to the independence of carbon diffusion coefficient on composition. This is the case according to various studies, (*e.g.* see [4] and references therein).

For linear dependence of the carbon chemical potential on composition

$$\mu_C = A + B\, x_C$$

that takes place in austenite (10) we have

$$\frac{\partial \mu_C}{\partial \ln x_C} = B\, x_C$$

and kinetic equation for chemical potential can be written in the form

$$\frac{\partial \mu_C}{\partial t} = D_C \left(\Delta \mu_C + \frac{|\nabla \mu_C|^2}{B\, x_C} \right) \qquad (12)$$

The mobility of carbon takes a form $\Gamma_C = D_C/(B/,x_C)$ and carbon diffusion coefficient appears to be proportional to composition. The strong compositional dependence of the coefficient of carbon diffusion in austenite has been previously found in many experiments, see [5] and references therein.

We have solved numerically the one-dimensional kinetic equation for carbon redistribution between ferrite and austenite at 400° according to Eqs.(14) and (16). The 'tracer' diffusion coefficient for carbon in ferrite and austenite was 2.27×10^{-13} and 8.432×10^{-15} m^2/s, respectively. The evolution of chemical potential and carbon concentration is shown in Figs.3 and 4.

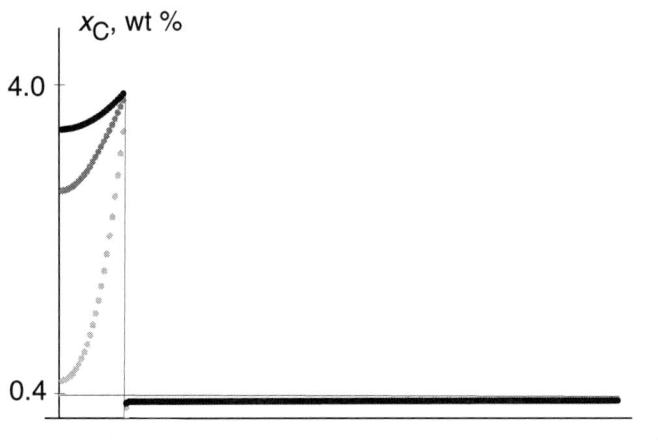

Figure 4: Evolution of the carbon composition in ferrite/austenite bilayer at 400°C

References

[1] C.P. Flynn, *Point Defects and Diffusion*, Clarendon Press, Oxford, 1972, ch.8

[2] M. Hillert, L. Höglund and J. Agren, *Acta Metal. Mater.* **41**, 1951 (1993)

[3] R. Reed, J. Goldak and E. Hughes, Phase Diagrams as a minimisation problem

[4] J. Agren, Acta Metall. **30**, 841 (1982)

[5] S.A. Mujahid and H.K.D.H. Bhadeshia, *Acta Metal. Mater.* **40**, 389 (1992)

SUBJECT INDEX

Aboav-Weaire's law, 15-22
active material, 49, 50
adaptive mesh refinement, 205-214
Al-Li alloys
　congruent ordering, 275, 276
　d' precipitation, 269, 270, 273, 274, 277, 278
Allen-Cahn equation, 355-365
allotriomorph, 174
amorphous, 181
anisotropic surface free energy, 135-148
anisotropy, 101, 135-148, 149-159, 161-164, 168, 225-244
antiphase boundary, 225-244, 245-268
antiphase domain boundary, 39-48
APB, 173, 178-180
applied stress effect, 103
austenite, 59-69
autonomy, 173, 174

B2 lattice, 178, 179
B2 order, 39-48
Bain strain, 105
bifurcation diagram
　particle shape, 125-133
binary alloys, 87-100, 355-365
boolean logic, 183
broken symmetry, 173, 174

Cahn-Hilliard equation, 355-365, 367-378
capillarity, 176
cell dynamical system (CDS), 355-365
chemical potential, 149-159, 175, 181, 379-385
chemical vapor deposition, 327-343
CIGM, 175, 181, 182
cluster approximation, 39-48
coherency strain, 101, 173-176
comparison lattice, 71-85
compensator, 327-343

computation, 135-148
computational materials science, 1-14
computer simulation, 15-22
continuum formulation, 225-244
crystalline surface free energy, 135-148
crystals
　small-scale, 24, 28, 29
cubic anisotropy, 87-100
curvature dependent motion, 161, 162
curvature, 175, 176

degeneracy, 173-175, 181, 183
degree of contact between grain boundary and second phase particles, 31, 33-35, 37
dendrite, 176, 181
dendritic growth, 195-204
diffuse interface, 245-268
diffusion potential, 125-133
diffusive ordering, 281-310
DIGM, 173
dilatational misfit strain, 101
dipole, 179, 180
discontinuous precipitation, 175, 176, 181, 182
discrete atom method (DAM), 102
dislocations
　line energy, 164
　line tension, 164
　loops, 161-168
dissipation principle, 173-175, 178, 182, 183
driving forces, 149-159

edge energy
　effects on crystal shape, 24-29
　evidence for, 24
　models for, 24
　reduction by corner truncation, 26-28
elastic effects, 87-100

elastic energy
    single particle, 125-133
elastic misfit, 87-100
equilibrium, 149-159
equipartition, 173, 177
euler equation, 178
eutectoid, 174

face-centered-cubic, 225-244
face-centered-cubic alloy, 245-268
feedback control, 327-343
finite media, 187-194
fluctuations, 39-48
fluid flow, 327-343
four-group, 173, 183
free energy, 39-48

Gibbs-Duhem, 175, 181
Ginzburg-Landau theory, 87-100
gradient energy, 176
grain growth, 15-22, 51, 55, 215-223
    Monte Carlo simulation, 31, 32, 34, 35, 37
    Zener pinning, 31-33, 37
grain side correlations, 15-22
grain side distribution, 15-22
grain size correlations, 15-22
grain size distribution, 15-22

Hamilton's principle, 176
hysteresis, 55
hysteretic behavior, 49
H∞, 327-343

inhomogeneity, 101
interface boundary, 245-268
interface dynamics, 355-365

kinetics, 321-326

lammelar fault, 174
latent heat, 181
lattice gas, 245-268
lattice misfit, 87-100
LFM, 176
long range order, 39-48
LQR, 327-343

magnetoelastic interactions, 49
magnetostriction, 53
magnetostrictive, 50, 53-55
Markov chains, 281-310
martensite, 59-69
martensitic material, 50
master equation, 281-310
mean curvature, 135-148, 149-159
mean field, 281-310
mean-field equation, 321-326
mean-field theory, 245-268
mean lattice parameter, 71-85
mechanical equivalent of heat, 177
Melzak's conjecture, 28
metastable, 367-378
microscopic diffusion equations
    Euler's equation, 272
    langevin noise, 269-271
microscopic reversability, 178, 179, 182, 183
microscopic solvability, 174
microstructure, 1-14, 49, 50, 53, 87-100, 379-385
minimal surfaces, 1-14
minmax, 327-343
misfit dislocation, 175
misfit strain, 125-133
mobility, 177, 179, 181
modulated patterns, 87-100
Monte Carlo method, 39-48, 103
morphological development
    cuboid-to-plate, 125-133
    numerical calculation, 125-133
    sphere-to-cuboid, 125-133
    three-dimensional, 125-133
motion by mean curvature, 135-148
motion by surface diffusion, 135-148
motion with triple junctions, 135-148
Mullins and Sekerlta, 176, 181
multicomponent systems, 149-159
multiple order parameter, 225-244, 245-268
multipoint correlations, 71-85

nickel-base superalloy, 102
niobium alloy, 102

non-equilibrium, 281-310
non-equilibrium alloy, 321-326
nonlinear analysis, 49
nonlinear deposition, 327-343
non-linear diffusion, 379-385
nonlinear Galerkin method, 327-343
nonlinear material, 50
nonuniform media, 345-354
nucleation, 187-194
    microstructure, 274
    non-classical, 272, 274, 276-278
    nuclei profiles, 275-278

onsager coupling, 173, 177, 181, 182
order-disorder, 178
order-disorder transition, 245-268
order parameter, 59-69, 173, 177-180, 182
oxidation, 345-354

p-polarized reflectance spectroscopy, 327-343
partially-stabilized zirconia, 102
partition coefficient, 181
path entropy, 177
pattern, 173, 174, 183
pattern formation, 345-354
periodic minimal surface, 39-48
phase evolution, 187-194
phase field, 173, 174, 176, 177, 181, 205-214
phase-field modeling, 195-204
phase-field theory, 245-268
phase kinetics, 281-310
phase separation, 376-378
phase transformation, 49
phenomenology, 173, 182
pinning of domains, 87-100
poisson voronoi structures, 15-22
pseudostable, 39-48
pure shear misfit strain, 107

quadrijunction, 215-223
quasiparticle, 174, 183

random alloy, 71-85
reaction-diffusion, 345-354

reduced basis elements method, 327-343
renormalization, 183
renormalization-group, 355-365
rigidity, 173, 174, 183

saddle point, 39-48
scaling, 15-22, 183
segregation, 179
self-organization, 173, 174
shape memory alloy, 49, 50
short range order, 39-48
simulation result
    Fe-Mo, 118-124
        coarsening, 118
        coherent binodal line, 118
        coherent spinodal line, 118
        elastic anisotropic factor, 118, 121
        elastic interaction energy, 118
        lattice mismatch, 118, 119
        modulated structure, 118
soldering, 1-14
solid-solid phase transitions, 59-69
solidification, 195-204
spinodal decomposition, 87-100, 173, 175, 182, 321-326, 355-365, 367-378
spontaneous pattern formation, 174, 183
statistical mechanics
    boundary tension, 316, 317
    cluster variation method, 312-318
    cooperative phenomena correlation, 312-318
    diffusion, 315, 316
    path probability method, 315
    phase diagram, 312-313
    spinodal decomposition, 313, 314
stochastic control, 281-310
stochastic variational principle, 71-85
surface energy anisotropy, 245-268
surface evolver, 15-22
    applications, 28-29
surface free energy, 149-159
surface motion, 135-148
surface tension, 181

Terfenol-D, 49, 51
ternary alloy, 367-378
ternary diffusion path, 175
tetragonal misfit strain, 105
theory
    diffusion equation, 112-117
        chemical free energy, 113
    interfacial energy, 114
        diffusion potential, 112
        elastic strain energy, 115
    mobility, 113
        interchange energy, 116
thermodynamics, 149-159, 281-310
time evolution, 15-22
time-reversal invariance, 173, 183
topological defect, 174
transformation path, 54, 55
transformation strain, 51
transient states, 321-326
trapping, 178 181

twin, 50-54
twinned, 59-69
twinned dendritic, 50, 52
twinning equation, 52, 53
two-dimensional grain growth, 15-22
two-phase solid, 215-223

universality, 355-365

vacancy, 173, 179, 180, 182
variant arrangement, 50
variational method for gradient flow, 135-148

wavenumber selection, 175
weighted mean curvature, 135-148, 149-159
Wulff shape, 24-26

Young measure, 50, 54

AUTHOR INDEX

Almgren, A. S., 205
Almgren, R. F., 205
Anthony, L., 39

Bachmann, K. J., 327
Bangia, A. K., 345
Banks, H. T., 327
Bär, M., 345
Beiden, S. V., 321
Brandon, D., 59
Braun, R. J., 225

Cahn, J. W., 149, 225
Carter, W. C., 1
Cerezo, A., 245
Chen, L. Q., 205, 269
Choy, J. H., 101

Dietz, N., 327
Dobretsov, V. Yu., 321

Eyre, D. J., 367

Fan, D., 215
Fradkin, M. A., 379
Fultz, B., 39

Gao, J., 31
Goldak, J., 379
Gosling, T. J., 71

Haas, G., 345
Hackney, S. A., 101
Hagedorn, J., 225
Hunderi, O., 15
Hyde, J. M., 245

Ito, K., 327

James, R., 49

Karma, A., 195
Kelley, M. J., 23
Kevrekidis, I. G., 345
Kikuchi, R., 311
Kinderlehrer, D., 49
Kirkaldy, J. S., 173
Koyama, T., 111

Lee, J. K., 101

Ma, L., 49
Makarov, S. D., 187
Marthinsen, K., 15
McFadden, G. B., 225
Miller, M. K., 245
Miyazaki, T., 111

Onuki, A., 87
Oono, Y., 355

Patterson, B. R., 31
Poduri, R., 269

Rajan, K., 161
Rappel, W.-J., 195
Reed, R. C., 379
Rogers, R. C., 59
Rotermund, H.-H., 345
Rumyantsev, E. L., 187
Ryum, N., 15

Schwendeman, D., 161
Scroggs, J. S., 327
Setna, R. P., 245
Shur, V. Ya., 187
Simmons, J. A., 281
Smith, G. D. W., 245

Taylor, J. E., 135, 149
Thompson, M. E., 125
Thompson, R. G., 31
Tran, H. T., 327

Vaks, V. G., 321
Voorhees, P. W., 125

Wheeler, A. A., 225
Willis, J. R., 71